Fundamentals of Matrix-Analytic Methods

Qi-Ming He

Fundamentals of Matrix-Analytic Methods

Springer

Qi-Ming He
Department of Management Sciences
University of Waterloo
Waterloo, ON, Canada

ISBN 978-1-4899-9002-0 ISBN 978-1-4614-7330-5 (eBook)
DOI 10.1007/978-1-4614-7330-5
Springer New York Heidelberg Dordrecht London

Printed on acid-free paper

Springer is part of Springer Science+Business Media (www.springer.com)

To my parents
To Rongjing Feng, Zechuan He,
and Patrick Zebing He

Preface

This is a textbook focusing on *matrix-analytic methods*. In this book, the basic theory is accompanied by a large number of examples and exercises. The book is suitable for senior undergraduates and graduate students in science, engineering, management, business, and finance. In addition, a number of recent developments are collected. Thus, the book can also be of value to researchers and practitioners.

Marcel F. Neuts pioneered matrix-analytic methods in the study of queueing models in the 1970s. Since then, matrix-analytic methods have become an indispensable tool in stochastic modeling and have found applications in the analysis and design of manufacturing systems, telecommunications networks, risk/insurance models, reliability models, and inventory and supply chain systems. The power and popularity of matrix-analytic methods come from their flexibility in stochastic modeling, capacity for analytic exploration, natural algorithmic thinking, and tractability in numerical computation.

Some Existing Work on Matrix-Analytic Methods

A number of monographs on matrix-analytic methods have been published. For a summary of earlier work on matrix-analytic methods, we refer to Neuts (1981, 1989) and references therein. We refer to Latouche and Ramaswami (1999) for the developments in the 1980s and 1990s. For specialized subjects, we refer to (i) Ost (2001) for matrix-analytic methods and telecommunications networks; (ii) Grassmann (2000) and Bini, Latouche, and Meini (2005) for matrix-analytic methods and numerical computation; (iii) Breuer and Baum (2005) for matrix-analytic methods and queueing theory; (iv) Tian and Zhang (2006) for matrix-analytic methods and vacation queues; (v) Artalejo and Gomez-Corral (2008) for matrix-analytic methods and retrial queues; (vi) Alfa (2010) for matrix-analytic methods and discrete time queues; and (vii) Li (2010) for matrix-analytic methods, Wiener-Hopf factorization, and structured Markov chains.

Since 1995, seven international conferences on matrix-analytic methods in stochastic models have been held. Seven conference proceedings have been published: Chakravarthy and Alfa (1996); Alfa and Chakravarthy (1998); Latouche and Taylor (2000, 2002); Bini, Latouche, and Meini (2005); He, Takine, Taylor, and Zhang (2008); and Latouche, Ramaswami, Sethuraman, Sigman, Squillante, and Yao (2013). The proceedings represent the developments of matrix-analytic methods and their applications in different time periods.

Features of the Book

This textbook covers three fundamental aspects of matrix-analytic methods and two of their important applications areas. Each chapter in the book begins with basic concepts and well-known models in stochastic processes, followed by concepts and models in matrix-analytic methods. The book takes a constructive approach to introduce stochastic models, which is intuitive, easy to identify and interpret parameters and variables, and mathematically rigorous. The fact that the approach is mathematically rigorous is especially useful for engineering students to develop advanced solution skills.

The book focuses on the basic theory, probabilistic and intuitive interpretations of results, and computational methods. For key results whose proof requires advanced mathematics, the book presents them without a proof, and reference(s) where the original proof can be found are provided. Furthermore, probabilistic and intuitive interpretations to all the basic results are given, in order to help readers gain in-depth understanding of the subjects. A large collection of computational methods in matrix-analytic methods are included.

There are a large number of exercises in the book. Some exercises are for review. They are given for readers to review basic calculus, linear algebra, probability theory, and matrix theory. Some exercises are very simple and are intended for understanding fundamental concepts and ideas. Some exercises are typical and are intended for practicing the use of the theory developed. Finally, some exercises are for extensions of the theory and are presented for information. Exercises of this type are difficult and are marked by an asterisk.

Topics and Organization of the Book

The selection of the topics reflects my preference, and is driven by my desire to include some topics that have not been in previous monographs. The book consists of a theoretical part and an application part. The theoretical part includes Chaps. 1, 2, and 3. Chapter 1 defines the exponential distribution and, from there, introduces phase-type distributions. Chapter 2 defines the Poisson process and, from there, introduces Markovian arrival processes. Chapter 3 defines the birth-and-death process and, from there, introduces Markov chains of the *QBD*, *GI*/*M*/1, and *M*/*G*/1 type. The application part includes Chaps. 4 and 5. Chapter 4 defines the *M*/*M*/1 queue and, from there, introduces more complex queueing models that can be analyzed using matrix-analytic methods. Chapter 5 focuses on inventory and supply chain models.

Reading the Book

Readers are expected to have a solid background in basic probability theory, Markov chains, Poisson processes, and matrix theory. In each chapter, a list of textbooks is provided for background reading. Readers should read the first parts of Chaps. 1, 2, and 3 to learn the basic theory of matrix-analytic methods. Sections 1.1, 1.2, and 1.3 cover the essentials of phase-type distributions. Sections 2.1, 2.2, 2.3, and 2.4 cover the essentials of Markovian arrival processes. Sections 3.1, 3.2, 3.3, 3.4, and 3.5 cover the essentials of structured Markov chains. Other sections in these chapters cover topics that are not normally discussed in books about matrix-analytic methods. In particular, Sect. 1.4 introduces multivariate phase-type distributions. Section 1.5 presents the EM-algorithm for parameter estimation of phase-type distributions. Section 2.5 introduces marked Markovian arrival processes. Sections 3.6, 3.7 and 3.8 define and analyze Markov chains with a tree structure. Section 3.9 deals with tail asymptotics of Markov chains with infinitely many background phases.

The first part in Chaps. 4 and 5 deal with simple and typical stochastic models in queueing theory and inventory management, respectively. The remaining parts of the chapters investigate advanced models. In particular, Sects. 4.6 and 4.9 introduce and investigate queues with multiple types of customers. Sections 5.3 and 5.4 introduce inventory management systems and develop algorithms for computing the optimal policy, which is not normally covered in books on matrix-analytic methods.

References

Alfa AS (2010) Queueing theory for telecommunications: discrete time modelling of a single node system. Springer, New York

Alfa AS, Chakravarthy SR (1998) Advances in matrix analytic methods for stochastic models. Notable Publications, New Jersey

Artalejo JR, Gomez-Corral A (2008) Retrial queueing systems: a computational approach. Springer, Berlin

Bini D, Meini B, Latouche G (2005) Stochastic models: fifth international conference on matrixanalytic methods, 21 (No 2–3)

Bini DA, Latouche G, Meini B (2005) Numerical methods for structured Markov chains. Oxford University Press, Oxford

Breuer L, Baum D (2005) An introduction to queueing theory and matrix-analytic methods. Springer, Dordrecht

Chakravarthy SR, Alfa AS (1996) Matrix-analytic methods in stochastic models. Marcel Dekker, New York

Grassmann WK (2000) Computational probability. Kluwer, Massachusetts

He QM, Takine T, Taylor P, Zhang HQ (2008) Annals of operations research: matrix-analytic methods in stochastic models 160

Latouche G, Taylor (2000) Advances in algorithmic methods for stochastic models. Notable Publications, New Jersey

Latouche G, Taylor (2002) Matrix-analytic methods: theory and applications. World Scientific, New Jersey

Latouche G, Ramaswami V (1999) Introduction to matrix analytic methods in stochastic modeling. ASA & SIAM, Philadelphia

Latouche G, Ramaswami V, Sethuraman J, Sigman K, Squillante MS, Yao DD (2013) Matrix-analytic methods in stochastic models. Springer proceedings in mathematics & statistics. Springer, New York

Li QL (2010) Constructive computation in stochastic models with applications: the RG-factorizations. Tsinghua University Press/Springer, Beijing/Berlin/Heidelberg

Neuts MF (1981) Matrix-geometric solutions in stochastic models – an algorithmic approach. The Johns Hopkins University Press, Baltimore

Neuts MF (1989) Structured stochastic matrices of M/G/1 type and their applications. Marcel Dekker, New York

Ost A (2001) Performance of communication systems: a model-based approach with matrix-geometric methods. Springer, New York

Tian NS, Zhang ZG (2006) Vacation queueing models: theory and applications. Springer, New York

Acknowledgments

This book was possible only with the guidance and support from my professional mentors. Professor Guanghui Hsu was my first Ph.D. supervisor, who introduced queueing theory and matrix-analytic methods to me. Dr. Marcel F. Neuts was my post-doctoral supervisor, who guided and supported me in my research. Drs. Elizabeth M. Jewkes and John Buzacott were my second Ph.D. supervisors, who guided and supported me in pursuing my career in industrial engineering and management sciences. I am grateful to all of them.

I have also benefitted from my friendship and collegial relationships with colleagues. Drs. Attahiru S. Alfa, Yigal Gerchak, Eldon Gunn, and David Stanford guided me in research through collaboration. Drs. Hui Li, Hanqin Zhang, Yiqiang Zhao, and Xiaobo Zhao had to correct mistakes that I made in research in order for us to complete our projects and papers. I am indebted to them for their support and encouragement in my career and in writing this book. I want to thank Drs. Barbara Margolius, Zhe George Zhang, and V. Ramaswami for reading some chapters of the book. I also want to thank Qishu Cai, Zurun Xu, Hao Zhang, and Tiffany A. Matuk for their careful reading of the book. A special thanks goes to Page Burton for carrying out the enormous task of proofreading and correcting. Without their help, the book would not be in its current form. The remaining errors in the book are my responsibility, though.

I started to write this book in 2007, when I was visiting Tsinghua University in Beijing. I offered a course called "Matrix-analytic methods and their applications" to a group of graduate students with a mathematical or engineering background. Following that, in 2008, 2009, and 2011, I offered the same course to graduate students with a mathematical, engineering, or business background at the University of Science and Technology of China, University of Waterloo, and

Tsinghua University, respectively. The four teaching opportunities gave me the chance to develop my lecture notes on matrix-analytic methods, which evolved into this book. I would like to thank the three universities and the talented students in my classes for their support of this book.

Qi-Ming He
Waterloo, ON, Canada

Contents

Chapter 1
From the Exponential Distribution to Phase-Type Distributions

Abstract This chapter introduces phase-type distributions. Topics covered in this chapter are: (i) the exponential distribution; (ii) definitions of phase-type distributions; (iii) closure properties of phase-type distributions; (iv) *PH*-representations; (v) multivariate phase-type distributions; and (vi) parameter estimation and fitting of phase-type distributions.

Probability distributions have been studied extensively and used widely in science and engineering. Well-known probability distributions include *normal*, *lognormal*, *uniform*, *exponential*, *Erlang*, *Gamma*, *Beta*, *Weibull*, *Cauchy*, *Pareto*, *phase-type*, *chi-squared*, *F*, *Student's t*, *binomial*, *negative binomial*, *geometric*, *hypergeometric*, and *Poisson* distributions. Those probability distributions possess certain interesting properties that appeal to some applications, which make them popular in many branches of science and engineering. This chapter focuses on the exponential distribution and phase-type distributions, whose popularity in stochastic modeling is mainly due to their memoryless or partial memoryless property.

The exponential distribution possesses the so-called memoryless property, which often leads to the introduction of Markov chains in the study of stochastic systems. Phase-type distributions can be defined on Markov chains and have a partial memoryless property, which also often leads to Markovian models that are analytically and algorithmically tractable. In addition to the (partial) memoryless property, the exponential distribution and phase-type distributions have a number of other interesting properties as well, many of which will be explored in this chapter. For instance, the set of phase-type random variables is closed under a number of operations such as "min", "max", and "+", which is useful for stochastic modeling. Phase-type distributions have matrix representations that are not unique. Some forms of the matrix representations, such as the Coxian representations, are suitable for theoretical analysis and numerical computation. Furthermore, phase-type distributions constitute a versatile class of distributions that can approximate arbitrarily closely any probability distribution defined on the nonnegative real line. Thus, phase-type distributions are used in most stochastic models investigated in this book.

Q.-M. He, *Fundamentals of Matrix-Analytic Methods*,

DOI 10.1007/978-1-4614-7330-5_1,

While we explicitly define all the concepts used in this book, we also provide readers with references where classical concepts are covered. We refer to Ross (2010) for basic concepts in probability theory: *random variable, probability distribution function, density function, mathematical expectation* (or *expectation/mean/average*), moment, *variance, standard deviation, coefficient of variation, co-variance, correlation, independence, conditional probability, probability generating function, Laplace-Stieltjes transform, joint probability distribution, equilibrium distribution*, etc. We refer to Asmussen (2003) and Ross (2010) for basic concepts on Markov chains: *state, state space, infinitesimal generator* (*Q-matrix*), *transition rate, transition probability matrix, absorption state, first passage time*, etc. We refer to Minc (1988) for basic concepts in matrix theory, especially on *nonnegative matrices* and *M-matrices*.

1.1 The Exponential Distribution

In this section, the exponential distribution is defined and some basic properties associated with the exponential distribution are presented.

Definition 1.1.1 A nonnegative *random variable X* has an *exponential distribution* if its *probability distribution function* (i.e., the *cumulative distribution function* (CDF)) is given as

$$F(t) = P\{X \le t\} = 1 - \exp(-\lambda t) \equiv 1 - \sum_{n=0}^{\infty} \frac{(-\lambda)^n t^n}{n!}, \quad t \ge 0, \tag{1.1}$$

where λ is a positive real number. We shall call X an *exponential random variable* with parameter λ.

Taking the derivative of the distribution function $F(t)$, the *probability density function* of the exponential distribution is obtained as

$$f(t) = \frac{\mathrm{d}F(t)}{\mathrm{d}t} = \lambda \exp(-\lambda t), \quad t \ge 0. \tag{1.2}$$

Exercise 1.1.1 Show that $\int_0^\infty \lambda e^{-\lambda t} \mathrm{d}t = 1$ using integration by parts. Explain probabilistically (without any calculation) why the integration must be one. (Note: We also write $\exp(-\lambda t)$ as $e^{-\lambda t}$ in this book.)

Density functions of exponential distributions with $\lambda = 1$ and $\lambda = 5$ are plotted in Fig. 1.1, where the two density functions demonstrate a similar "exponential" shape.

Exercise 1.1.2 Plot the density functions of exponential distributions with $\lambda = 0.2$ and $\lambda = 10$, respectively.

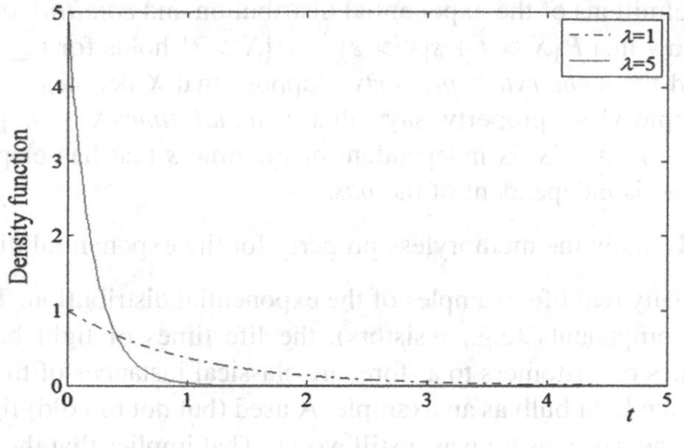

Fig. 1.1 Density functions of exponential distributions

Exercise 1.1.3 Show that the *moments* of an exponential distribution with parameter λ are given as $E[X^n] \equiv \int_0^\infty t^n \mathrm{d}F(t) = n!/\lambda^n$, for $n = 0, 1, 2, \ldots$. (Hint: Use integration by parts and the *induction method*. Also note: $\mathrm{d}(\exp(-\lambda t))/\mathrm{d}t = -\lambda\exp(-\lambda t)$ and $\mathrm{d}(t^n)/\mathrm{d}t = nt^{n-1}$.)

From Exercise 1.1.3, the *mean*, *variance* (*Var*), and *coefficient of variation* (*cv*) of an exponential random variable can be obtained.

Proposition 1.1.1 *For exponential random variable X with parameter λ, we have*

$$E[X] = \frac{1}{\lambda};$$
$$Var(X) \equiv E\left[(X - E[X])^2\right] = E[X^2] - (E[X])^2 = \frac{1}{\lambda^2}; \tag{1.3}$$
$$cv(X) \equiv \frac{\sqrt{Var(X)}}{E[X]} = \sqrt{Var\left(\frac{X}{E[X]}\right)} = 1.$$

We remark that (i) the fact that the mean of X is given by $1/\lambda$ provides a simple method for estimating λ; and (ii) the fact that $cv(X) = 1$ for all exponential distributions limits the use of exponential distributions significantly. The sole parameter λ of the exponential distribution can be interpreted explicitly under different circumstances. For instance, λ is interpreted as the failure rate in reliability theory (see Exercise 1.1.10), while it represents the arrival rate of the *Poisson process* (see Sect. 2.1 in Chapt. 2).

Note: In the literature, the mathematical expectation of a random variable is called the mean, expectation, expected value, or average. In this book, we shall mainly use "mean" and "expectation", which one to be used depends on the context and the area of interest.

From the definitions of the exponential distribution and *conditional probability*, it is easy to show that $P\{X > t + s|X > s\} = P\{X > t\}$ holds for $t \geq 0$ and $s \geq 0$, which is called the *memoryless property*. Suppose that X denotes the time. Intuitively, the memoryless property says that *residual time* $X - s$, given $X > s$, denoted as $X - s \mid X > s$, is independent of the time s that has elapsed. That is, the residual time is independent of the *past*.

Exercise 1.1.4 Show the memoryless property for the exponential distribution.

There are many real life examples of the exponential distribution. The life times of electronic components (e.g., resistors), the life times of light bulbs, and the interarrival times of customers to a store, are classical instances of the exponential distribution. Take light bulb as an example. A used (but not too old) light bulb may be as good as a new one, as long as it still works. That implies that the life time of a light bulb may possess the memoryless property. Consequently, the life time distribution of a light bulb can be approximated by the exponential distribution.

The memoryless property is the most important property of the exponential distribution, due to its usefulness in stochastic modeling and its uniqueness to the exponential distribution amongst all continuous random variables.

Theorem 1.1.1 *Exponential random variables are the only nonnegative, nonzero, and finite continuous random variables that possess the memoryless property.*

Proof. By Exercise 1.1.4, an exponential random variable has the memoryless property. On the other hand, suppose that a continuous random variable X on the nonnegative real line possesses the memoryless property. Then we must have $P\{X > t + s\} = P\{X > t\}P\{X > s\}$ for any $t, s \geq 0$. It can be shown that $P\{X > t\} > 0$ and $P\{X > t\} < 1$ for all $t \geq 0$. In fact, if $P\{X > t_0\} = 0$ for some $t_0 \geq 0$, by the definition of probability, we have $P\{X > t\} = 0$ for all $t \geq t_0$; and by the memoryless property, we obtain $P\{X > t_0/2^n\} = 0$ for $n = 0, 1, 2, \ldots$, which implies $P\{X > t\} = 0$ for all $t > 0$. Then $X = 0$ with probability one, which contradicts the "nonzero" assumption on X. If $P\{X > t_1\} = 1$ for some $t_1 > 0$, then $P\{X > t\} \geq P\{X > t_1\} = 1$ for all $t \leq t_1$. Applying the memoryless property, we obtain $P\{X > t\} = 1$ for all t. This implies $X = \infty$ with probability one, which contradicts the assumption of finiteness on X.

Let $g(t) = \log(P\{X > t\})$, $t \geq 0$. Then we have $g(t + s) = g(t) + g(s)$, which leads to $g(mt) = mg(t)$, $g(1/m) = g(1)/m$, and $g(n/m) = ng(1)/m$, for any positive integer m and nonnegative integer n. Since any irrational number can be approximated by a sequence of rational numbers, due to the continuity of functions $P\{X > t\}$ and $\log(t)$, we have $g(t) = tg(1)$, for any $t \geq 0$. Define $\lambda = -g(1)$. Then we obtain $P\{X > t\} = \exp(-\lambda t)$ for $t \geq 0$. Since $0 < P\{X > 1\} < 1$, we must have $g(1) < 0$ and, consequently, $\lambda > 0$. This completes the proof of Theorem 1.1.1.

Exercise 1.1.5 Assume that X_1 and X_2 are *independent* exponential random variables with parameters λ_1 and λ_2, respectively. Define $X = \min\{X_1, X_2\}$, i.e., the minimum of X_1 and X_2. Show that X has an exponential distribution with parameter $\lambda_1 + \lambda_2$. (Hint: $P\{\min\{X_1, X_2\} > t\} = P\{X_1 > t\}P\{X_2 > t\}$).

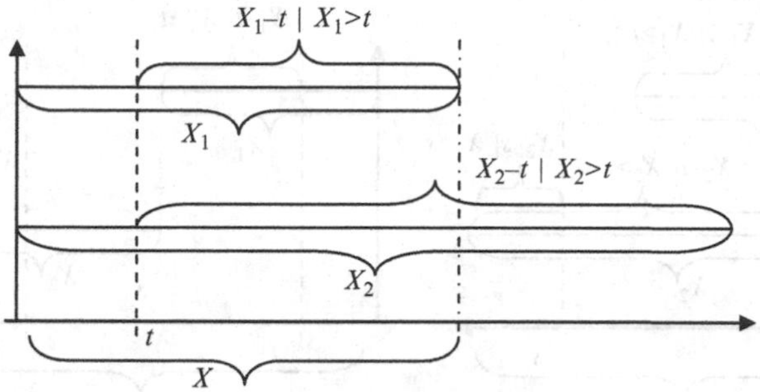

Fig. 1.2 A plot of $X = \min\{X_1, X_2\}$ and its residual at time t

The memoryless property can be used to explain that random variable X defined in Exercise 1.1.5 has an exponential distribution. Since both X_1 and X_2 possess the memoryless property, if $t < \min\{X_1, X_2\}$, residual times $X_1 - t \mid X_1 > t$ and $X_2 - t \mid X_2 > t$ have the same exponential distributions as X_1 and X_2, respectively. Since $X - t \mid X > t = \min\{X_1 - t \mid X_1 > t, X_2 - t \mid X_2 > t\}$, the random variable $X - t \mid X > t$ has the same distribution as X. Therefore, the random variable X possesses the memoryless property (see Fig. 1.2). According to Theorem 1.1.1, X must be exponentially distributed. Intuitively, we can consider the forecast for X at different time epochs. The forecast for X at time zero (i.e., looking forward from time zero) is probabilistically equivalent to the forecast for X at t, given $X > t$. At time 0, X is the minimum of two independent exponential random variables. At time t, given $X > t$, $X - t$ is still the minimum of two independent exponential random variables (with the same parameters as the previous two). Thus, X has the memoryless property.

We remark that the intuition used here to identify the memoryless property will be used repeatedly to observe the Markovian property of Markov chains throughout this book.

Exercise 1.1.6 Assume that X_1 and X_2 are independent exponential random variables with parameters λ_1 and λ_2, respectively. Find the probability distribution functions and density functions of $Y = \max\{X_1, X_2\}$ (i.e., the maximum of X_1 and X_2) and $Z = X_1 + X_2$. Do random variables Y and Z possess the memoryless property? Explain your conclusions intuitively (without any calculation). (Hint: Compare the distributions of the residuals at times t and s, i.e., $Y - t \mid Y > t$, $Y - s \mid Y > s$, $Z - t \mid Z > t$, and $Z - s \mid Z > s$, plotted in Fig. 1.3.)

While Exercise 1.1.5 shows that the set of exponential random variables is closed under the "*min*" operation, Exercise 1.1.6 indicates that the set is not closed under the "*max*" and "+" operations.

Exercise 1.1.5 can be generalized as follows.

Proposition 1.1.2 *Let* $\{X_j, j = 1,\ldots, n\}$ *be independent exponential random variables with parameters* $\{\lambda_j,\ j = 1,\ \ldots,\ n\}$, *respectively. Then* $X = \min\{X_1,\ \ldots,\ X_n\}$ *is exponentially distributed with parameter* $\lambda_1 + \ldots + \lambda_n$.

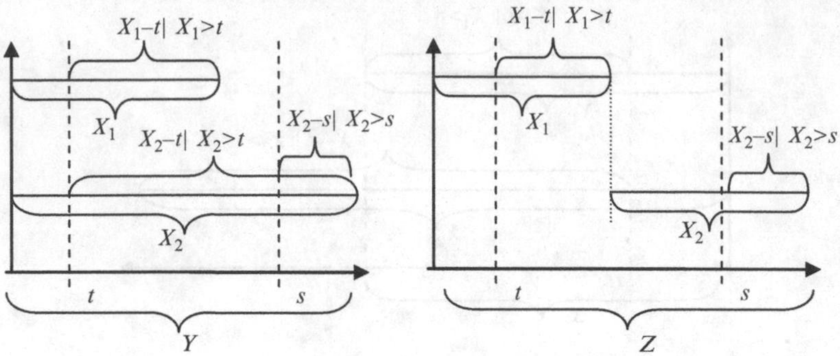

Fig. 1.3 Plots of $Y = \max\{X_1, X_2\}$ and $Z = X_1 + X_2$

Proof. We use the induction method to prove the proposition. By Exercise 1.1.5, the result holds for $n = 2$. Suppose that the result holds for $n = k$. For $n = k + 1$, $X = \min\{X_1, \ldots, X_{k+1}\} = \min\{\min\{X_1, \ldots, X_k\}, X_{k+1}\}$. By the induction assumption, $\min\{X_1, \ldots, X_k\}$ has an exponential distribution with parameter $\lambda_1 + \ldots + \lambda_k$. Again by Exercise 1.1.5, X has an exponential distribution with parameter $(\lambda_1 + \ldots + \lambda_k) + \lambda_{k+1} = \lambda_1 + \ldots + \lambda_{k+1}$. Therefore, the result holds for any positive integer n. This completes the proof of Proposition 1.1.2.

Exercise 1.1.7 Let X_1 and X_2 be independent exponential random variables with parameters λ_1 and λ_2, respectively. Show that $P\{X_1 \le X_2\} = \lambda_1/(\lambda_1 + \lambda_2)$. (Hint: Use conditional probability to obtain $P\{X_1 \le X_2\} = \int_0^\infty P\{X_1 \le X_2 | X_1 = s\} \mathrm{d}P\{X_1 \le s\} = \int_0^\infty P\{X_2 \ge s\} \mathrm{d}P\{X_1 \le s\}$.)

Exercise 1.1.7 can be generalized as follows.

Proposition 1.1.3 *Let* $\{X_j, j = 1, \ldots, n\}$ *be independent exponential random variables with parameters* $\{\lambda_j, j = 1, \ldots, n\}$*, respectively. Then* $P\{X_1 = \min\{X_1, \ldots, X_n\}\} = \lambda_1/(\lambda_1 + \ldots + \lambda_n)$.

Exercise 1.1.8 Use Proposition 1.1.2 and Exercise 1.1.7 to prove Proposition 1.1.3.

Propositions 1.1.2 and 1.1.3 play a key role in the introduction of continuous time Markov chains for stochastic systems in which exponential distributions are involved. We shall use the two propositions repeatedly throughout this book.

The *Laplace-Stieltjes transform* (LST) is a useful tool in the study and applications of probability distributions. There is a one-to-one relationship between LSTs of probability distribution functions and random variables. For exponential random variable X with parameter λ, we have, for $s \ge 0$,

$$\begin{aligned} f^*(s) &\equiv E[e^{-sX}] \equiv \int_0^\infty e^{-st} \mathrm{d}P\{X < t\} \\ &= \int_0^\infty e^{-st} \lambda e^{-\lambda t} \mathrm{d}t = \lambda \int_0^\infty e^{-(\lambda+s)t} \mathrm{d}t = \frac{\lambda}{\lambda + s}. \end{aligned} \tag{1.4}$$

The function $f^*(s)$ is called the LST of X. It is true that any random variable with an LST given by Eq. (1.4) has an exponential distribution with parameter λ.

Exponential distributions can be used in the construction of more complicated probability distributions. Exercise 1.1.9 presents a simple and well-known example. An *Erlang distribution* with parameters (n, λ) is defined as a probability distribution with density function

$$f_{(n,\lambda)}(t) = \frac{t^{n-1}\lambda^n}{(n-1)!}\exp(-\lambda t), \quad t \geq 0. \tag{1.5}$$

Exercise 1.1.9 Assume that $\{X_j, j = 1, \ldots, n\}$ are independent exponential random variables with the same parameter λ. Show that $X = X_1 + \ldots + X_n$ has an Erlang distribution with parameters (n, λ) using two methods: (i) the induction method; and (ii) LSTs of X and the Erlang distribution defined by Eq. (1.5) (i.e., LST $= \int_0^\infty e^{-st} f_{(n,\lambda)}(t)dt$). (Hint: Find the probability distribution function of the Erlang distribution from Eq. (1.5). For independent random variables, the LST of their sum equals the product of their individual LSTs.)

Based on Exercise 1.1.9, an Erlang random variable can be defined as the sum of a finite number of independent and identically distributed exponential random variables. This relationship provides great flexibility in the use of Erlang distributions. For instance, it allows us to immediately determine the mean and variance of the Erlang distribution as n/λ and n/λ^2, respectively.

Exercise 1.1.10 For an exponential distribution with parameter λ, show that the *failure rate* at time t, defined as $f(t)/(1 - F(t))$, is given by λ. Explain the result intuitively by linking it to the memoryless property.

Note that $P\{t < X < t + \delta t \mid X > t\} \approx \delta t\, f(t)/(1 - F(t))$, which, in reliability theory, is interpreted as the expected number of failures in $(t, t + \delta t)$. Then $P\{t < X < t + \delta t \mid X > t\}/\delta t \to f(t)/(1 - F(t))$, as $\delta t \to 0$, is called the failure rate at time t (i.e., the number of failures per unit time at time t).

Proposition 1.1.4 *Let $\{X_n, n = 1, 2, \ldots\}$ be independent exponential random variables with the same parameter λ. Assume that random variable N, independent of $\{X_n, n = 1, 2, \ldots\}$, has a* geometric distribution *with parameter p on positive integers $\{1, 2, \ldots\}$, i.e., $P\{N = n\} = (1 - p)p^{n-1}$, $n = 1, 2, \ldots$. Define $Y = \sum_{n=1}^{N} X_n$. Then Y has an exponential distribution with parameter* $(1 - p)\lambda$.

Proof. The proof is done by routine calculations of the LST of Y. Conditioning on N, we obtain

$$\begin{aligned}
E[\exp(-sY)] &= \sum_{n=1}^{\infty} P\{N=n\}E[\exp(-s(X_1+\ldots+X_N))|N=n] \\
&= \sum_{n=1}^{\infty}(1-p)p^{n-1}E[\exp(-s(X_1+\ldots+X_n))] \\
&= \sum_{n=1}^{\infty}(1-p)p^{n-1}(E[\exp(-sX_1)])^n \qquad (1.6)\\
&= \sum_{n=1}^{\infty}(1-p)p^{n-1}\left(\frac{\lambda}{s+\lambda}\right)^n \\
&= \frac{(1-p)\lambda}{s+(1-p)\lambda},
\end{aligned}$$

which is the LST of an exponential distribution with parameter $(1-p)\lambda$. Due to the one-to-one relationship between LSTs and random variables, Y has an exponential distribution with parameter $(1-p)\lambda$. This completes the proof of Proposition 1.1.4.

Intuitively, if $Y > t$, by the memoryless property of X_n (see Exercise 1.1.4) and the memoryless property of N (see Exercise 1.1.15), $Y - t \mid Y > t$ has the same distribution as Y (whatever the distribution is). Since the exponential distribution is the only continuous distribution that possesses the memoryless property (Theorem 1.1.1), random variable Y must have an exponential distribution. The parameter of Y can be obtained as follows. Random variable Y can be considered as the total time until an exponential random variable ends and the next exponential random variable fails to begin. At any time t with $Y > t$, the failure rate λ of X_n can be decomposed into two parts: $p\lambda$ and $(1-p)\lambda$. With rate $p\lambda$, an exponential random variable ends and the next exponential random variable is generated. Then the newly generated random variable is added to Y. With rate $(1-p)\lambda$, an exponential random variable ends and no more exponential random variable is generated. Then no more time is added to Y. Thus, the failure rate of Y is $(1-p)\lambda$. Consequently, the parameter of Y is $(1-p)\lambda$.

Commentary Since the exponential distribution is widely used in science and engineering, it is studied in many textbooks. We refer to Ross (2010) for more details on the exponential distribution as well as basic tools such as LST and probability generating functions in stochastic models.

Additional Exercises and Extensions

Exercise 1.1.11 Use the LST of an exponential random variable to obtain its moments given in Exercise 1.1.3. (Hint: $E[X^n] = (-1)^n (f^*(s))^{(n)}\Big|_{s=0}$, where $(f^*(s))^{(n)}$ is the n-th derivative of $f^*(s)$.)

Exercise 1.1.12 Compute the first twenty moments of two exponential random variables with parameters $\lambda = 0.2$ and $\lambda = 2$, respectively. Comment on the results.

Exercise 1.1.13 Let X_1 and X_2 be independent exponential random variables with parameters $\lambda_1 = 1$ and $\lambda_2 = 2$, respectively. (Note: This exercise provides a good opportunity to practice some of the basic concepts in probability theory.)

(i) Find the distribution and density functions of random variables $\min\{X_1, X_2\}$, $\max\{X_1, X_2\}$, and $X_1 + X_2$. Give all the details.
(ii) Plot the density functions of $\min\{X_1, X_2\}$, $\max\{X_1, X_2\}$, and $X_1 + X_2$.
(iii) Find the LSTs of $\min\{X_1, X_2\}$, $\max\{X_1, X_2\}$, and $X_1 + X_2$.

Exercise 1.1.14 Let X be an exponential random variable with parameter λ. Show that cX has an exponential distribution with parameter λ/c, for $c > 0$.

Exercise 1.1.15 Show that a geometric distribution with parameter p on positive integer $\{1, 2, \ldots\}$, i.e., $P\{N = n\} = p^{n-1}(1 - p)$, $n = 1, 2, \ldots$, possesses the memoryless property, i.e., $P\{N = n + k \mid N > k\} = P\{N = n\}$ for n, $k = 1, 2, \ldots$. In addition, show that the *probability generating function* of N is given by $E[z^N] \equiv \sum_{n=1}^{\infty} z^n P\{N = n\} = (1 - p)z/(1 - pz)$.

Exercise 1.1.16 Let $\{X_n, n = 1, 2, \ldots\}$ be independent random variables with common LST $f^*(s)$. Define $Y = \sum_{n=1}^{N} X_n$.

(i) Assume that, independent of $\{X_n, n = 1, 2, \ldots\}$, random variable N has a geometric distribution with parameter p on positive integers $\{1, 2, \ldots\}$. Show that $E[e^{-sY}] = (1 - p)f^*(s)/(1 - pf^*(s))$.
(ii) Assume that, independent of $\{X_n, n = 0, 1, 2, \ldots\}$, random variable N has a *Poisson distribution* with parameter λ (>0), i.e.,

$$P\{N = n\} = e^{-\lambda}\lambda^n/n!, \quad \text{for } n = 0, 1, 2, \ldots. \tag{1.7}$$

Show that $E[e^{-sY}] = \exp(-\lambda(1 - f^*(s)))$.
(iii) Assume that, independent of $\{X_n, n = 1, 2, \ldots\}$, random variable N has a *binomial distribution* with parameters p and K, i.e.,

$$P\{N = n\} = \frac{K!}{n!(K - n)!} p^n (1 - p)^{K-n}, \quad \text{for } n = 0, 1, 2, \ldots, K. \tag{1.8}$$

Show that $E[e^{-sY}] = (pf^*(s) + 1 - p)^K$.
(iv) Based on the above three cases, find a general relationship between the LST of X_n, the probability generating function of N, and the LST of Y.

Exercise 1.1.17 (*Bernoulli distribution* and *binomial distribution*) Let $\{X_1, X_2, \ldots, X_n, \ldots\}$ be independent Bernoulli random variables with common parameter p, i.e., $P\{X_n = 1\} = 1 - P\{X_n = 0\} = p$, for $n = 1, 2, \ldots$.

(i) Let X be a binomial random variable with parameters (n, p). Show that $X = X_1 + X_2 + \ldots + X_n$.
(ii) Let X be a geometric distribution with parameter p. Then $X = \min_{n \geq 1}\{n: X_n = 0\}$.

Part (ii) in Exercise 1.1.17 has been used in the intuitive interpretation of Proposition 1.1.4. Both relationships in Exercise 1.1.17 will be used repeatedly to prove and/or interpret results presented in this book.

Exponential functions, for scalars and matrices, are used throughout the book. The following two exercises illustrate elementary properties of exponential functions.

Exercise 1.1.18 Based on the definition of exponential function $\exp(ct)$ (see Eq. (1.1)), prove the following basic properties mathematically: (i) $de^{ct}/dt = ce^{ct}$; and (ii) $e^{(a+b)t} = e^{at}e^{bt}$. (Hint: You need to consider the issue of convergence of infinite summations and use the *binomial formula* for $(a + b)^n$).

Exercise 1.1.19 The exponential function $\exp(t)$ can be defined by $e^t = \lim_{x\to 0}(1 + xt)^{1/x}$. Use this definition to show $e^{(a+b)t} = e^{at}e^{bt}$.

1.2 Phase-Type Distributions: Definitions and Basic Properties

Phase-type distributions were introduced in Neuts (1975) as a generalization of the exponential distribution. In this section, we give an algebraic definition and a probabilistic definition of phase-type distributions, and present some basic properties.

Definition 1.2.1 A nonnegative random variable X has a *phase-type distribution* (*PH*-distribution) if its distribution function is given by

$$F(t) = P\{X \leq t\} = 1 - \boldsymbol{\alpha}\exp(Tt)\mathbf{e} \equiv 1 - \boldsymbol{\alpha}\left(\sum_{n=0}^{\infty}\frac{t^n}{n!}T^n\right)\mathbf{e}, \quad t \geq 0, \tag{1.9}$$

where

(i) $\mathbf{e}$ is the column vector with all elements being one;
(ii) $\boldsymbol{\alpha}$ is a *substochastic* vector of order m, i.e., $\boldsymbol{\alpha}$ is a row vector, all elements of $\boldsymbol{\alpha}$ are nonnegative, and $\boldsymbol{\alpha}\mathbf{e} \leq 1$, where m is a positive integer; and
(iii) T is a *subgenerator* of order m, i.e., T is an $m\times m$ matrix such that (1) all diagonal elements are negative; (2) all off-diagonal elements are nonnegative; (3) all row sums are non-positive; and (4) T is invertible.

We shall call T a *PH-generator*. The 2-tuple $(\boldsymbol{\alpha}, T)$ is called a *phase-type representation* (*PH-representation*) of order m for the *PH*-distribution.

Example 1.2.1 The following are four *PH*-distributions with *PH*-representations:

$$\boldsymbol{\alpha}_1 = (0.2,\ 0.8),\ T_1 = \begin{pmatrix} -2 & 1 \\ 0.5 & -10 \end{pmatrix};$$

$$\boldsymbol{\alpha}_2 = (0.1,\ 0,\ 0.9),\quad T_2 = \begin{pmatrix} -15 & 0 & 10 \\ 0 & -2 & 0 \\ 0 & 2 & -2 \end{pmatrix};$$

$$\boldsymbol{\alpha}_3 = (0,\ 0.1,\ 0.8,\ 0.1),\quad T_3 = \begin{pmatrix} -20 & 0 & 0 & 15 \\ 20 & -20 & 0 & 0 \\ 0 & 0.5 & -1 & 0.5 \\ 0 & 0 & 0 & -1 \end{pmatrix};\ \text{and} \tag{1.10}$$

$$\boldsymbol{\alpha}_4 = (0,\ 0.2,\ 0,\ 0,\ 0.8),\quad T_4 = \begin{pmatrix} -15 & 0 & 0 & 0 & 0 \\ 15 & -15 & 0 & 0 & 0 \\ 0 & 0 & -4 & 0 & 0 \\ 0 & 0 & 2 & -2 & 0 \\ 0 & 0 & 0 & 1 & -1 \end{pmatrix}.$$

Proposition 1.2.1 *For PH-representation* ($\boldsymbol{\alpha}$, *T*) *given in Definition* 1.2.1, *the function F*(*t*) *defined in Eq.* (1.9) *is a probability distribution and its density function is given by*

$$f(t) = \frac{dF(t)}{dt} = \boldsymbol{\alpha}\exp(Tt)\mathbf{T}^0,\quad t \ge 0, \tag{1.11}$$

where $\mathbf{T}^0 = -T\mathbf{e}$.

Proof. The function *f*(*t*) is obtained by taking the derivative of the function *F*(*t*). Taking the derivative of the matrix exponential function exp(*Tt*) is similar to that of the scalar case. However, a rigorous mathematical justification is required and has been established in the existing literature (see Minc (1988)). For more details about the matrix exponential function exp(*Tt*), see Minc (1988) and Exercises 1.2.22, 1.2.23, and 1.2.24. To prove Eq. (1.11), we have, for $t \ge 0$,

$$\begin{aligned}\frac{d\exp(Tt)}{dt} &= \frac{d}{dt}\left(\sum_{n=0}^{\infty}\frac{T^n t^n}{n!}\right) \\ &= \sum_{n=0}^{\infty}\frac{d}{dt}\left(\frac{T^n t^n}{n!}\right) = \sum_{n=1}^{\infty}\frac{T^n t^{n-1}}{(n-1)!} = \exp(Tt)T,\end{aligned} \tag{1.12}$$

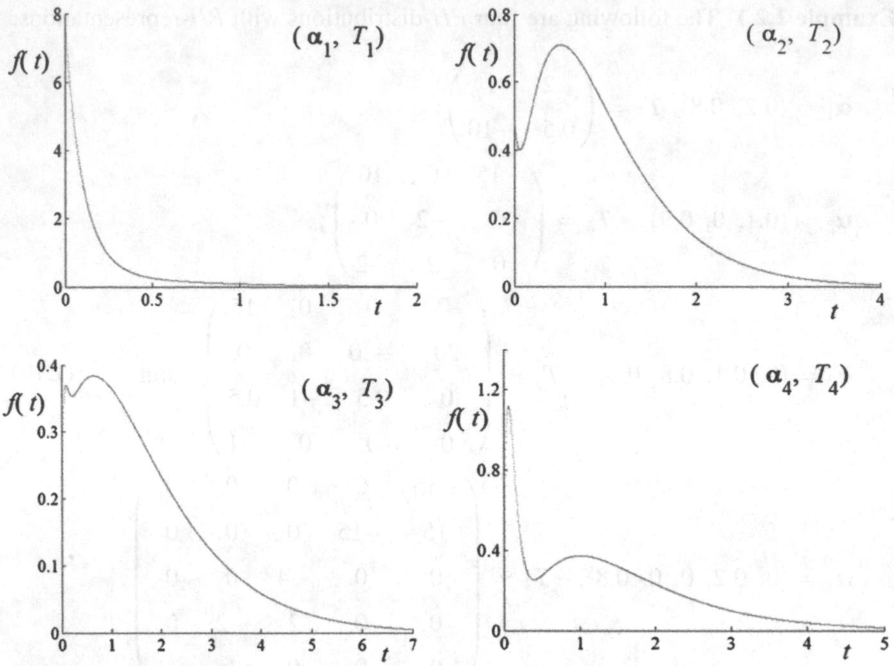

Fig. 1.4 Density functions of *PH*-distributions in Example 1.2.1

where the exchange of differentiation and summation and the convergence of the summation can be justified. Details are omitted. Premultiplying by $\boldsymbol{\alpha}$ and postmultiplying by $-\mathbf{e}$ on both sides of Eq. (1.12), Eq. (1.11) is obtained.

Rewrite $f(t) = \boldsymbol{\alpha}\exp(Tt)\mathbf{T}^0 = e^{-ct}\boldsymbol{\alpha}\exp((cI+T)t)\mathbf{T}^0$, where c is greater than or equal to the absolute value of any diagonal element of T. Then $cI + T$ and $\exp((cI + T)t)$ are nonnegative matrices. Since row sums of T are nonpositive, $\mathbf{T}^0$ is nonnegative. Consequently, $f(t)$ is nonnegative and $F(t)$ is nondecreasing in t. Since $F(0) = 1 - \boldsymbol{\alpha}\mathbf{e} \geq 0$, $F(t)$ is a nonnegative and non-decreasing function. Since $-T$ is an M-matrix (see Berman and Plemmons (1979), Minc (1988), and Exercise 1.2.24), all its eigenvalues have a positive real part. Thus, all eigenvalues of T have a negative real part and, consequently, $\exp(Tt)$ converges to a matrix with all elements zero if $t \to \infty$. This implies that $F(t)$ increases to one if $t \to \infty$, and, consequently, $F(t)$ is a probability distribution function with density function $f(t)$. This completes the proof of Proposition 1.2.1.

Density functions of the four *PH*-distributions in Example 1.2.1 are plotted in Fig. 1.4.

Exercise 1.2.1 Re-plot the density functions of the *PH*-distributions given in Example 1.2.1.

As shown in Fig. 1.4, the four density functions are quite different. In fact, by changing the parameters of *PH*-representations of a small order (e.g., $m = 2, 3, 4$, or 5), we can generate a broad range of probability distributions with different features. Example 1.2.1 indicates that the set of *PH*-distributions can be large and versatile, which is confirmed formally by the following theorem.

Theorem 1.2.1 *The set of PH-distributions is dense in the set of probability distributions on the nonnegative half-line.*

It is well-known that the set of generalized Erlang distributions (to be defined in Example 1.2.3) is dense in the set of all probability distributions on the nonnegative half-line. Since the set of generalized Erlang distributions is a subset of *PH*-distributions, Theorem 1.2.1 follows. A detailed proof of Theorem 1.2.1 can be found, for example, in Asmussen (2000).

Theorem 1.2.1 implies that *PH*-distributions can approximate any probability distribution on the nonnegative half-line, providing one of the main reasons that they are used so widely in stochastic modeling. The probabilistic interpretation associated with the *PH*-representations of *PH*-distributions presents another major reason. To explain, we give a probabilistic definition of *PH*-distributions, which is the first definition of *PH*-distributions given in Neuts (1975). First, we use the following example and exercise to review *continuous time Markov chains* that are utilized in the definition of *PH*-distributions.

Example 1.2.2 Consider a continuous time Markov chain $\{I(t), t \geq 0\}$ with five states $\{1, 2, 3, 4, 5\}$ and *infinitesimal generator* (i.e., Q-matrix)

$$Q = \begin{matrix} 1 \\ 2 \\ 3 \\ 4 \\ 5 \end{matrix} \begin{pmatrix} -20 & 0 & 0 & 15 & 5 \\ 20 & -20 & 0 & 0 & 0 \\ 0 & 0.5 & -1 & 0.5 & 0 \\ 0 & 0 & 0 & -1 & 1 \\ 0 & 0 & 0 & 0 & 0 \end{pmatrix}. \tag{1.13}$$

Note that an infinitesimal generator is a finite/infinite matrix, for which (1) all off-diagonal elements are nonnegative, (2) all diagonal elements are negative or zero, and (3) all row sums are zero.

By the definition of continuous time Markov chains, the *sojourn times* in states $\{1, 2, 3, 4, 5\}$ of $\{I(t), t \geq 0\}$ are exponentially distributed with parameters 20, 20, 1, 1, and 0, respectively. State 5 is special since the parameter of its sojourn time is zero, which implies that there is no transition of state once the Markov chain enters state 5. Consequently, the Markov chain will remain in state 5 forever. We call state 5 an *absorption state*.

Let $p_{i,j}(t)$ be the probability that the Markov chain is in state j at time t, given $I(0) = i$, for $i, j = 1, 2, \ldots, 5$. Let $P(t) = (p_{i,j}(t))$, a 5×5 matrix. By the general theory of continuous time Markov chain, we have $P(t) = \exp(Qt)$, for $t \geq 0$. If $I(0)$ has a probability distribution $\boldsymbol{\gamma}$, then the distribution of $I(t)$ is given by $\boldsymbol{\gamma}\exp(Qt)$, for $t \geq 0$.

Note that we must have $\boldsymbol{\gamma}\mathbf{e} = 1$. We also have $\boldsymbol{\gamma}\exp(Qt)\mathbf{e} = 1$, which can be shown algebraically (since $Q\mathbf{e} = 0$ and $\boldsymbol{\gamma}\mathbf{e} = 1$) and probabilistically (since the Markov chain has to be in one of the five states at time t). (Note: throughout the paper, "0" is also used for vectors and matrices with all elements being zero.)

Exercise 1.2.2 Draw a *sample path* of the Markov chain defined in Example 1.2.2. What is the probability that the Markov chain is in state 5 at time $t = 0.5$, 1, and 15, given that $I(0) = 2$?

In Example 1.2.2, state 5 is an absorption state, i.e., once the Markov chain enters state 5, it stays there forever. Thus, the time until the Markov chain enters state 5 is well defined, and is called the *absorption time* of state 5. The answer to Exercise 1.2.2 actually leads to the distribution of the absorption time of state 5, given that the Markov chain is initially in state 2.

In general, define a continuous time Markov chain $\{I(t), t \geq 0\}$ with $m + 1$ states $\{1, 2, \ldots, m, m + 1\}$ and infinitesimal generator

$$Q = \begin{pmatrix} T & \mathbf{T}^0 \\ 0 & 0 \end{pmatrix}, \tag{1.14}$$

where T is a *PH*-generator (see Definition 1.2.1). Since the row sums of an infinitesimal generator are zero, we must have $\mathbf{T}^0 = -T\mathbf{e}$. Since the total transition rate of state $m + 1$ is zero, state $m + 1$ is an *absorption state*. Define

$$X = \min\{t:\ I(t) = m + 1, t \geq 0\}, \tag{1.15}$$

which is the absorption time of state $m + 1$. Then $P\{X < t\}$ can be interpreted as the probability that the Markov chain has been absorbed into state $m + 1$ before time t.

For the Markov chain defined in Example 1.2.2 (see Eq. (1.13)), state 5 is an absorption state. The absorption time of state 5 has a *PH*-distribution with $m = 4$, and

$$T = \begin{pmatrix} -20 & 0 & 0 & 15 \\ 20 & -20 & 0 & 0 \\ 0 & 0.5 & -1 & 0.5 \\ 0 & 0 & 0 & -1 \end{pmatrix}, \quad \mathbf{T}^0 = \begin{pmatrix} 5 \\ 0 \\ 0 \\ 1 \end{pmatrix}. \tag{1.16}$$

Note that T in Eq. (1.16) is T_3 in Example 1.2.1. Figure 1.5 shows two sample paths of the corresponding X and $\{I(t), t \geq 0\}$, which are generated using initial probability vector $\boldsymbol{\alpha}_3$ given in Example 1.2.1.

Given an initial distribution $(\boldsymbol{\alpha}, 1 - \boldsymbol{\alpha}\mathbf{e})$ for the Markov chain $\{I(t), t \geq 0\}$, the probability distribution of $I(t)$ is given by $(\boldsymbol{\alpha}, 1 - \boldsymbol{\alpha}\mathbf{e})P(t)$, which can be evaluated as

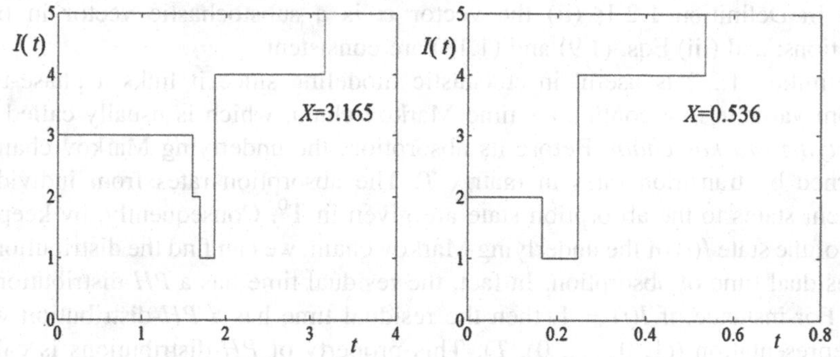

Fig. 1.5 Two sample paths of $\{I(t), t \geq 0\}$ for $(\boldsymbol{\alpha}_3, T_3)$ in Example 1.2.1

$$
\begin{aligned}
(\boldsymbol{\alpha}, 1-\boldsymbol{\alpha}\mathbf{e}) \exp\left(\begin{pmatrix} T & \mathbf{T}^0 \\ 0 & 0 \end{pmatrix} t\right)
&= (\boldsymbol{\alpha}, 1-\boldsymbol{\alpha}\mathbf{e}) \sum_{n=0}^{\infty} \frac{t^n}{n!} \begin{pmatrix} T & \mathbf{T}^0 \\ 0 & 0 \end{pmatrix}^n \\
&= (\boldsymbol{\alpha}, 1-\boldsymbol{\alpha}\mathbf{e}) \begin{pmatrix} \sum_{n=0}^{\infty} \frac{t^n}{n!} T^n & \sum_{n=1}^{\infty} \frac{t^n}{n!} T^{n-1}\mathbf{T}^0 \\ 0 & 1 \end{pmatrix} \\
&= (\boldsymbol{\alpha}, 1-\boldsymbol{\alpha}\mathbf{e}) \begin{pmatrix} \exp(Tt) & \mathbf{e} - \exp(Tt)\mathbf{e} \\ 0 & 1 \end{pmatrix} \\
&= (\boldsymbol{\alpha} \exp(Tt), \;\; 1 - \boldsymbol{\alpha} \exp(Tt)\mathbf{e}).
\end{aligned} \tag{1.17}
$$

The second term in the last line in Eq. (1.17) gives the probability that the Markov chain is in state $m + 1$ at time t, i.e., probability $P\{X \leq t\}$. Thus, we obtain

$$
P\{X \leq t\} = P\{I(t) = m+1\} = 1 - \boldsymbol{\alpha} \exp(Tt)\mathbf{e}, \quad t \geq 0. \tag{1.18}
$$

Exercise 1.2.3 Explain the elements in matrix $\exp(Tt)$ and vector $\mathbf{e} - \exp(Tt)\mathbf{e}$ probabilistically.

Equation (1.18) leads to the following definition of *PH*-distributions.

Definition 1.2.2 Assume that the continuous time Markov chain $\{I(t), t \geq 0\}$ defined by Eq. (1.14) will be absorbed into state $m + 1$ with probability one. A *phase-type random variable X* is defined as the absorption time of state $m + 1$ of the continuous time Markov chain $\{I(t), t \geq 0\}$, given that the initial distribution of the Markov chain is $(\boldsymbol{\alpha}, 1 - \boldsymbol{\alpha}\mathbf{e})$. The tuple $(\boldsymbol{\alpha}, T)$ is a *PH*-representation of X.

Definitions 1.2.1 and 1.2.2 are equivalent since (i) the matrix T for the Markov chain $\{I(t), t \geq 0\}$ with an absorption state satisfies conditions given in Definition 1.2.1; (ii) the vector $\boldsymbol{\alpha}$ is a substochastic vector in both definitions; and (iii) Eqs. (1.9) and (1.18) are consistent.

Definition 1.2.2 is useful in stochastic modeling since it links a phase-type random variable to a continuous time Markov chain, which is usually called the *underlying Markov chain*. Before its absorption, the underlying Markov chain is governed by transition rates in matrix T. The absorption rates from individual transient states to the absorption state are given in $\mathbf{T}^0$. Consequently, by keeping track of the state $I(t)$ of the underlying Markov chain, we can find the distribution of the residual time of absorption. In fact, the residual time has a *PH*-distribution as well. For instance, if $I(t) = 1$, then the residual time has a *PH*-distribution with *PH*-representation ((1, 0, …, 0), T). This property of *PH*-distributions is called the *partial memoryless* property based on the state of the underlying Markov chain. The probabilistic interpretation of *PH*-distributions described in Definition 1.2.2 is key for the applications of *PH*-distributions in stochastic modeling. In later chapters, whenever *PH*-distributions are utilized, the state variable $I(t)$ will be used as an ancillary variable in the introduction of a Markov chain for the stochastic model of interest.

Exercise 1.2.4 Consider the *PH*-distribution $(\boldsymbol{\alpha}_3, T_3)$ in Example 1.2.1. (i) Suppose that at time $t = 1.5$, the underlying Markov chain is in state 2. Find the distribution of the residual time (i.e., the distribution of conditional random variable $X - 1.5 \mid \{X > 1.5, I(1.5) = 2\}$), and plot its density function. (ii) Find the distribution of $X - 10 \mid \{X > 10, I(10) = 2\}$. (iii) Find the distribution of $X - 10 \mid \{I(10) = 2\}$.

We remark that, traditionally, a state of the underlying Markov chain of a *PH*-distribution is called a *phase*. Thus, in the rest of the book, we shall use "phase" and "state" interchangeably.

Example 1.2.3 presents *PH*-representations of several well-known probability distributions that are special cases of *PH*-distributions. The *PH*-representations given in Example 1.2.3 can be verified by direct calculations using Definition 1.2.1 or by applying the probabilistic interpretation provided in Definition 1.2.2.

Example 1.2.3 Here are some well-known *PH*-distributions and their corresponding *PH*-representations.

(1) The exponential distribution $F(t) = 1 - \exp(-\lambda t)$, $t \geq 0:$ $\boldsymbol{\alpha} = 1$, $T = -\lambda$, and $m = 1$.

(2) Erlang distribution $F(t) = 1 - \sum_{j=0}^{m-1} e^{-\lambda t}(\lambda t)^j/j!$, $t \geq 0$, for $m \geq 1$ and $\lambda > 0$:

$$\left(\boldsymbol{\alpha} = (0, \ldots, 0,\ 1), \quad T = E(m, \lambda) = \begin{pmatrix} -\lambda & & & \\ \lambda & -\lambda & & \\ & \ddots & \ddots & \\ & & \lambda & -\lambda \end{pmatrix}_{m \times m} \right). \tag{1.19}$$

(3) *Generalized Erlang distribution* $F(t) = 1 - \sum_{k=1}^{m} \alpha_k \left(\sum_{j=0}^{k-1} e^{-\lambda t} (\lambda t)^j / j! \right)$, $t \geq 0$, for $\lambda > 0$,

$$\left(\boldsymbol{\alpha} = (\alpha_1, \ldots, \alpha_m), \quad T = \begin{pmatrix} -\lambda & & & \\ \lambda & -\lambda & & \\ & \ddots & \ddots & \\ & & \lambda & -\lambda \end{pmatrix}_{m \times m} \right), \tag{1.20}$$

where $\boldsymbol{\alpha}$ is a substochastic vector.

(4) *Coxian distribution* $F(t) = 1 - \sum_{k=1}^{m} \alpha_k \left(\sum_{j=1}^{k} e^{-\lambda_j t} \prod_{i=1:i \neq j}^{k} \left(\lambda_i / (\lambda_i - \lambda_j) \right) \right)$, $t \geq 0$:

$$\left(\boldsymbol{\alpha} = (\alpha_1, \ldots, \alpha_m), \quad T = \begin{pmatrix} -\lambda_1 & & & \\ \lambda_2 & -\lambda_2 & & \\ & \ddots & \ddots & \\ & & \lambda_m & -\lambda_m \end{pmatrix}_{m \times m} \right), \tag{1.21}$$

where $\boldsymbol{\alpha}$ is a substochastic vector and $\lambda_j > 0, j = 1, \ldots, m$. Note that $\{\lambda_j, j = 1, \ldots, m\}$ must be different positive real numbers in the analytic expression of $F(t)$. The corresponding *PH*-representation does not impose the constraint on the parameters, and is valid for all positive $\{\lambda_j, j = 1, \ldots, m\}$, which is a key advantage using *PH*-representations. The *PH*-representation in Eq. (1.21) is called a *Coxian representation*.

Note: The missing elements in matrix T are all zeros. Throughout this book, if an element or a block of elements is missing, then it is zero or a block of zeros.

For part (1) of Example 1.2.3, the exponential distribution can be viewed as the absorption time of a continuous time Markov chain with two phases: a transient phase with sojourn time exponentially distributed with parameter λ; and an absorption phase. For part (2), based on Exercise 1.1.9, the Erlang random variable is the sum of m independent exponential random variables with common parameter λ. Then we introduce an underlying Markov chain with m transient phases and an absorption phase as in Eq. (1.19). The sojourn times in the transient phases are independent exponential random variables with common parameter λ. If the underlying process begins in phase m, it will enter the absorption phase after going through all m transient phases. Consequently, the absorption time is the sum of m independent exponential random variables with common parameter λ. Parts (3) and (4) can be shown similarly.

Example 1.2.3 shows that the *PH*-distribution is a natural generalization of the exponential, Erlang (1917), and Coxian (Cox (1955a, b)) distributions.

Exercise 1.2.5 Prove the *PH*-representations given in parts (3) and (4) of Example 1.2.3 for the generalized Erlang distribution and the Coxian distribution, respectively. (Hint: Both the generalized Erlang random variables and Coxian random variables can be viewed as the mixed sum of independent exponential random variables.)

Two basic properties of *PH*-distributions are given in the following proposition.

Proposition 1.2.2 *The following properties hold for PH-random variable X with PH-representation* $(\boldsymbol{\alpha}, T)$.

(i) *The LST of X is* $f^*(s) = 1 - \boldsymbol{\alpha}\mathbf{e} + \boldsymbol{\alpha}(sI - T)^{-1}\mathbf{T}^0$, *for* $s \geq 0$, *where I is the identity matrix.*
(ii) $E[X^n] = (-1)^n n! \boldsymbol{\alpha} T^{-n}\mathbf{e}$, for $n = 1, 2, \ldots$.

Proof. First note $F(0-) = \lim_{t\uparrow 0} F(t) = 0$ and $F(0+) = \lim_{t\downarrow 0} F(t) = 1 - \boldsymbol{\alpha}\mathbf{e}$. Part (i) is obtained as follows:

$$\begin{aligned} f^*(s) = E[e^{-sX}] &= \int_0^\infty e^{-st}\mathrm{d}F(t) \\ &= \int_0^{0+} e^0 \mathrm{d}F(t) + \int_{0+}^\infty e^{-st}\mathrm{d}F(t) \\ &= F(0+) - F(0-) + \int_{0+}^\infty e^{-st}\mathrm{d}F(t) \\ &= 1 - \boldsymbol{\alpha}\mathbf{e} + \int_{0+}^\infty e^{-st} f(t)\mathrm{d}t \\ &= 1 - \boldsymbol{\alpha}\mathbf{e} + \int_{0+}^\infty e^{-st}\boldsymbol{\alpha}\exp(Tt)\mathbf{T}^0\mathrm{d}t \\ &= 1 - \boldsymbol{\alpha}\mathbf{e} + \boldsymbol{\alpha}\int_{0+}^\infty \exp((-sI + T)t)\mathrm{d}t\mathbf{T}^0, \end{aligned} \tag{1.22}$$

which leads to the expected result since $-sI + T$ is invertible for $s \geq 0$.

Part (ii) can be obtained by using Part (i) and the induction method as follows. By taking the *n*-th derivative of the LST of *X*, we obtain

$$\begin{aligned} E[X^n] &= (-1)^n (f^*(s))^{(n)}\Big|_{s=0} \\ &= (-1)^n \left((-1)^{n-1}(n-1)!\boldsymbol{\alpha}(sI - T)^{-n}\mathbf{T}^0\right)^{(1)}\Big|_{s=0} \\ &= (-1)^n \left((-1)^n n!\boldsymbol{\alpha}(sI - T)^{-n-1}\mathbf{T}^0\right)\Big|_{s=0} \\ &= (-1)^n n!\boldsymbol{\alpha}T^{-n}(-T^{-1}\mathbf{T}^0). \end{aligned} \tag{1.23}$$

Since $-T^{-1}\mathbf{T}^0 = \mathbf{e}$, Eq. (1.23) leads to the expected result. This completes the proof of Proposition 1.2.2.

We note that, in probability theory, computing moments of a continuous distribution involves integration (i.e., $E[X^n] = \int_0^\infty x^n \mathrm{d}F(x)$). Using *PH*-distributions transforms integration into matrix calculations, which can be done effectively and efficiently. This is another advantage of using *PH*-distributions in stochastic modeling.

A fundamental question about *PH*-distributions is: what kind of probability distribution are *PH*-distributions? The following theorem gives a complete answer in terms of LST, which is useful in both theoretical study and computation of *PH*-distributions.

Theorem 1.2.2 (O'Cinneide (1990a)) *A probability distribution on the nonnegative half-line is a PH-distribution if and only if it is either the point mass at zero or (a) it has a strictly positive continuous density on the positive real numbers, and (b) it has a rational Laplace-Stieltjes transform with a unique pole of maximal real part.*

We note that advanced mathematical techniques are required to prove Theorem 1.2.2 and a number of other results presented in this book. We choose not to include proofs for them. Readers are referred to the original papers/books for proofs.

Commentary We recommend Neuts (1981), Latouche and Ramaswami (1999), and Asmussen (2000) for further reading on basic properties and applications of *PH*-distributions. In particular, Chapter 2 in Neuts (1981) gives a comprehensive treatment of *PH*-distributions. For advanced characterizations of *PH*-distributions, we refer readers to a serial of papers by O'Cinneide (1989, 1990a, 1991a, b, 1993, and 1999).

Additional Exercises and Extensions

Exercise 1.2.6 (1) Use part (i) in Proposition 1.2.2 to find the LST for the Coxian distribution defined in Eq. (1.21). (2) Use the probabilistic interpretation of *PH*-representations to find the LST for the Coxian distribution defined in Eq. (1.21).

Exercise 1.2.7 Compute the first 20 moments of the *PH*-distributions given in Example 1.2.1. Comment on the results.

Exercise 1.2.8 Assume that $(\boldsymbol{\alpha}_1, T_1)$ and $(\boldsymbol{\alpha}_2, T_2)$ are Erlang distributions with parameters $\{m_1 = 5,\ \lambda_1 = 2\}$ and $\{m_2 = 5,\ \lambda_2 = 0.2\}$, respectively. Define $\boldsymbol{\alpha} = (0.5\boldsymbol{\alpha}_1, 0.5\boldsymbol{\alpha}_2)$ and

$$T = \begin{pmatrix} T_1 & 0 \\ 0 & T_2 \end{pmatrix}. \tag{1.24}$$

It is easy to verify that $(\boldsymbol{\alpha}, T)$ is a *PH*-representation. The density function of the corresponding *PH*-distribution is plotted in Fig. 1.6a, which has two peak points. Find a *PH*-distribution with a density function similar to Fig. 1.6b, which has three peak points.

Exercise 1.2.9 Show that the probability distribution with a density function $f(t) = c(t-1)^2\exp(-2t)$, for $t \geq 0$, where c is a normalization factor, is not a *PH*-distribution. (Hint: See Theorem 1.2.2)

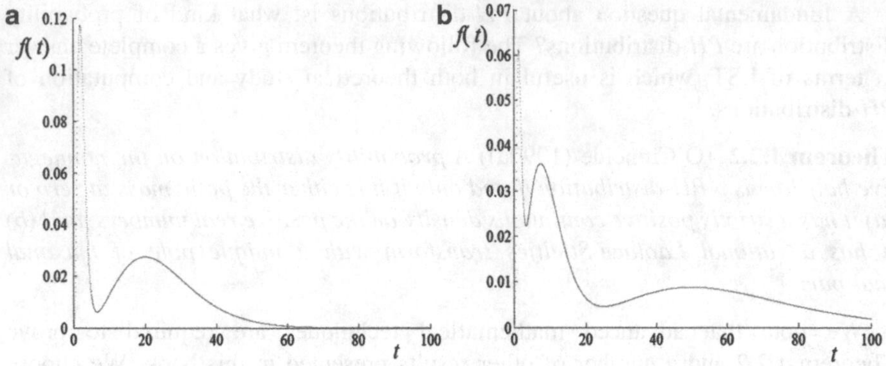

Fig. 1.6 Density functions for Exercise 1.2.8

Exercise 1.2.10 Consider the matrix representation

$$\boldsymbol{\alpha} = (0.1, 0.2, 0.7), \quad T = \begin{pmatrix} -15 & 2 & 0 \\ 0 & -2 & 2 \\ 0 & 5 & -5 \end{pmatrix}, \tag{1.25}$$

Is $(\boldsymbol{\alpha}, T)$ a *PH*-representation? Why or why not? Plot $F(t) = 1-\boldsymbol{\alpha}\exp(Tt)\mathbf{e}$, for $t \geq 0$.

Exercise 1.2.11 Assume that T and S are both *PH*-generators of the same order. Show that (i) $T + S$ is also a *PH*-generator; and (ii) cT is a *PH*-generator for $c > 0$. (Note: The results in this exercise imply that the set of all *PH*-generators of the same order form a *convex set* (Rockafellar (1970)).)

Exercise 1.2.12 Assume that X has a *PH*-distribution with matrix representation $(\boldsymbol{\alpha}, T)$. Let c be a positive real number. Show that cX has a *PH*-distribution. Find a *PH*-representation for cX.

Exercise 1.2.12 indicates that it is easy to find a *PH*-distribution with any given mean by normalizing an existing *PH*-distribution.

Exercise 1.2.13 Assume that X_1 and X_2 are two *PH*-distributions with *PH*-representations $(\boldsymbol{\alpha}_1, T_1)$ and $(\boldsymbol{\alpha}_2, T_2)$, respectively. Define $X = X_1$, w.p. p; X_2, w.p. $1 - p$. Show that X has a *PH*-distribution and find its *PH*-representation. (Note: "w.p." means "with probability". This exercise indicates that the set of *PH*-distributions are closed under convex combination, i.e., $pF_1(t) + (1-p)F_2(t)$, where $F_1(t)$ and $F_2(t)$ are the probability distributions of X_1 and X_2, respectively.)

Exercise 1.2.14 Let $(\mathbf{e}(m), E(m, \lambda))$ be the *PH*-representation of an Erlang distribution with parameters m and λ (see Eq. (1.19)), where $\mathbf{e}(m) = (0, \ldots, 0, 1)$. Find the mean and variance of the Erlang distribution $(\mathbf{e}(m), E(m, \lambda))$. Use the *Central Limit Theorem* in probability theory (Ross (2010)) to show that the sequence of distributions $\{(\mathbf{e}(m), E(m, \lambda m)), m = 1, 2, \ldots\}$ converges to constant $1/\lambda$.

Exercise 1.2.15 Find a *PH*-distribution that approximates a probability distribution with $P\{X = 1\} = 0.3$ and $P\{X = 5\} = 0.7$. Plot the density function of your *PH*-distribution. (Hint: Use results in Exercises 1.2.13 and 1.2.14.)

Exercise 1.2.16 Assume that X has a *PH*-distribution with *PH*-representation $(\boldsymbol{\alpha}, T)$. Show that $X - t \mid X \geq t$ has a *PH*-distribution with *PH*-representation $(\boldsymbol{\alpha}\exp(Tt)/(\boldsymbol{\alpha}\exp(Tt)\mathbf{e}), T)$, for $t \geq 0$.

Exercise 1.2.17 Assume that X has a *PH*-distribution with *PH*-representation $(\boldsymbol{\alpha}, T)$. Define $Y = \max\{0, X - t\}$, for $t \geq 0$. Find the probability $P\{Y = 0\}$. Show that Y has a *PH*-distribution with *PH*-representation $(\boldsymbol{\alpha}\exp(Tt), T)$.

Exercise 1.2.18 Consider *PH*-representation $(\boldsymbol{\alpha}, T)$. The eigenvalue η of T with the largest real part is real and negative (Berman and Plemmons (1979) and Minc (1988)). Show that $F(t) = ct^k\exp(-\eta t) + o(t^k\exp(-\eta t))$, as $t \to \infty$, for some $c > 0$ and $k \geq 0$. If T is *irreducible*, show that $k = 0$.

(i) Find the approximation of $F(t)$ for the Erlang distribution with parameters (λ, n).
(ii) For each *PH*-distribution in Example 1.2.1, find the approximation of $F(t)$ numerically (i.e., find η, c, and k).

Exercise 1.2.18 indicates that *PH*-distributions have an *exponential tail* and the *decay rate* is η.

Exercise 1.2.19 A four-state continuous time Markov chain $\{I(t), t \geq 0\}$ is defined as follows: $I(0)$ has a distribution (0.2, 0.5, 0.15, 0.15) and infinitesimal generator Q is given by

$$Q = \begin{matrix} 1 \\ 2 \\ 3 \\ 4 \end{matrix} \begin{pmatrix} -15 & 2 & 1 & 12 \\ 0 & -2 & 2 & 0 \\ 2 & 1 & -5 & 2 \\ 1 & 1.5 & 0 & -2.5 \end{pmatrix}. \tag{1.26}$$

Define Y as the time that the Markov chain enters states $\{2, 4\}$ for the first time. Find the distribution of the random variable Y.

Exercise 1.2.20 The *PH*-generators in Example 1.2.3 are lower triangular matrices. Find *PH*-representations with upper triangular *PH*-generators for those *PH*-distributions.

Exercise 1.2.21 Consider matrix representation

$$\boldsymbol{\alpha} = (1.125, -0.125), \quad T = \begin{pmatrix} -2 & 0 \\ 5 & -5 \end{pmatrix}. \tag{1.27}$$

Is $(\boldsymbol{\alpha}, T)$ a *PH*-representation? Plot the distribution function $F(t)$ for $(\boldsymbol{\alpha}, T)$ defined in Eq. (1.27). (Note: $F(t)$ of $(\boldsymbol{\alpha}, T)$ is actually a *PH*-distribution function. In Exercise 1.4.10, an equivalent *PH*-representation is found for $(\boldsymbol{\alpha}, T)$.)

Exercise 1.2.22 (a) Provide a complete mathematical proof of Eq. (1.12) (Hint: Convergence of infinite sum has to be considered.) (b) By the definition of matrix exponential function (see Eq. (1.9)), show that $\exp(Tt) = e^{-ct}\exp((cI + T)t)$ for any square matrix T and real numbers c and t.

Exercise 1.2.23 Use the definition given in Eq. (1.9) to show that $\exp(T(t + s)) = \exp(Tt)\exp(Ts)$.

Matrix A is called an M-matrix if (i) all off-diagonal elements are less than or equal to zero; and (ii) all eigenvalues have positive real parts. See Berman and Plemmons (1979) and Minc (1988) for detailed coverage of the M-matrix.

Exercise 1.2.24* Assume that T is a PH-generator. Verify that $-T$ is an M-matrix. (Note: This result has been used in the proofs of Proposition 1.2.1 and Exercise 1.2.18.) Since $-T$ is an M-matrix, all eigenvalues of $-T$ have a positive real part. Use the *Jordan canonical form* of matrix (Lancaster and Tismenetsky (1985)) to show $\int_0^\infty \exp(Tt)dt = -T^{-1}$.

1.3 Phase-Type Distributions: Closure Properties

In this section, we show that the set of PH-distributions is closed under a number of operations (e.g., "*min*", "*max*", "+"). The closure properties demonstrate the mathematical maneuverability of PH-distributions in stochastic modeling, which plays a key role in their applications. First, we consider a Markov chain approach to do Exercises 1.1.5, 1.1.6, and 1.1.9.

Example 1.3.1 Assume that X_1 and X_2 are independent exponential random variables with parameters λ_1 and λ_2, respectively. Show that $\min\{X_1, X_2\}$, $\max\{X_1, X_2\}$, and $X_1 + X_2$ have PH-distributions. Find a PH-representation for each.

Recall that there is a two-phase continuous time Markov chain associated with X_1 (see Example 1.2.3). Let $I_1(t)$ be the phase of the underlying Markov chain at time t. The infinitesimal generator of two-phase underlying Markov chain $\{I_1(t), t \geq 0\}$ is given by

$$Q_1 = \begin{matrix} 1 \\ 2 \end{matrix} \begin{pmatrix} -\lambda_1 & \lambda_1 \\ 0 & 0 \end{pmatrix}, \tag{1.28}$$

where phase 2 is the absorption phase. Similarly, $I_2(t)$ is defined for the two-phase underlying Markov chain associated with X_2. The two Markov chains $\{I_1(t), t \geq 0\}$ and $\{I_2(t), t \geq 0\}$ are independent.

To find the distributions of $\max\{X_1, X_2\}$ and $\min\{X_1, X_2\}$, we define a new Markov chain $\{I(t) = (I_1(t), I_2(t)), t \geq 0\}$. Since the processes $\{I_1(t), t \geq 0\}$ and

$\{I_2(t), t \geq 0\}$ are independent, $\{I(t), t \geq 0\}$ is a continuous time Markov chain with state space $\{(1, 1), (1, 2), (2, 1), (2, 2)\}$ and infinitesimal generator

$$Q = \begin{matrix} (1,1) \\ (1,2) \\ (2,1) \\ (2,2) \end{matrix} \begin{pmatrix} -(\lambda_1+\lambda_2) & \lambda_2 & \lambda_1 & 0 \\ 0 & -\lambda_1 & 0 & \lambda_1 \\ 0 & 0 & -\lambda_2 & \lambda_2 \\ 0 & 0 & 0 & 0 \end{pmatrix}. \tag{1.29}$$

(i) It is clear that $\max\{X_1, X_2\}$ is the absorption time of state (2, 2), given that the Markov chain $\{I(t), t \geq 0\}$ starts in state (1, 1). Thus, $\max\{X_1, X_2\}$ has a *PH*-distribution with *PH*-representation $\boldsymbol{\alpha} = (1, 0, 0)$ and

$$T = \begin{pmatrix} -(\lambda_1+\lambda_2) & \lambda_2 & \lambda_1 \\ 0 & -\lambda_1 & 0 \\ 0 & 0 & -\lambda_2 \end{pmatrix}. \tag{1.30}$$

(ii) It is also easy to see that $\min\{X_1, X_2\}$ is the first passage time into one of states $\{(1, 2), (2, 1), (2, 2)\}$, given that the Markov chain $\{I(t), t \geq 0\}$ starts in state (1,1). Thus, we can consolidate states $\{(1, 2), (2, 1), (2, 2)\}$ into a single *super* absorption state. The absorption time is $\min\{X_1, X_2\}$ and has an exponential distribution with *PH*-representation $\boldsymbol{\alpha} = (1)$ and $T = (-(\lambda_1 + \lambda_2))$.

(iii) For $X_1 + X_2$, the process $\{I_1(t), t \geq 0\}$ is initialized at $t = 0$. Define $J(t) = I_1(t) = 1$ for $t < X_1$. At $t = X_1$, i.e., when the process $\{I_1(t), t \geq 0\}$ enters its absorption phase, initialize $\{I_2(t), t \geq 0\}$ and set $J(t) = 2$. If $I_2(t) = 2$, set $J(t) = 3$. The process $\{J(t), t \geq 0\}$ stays in state 3, once it enters state 3. It is clear that $\{J(t), t \geq 0\}$ is a continuous time Markov chain. Then it is easy to see that $X_1 + X_2$ is the time until absorption into state 3 for the Markov chain $\{J(t), t \geq 0\}$. Thus, $X_1 + X_2$ has a *PH*-distribution with *PH*-representation $\boldsymbol{\alpha} = (1, 0)$ and

$$T = \begin{pmatrix} -\lambda_1 & \lambda_1 \\ 0 & -\lambda_2 \end{pmatrix}. \tag{1.31}$$

It is easy to see $X_1 + X_2$ is a Coxian random variable (see Example 1.2.3).

Example 1.3.1 can be generalized as follows.

Example 1.3.2 Consider *PH*-random variables X_1 and X_2 with *PH*-representations

$$\left(\boldsymbol{\alpha} = (1,\ 0),\quad T = \begin{pmatrix} -2 & 2 \\ 5 & -10 \end{pmatrix}\right) \quad \text{and}$$
$$\left(\boldsymbol{\beta} = (0.2,\ 0.8),\quad S = \begin{pmatrix} -1 & 1 \\ 3 & -4 \end{pmatrix}\right), \tag{1.32}$$

respectively. Denote by $\{I_1(t), t \geq 0\}$ and $\{I_2(t), t \geq 0\}$ the underlying Markov chains associated with the two *PH*-representations. Both Markov chains have three phases $\{1, 2, 3\}$. We define $\{I(t) = (I_1(t), I_2(t)), t \geq 0\}$, which is a Markov chain with phases $\{(1, 1), (1, 2), (1, 3), (2, 1), (2, 2), (2, 3), (3, 1), (3,2), (3,3)\}$ and infinitesimal generator

$$
\begin{array}{c} (1,1) \\ (1,2) \\ (2,1) \\ (2,2) \\ (3,1) \\ (3,2) \\ (1,3) \\ (2,3) \\ (3,3) \end{array}
\overset{\begin{array}{ccccccccc} (1,\ 1) & (1,\ 2) & (2,1) & (2,2) & (3,1) & (3,2) & (1,3) & (2,3) & (3,3) \end{array}}{
\begin{pmatrix}
-2-1 & 1 & 2 & 0 & 0 & 0 & 0 & 0 & 0 \\
3 & -2-4 & 0 & 2 & 0 & 0 & 1 & 0 & 0 \\
5 & 0 & -10-1 & 1 & 5 & 0 & 0 & 0 & 0 \\
0 & 5 & 3 & -10-4 & 0 & 5 & 0 & 1 & 0 \\
0 & 0 & 0 & 0 & -1-0 & 1 & 0 & 0 & 0 \\
0 & 0 & 0 & 0 & 3 & -4-0 & 0 & 0 & 1 \\
0 & 0 & 0 & 0 & 0 & 0 & -2-0 & 2 & 0 \\
0 & 0 & 0 & 0 & 0 & 0 & 5 & -10-0 & 5 \\
0 & 0 & 0 & 0 & 0 & 0 & 0 & 0 & 0+0
\end{pmatrix}}.
\tag{1.33}
$$

The elements in the above infinitesimal generator can be obtained by the definition of transition rates. For instance, the total transition rate for phase (2, 2) is 10 + 4. At rate 5, the process jumps from phase (2, 2) to (1, 2); at rate 3 to (2, 1); at rate 5 to (3, 2); at rate 1 to (2, 3); and at rate 0 to all other phases. The infinitesimal generator in Eq. (1.33) can be rewritten in the following matrix form:

$$
\begin{pmatrix}
T \otimes I + I \otimes S & (-T\mathbf{e}) \otimes I & I \otimes (-S\mathbf{e}) & 0 \\
0 & S & 0 & -S\mathbf{e} \\
0 & 0 & T & -T\mathbf{e} \\
0 & 0 & 0 & 0
\end{pmatrix},
\tag{1.34}
$$

where notation "$\otimes$" denotes the *Kronecker product* of two matrices. For matrices $A = (a_{i,j})$ and B, $A \otimes B$ is defined as $(a_{i,j}B)$. For T and S given in Eq. (1.32), we have

$$
\begin{aligned}
T \otimes I + I \otimes S &= \begin{pmatrix} -2 & 0 & 2 & 0 \\ 0 & -2 & 0 & 2 \\ 5 & 0 & -10 & 0 \\ 0 & 5 & 0 & -10 \end{pmatrix} + \begin{pmatrix} -1 & 1 & 0 & 0 \\ 3 & -4 & 0 & 0 \\ 0 & 0 & -1 & 1 \\ 0 & 0 & 3 & -4 \end{pmatrix} \\
&= \begin{array}{c} (1,1) \\ (1,2) \\ (2,1) \\ (2,2) \end{array}
\overset{\begin{array}{cccc} (1,1) & (1,2) & (2,1) & (2,2) \end{array}}{
\begin{pmatrix} -2-1 & 0+1 & 2+0 & 0+0 \\ 0+3 & -2-4 & 0+0 & 2+0 \\ 5+0 & 0+0 & -10-1 & 0+1 \\ 0+0 & 5+0 & 0+3 & -10-4 \end{pmatrix}},
\end{aligned}
\tag{1.35}
$$

which is the northwestern corner of the infinitesimal generator given in Eq. (1.33). Note that the size of the identity matrix I depends on the context.

Exercise 1.3.1 For matrices A, B, C, and D, show that (i) $(A \otimes B)(C \otimes D) = (AC) \otimes (BD)$; (ii) $(A + B) \otimes C = A \otimes C + B \otimes C$; and (iii) $\boldsymbol{\alpha}\exp(At)\mathbf{e}\boldsymbol{\beta}\exp(Bt)\mathbf{e} = (\boldsymbol{\alpha} \otimes \boldsymbol{\beta})\exp((A \otimes I + I \otimes B)t)(\mathbf{e} \otimes \mathbf{e})$. Assume that all matrix operations in these expressions are valid.

Exercise 1.3.2 Show that $T \otimes I + I \otimes S$ is a *PH*-generator for any two *PH*-generators T and S.

The initial distribution of the Markov chain $\{I(t), t \geq 0\}$ is $(\alpha_1\beta_1, \alpha_1\beta_2, \alpha_2\beta_1, \alpha_2\beta_2)$ on states $\{(1, 1), (1, 2), (2, 1), (2, 2)\}$, respectively, which can be written as $\boldsymbol{\alpha} \otimes \boldsymbol{\beta}$; for states (1, 3), (2, 3), (3, 1), (3, 2), (3, 3), the probabilities that the Markov chain is initially in them are zero.

(i) It is clear that $\max\{X_1, X_2\}$ is the absorption time of the phase (3, 3) of $\{I(t), t \geq 0\}$, given that the initial distribution is $(\boldsymbol{\alpha} \otimes \boldsymbol{\beta}, 0, 0, 0, 0, 0) = (0.2, 0.8, 0, 0, 0, 0, 0, 0, 0)$. Therefore, $\max\{X_1, X_2\}$ has a *PH*-distribution with *PH*-representation: (0.2, 0.8, 0, 0, 0, 0, 0, 0) and

$$\begin{pmatrix} T \otimes I + I \otimes S & (-T\mathbf{e}) \otimes I & I \otimes (-S\mathbf{e}) \\ 0 & S & 0 \\ 0 & 0 & T \end{pmatrix}. \tag{1.36}$$

(ii) It is also clear that $\min\{X_1, X_2\}$ is the time until the Markov chain $\{I(t), t \geq 0\}$ enters any phase in the subset $\{(1, 3), (2, 3), (3, 1), (3, 2), (3, 3)\}$, given that the initial distribution is $(\boldsymbol{\alpha} \otimes \boldsymbol{\beta}, 0, 0, 0, 0, 0)$. Therefore, $\min\{X_1, X_2\}$ has a *PH*-distribution with *PH*-representation $(\boldsymbol{\alpha} \otimes \boldsymbol{\beta}, T \otimes I + I \otimes S)$.

(iii) For $X_1 + X_2$, we initialize the underlying Markov chain $\{I_1(t), t \geq 0\}$ of X_1 at time zero. As soon as $\{I_1(t), t \geq 0\}$ enters its absorption phase 3, we initialize the underlying Markov chain $\{I_2(t), t \geq 0\}$ of X_2. Then $X_1 + X_2$ is the time until the second underlying Markov chain enters its absorption phase, which has a *PH*-distribution with *PH*-representation

$$\left((\boldsymbol{\alpha}, 0), \begin{pmatrix} T & (-T\mathbf{e})\boldsymbol{\beta} \\ 0 & S \end{pmatrix} \right) = \left((1, 0, 0, 0), \begin{pmatrix} -2 & 2 & 0 & 0 \\ 5 & -10 & 1 & 4 \\ 0 & 0 & -1 & 1 \\ 0 & 0 & 3 & -4 \end{pmatrix} \right). \tag{1.37}$$

Note that the (i, j)-th element of matrix $(-T\mathbf{e})\boldsymbol{\beta}$ is the transition rate for which the Markov chain $\{I_1(t), t \geq 0\}$ enters its absorption phase from its transient phase i and the Markov chain $\{I_2(t), t \geq 0\}$ is initialized in its transient phase j.

Now, we are ready to generalize the results in Exercise 1.1.5 and Proposition 1.1.2.

Proposition 1.3.1 *Assume that X_1 has a PH-distribution with PH-representation $(\boldsymbol{\alpha}, T)$, X_2 has a PH-distribution with PH-representation $(\boldsymbol{\beta}, S)$, and X_1 and X_2 are independent. Then $X = \min\{X_1, X_2\}$ has a PH-distribution with PH-representation*

$$(\boldsymbol{\alpha}\otimes\boldsymbol{\beta},\quad T\otimes I+I\otimes S).\tag{1.38}$$

Proof. We give two proofs to the proposition: one algebraic and the other probabilistic. The first proof is done by routine calculations based on Exercises 1.3.1 and 1.3.2. For completeness, we provide some necessary details. Note that $(ABC)\otimes(DEF)=(A\otimes D)(B\otimes E)(C\otimes F)$ if all the matrix multiplications are valid. We have

$$\begin{aligned}
P\{\min(X_1,X_2)\} &= P\{X_1>t\}P\{X_2>t\}\\
&= (\boldsymbol{\alpha}\exp(Tt)\mathbf{e})(\boldsymbol{\beta}\exp(St)\mathbf{e})\\
&= (\boldsymbol{\alpha}\exp(Tt)\mathbf{e})\otimes(\boldsymbol{\beta}\exp(St)\mathbf{e})\\
&= (\boldsymbol{\alpha}\otimes\boldsymbol{\beta})(\exp(Tt)\otimes\exp(St))(\mathbf{e}\otimes\mathbf{e})\\
&= (\boldsymbol{\alpha}\otimes\boldsymbol{\beta})\left(\sum_{k=0}^{\infty}\sum_{n=0}^{\infty}\frac{t^{k+n}T^k\otimes S^n}{k!n!}\right)\mathbf{e}\\
&= (\boldsymbol{\alpha}\otimes\boldsymbol{\beta})\left(\sum_{v=0}^{\infty}\frac{t^v}{v!}\sum_{n=0}^{v}\frac{v!(T^v\otimes S^{v-n})}{n!(v-n)!}\right)\mathbf{e}\\
&= (\boldsymbol{\alpha}\otimes\boldsymbol{\beta})\left(\sum_{k=0}^{\infty}\frac{t^k}{k!}(T\otimes I+I\otimes S)^k\right)\mathbf{e}\\
&= (\boldsymbol{\alpha}\otimes\boldsymbol{\beta})\exp((T\otimes I+I\otimes S)t)\mathbf{e}.
\end{aligned}\tag{1.39}$$

Note that $T\otimes S=(TI)\otimes(IS)=(T\otimes I)(I\otimes S)=(I\otimes S)(T\otimes I)$. It is easy to verify that both $\boldsymbol{\alpha}\otimes\boldsymbol{\beta}$ and $T\otimes I+I\otimes S$ satisfy the conditions for a *PH*-representation (Definition 1.2.1). This completes the algebraic proof of Proposition 1.3.1.

The second proof is based on the probabilistic interpretation of *PH*-representations. Mainly, we utilize the underlying Markov chains associated with *PH*-representations. To show that $\min\{X_1, X_2\}$ is a *PH*-distribution, we simply construct an underlying Markov chain for which $\min\{X_1, X_2\}$ is the absorption time of certain states.

Denote by $\{I_1(t),\ t\geq 0\}$ and $\{I_2(t),\ t\geq 0\}$ the underlying continuous time Markov chains associated with *PH*-representations $(\boldsymbol{\alpha}, T)$ and $(\boldsymbol{\beta}, S)$, respectively. Define $I(t)=(I_1(t),\ I_2(t))$. Since $\{I_1(t),\ t\geq 0\}$ and $\{I_2(t),\ t\geq 0\}$ are independent Markov chains, it is easy to see that $\{I(t), t\geq 0\}$ is a continuous time Markov chain with state space $\{1, 2, \ldots, m_1+1\}\times\{1, 2, \ldots, m_2+1\}$, where m_1+1 and m_2+1 are the numbers of phases of Markov chains $\{I_1(t),\ t\geq 0\}$ and $\{I_2(t),\ t\geq 0\}$, respectively. It is easy to see that X is the first time that the Markov chain $\{I(t),\ t\geq 0\}$ leaves the set of phases $\{1, 2, \ldots, m_1\}\times\{1, 2, \ldots, m_2\}$, or, the first time that the Markov chain enters the set of phases $\{m_1+1\}\times\{1, 2, \ldots, m_2+1\}$ and $\{1, 2, \ldots, m_1+1\}\times\{m_2+1\}$, which are now lumped together to form a *super* absorption phase. Thus, we have shown that $X=\min\{X_1,\ X_2\}$ has a

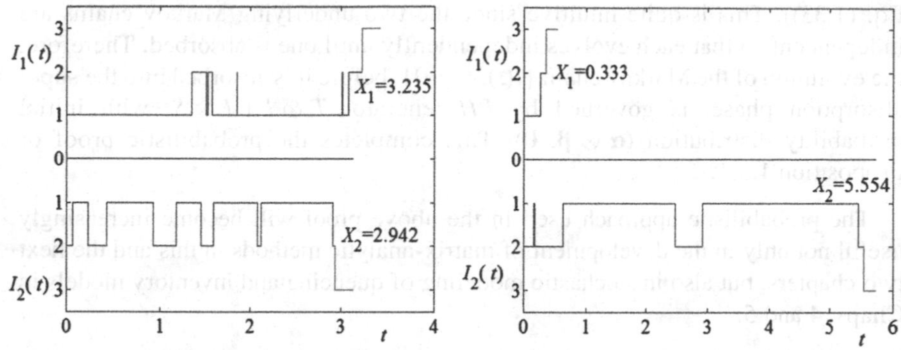

Fig. 1.7 Two sample paths for $\{I_1(t), t \geq 0\}$ and $\{I_2(t), t \geq 0\}$ of Proposition 1.3.1

PH-distribution. Figure 1.7 illustrates sample paths of the two Markov chains, which are generated by using the *play method* defined in Exercise 1.3.16.

Next, we find a *PH*-representation for *X*. We need to find the initial probability distribution of the Markov chain $\{I(t), t \geq 0\}$ and the transition rates among the phases $\{1, 2, \ldots, m_1\} \times \{1, 2, \ldots, m_2\}$. Since the initial distributions of $\{I_1(t), t \geq 0\}$ and $\{I_2(t), t \geq 0\}$ are $\boldsymbol{\alpha}$ and $\boldsymbol{\beta}$, the probability that the Markov chain $\{I(t), t \geq 0\}$ is in state (i, j) at $t = 0$ is $\alpha_i\beta_j$, for $i = 1, 2, \ldots, m_1$ and $j = 1, 2, \ldots, m_2$. That is, the initial distribution on $\{1, 2, \ldots, m_1\} \times \{1, 2, \ldots, m_2\}$ is given by $\boldsymbol{\alpha} \otimes \boldsymbol{\beta}$. The sojourn time of the Markov chain $\{I_1(t), t \geq 0\}$ in phase *i* is exponentially distributed with parameter $-t_{i,i}$, and the sojourn time of the Markov chain $\{I_2(t), t \geq 0\}$ in phase *j* is exponentially distributed with parameter $-s_{j,j}$. Note $T = (t_{i,j})$ and $S = (s_{i,j})$. By the memoryless property, we know that the residual sojourn time is exponentially distributed with the same parameter in each phase for each Markov chain. Thus, the sojourn time of the Markov chain $\{I(t), t \geq 0\}$ in phase (i, j) is the minimum of two independent exponential random variables with parameters $-t_{i,i}$ and $-s_{j,j}$, respectively. By Proposition 1.1.2, the sojourn time in phase (i, j) has an exponential distribution with parameter $-t_{i,i}-s_{j,j}$. By Proposition 1.1.3, the next transition can come from $\{I_1(t), t \geq 0\}$ with probability $-t_{i,i}/(-t_{i,i}-s_{j,j})$ or come from $\{I_2(t), t \geq 0\}$ with probability $-s_{j,j}/(-t_{i,i}-s_{j,j})$. Furthermore, by Proposition 1.1.3, the individual transition rates are

$$
\begin{aligned}
&\frac{(t_{i,1}, \cdots, t_{i,i-1}, t_{i,i+1}, \cdots, t_{i,m_1}, \mathbf{T}_i^0)}{(-t_{i,i})}\left(\frac{-t_{i,i}}{-t_{i,i}-s_{j,j}}\right)(-t_{i,i}-s_{j,j}) \\
&= (t_{i,1}, \cdots, t_{i,i-1}, t_{i,i+1}, \cdots, t_{i,m_1}, \mathbf{T}_i^0),
\end{aligned}
\tag{1.40}
$$

for transitions from phase (i, j) to (k, j), $i \neq k$, and are equal to $(s_{j,1}, \cdots, s_{j,j-1}, s_{j,j+1}, \cdots, s_{j,m_2}, \mathbf{S}_j^0)$, for transitions from phase (i, j) to (i, l), $l \neq j$. Note, by the definition of *PH*-representations, $\mathbf{T}^0 = -T\mathbf{e}$ and $\mathbf{S}^0 = -S\mathbf{e}$. Thus, the transitions of $\{I(t), t \geq 0\}$ are governed by $T \otimes I + I \otimes S$ before absorption (for example, see

Eq. (1.33)). This is quite intuitive since the two underlying Markov chains are independent so that each evolves independently until one is absorbed. Therefore, the evolution of the Markov chain $\{I(t), t \geq 0\}$, before it is absorbed into the super absorption phase, is governed by PH-generator $T \otimes I + I \otimes S$, with initial probability distribution $(\boldsymbol{\alpha} \otimes \boldsymbol{\beta}, 0)$. This completes the probabilistic proof of Proposition 1.3.1.

The probabilistic approach used in the above proof will become increasingly useful not only in the development of matrix-analytic methods in this and the next two chapters, but also in stochastic modeling of queueing and inventory models in Chaps. 4 and 5.

Exercise 1.3.3 For the *PH*-distributions defined in Proposition 1.3.1, explain intuitively that the 2-tuple $(\boldsymbol{\beta} \otimes \boldsymbol{\alpha}, S \otimes I + I \otimes T)$ is also a *PH*-representation of $X = \min\{X_1, X_2\}$.

Exercise 1.3.4 (Generalization of Proposition 1.3.1 to $X = \min\{X_1, \ldots, X_n\}$) Assume that $\{X_1, \ldots, X_n\}$ are independent and have *PH*-distributions. Show that $X = \min\{X_1, \ldots, X_n\}$ has a *PH*-distribution. Find a *PH*-representation for X. Design a simple recursive algorithm to find a *PH*-representation of $\min\{X_1, \ldots, X_n\}$. Test your program with the four *PH*-distributions in Example 1.2.1. (Hint: Let $(\boldsymbol{\alpha}_k, T_k)$ be a *PH*-representation of X_k. Let $(\boldsymbol{\beta}_k, S_k)$ be a *PH*-representation of $\min\{X_1, \ldots, X_k\}$. Then $(\boldsymbol{\beta}_{k+1}, S_{k+1})$ can be obtained from $(\boldsymbol{\beta}_k, S_k)$ and $(\boldsymbol{\alpha}_{k+1}, T_{k+1})$ by Proposition 1.3.1.)

In Exercise 1.1.6, it has been shown that the set of exponential distributions is not closed under the "*max*" and "+" operations. Next, in Propositions 1.3.2 and 1.3.3, it is shown that the set of *PH*-distributions is closed under both operations.

Proposition 1.3.2 *Assume that X_1 has a PH-distribution with PH-representation $(\boldsymbol{\alpha}, T)$, X_2 has a PH-distribution with PH-representation $(\boldsymbol{\beta}, S)$, and X_1 and X_2 are independent. Then $X = X_1 + X_2$ has a PH-distribution with PH-representation*

$$\left((\boldsymbol{\alpha}, (1 - \boldsymbol{\alpha}\mathbf{e})\boldsymbol{\beta}), \quad \begin{pmatrix} T & \mathbf{T}^0\boldsymbol{\beta} \\ 0 & S \end{pmatrix} \right) \tag{1.41}$$

Proof. An algebraic proof based on routine calculations is given first. By contrast with the proof of Proposition 1.3.1, we calculate the LSTs of random variables. The proof is based on the well-known fact that there is a one-to-one relationship between LSTs and probability distributions. First, it is easy to verify that the representation given in Eq. (1.41) satisfies the conditions for a *PH*-representation. By part (i) of Proposition 1.2.2, the LST of the *PH*-representation given in Eq. (1.41) can be calculated as

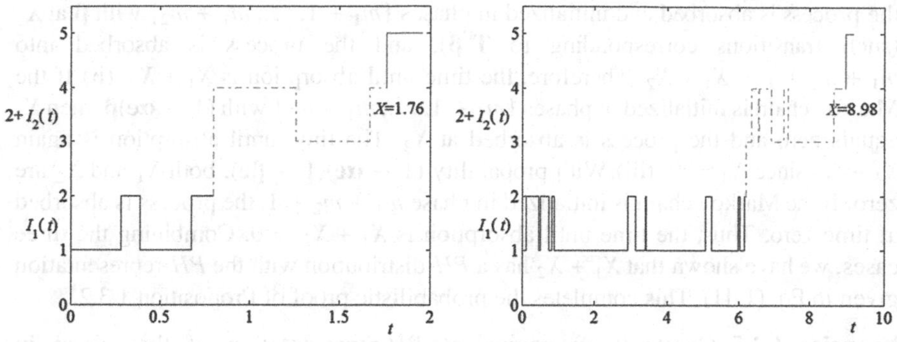

Fig. 1.8 Sample paths of $\{I_1(t), t \geq 0\}$ and $\{I_2(t), t \geq 0\}$ of Proposition 1.3.2

$$
\begin{aligned}
&1-(\boldsymbol{\alpha},\,(1-\boldsymbol{\alpha}\mathbf{e})\boldsymbol{\beta})\mathbf{e}+(\boldsymbol{\alpha},\,(1-\boldsymbol{\alpha}\mathbf{e})\boldsymbol{\beta})\left(sI-\begin{pmatrix} T & \mathbf{T}^0\boldsymbol{\beta} \\ 0 & S\end{pmatrix}\right)^{-1}\left(-\begin{pmatrix} T & \mathbf{T}^0\boldsymbol{\beta} \\ 0 & S\end{pmatrix}\mathbf{e}\right)\\
&\quad=(1-\boldsymbol{\alpha}\mathbf{e})(1-\boldsymbol{\beta}\mathbf{e})\\
&\qquad+(\boldsymbol{\alpha},(1-\boldsymbol{\alpha}\mathbf{e})\boldsymbol{\beta})\begin{pmatrix}(sI-T)^{-1} & (sI-T)^{-1}\mathbf{T}^0\boldsymbol{\beta}(sI-S)^{-1}\\ 0 & (sI-S)^{-1}\end{pmatrix}\begin{pmatrix}(1-\boldsymbol{\beta}\mathbf{e})\mathbf{T}^0\\ \mathbf{S}^0\end{pmatrix}\\
&\quad=(1-\boldsymbol{\alpha}\mathbf{e})(1-\boldsymbol{\beta}\mathbf{e})+(1-\boldsymbol{\beta}\mathbf{e})\boldsymbol{\alpha}(sI-T)^{-1}\mathbf{T}^0+(1-\boldsymbol{\alpha}\mathbf{e})\boldsymbol{\beta}(sI-S)^{-1}\mathbf{S}^0\\
&\qquad+\boldsymbol{\alpha}(sI-T)^{-1}\mathbf{T}^0\boldsymbol{\beta}(sI-S)^{-1}\mathbf{S}^0\\
&\quad=\left(1-\boldsymbol{\alpha}\mathbf{e}+\boldsymbol{\alpha}(sI-T)^{-1}\mathbf{T}^0\right)\left(1-\boldsymbol{\beta}\mathbf{e}+\boldsymbol{\beta}(sI-S)^{-1}\mathbf{S}^0\right)\\
&\quad=E[\exp(-sX_1)]E[\exp(-sX_2)]\\
&\quad=E[\exp(-s(X_1+X_2))].
\end{aligned}
\tag{1.42}
$$

The last equality holds due to the independence of the two random variables. This completes the algebraic proof of Proposition 1.3.2.

Now, we give a probabilistic proof for Proposition 1.3.2. We introduce a continuous time Markov chain $\{I(t), t \geq 0\}$ as follows: (i) it has state space $\{1, 2, \ldots, m_1, m_1 + 1, m_1 + 2, \ldots, m_1+ m_2, m_1 + m_2 + 1\}$; (ii) transitions among phases $\{1, 2, \ldots, m_1\}$ are governed by T, transitions among $\{m_1 + 1, \ldots, m_1+ m_2\}$ are governed by S, transitions from $\{1, 2, \ldots, m_1\}$ to $\{m_1 + 1, \ldots, m_1 + m_2\}$ are governed by $\mathbf{T}^0\boldsymbol{\beta}$, and there is no transition from phases $\{m_1 + 1, \ldots, m_1 + m_2\}$ to $\{1, 2, \ldots, m_1\}$; (iii) transitions from $\{1, 2, \ldots, m_1\}$ to phase $m_1 + m_2 + 1$ are governed by $\mathbf{T}^0(1 - \boldsymbol{\beta}\mathbf{e})$; (iv) transitions from $\{m_1 + 1, \ldots, m_1 + m_2\}$ to phase $m_1 + m_2 + 1$ are governed by $\mathbf{S}^0$; and (v) phase $m_1 + m_2 + 1$ is an absorption phase. See two sample paths of $\{I(t), t \geq 0\}$ in Fig. 1.8.

Suppose that the initial distribution of Markov chain $\{I(t), t \geq 0\}$ is $(\boldsymbol{\alpha}, (1 - \boldsymbol{\alpha}\mathbf{e})\boldsymbol{\beta}, (1 - \boldsymbol{\alpha}\mathbf{e})(1 - \boldsymbol{\beta}\mathbf{e}))$. Then the Markov chain shall be absorbed into $m_1 + m_2 + 1$ in three ways. (i) If the Markov chain is initialized in phases $\{1, 2, \ldots, m_1\}$ with $\boldsymbol{\alpha}$, the process is absorbed and initialized in phases $\{m_1 + 1, \ldots, m_1 + m_2\}$ with $\boldsymbol{\beta}$ at X_1 (such transitions corresponding to $\mathbf{T}^0\boldsymbol{\beta}$), and the process is absorbed into $m_1 + m_2 + 1$ at $X_1 + X_2$. Therefore, the time until absorption is $X_1 + X_2$. (ii) If the Markov chain is initialized in phases $\{m_1 + 1, \ldots, m_1 + m_2\}$ with $(1 - \boldsymbol{\alpha}\mathbf{e})\boldsymbol{\beta}$, then X_1 equals zero and the process is absorbed at X_2. The time until absorption is again $X_1 + X_2$ since $X_1 = 0$. (iii) With probability $(1 - \boldsymbol{\alpha}\mathbf{e})(1 - \boldsymbol{\beta}\mathbf{e})$, both X_1 and X_2 are zero. If the Markov chain is initialized in phase $m_1 + m_2 + 1$, the process is absorbed at time zero. Thus, the time until absorption is $X_1 + X_2 = 0$. Combining the three cases, we have shown that $X_1 + X_2$ has a *PH*-distribution with the *PH*-representation given in Eq. (1.41). This completes the probabilistic proof of Proposition 1.3.2.

Exercise 1.3.5 Construct an equivalent *PH*-representation of that given in Eq. (1.41) by first considering the representation of X_2.

Exercise 1.3.6 (Generalization of Proposition 1.3.2 to $X_1 + \ldots + X_n$) Assume that $X_1, \ldots,$ and X_n are independent and have *PH*-distributions. Show that $X_1 + \ldots + X_n$ has a *PH*-distribution. Design a simple recursive algorithm to find a *PH*-representation for $X_1 + \ldots + X_n$. Test your algorithm with the four *PH*-representations given in Example 1.2.1.

Proposition 1.3.3 *Assume that X_1 has a PH-distribution with PH-representation $(\boldsymbol{\alpha}, T)$, X_2 has a PH-distribution with PH-representation $(\boldsymbol{\beta}, S)$, and X_1 and X_2 are independent. Then $X = \max\{X_1, X_2\}$ has a PH-distribution with PH-representation*

$$\left((\boldsymbol{\alpha}\otimes\boldsymbol{\beta},\ (1-\boldsymbol{\alpha}\mathbf{e})\boldsymbol{\beta},\ (1-\boldsymbol{\beta}\mathbf{e})\boldsymbol{\alpha}),\ \begin{pmatrix} T\otimes I + I\otimes S & \mathbf{T}^0\otimes I & I\otimes\mathbf{S}^0 \\ 0 & S & 0 \\ 0 & 0 & T \end{pmatrix} \right). \tag{1.43}$$

Proof. The first proof is based on $P\{\max\{X_1, X_2\} \le t\} = P\{X_1 \le t\}P\{X_2 \le t\}$ for independent random variables X_1 and X_2. It is easy to check that the representation in Eq. (1.43) is a *PH*-representation of a *PH*-random variable, to be called X. The distribution function of X can be calculated routinely as follows, for $t > 0$,

$$\begin{aligned} P\{X \le t\} &= 1 - (\boldsymbol{\alpha}\otimes\boldsymbol{\beta},\ (1-\boldsymbol{\alpha}\mathbf{e})\boldsymbol{\beta},\ (1-\boldsymbol{\beta}\mathbf{e})\boldsymbol{\alpha}) \exp\left(\begin{pmatrix} T\otimes I + I\otimes S & \mathbf{T}^0\otimes I & I\otimes\mathbf{S}^0 \\ 0 & S & 0 \\ 0 & 0 & T \end{pmatrix} t \right) \mathbf{e} \\ &= 1 - (\boldsymbol{\alpha}\otimes\boldsymbol{\beta},\ (1-\boldsymbol{\alpha}\mathbf{e})\boldsymbol{\beta},\ (1-\boldsymbol{\beta}\mathbf{e})\boldsymbol{\alpha}) \begin{pmatrix} \exp(Tt)\mathbf{e}\otimes(\mathbf{e}-\exp(St)\mathbf{e}) + \mathbf{e}\otimes\exp(St)\mathbf{e} \\ \exp(St)\mathbf{e} \\ \exp(Tt)\mathbf{e} \end{pmatrix} \\ &= (1-\boldsymbol{\alpha}\exp(Tt)\mathbf{e})(1-\boldsymbol{\beta}\exp(St)\mathbf{e}) \\ &= P\{X_1 \le t\}P\{X_2 \le t\} = P\{\max\{X_1, X_2\} \le t\}. \end{aligned} \tag{1.44}$$

Note that the results in Examples 1.3.3 and 1.3.4 at the end of this section are used in Eq. (1.44). Consequently, we must have $X = \max\{X_1, X_2\}$. This completes the algebraic proof of Proposition 1.3.3.

Denote by $\{I_1(t),\ t \geq 0\}$ and $\{I_2(t),\ t \geq 0\}$ the underlying continuous time Markov chains associated with *PH*-representations ($\boldsymbol{\alpha}$, T) and ($\boldsymbol{\beta}$, S), respectively. Similar to the probabilistic proof of Proposition 1.3.1, define $I(t) = (I_1(t), I_2(t))$. It is easy to see that X is the first time that the Markov chain $\{I(t),\ t \geq 0\}$ enters phase $(m_1 + 1, m_2 + 1)$. Thus, $X = \max\{X_1, X_2\}$ has a *PH*-distribution. The initial distribution of $\{I(t),\ t \geq 0\}$ is given by $(\boldsymbol{\alpha} \otimes \boldsymbol{\beta}, (1 - \boldsymbol{\alpha}\mathbf{e})\boldsymbol{\beta}, (1 - \boldsymbol{\beta}\mathbf{e})\boldsymbol{\alpha}, (1 - \boldsymbol{\alpha}\mathbf{e})(1 - \boldsymbol{\beta}\mathbf{e}))$ by conditioning on the initial phases of $\{I_1(t),\ t \geq 0\}$ and $\{I_2(t),\ t \geq 0\}$. By the probabilistic proof of Proposition 1.3.1, the infinitesimal generator of $\{I(t),\ t \geq 0\}$ is given by

$$\begin{array}{r} \{1,2,\ldots,m_1\} \times \{1,2,\ldots,m_2\} \\ \{m_1+1\} \times \{1,2,\ldots,m_2\} \\ \{1,2,\ldots,m_1\} \times \{m_2+1\} \\ \{m_1+1\} \times \{m_2+1\} \end{array} \begin{pmatrix} T \otimes I + I \otimes S & \mathbf{T}^0 \otimes I & I \otimes \mathbf{S}^0 & 0 \\ 0 & S & 0 & \mathbf{S}^0 \\ 0 & 0 & T & \mathbf{T}^0 \\ 0 & 0 & 0 & 0 \end{pmatrix}. \tag{1.45}$$

The *PH*-representation for X, given in Eq. (1.43), is obtained by Definition 1.2.2. This completes the proof of Proposition 1.3.3.

We present a slightly different way to construct the *PH*-representation for $X = \max\{X_1, X_2\}$ (see Fig. 1.7 for two sample paths of X). We define a continuous time Markov chain $\{I(t),\ t \geq 0\}$ with state space $\{\{1, 2, \ldots, m_1\} \times \{1, 2, \ldots, m_2\}\} \cup \{\{1, 2, \ldots, m_1\} \times \{m_2 + 1\}\} \cup \{\{m_1 + 1\} \times \{1, 2, \ldots, m_2\}\} \cup \{\Delta\}$, where $\{1, 2, \ldots, m_1\}$ are the transient phases associated with X_1, $\{1, 2, \ldots, m_2\}$ are the transient phases associated with X_2, and $\Delta = \{m_1 + 1\} \times \{m_2 + 1\}$. The transitions within $\{1, 2, \ldots, m_1\} \times \{1, 2, \ldots, m_2\}$ are governed by $T \otimes I + I \otimes S$ (since $I(t) = (I_1(t), I_2(t))$ in this case); transitions within $\{1, 2, \ldots, m_1\} \times \{m_2 + 1\}$ are governed by T; transitions within $\{m_1 + 1\} \times \{1, 2, \ldots, m_2\}$ are governed by S; transitions from $\{1, 2, \ldots, m_1\} \times \{1, 2, \ldots, m_2\}$ to $\{1, 2, \ldots, m_1\} \times \{m_2 + 1\}$ are governed by $I \otimes \mathbf{S}^0$, transitions from $\{1, 2, \ldots, m_1\} \times \{1, 2, \ldots, m_2\}$ to $\{m_1 + 1\} \times \{1, 2, \ldots, m_2\}$ are governed by $\mathbf{T}^0 \otimes I$, and Δ is an absorption phase. On the other hand, we write $X = \min\{X_1, X_2\} + (\max\{X_1, X_2\} - \min\{X_1, X_2\})$. The first part $\min\{X_1, X_2\}$ of X corresponds to $(\boldsymbol{\alpha} \otimes \boldsymbol{\beta}, T \otimes I + I \otimes S)$ (see the proof of Proposition 1.3.1). At $t = \min\{X_1, X_2\}$, the process $\{I(t),\ t \geq 0\}$ may go into a phase in $\{1, 2, \ldots, m_1\} \times \{m_2 + 1\}$ or $\{m_1 + 1\} \times \{1, 2, \ldots, m_2\}$, depending on which process, $\{I_1(t),\ t \geq 0\}$ or $\{I_2(t),\ t \geq 0\}$, has been absorbed. Then $\{I(t),\ t \geq 0\}$ will stay in the set of phases for some extra time denoted by $\max\{X_1, X_2\} - \min\{X_1, X_2\}$, until absorption into Δ. If $X_1 = 0$ or $X_2 = 0$, then $\{I(t),\ t \geq 0\}$ gets into process $\{I_2(t),\ t \geq 0\}$ or $\{I_1(t),\ t \geq 0\}$ with an initial distribution $(1 - \boldsymbol{\alpha}\mathbf{e})\boldsymbol{\beta}$ or $(1 - \boldsymbol{\beta}\mathbf{e})\boldsymbol{\alpha}$, and stays there until absorption into Δ. Combining all the cases, we have shown that Eq. (1.43) gives a *PH*-representation of $\max\{X_1, X_2\}$.

Exercise 1.3.7 For the *PH*-distributions defined in Proposition 1.3.3, explain intuitively that the *PH*-representation given in Eq. (1.46) is also a *PH*-representation of $X = \max\{X_1, X_2\}$.

$$\left((\boldsymbol{\beta} \otimes \boldsymbol{\alpha},\ (1-\boldsymbol{\beta}\mathbf{e})\boldsymbol{\alpha},\ (1-\boldsymbol{\alpha}\mathbf{e})\boldsymbol{\beta}),\ \begin{pmatrix} S \otimes I + I \otimes T & \mathbf{S}^0 \otimes I & I \otimes \mathbf{T}^0 \\ 0 & T & 0 \\ 0 & 0 & S \end{pmatrix} \right). \tag{1.46}$$

Exercise 1.3.8 (Generalization of Proposition 1.3.3 to $\max\{X_1, \ldots, X_n\}$) Assume that $X_1, \ldots,$ and X_n are independent and have *PH*-distributions. Show that $\max\{X_1, \ldots, X_n\}$ has a *PH*-distribution. Design a simple recursive algorithm to find a *PH*-representation for $\max\{X_1, \ldots, X_n\}$. Test your algorithm with the four *PH*-representations given in Example 1.2.1.

Proposition 1.1.4 shows that the sum of a geometrical number of independent exponential random variables with the same parameter is exponentially distributed. That result is generalized to *PH*-distributions as follows.

Proposition 1.3.4 *Assume that N has a geometric distribution with parameter p* $(0 < p < 1)$, *i.e.,* $P\{N = n\} = p^{n-1}(1-p)$, $n = 1, 2, \ldots$. *Assume that* $\{X_n,\ n = 1, 2, \ldots\}$ *has a common PH-distribution with PH-representation* $(\boldsymbol{\alpha}, T)$ *with* $\boldsymbol{\alpha}\mathbf{e} = 1$. *All random variables are independent. Then random variable* $Y = \sum_{n=1}^{N} X_n$ *has a PH-distribution with PH-representation* $(\boldsymbol{\alpha},\ T + p\mathbf{T}^0\boldsymbol{\alpha})$.

Proof. The following proof is based on LST and routine calculations. First, we note:

$$\begin{aligned} P\{Y \le t\} &= \sum_{n=1}^{\infty} P\{Y \le t | N = n\} P\{N = n\} \\ &= \sum_{n=1}^{\infty} P\{X_1 + \ldots + X_n \le t\} P\{N = n\}. \end{aligned} \tag{1.47}$$

Then we have

$$\begin{aligned} & E[\exp(-sY)] \\ &= \sum_{n=1}^{\infty} p^{n-1}(1-p) E[\exp(-s(X_1 + \ldots + X_n))] \\ &= \sum_{n=1}^{\infty} p^{n-1}(1-p)(E[\exp(-sX_1)])^n \\ &= \sum_{n=1}^{\infty} p^{n-1}(1-p)\left(\boldsymbol{\alpha}(sI - T)^{-1}\mathbf{T}^0\right)^n \\ &= \boldsymbol{\alpha}\left(\sum_{n=1}^{\infty} p^{n-1}\left((sI - T)^{-1}\mathbf{T}^0\boldsymbol{\alpha}\right)^{n-1}\right)(sI - T)^{-1}\mathbf{T}^0(1-p) \\ &= \boldsymbol{\alpha}\left(I - p(sI - T)^{-1}\mathbf{T}^0\boldsymbol{\alpha}\right)^{-1}(sI - T)^{-1}\mathbf{T}^0(1-p) \\ &= \boldsymbol{\alpha}\left(sI - T - p\mathbf{T}^0\boldsymbol{\alpha}\right)^{-1}\mathbf{T}^0(1-p). \end{aligned} \tag{1.48}$$

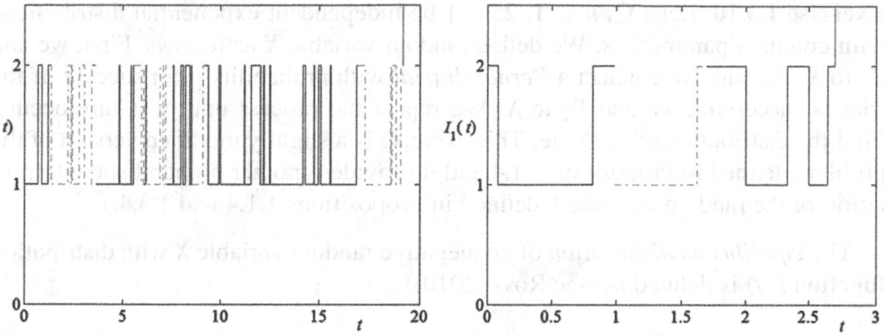

Fig. 1.9 Two sample paths of $\{I_1(t), t \geq 0\}$ for Proposition 1.3.4

The proposition is proved by noting $\mathbf{T}^0 = -T\mathbf{e}$ and $(T + p\mathbf{T}^0\boldsymbol{\alpha})\mathbf{e} = -(1-p)\mathbf{T}^0$. This completes the proof of Proposition 1.3.4.

We also offer a probabilistic interpretation of $(\boldsymbol{\alpha}, T + p\mathbf{T}^0\boldsymbol{\alpha})$. First note that the sum of a geometric number of *PH*-random variables can be obtained by summing up independent *PH*-random variables one at a time until we decide to terminate the summation process. After adding a *PH*-random variable to Y, we stop the process with probability $1 - p$ and continue with probability p. Consider the underlying Markov chain $\{I_1(t), t \geq 0\}$ for $(\boldsymbol{\alpha}, T)$. We modify the Markov chain $\{I_1(t), t \geq 0\}$ as follows. When it is absorbed, with probability p, we start the process again with the same initial probability distribution $\boldsymbol{\alpha}$. Thus, if the Markov chain is in phase i, then there are three cases at its transitions: (1) transiting to another phase j with rate $t_{i,j}$; (2) being absorbed but immediately going into phase k with rate $(\mathbf{T}^0)_i p\alpha_k$, where k can be i; and (3) being absorbed and the process is terminated with rate $(\mathbf{T}^0)_i(1-p)$. Combining the three cases, we have constructed a Markov chain with transition matrix $T + p\mathbf{T}^0\boldsymbol{\alpha}$ on the m transient phases. If the initial distribution of the Markov chain is $(\boldsymbol{\alpha}, 0)$, then the time until absorption is the sum of a geometric number of independent *PH*-random variables with a common *PH*-representation $(\boldsymbol{\alpha}, T)$. See Fig. 1.9 for two sample paths of the absorption process. (Note: Since geometric distributions possess the memoryless property and *PH*-distributions possess the partial memoryless property, given the current phase, when the process will be absorbed is independent of how many time the summation process has survived.)

Exercise 1.3.9 For a *PH*-distribution with *PH*-representation $(\boldsymbol{\alpha}, T)$ given as

$$\boldsymbol{\alpha} = (0,\ 0.2,\ 0.8,\ 0), \quad T = \begin{pmatrix} -20 & 0 & 0 & 15 \\ 20 & -20 & 0 & 0 \\ 0 & 0.5 & -1 & 0.5 \\ 0 & 0 & 0 & -1 \end{pmatrix}, \tag{1.49}$$

and $p = 0.4$, find the distribution of Y defined in Proposition 1.3.4. Plot the density functions of $(\boldsymbol{\alpha}, T)$ and Y.

Exercise 1.3.10 Let $\{X_n, n = 1, 2, \ldots\}$ be independent exponential distributions with common parameter λ. We define random variable X as follows. First, we add X_1 to X. Second, we conduct a *Bernoulli trial* with probability p for success. If the trial is successful, we add X_2 to X. We repeat the process until a failure occurs. Find the distribution of X. (Note: This exercise is a slightly modified version of the problem defined in Proposition 1.1.4 and it provides another probabilistic interpretation of the random variable Y defined in Propositions 1.1.4 and 1.3.4.)

The *equilibrium distribution* of nonnegative random variable X with distribution function $F(t)$ is defined as (see Ross (2010))

$$F_e(t) = \frac{1}{E[X]}\int_0^t (1 - F(x))\mathrm{d}x, \quad t \geq 0. \tag{1.50}$$

The term equilibrium distribution comes from the theory of *renewal process* (see Ross (2010)), which is a *counting process* for which the *interarrival times* of events are independent random variables with common distribution function $F(t)$. The function $F_e(t)$ is the distribution of the residual time of the renewal process at an arbitrary time.

A *PH-renewal process* is a renewal process with a *PH*-interarrival time $(\boldsymbol{\alpha}, T)$ of order m. Let $\{I(t), t \geq 0\}$ be the underlying Markov chain associated with $(\boldsymbol{\alpha}, T)$. From the renewal process point of view, as soon as the underlying Markov chain is absorbed into phase $m + 1$, it is reinitialized in one of the transient phases with $\boldsymbol{\alpha}$. Thus, for the renewal process, the absorption phase $m + 1$ is invisible. We define $I_r(t)$ as the phase of the underlying Markov chain at time t. It is easy to see that $\{I_r(t), t \geq 0\}$ is a Markov chain with state space $\{1, 2, \ldots, m\}$. For $\{I_r(t), t \geq 0\}$, its phase can go from i to j in two ways: (1) $I(t)$ transits from i to j (with transition rate $t_{i,j}$), for $i \neq j$; and (2) $I(t)$ transits from i to the absorption phase $m + 1$ and is immediately reinitialized in phase j (with transition rate $(\mathbf{T}^0)_i(\boldsymbol{\alpha})_j$). Then the total transition rate from phase i to phase j of $\{I_r(t), t \geq 0\}$ is given by $t_{i,j} + (\mathbf{T}^0)_i(\boldsymbol{\alpha})_j$. In matrix form, the infinitesimal generator of $\{I_r(t), t \geq 0\}$ is given by $T + \mathbf{T}^0\boldsymbol{\alpha}$.

For *PH*-representation $(\boldsymbol{\alpha}, T)$, define $Q = T + \mathbf{T}^0\boldsymbol{\alpha}$. If Q is *irreducible*, we call $(\boldsymbol{\alpha}, T)$ an *irreducible PH-representation*.

Lemma 1.3.1 *Assume that X has a PH-distribution with irreducible PH-representation $(\boldsymbol{\alpha}, T)$ and $\boldsymbol{\alpha}\mathbf{e} = 1$. Let $\boldsymbol{\theta}$ satisfy $\boldsymbol{\theta}Q = 0$ and $\boldsymbol{\theta}\mathbf{e} = 1$. Then $\boldsymbol{\theta} = -\boldsymbol{\alpha}T^{-1}/E[X]$ and $E[X] = 1/(\boldsymbol{\theta}\mathbf{T}^0)$.*

Proof. Since Q is irreducible and finite in size, $\boldsymbol{\theta}$ is the unique *stationary distribution* of Q (e.g., Berman and Plemmons (1979)). By $\boldsymbol{\theta}Q = 0$, we obtain $\boldsymbol{\theta}T + (\boldsymbol{\theta}\mathbf{T}^0)\boldsymbol{\alpha} = 0$, which leads to $\boldsymbol{\theta} = -(\boldsymbol{\theta}\mathbf{T}^0)\boldsymbol{\alpha}T^{-1}$. Multiplying by $\mathbf{e}$ on both sides of $\boldsymbol{\theta} = -(\boldsymbol{\theta}\mathbf{T}^0)\boldsymbol{\alpha}T^{-1}$, we obtain $1 = -(\boldsymbol{\theta}\mathbf{T}^0)\boldsymbol{\alpha}T^{-1}\mathbf{e} = (\boldsymbol{\theta}\mathbf{T}^0)E[X]$. This completes the proof of Lemma 1.3.1.

The equilibrium distribution of a *PH*-distribution is a *PH*-distribution with an explicit *PH*-representation.

Proposition 1.3.5 *Assume that X has a PH-distribution with irreducible PH-representation* ($\boldsymbol{\alpha}$, *T*) *and* $\boldsymbol{\alpha}\mathbf{e} = 1$. *Let* $\boldsymbol{\theta}$ *satisfy* $\boldsymbol{\theta}Q = 0$ *and* $\boldsymbol{\theta}\mathbf{e} = 1$. *Then* ($\boldsymbol{\theta}$, *T*) *is a PH-representation of the equilibrium distribution of X.*

Proof. The proposition can be proved by routine calculations:

$$\begin{aligned}\frac{1}{E[X]}\int_0^t (1 - F(x))\mathrm{d}x &= \frac{1}{E[X]}\int_0^t \boldsymbol{\alpha}\exp(Tx)\mathbf{e}\mathrm{d}x \\ &= \frac{1}{E[X]}\boldsymbol{\alpha}T^{-1}(\exp(Tt) - I)\mathbf{e} \\ &= 1 - \left(-\frac{\boldsymbol{\alpha}T^{-1}}{E[X]}\right)\exp(Tt)\mathbf{e}.\end{aligned} \tag{1.51}$$

By Lemma 1.3.1, we obtain the expected result. This completes the proof of Proposition 1.3.5.

Proposition 1.3.5 can be interpreted probabilistically as follows. For the Markov chain $\{I_r(t),\ t \geq 0\}$ with infinitesimal generator Q, its stationary distribution is given by $\boldsymbol{\theta} = (\theta_1, \ldots, \theta_m)$. That is $\lim_{t\to\infty} P\{I_r(t) = j\} = \theta_j$, for $j = 1, 2, \ldots, m$. The equilibrium distribution is the distribution of the residual time at an arbitrary epoch. At an arbitrary epoch, the phase distribution should be $\boldsymbol{\theta}$ and, consequently, the time until the next absorption should be a *PH*-distribution with *PH*-representation ($\boldsymbol{\theta}$, *T*).

Exercise 1.3.11 For the *PH*-distribution defined in Exercise 1.3.9, find a *PH*-representation for its equilibrium distribution and plot the corresponding density function.

Exercise 1.3.12 Find the equilibrium distribution of a renewal process with exponential interarrival times, also called a Poisson process (see Sect. 2.1). Find the equilibrium distribution of a renewal process with Erlang interarrival times.

Commentary There are many other closure properties associated with *PH*-distributions (see Neuts (1981, 1992) and Latouche and Ramaswami (1999)). Closure properties of *PH*-distributions can be extended to stochastic systems and their performance measures, e.g., (i) the waiting time in the *GI*/*PH*/1 queue (Sengupta (1989) and Asmussen and O'Cinneide (1998)); (ii) ruin probability for perturbed risk processes (Asmussen (1995)); and (iii) reliability systems (Assaf and Levikson (1982)).

Table 1.1 summarizes basic facts about the exponential distribution and *PH*-distributions.

Additional Exercises and Extensions

Exercise 1.3.13 The random variable *Y* defined in Proposition 1.3.4 has the following alternative *PH*-representation

Table 1.1 Comparison of the exponential and *PH*- distributions

	Exponential	Phase-type
Parameter(s)	λ	$(m, \boldsymbol{\alpha}, T)$
CDF	$1-\exp(-\lambda t)$	$1-\boldsymbol{\alpha}\exp(Tt)\mathbf{e}$
PDF	$\lambda\exp(-\lambda t)$	$-\boldsymbol{\alpha}\exp(Tt)T\mathbf{e}$
Mean	$1/\lambda$	$-\boldsymbol{\alpha}T^{-1}\mathbf{e}$
Variance	$1/\lambda^2$	$2\boldsymbol{\alpha}T^{-2}\mathbf{e} - (\boldsymbol{\alpha}T^{-1}\mathbf{e})^2$
cv	1	$\sqrt{2\boldsymbol{\alpha}T^{-2}\mathbf{e}/(\boldsymbol{\alpha}T^{-1}\mathbf{e})^2 - 1}$
Properties	Memoryless	(1) Partial memoryless (2) Approximate any distribution on $[0, \infty]$
Closure properties	(1) min (2) Geometric sum	(1) min; (2) max; (3) sum (4) Geometric sum

$$\left((\boldsymbol{\alpha}, 0), \quad \begin{pmatrix} T & p\mathbf{T}^0\boldsymbol{\alpha} \\ p\mathbf{T}^0\boldsymbol{\alpha} & T \end{pmatrix}\right), \tag{1.52}$$

which can be proved as follows. In part (i) of Proposition 1.2.2, the inverse matrix in the LST satisfies

$$\begin{pmatrix} X_1 & X_2 \\ X_2 & X_1 \end{pmatrix}\begin{pmatrix} sI - T & -p\mathbf{T}^0\boldsymbol{\alpha} \\ -p\mathbf{T}^0\boldsymbol{\alpha} & sI - T \end{pmatrix} = \begin{pmatrix} I & 0 \\ 0 & I \end{pmatrix}, \tag{1.53}$$

which leads to $(X_1 + X_2)(sI - T - p\mathbf{T}^0\boldsymbol{\alpha}) = I$. Then the LST of the given *PH*-representation is obtained as

$$\begin{aligned}(\boldsymbol{\alpha}, 0)\begin{pmatrix} sI - T & -p\mathbf{T}^0\boldsymbol{\alpha} \\ -p\mathbf{T}^0\boldsymbol{\alpha} & sI - T \end{pmatrix}^{-1}\begin{pmatrix} (1-p)\mathbf{T}^0 \\ (1-p)\mathbf{T}^0 \end{pmatrix} \\ = \boldsymbol{\alpha}(X_1 + X_2)(1-p)\mathbf{T}^0 \\ = \boldsymbol{\alpha}\left(sI - (T + p\mathbf{T}^0\boldsymbol{\alpha})\right)^{-1}(1-p)\mathbf{T}^0.\end{aligned} \tag{1.54}$$

(1) Show that

$$\left(\frac{1}{2}(\boldsymbol{\alpha}, \boldsymbol{\alpha}), \quad \begin{pmatrix} T & p\mathbf{T}^0\boldsymbol{\alpha} \\ p\mathbf{T}^0\boldsymbol{\alpha} & T \end{pmatrix}\right) \tag{1.55}$$

is also a *PH*-representation of *Y*.

(2) Find more alternative *PH*-representations for *Y*.

Exercise 1.3.14 In Proposition 1.3.4, show that

$$\left(\frac{\boldsymbol{\alpha}}{1 - p(1 - \boldsymbol{\alpha}\mathbf{e})}, \quad T + \mathbf{T}^0\boldsymbol{\alpha}\frac{p}{1 - p(1 - \boldsymbol{\alpha}\mathbf{e})}\right) \tag{1.56}$$

is a *PH*-representation for *Y* if the condition $\boldsymbol{\alpha}\mathbf{e} = 1$ is removed.

Exercise 1.3.15 Assume that $\{X_n, n = 1, 2, \ldots\}$ has a common *PH*-distribution with *PH*-representation $(\boldsymbol{\alpha}, T)$ and $\boldsymbol{\alpha}\mathbf{e} = 1$. Assume that N has a *discrete PH-distribution* with *PH*-representation $(\boldsymbol{\beta}, S)$, where $\boldsymbol{\beta}$ is a *stochastic vector* (i.e., $\boldsymbol{\beta} \geq 0$ and $\boldsymbol{\beta}\mathbf{e} = 1$) and S is a *substochastic matrix* (i.e., $S \geq 0$ and $S\mathbf{e} \leq \mathbf{e}$), i.e., $P\{N = n\} = \boldsymbol{\beta} S^{n-1}(\mathbf{e} - S\mathbf{e})$, $n = 1, 2, \ldots$. All random variables are independent. Show that $Y = \sum_{n=1}^{N} X_n$ has a *PH*-distribution with *PH*-representation $\left(\boldsymbol{\alpha} \otimes \boldsymbol{\beta},\quad T \otimes I + (\mathbf{T}^0\boldsymbol{\alpha}) \otimes S\right)$. Find a *PH*-representation of Y if the condition $\boldsymbol{\alpha}\mathbf{e} = 1$ is removed. (Note: The problems in Propositions 1.1.4, 1.3.4, and this exercise are equivalent to showing that $G(t) = \sum_{k=1}^{\infty} p_k F^{*(k)}(t)$ is a *PH*-distribution, where $\{p_k, k = 1, 2, \ldots\}$ is a discrete *PH*-distribution, $F(t)$ is a continuous *PH*-distribution, and $F^{*(k)}(t)$ is the *k-th convolution* of $F(t)$.)

Note: The convolution of $F(t)$ is defined as: $F^{*(1)}(t) = F(t)$, and, for $k \geq 1$,

$$F^{*(k+1)}(t) = \int_0^t F^{*(k)}(t - x)\mathrm{d}F(x). \tag{1.57}$$

Simulation (Law and Kelton (2000)) of Markov chains can help us understand and visualize the absorption process associated with *PH*-representations. To simulate a *PH*-random variable, its underlying Markov chain can be utilized. The key steps are to generate the sojourn time in each phase for each visit and to identify the next phase to be visited.

Exercise 1.3.16 Use simulation methods to generate sample paths for *PH*-distributions with *PH*-representation $(\boldsymbol{\alpha}, T)$. The simulation program for the underlying continuous time Markov chain can be built in the following steps.

1. Use $X_i = -\log(U)/(-t_{i,i})$ to generate the exponential sojourn time in phase i, where U is the uniform random variable on [0, 1].
2. Use $\boldsymbol{\alpha}$ to generate the initial phase.
3. Use matrix $P = [-(\mathrm{diag}(t_{1,1}, \ldots, t_{m,m}))^{-1}(T, \mathbf{T}^0) + (I, 0)]J$ to determine the next phase, where $\mathrm{diag}(t_{1,1}, \ldots, t_{m,m})$ is a diagonal matrix with elements $\{t_{1,1}, \ldots, t_{m,m}\}$ on the diagonal, J is a matrix with elements in the upper triangular part and on the diagonal being one, and the lower triangular part zero. Note that P is an $m\times(m + 1)$ matrix. Each row of P gives the cumulative distribution function of a probability distribution on $\{1, 2, \ldots, m, m + 1\}$. If the next phase is $m + 1$, the process is terminated.
4. Add up X_i for all transient phases until absorption to generate the phase-type random variable X. (Note: This method is referred to as the *play method* in the literature. Evaluating the density function to generate a *PH*-random variable X is simple but computationally expensive.)

We exemplify the closure properties of *PH*-distributions with a few applications.

Exercise 1.3.17 A barber shop has two barbers who are serving customers at rates $\mu_1 = 2$ person/hour and $\mu_2 = 1.5$ person/hour, respectively (i.e., the service times are exponentially distributed). Assume all customers are served on a first-come-first-served basis. When a customer arrives at the barber shop, he/she finds two customers in service and two more waiting. Show that the waiting time of the new customer has a *PH*-distribution. Find a *PH*-representation for the waiting time. Calculate the mean waiting time.

Exercise 1.3.18 (Exercise 1.3.17 continued) In Exercise 1.3.17, assume that the service times of the two barbers have *PH*-distributions with *PH*-representations $(\boldsymbol{\alpha}, T)$ and $(\boldsymbol{\beta}, S)$, respectively. We also assume $\boldsymbol{\alpha}\mathbf{e} = 1$ and $\boldsymbol{\beta}\mathbf{e} = 1$. Show probabilistically that the waiting time of a customer, who finds two waiting customers and two customers in service at arrival, has a *PH*-distribution with the following *PH*-representation:

$$\left((\boldsymbol{\gamma},\ 0,\ 0),\quad \begin{pmatrix} T\otimes I+I\otimes S & \mathbf{T}^0\boldsymbol{\alpha}\otimes I+I\otimes \mathbf{S}^0\boldsymbol{\beta} & 0 \\ 0 & T\otimes I+I\otimes S & \mathbf{T}^0\boldsymbol{\alpha}\otimes I+I\otimes \mathbf{S}^0\boldsymbol{\beta} \\ 0 & 0 & T\otimes I+I\otimes S \end{pmatrix} \right), \tag{1.58}$$

where $\boldsymbol{\gamma}$ is for the distribution of the phases of the two services at the arrival epoch. If conditions $\boldsymbol{\alpha}\mathbf{e} = 1$ and $\boldsymbol{\beta}\mathbf{e} = 1$ are removed, show that the waiting time of the new customer has a *PH*-distribution with the following *PH*-representation:

$$\left((\boldsymbol{\gamma},\ 0,\ 0),\quad \begin{pmatrix} T\otimes I+I\otimes S & \mathbf{T}^0\boldsymbol{\alpha}\otimes I+I\otimes \mathbf{S}^0\boldsymbol{\beta} & (1-\boldsymbol{\alpha}\mathbf{e})\mathbf{T}^0\boldsymbol{\alpha}\otimes I+(1-\boldsymbol{\beta}\mathbf{e})I\otimes \mathbf{S}^0\boldsymbol{\beta} \\ 0 & T\otimes I+I\otimes S & \mathbf{T}^0\boldsymbol{\alpha}\otimes I+I\otimes \mathbf{S}^0\boldsymbol{\beta} \\ 0 & 0 & T\otimes I+I\otimes S \end{pmatrix} \right). \tag{1.59}$$

Exercise 1.3.19 An electronic system consists of a working unit and a *cold standby* unit (i.e., no failure in standby status). The two units are identical. The *time to failure* for the working unit has an exponential distribution with parameter $\lambda = 0.1$/h. If the working unit fails, the standby unit (if there is one) is put into work. The failed unit is sent to a repair shop for repair. The repair time has an exponential distribution with parameter $\mu = 1.5$/h. Repaired units are as good as new. If the repair completes before the working unit fails, the repaired unit becomes a cold standby unit; otherwise, the system fails. Assume that both units are in good condition at time zero. Show that the *time to system failure* (i.e., both units fail for the first time) has a *PH*-distribution. Find a *PH*-representation for that *PH*-distribution.

Exercise 1.3.20 (Exercise 1.3.19 continued) In Exercise 1.3.19, we assume that the time to failure and the repair time of each unit both have *PH*-distributions with *PH*-representations $(\boldsymbol{\alpha}, T)$ and $(\boldsymbol{\beta}, S)$, respectively. Assume that the two units are new at time zero. Show that the time to system failure has a *PH*-distribution with the following *PH*-representation

$$\left((\boldsymbol{\alpha},\ (1-\boldsymbol{\alpha}\mathbf{e})\boldsymbol{\alpha}\otimes\boldsymbol{\beta}),\ \begin{pmatrix} T+\mathbf{T}^0\boldsymbol{\alpha}(1-\boldsymbol{\beta}\mathbf{e}) & \mathbf{T}^0\boldsymbol{\alpha}\otimes\boldsymbol{\beta} \\ I\otimes\mathbf{S}^0 & T\otimes I+I\otimes S \end{pmatrix} \right). \tag{1.60}$$

Exercise 1.3.21 A system consists of three components A, B, and C. Components A and B form a *parallel* subsystem, denoted as (A, B). The subsystem (A, B) and component C form a *serial* system. Suppose the times to failure of the three components are exponentially distributed with parameters $\lambda_A = 0.05$/h, $\lambda_B = 0.01$/h, and $\lambda_C = 0.1$/h, respectively. Let X_A, X_B, and X_C be the times to failure for the three components. Show that the time to system failure (i.e., all components fail for the first time) has a *PH*-distribution. Find a *PH*-representation for that *PH*-distribution. Calculate the mean time to system failure.

Exercise 1.3.22 (Exercise 1.3.21 continued) In Exercise 1.3.21, assume that the times to failure for components A, B, and C have *PH*-distributions with *PH*-representations $(\boldsymbol{\alpha}_A, T_A)$, $(\boldsymbol{\alpha}_B, T_B)$, and $(\boldsymbol{\alpha}_C, T_C)$, respectively. Assume that all three components are new at time zero and $\boldsymbol{\alpha}_A\mathbf{e} = \boldsymbol{\alpha}_B\mathbf{e} = \boldsymbol{\alpha}_C\mathbf{e} = 1$. Show that the time to system failure has a *PH*-distribution with the following *PH*-representation

$$\left((\boldsymbol{\alpha}_A\otimes\boldsymbol{\alpha}_B,\ 0,\ 0)\otimes\boldsymbol{\alpha}_C,\ \begin{pmatrix} T_A\otimes I+I\otimes T_B & \mathbf{T}_A^0\otimes I & I\otimes\mathbf{T}_B^0 \\ 0 & T_B & 0 \\ 0 & 0 & T_A \end{pmatrix}\otimes I+I\otimes T_C \right). \tag{1.61}$$

Find a *PH*-representation for the time to system failure if the condition $\boldsymbol{\alpha}_A\mathbf{e} = \boldsymbol{\alpha}_B\mathbf{e} = \boldsymbol{\alpha}_C\mathbf{e} = 1$ is removed.

Exercise 1.3.23 Assume that X has a *PH*-distribution with matrix representation $(\boldsymbol{\alpha}, T)$ of order m. If the underlying Markov chain is in phase i, an amount of money $\$h_i$ (>0) is made per unit time, for $i = 1, 2, \ldots, m$. Denote by Y the total amount of money made in $[0, X]$. Show that Y has a *PH*-distribution with matrix representation $(\boldsymbol{\alpha}, \text{diag}(1/h_1, \ldots, 1/h_m)T)$. (Hint: Consider time *transformation* in each phase. See Exercise 1.1.14.)

Exercise 1.3.24 (Exercise 1.3.23 continued) (Time transformation) Consider *PH*-distribution X with $\boldsymbol{\alpha} = (0.2,\ 0.8)$ and

$$T = \begin{pmatrix} -2 & 1 \\ 2 & -3 \end{pmatrix}. \tag{1.62}$$

Assume $h_1 = \$1$ and $h_2 = \$5$. Find the distribution of the total amount of money, denoted by Y, made in $[0, X]$. Find the mean and variance of Y. Plot the density functions of X and Y.

Random variable X, with distribution function $F(x)$, is *stochastically larger than* random variable Y, with distribution function $G(x)$, if $F(x) \leq G(x)$ for all real x.

Exercise 1.3.25 Assume that X has a PH-distribution with PH-representation ($\boldsymbol{\alpha}$, T). Show that ($\boldsymbol{\alpha}$, $T - \lambda I$) is a PH-representation for $\lambda \geq 0$. Denote by X_λ the random variable corresponding to ($\boldsymbol{\alpha}$, $T - \lambda I$). Show that X_λ is stochastically larger than X_μ, if $\lambda < \mu$. Show that $E[X_\lambda] \geq E[X_\mu]$. (Hint: Use $\exp((T - \lambda I)t) = e^{-\lambda t}\exp(Tt)$. See Exercise 1.2.22.)

For n random variables $\{X_1, X_2, \ldots, X_n\}$, their *k-th order statistic* is defined as the k-th smallest and denoted as $X_{[k]}$, for $k = 1, \ldots, n$. It is easy to see $X_{[1]} = \min\{X_1, X_2, \ldots, X_n\}$ and $X_{[n]} = \max\{X_1, X_2, \ldots, X_n\}$.

Exercise 1.3.26* Show that all the order statistics of n independent phase-type random variables have phase-type distributions. (Hint: The result can be shown using the following steps. Start with set $\{1, 2, \ldots, n\}$. Construct an underlying Markov chain with the (independent) underlying Markov chains of all random variables. By definition, $X_{[1]}$ is the time until one of the underlying Markov chain enters its absorption state. By Exercise 1.3.4, $X_{[1]}$ has a PH-distribution. After the first absorption, there are n possibilities for the remaining set of underlying Markov chains: $\{1, 2, \ldots, n-1\}$, $\{1, 2, \ldots, n-2, n\}$, …, $\{2, 3, \ldots, n\}$. In the remaining set, absorptions will occur. The time (from zero) to the second absorption is $X_{[2]}$. It is easy to see that $X_{[2]}$ also has a PH-distribution. This process can be repeated to define $X_{[k]}$, which shows that $X_{[k]}$ has a PH-distribution. The PH-representations become very large in size for large k.)

Exercise 1.3.27* Consider a shipment consolidation model in transportation. Demands accumulate over time. The amount of goods (in continuous unit) in each demand has a PH-distribution with PH-representation ($\boldsymbol{\alpha}$, T). As soon as the total accumulated amount reaches K, a shipment is dispatched with all accumulated goods. Then the next cycle begins with no goods initially. *Overshot* is defined as the total shipped amount less K (i.e., the shipped portion that is above K). Show that the overshot has a PH-distribution and find a PH-representation for it. (Hint: Consider a PH-renewal process with PH-renewal time ($\boldsymbol{\alpha}$, T). Overshot is then the residual time of the PH-random variable at time $t = K$).

Exercise 1.3.28* (Neuts (1981)) For the classical $M/G/1$ queue, the LST of the waiting time of an arbitrary customer is given by the well-known *Pollachek-Khinchine formula* (Asmussen (2003))

$$w^*(s) = \frac{(1-\rho)s}{s - \lambda + \lambda f^*(s)}, \tag{1.63}$$

where $0 < \rho < 1$, $\lambda > 0$, and $f^*(s)$ is the LST of the service time. The mean service time is $1/\mu$, and the traffic intensity is $\rho = \lambda/\mu$. Assume that the service time has a PH-distribution with irreducible PH-representation ($\boldsymbol{\alpha}$, T) and $\boldsymbol{\alpha}\mathbf{e} = 1$. Show that the waiting time has a PH-distribution with PH-representation ($\rho\boldsymbol{\theta}$, $T + \rho\mathbf{T}^0\boldsymbol{\theta}$), where $\boldsymbol{\theta}$ satisfies $\boldsymbol{\theta}(T + \mathbf{T}^0\boldsymbol{\alpha}) = 0$ and $\boldsymbol{\theta}\mathbf{e} = 1$. (Hint: Use Propositions 1.3.4 and 1.3.5.)

Next, we provide some technical details for the proof of Proposition 1.3.3, particularly for the proof of Eq. (1.44).

Example 1.3.3 Define

$$\begin{pmatrix} \mathbf{p}_1(t) \\ \mathbf{p}_2(t) \\ \mathbf{p}_3(t) \end{pmatrix} = \exp\left(\begin{pmatrix} T\otimes I+I\otimes S & \mathbf{T}^0\otimes I & I\otimes \mathbf{S}^0 \\ 0 & S & 0 \\ 0 & 0 & T \end{pmatrix} t\right)\mathbf{e}, \quad t\geq 0. \tag{1.64}$$

It is easy to see $\mathbf{p}_2(t) = \exp(St)\mathbf{e}$ and $\mathbf{p}_3(t) = \exp(Tt)\mathbf{e}$. To find $\mathbf{p}_1(t)$, we note that $\mathrm{d}(\exp(At))/\mathrm{d}t = A\exp(At)$. Taking derivatives of both sides of Eq. (1.64) yields

$$\frac{\mathrm{d}\mathbf{p}_1(t)}{\mathrm{d}t} = (T\otimes I+I\otimes S)\mathbf{p}_1(t) + (\mathbf{T}^0\otimes I)\mathbf{p}_2(t) + (I\otimes \mathbf{S}^0)\mathbf{p}_3(t). \tag{1.65}$$

Equation (1.65) leads to

$$\begin{aligned} &\frac{\mathrm{d}(\exp(-(T\otimes I+I\otimes S)t)\mathbf{p}_1(t))}{\mathrm{d}t} \\ &= \exp(-(T\otimes I+I\otimes S)t))\left((\mathbf{T}^0\otimes I)\mathbf{p}_2(t) + (I\otimes \mathbf{S}^0)\mathbf{p}_3(t)\right). \end{aligned} \tag{1.66}$$

Taking integrals of both sides of Eq. (1.66), noting $\mathbf{p}_1(0) = \mathbf{e}$, by routine calculations, we obtain

$$\mathbf{p}_1(t) = (\exp(Tt)\mathbf{e})\otimes\mathbf{e} + \mathbf{e}\otimes(\exp(St)\mathbf{e}) - (\exp(Tt)\mathbf{e})\otimes(\exp(St)\mathbf{e}). \tag{1.67}$$

Example 1.3.4 Define

$$\begin{aligned} &\begin{pmatrix} P_{1,1}(t) & P_{1,2}(t) & P_{1,3}(t) \\ P_{2,1}(t) & P_{2,2}(t) & P_{2,3}(t) \\ P_{3,1}(t) & P_{3,2}(t) & P_{3,3}(t) \end{pmatrix} \\ &= \exp\left(\begin{pmatrix} T\otimes I+I\otimes S & \mathbf{T}^0\otimes I & I\otimes \mathbf{S}^0 \\ 0 & S & 0 \\ 0 & 0 & T \end{pmatrix} t\right), \quad t\geq 0. \end{aligned} \tag{1.68}$$

By the definition of matrix exponential function (see Eq. (1.9)), it is easy to see $P_{2,1}(t) = P_{2,3}(t) = P_{3,1}(t) = P_{3,2}(t) = 0$, $P_{2,2}(t) = \exp(St)$, and $P_{3,3}(t) = \exp(Tt)$. By the proof in Eq. (1.39), we obtain $P_{1,1}(t) = \exp((T\otimes I + I\otimes S)t)$. For $P_{1,2}(t)$, we have the following calculations:

$$
\begin{aligned}
P_{1,2}(t) &= \sum_{n=1}^{\infty} \frac{t^n}{n!} \left(\sum_{i=0}^{n-1} (T \otimes I + I \otimes S)^i (\mathbf{T}^0 \otimes I) S^{n-1-i} \right) \\
&= \sum_{n=1}^{\infty} \frac{t^n}{n!} \left(\sum_{i=0}^{n-1} \left(\sum_{j=0}^{i} \frac{i!}{j!(i-j)!} T^j \otimes S^{i-j} \right) (\mathbf{T}^0 \otimes S^{n-1-i}) \right) \\
&= \sum_{i=0}^{\infty} \left(\sum_{n=i+1}^{\infty} \frac{t^n}{n!} \left(\sum_{j=0}^{i} \frac{i!}{j!(i-j)!} (T^j \mathbf{T}^0) \otimes S^{n-1-j} \right) \right) \\
&= \sum_{i=0}^{\infty} \left(\sum_{n=0}^{\infty} \frac{t^{n+i+1}}{(n+i+1)!} \left(\sum_{j=0}^{i} \frac{i!}{j!(i-j)!} (T^j \mathbf{T}^0) \otimes S^{n+i-j} \right) \right) \\
&= \sum_{n=0}^{\infty} \left(\sum_{i=0}^{\infty} \left(\sum_{j=0}^{i} \frac{t^{n+i+1}}{(n+i+1)!} \frac{i!}{j!(i-j)!} (T^j \mathbf{T}^0) \otimes S^{n+i-j} \right) \right) \\
&= \sum_{n=0}^{\infty} \left(\sum_{j=0}^{\infty} \left(\sum_{i=j}^{\infty} \frac{t^{n+i+1}}{(n+i+1)!} \frac{i!}{j!(i-j)!} (T^j \mathbf{T}^0) \otimes S^{n+i-j} \right) \right) \\
&= \sum_{n=0}^{\infty} \left(\sum_{j=0}^{\infty} \left(\sum_{i=0}^{\infty} \frac{t^{n+i+j+1}}{(n+i+j+1)!} \frac{(i+j)!}{j!i!} (T^j \mathbf{T}^0) \otimes S^{n+i} \right) \right) \\
&= \sum_{j=0}^{\infty} \frac{t^{j+1}}{j!} (T^j \mathbf{T}^0) \otimes \left(\sum_{n=0}^{\infty} \left(\sum_{i=0}^{\infty} \frac{t^{n+i} S^{n+i}}{(n+i+j+1)!} \frac{(i+j)!}{i!} \right) \right).
\end{aligned}
\tag{1.69}
$$

For fixed j, we have

$$
\sum_{n=0}^{\infty} \left(\sum_{i=0}^{\infty} \frac{t^{n+i} S^{n+i}}{(n+i+j+1)!} \frac{(i+j)!}{i!} \right) = \sum_{n=0}^{\infty} \frac{t^n S^n}{(n+j+1)!} \left(\sum_{i=0}^{n} \frac{(i+j)!}{i!} \right). \tag{1.70}
$$

By induction, it can be shown that

$$
\sum_{i=0}^{n} \frac{(i+j)!}{i!} = \frac{(n+j+1)!}{n!(j+1)}. \tag{1.71}
$$

Combining Eqs. (1.69), (1.70), and (1.71), we obtain

$$
\begin{aligned}
P_{1,2}(t) &= \sum_{j=0}^{\infty} \frac{t^{j+1}}{j!} (T^j \mathbf{T}^0) \otimes \left(\sum_{n=0}^{\infty} \frac{t^n S^n}{n!(j+1)} \right) \\
&= \sum_{j=0}^{\infty} \frac{t^{j+1}}{(j+1)!} (-T^{j+1} \mathbf{e}) \otimes \exp(St) \\
&= (\mathbf{e} - \exp(Tt)\mathbf{e}) \otimes \exp(St).
\end{aligned}
\tag{1.72}
$$

Similarly, we obtain $P_{1,3}(t) = \exp(Tt) \otimes (\mathbf{e} - \exp(St)\mathbf{e})$.

Exercise 1.3.29 In Examples 1.3.3 and 1.3.4, explain the expressions of functions $P_{1,2}(t)$, $P_{1,3}(t)$, and $\mathbf{p}_1(t)$, probabilistically. (Hint: $\mathbf{e} = \mathbf{e} - \exp(Tt)\mathbf{e} + \exp(Tt)\mathbf{e}$.)

1.4 Phase-Type Distributions: *PH*-Representations

Section 1.3 demonstrates the importance of *PH*-representations in the applications of *PH*-distributions. It also shows that the *PH*-representation of a given *PH*-distribution is not unique. In this section, we further investigate *PH*-representations and related properties. We develop algorithms to find simple equivalent representations (e.g., Coxian representations) for *PH*-distributions.

Example 1.4.1 The following *PH*-representations represent an exponential distribution with parameter $\lambda = 1$, which can be shown by using Definition 1.2.1.

(1) $\boldsymbol{\alpha} = (1,\ \ 0)$, $T = \begin{pmatrix} -1 & 0 \\ 1 & -5 \end{pmatrix}$. Since phase 2 can never be reached, the result follows.

(2) $\boldsymbol{\alpha} = (0.2,\ \ 0.8)$, $T = \begin{pmatrix} -2 & 1 \\ 2 & -3 \end{pmatrix}$. By $T\mathbf{e} = -\mathbf{e}$, it is easy to verify $\exp(Tt)\mathbf{e} = e^{-t}\mathbf{e}$. The result follows.

(3) $\boldsymbol{\alpha} = (0.5,\ \ 0.5)$, $T = \begin{pmatrix} -1 & 1 \\ 0 & -2 \end{pmatrix}$. By $\boldsymbol{\alpha}T = -\boldsymbol{\alpha}$, it is easy to verify $\boldsymbol{\alpha}\exp(Tt) = \boldsymbol{\alpha}e^{-t}$. The result follows.

Exercise 1.4.1 Show that $\boldsymbol{\alpha} = (2/3,\ 1/3)$ and $T = \begin{pmatrix} -4 & 4 \\ 4 & -10 \end{pmatrix}$ represent an exponential distribution with parameter $\lambda = 2$.

Two *PH*-representations are called *equivalent* if they represent the same probability distribution. Equivalent *PH*-representations can have a different number of phases and a different structure in their *PH*-generators. In general, equivalent *PH*-representations can be constructed by adding ancillary phases (e.g., Example 1.4.1), by rearranging phases, or by transformation. It is easy to see that any given *PH*-distribution has infinitely many equivalent *PH*-representations.

Exercise 1.4.2 The following methods construct infinitely many equivalent *PH*-representations for any *PH*-representation $(\boldsymbol{\alpha}, T)$.

(1) Show that $(\boldsymbol{\alpha}, T)$ and $((\boldsymbol{\alpha}, 0), \mathrm{diag}(T, S))$ are equivalent *PH*-representations, where S is a *PH*-generator and $\mathrm{diag}(T, S)$ represents a matrix with T and S on the (block) diagonal.
(2) Show that $(\boldsymbol{\alpha}, T)$ and $((\boldsymbol{\alpha}, \boldsymbol{\alpha}, \ldots, \boldsymbol{\alpha})/n, \mathrm{diag}(T, T, \ldots, T))$ are equivalent *PH*-representations for any positive integer n.

(3) Show that $(\boldsymbol{\alpha}, T)$ and $((c_1\boldsymbol{\alpha}, c_2\boldsymbol{\alpha}, \ldots, c_n\boldsymbol{\alpha}), \text{diag}(T, T, \ldots, T))$ are equivalent *PH*-representations, where $\{c_1, c_2, \ldots, c_n\}$ are nonnegative real numbers with a unit sum.

For *PH*-representation $(\boldsymbol{\alpha}, T)$ of order m, the integer m is called the *order of the representation*. Consider *PH*-representation

$$\left(\boldsymbol{\alpha} = (0.2, 0,8, 0), \quad T = \begin{pmatrix} -2 & 1.5 & 0 \\ 1 & -5 & 0 \\ 1 & 1 & -3 \end{pmatrix}\right). \tag{1.73}$$

It is easy to verify that the *PH*-representation is reducible (i.e., $T + \mathbf{T}^0\boldsymbol{\alpha}$ is not irreducible). By removing the third element in $\boldsymbol{\alpha}$, and the third row and third column in T, an equivalent *PH*-representation of order 2 is obtained.

Proposition 1.4.1 *If PH-representation $(\boldsymbol{\alpha}, T)$ is reducible, then an equivalent PH-representation of a smaller order can be found.*

Proof. Since $Q = T + \mathbf{T}^0\boldsymbol{\alpha}$ is reducible, by rearranging the phases, Q, T, and $\mathbf{T}^0\boldsymbol{\alpha}$ must have the structure

$$Q = \begin{pmatrix} Q_{1,1} & 0 \\ Q_{2,1} & Q_{2,2} \end{pmatrix}, \; T = \begin{pmatrix} T_{1,1} & 0 \\ T_{2,1} & T_{2,2} \end{pmatrix}, \; \mathbf{T}^0\boldsymbol{\alpha} = \begin{pmatrix} T^0_{1,1} & 0 \\ T^0_{1,2} & T^0_{2,2} \end{pmatrix} \tag{1.74}$$

where $Q_{1,1}$, $T_{1,1}$, and $T^0{}_{1,1}$ are square matrices of the same order. Partition $\mathbf{T}^0$ and $\boldsymbol{\alpha}$ accordingly as follows:

$$\mathbf{T}^0 = \begin{pmatrix} \mathbf{T}^0_1 \\ \mathbf{T}^0_2 \end{pmatrix}, \quad \boldsymbol{\alpha} = (\boldsymbol{\alpha}_1, \boldsymbol{\alpha}_2). \tag{1.75}$$

By eq. (1.75), $\mathbf{T}^0{}_1\boldsymbol{\alpha}_2 = 0$. Since T is a *PH*-generator, $T_{1,1}$ must be invertible. Since $\mathbf{T}^0{}_1 = -T_{1,1}\mathbf{e}$, we must have $\mathbf{T}^0{}_1 \neq 0$. Consequently, we must have $\boldsymbol{\alpha}_2 = 0$. Therefore, $(\boldsymbol{\alpha}, T)$ is equivalent to $(\boldsymbol{\alpha}_1, T_{1,1})$. This completes the proof of Proposition 1.4.1.

Proposition 1.4.1 implies that any given *PH*-representation has an equivalent irreducible *PH*-representation. The reason is that, if the *PH*-representation $(\boldsymbol{\alpha}_1, T_{1,1})$ defined in the proof of Proposition 1.4.1 is reducible, the same procedure can be applied to find a smaller equivalent *PH*-representation for $(\boldsymbol{\alpha}_1, T_{1,1})$. Eventually, an equivalent irreducible *PH*-representation can be found. Note that a *PH*-representation of order one is irreducible. Further, by combining Exercise 1.4.2 and Proposition 1.4.1, the following result can be obtained.

Exercise 1.4.3 Show that any given *PH*-distribution can have infinitely many irreducible *PH*-representations.

It is clear that, if the order of *PH*-representations is higher, then greater number of different probability distributions can be constructed. Suppose we want to find *PH*-distributions to approximate a probability distribution $F(t)$. To find a satisfactory

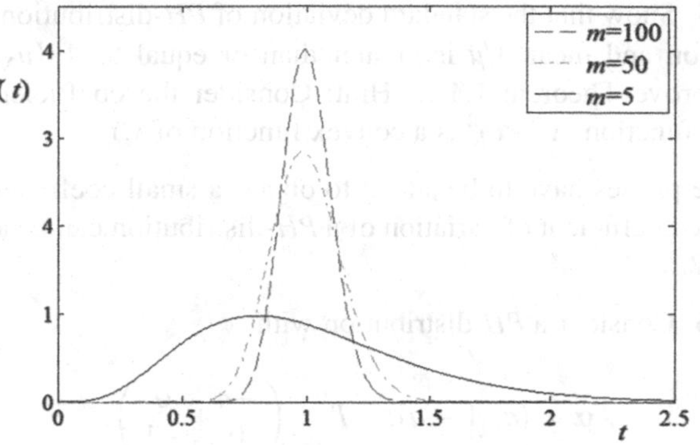

Fig. 1.10 Density functions of Erlang distributions with mean one

approximation, the order of *PH*-representations may have to be large. How large must the order be to produce a satisfactory approximation? For example, how can we find *PH*-distributions to approximate constant 1 (i.e., $F(t) = 0$, if $t < 1$; 1, if $t \geq 1$)? The following well-known result gives a lower bound on the order required with respect to the coefficient of variation (see Proposition 1.1.1) of probability distributions. It finds that the Erlang distribution of order *m* has the smallest coefficient of variation among all the *PH*-distributions with a *PH*-representation of order *m* or less.

Theorem 1.4.1 (Aldous and Shepp (1987)) *The coefficient of variation of a PH-random variable with an order m PH-representation is at least* $1/\sqrt{m}$, *and the order* m *Erlang distribution is the only one to attain the lower bound.*

Exercise 1.4.4 Determine how large an order is required for a *PH*-representation to have coefficient of variation 0.1? Re-plot the density functions of the Erlang distributions of orders 5, 50, and 100, and with mean one, as shown in Fig. 1.10. Plot the density functions of any randomly generated *PH*-representations of orders 5, 50, and 100, and with mean one. Calculate their coefficients of variation. (See Exercise 1.2.14.)

Theorem 1.4.1 can be strengthened as follows. For two probability distribution functions $F(x)$ and $G(x)$ corresponding to random variables X and Y, $F(x)$ *majorizes* $G(x)$ if $E[h(X)] \geq E[h(Y)]$ for every *convex function* $h(x)$. In the literature, Y is said to be smaller than X in the convex order (denoted as $Y \leq_{cx} X$) (see Shaked and Shanthikumar (2006)). Since both functions $h(x) = x$ and $g(x) = -x$ are convex, $Y \leq_{cx} X$ implies $E[Y] = E[X]$. The convex order implies that X is "more variable" than Y in the sense $E[\max(0, X - x)] \geq E[\max(0, Y - x)]$ and $E[\max(0, x - X)] \geq E[\max(0, x - Y)]$ hold for all x.

Theorem 1.4.2 (O'Cinneide (1991b)) *A PH-distribution with an order m PH-representation majorizes the order m Erlang distribution of the same mean.*

Exercise 1.4.5 Show that the standard deviation of *PH*-distributions with an order m representation and mean $1/\mu$ is greater than or equal to $1/(\mu\sqrt{m})$. Then use the result to prove Theorem 1.4.1. (Hint: Consider the coefficient of variation. Also note that function $(x - c)^2$ is a convex function of x.)

While more phases have to be added to obtain a small coefficient of variation, the value of the coefficient of variation of a *PH*-distribution can be arbitrarily large even at order 2.

Exercise 1.4.6 Consider a *PH*-distribution with

$$\boldsymbol{\alpha} = (a,\ 1 - a), \quad T = \begin{pmatrix} -a & 0 \\ 0 & -1 \end{pmatrix}, \tag{1.76}$$

for $0 < a < 1$. Show that $cv = \sqrt{2(1 - a + 1/a)(2 - a)^{-2} - 1}$. Show that cv approaches positive infinity if a approaches zero.

As indicated in Theorems 1.4.1 and 1.4.2, Erlang distributions play an important role in the set of *PH*-distributions. Equation (1.19) gives a simple matrix representation for an Erlang distribution. The matrix T in Eq. (1.19) has a *bi-diagonal structure*, which is amenable to both numerical computation and probabilistic interpretation. In fact, Eqs. (1.20) and (1.21) show that Coxian distributions, so defined, have a simple matrix representation, called *Coxian representation*, and a simple probabilistic interpretation as well. Many *PH*-representations have an equivalent Coxian representation of the same or a different order since they actually represent Coxian distributions. It is then interesting to find conditions under which a *PH*-representation has an equivalent Coxian representation, and to develop algorithms to find an equivalent Coxian representation. The first result of this kind is given in Cumani (1982) and Dehon and Latouche (1982). Let

$$S(\boldsymbol{\lambda}) = \begin{pmatrix} -\lambda_1 & & & \\ \lambda_2 & -\lambda_2 & & \\ & \ddots & \ddots & \\ & & \lambda_m & -\lambda_m \end{pmatrix}, \tag{1.77}$$

where $\boldsymbol{\lambda} = (\lambda_1, \lambda_2, \ldots, \lambda_m)$. For convenience, the elements of $\boldsymbol{\lambda}$ are arranged in descending order of their *modulus* (i.e., the absolute value of a real or complex number), unless specified otherwise.

For *PH*-generators T and S, if for any *PH*-representation $(\boldsymbol{\alpha}, T)$, there exists a *PH*-representation $(\boldsymbol{\beta}, S)$ such that $(\boldsymbol{\alpha}, T)$ and $(\boldsymbol{\beta}, S)$ are equivalent, then T is *PH-majorized* by S.

Theorem 1.4.3 (Cumani (1982) and Dehon and Latouche (1982)) *Let T be a (upper or lower) triangular PH-generator with diagonal elements $\{-\lambda_1, -\lambda_2, \ldots, -\lambda_m\}$ satisfying $\lambda_1 \geq \lambda_2 \geq \ldots \geq \lambda_m$. Then any PH-representation $(\boldsymbol{\alpha}, T)$ has an equivalent Coxian representation $(\boldsymbol{\beta}, S(\boldsymbol{\lambda}))$ of the order m, i.e., T is PH-majorized by $S(\boldsymbol{\lambda})$.*

Proof. Let $\mathbf{y} = (y_1, y_2, \ldots, y_m)$ and all elements of $\mathbf{y}$ be positive. Define $\mathbf{z} = (y_1, \ldots, y_{i-1}, y_{i+1}, y_i, y_{i+2}, \ldots, y_m)$, i.e., $\mathbf{z}$ is obtained by interchanging y_i and y_{i+1} in $\mathbf{y}$. Then $S(\mathbf{y})$ and $S(\mathbf{z})$ are two Coxian generators satisfying $S(\mathbf{y})P = PS(\mathbf{z})$, where

$$P = \begin{pmatrix} I & 0 & 0 & 0 \\ 0 & \begin{pmatrix} \frac{y_i}{y_{i+1}} & 1 - \frac{y_i}{y_{i+1}} \\ 0 & 1 \end{pmatrix} & & 0 \\ 0 & & & 0 \\ 0 & 0 & 0 & I \end{pmatrix}, \tag{1.78}$$

which has unit row sums. If $y_i \leq y_{i+1}$, the matrix P is nonnegative. Then any *PH*-representation $(\boldsymbol{\alpha}, S(\mathbf{y}))$ has an equivalent Coxian representation $(\boldsymbol{\alpha}P, S(\mathbf{z}))$. Consequently, $S(\mathbf{z})$ *PH*-majorizes $S(\mathbf{y})$. Repeating this process, we obtain that $S(\mathbf{x})$ *PH*-majorizes $S(\mathbf{y})$, where $\mathbf{x}$ is the vector with elements $\{y_1, y_2, \ldots, y_m\}$ in descending order. Let $\mathbf{w} = (w_1, w_2, \ldots, w_n)$, where all elements are positive. Since $S(\mathbf{y})(I, 0) = (I, 0)S((\mathbf{y}, \mathbf{w}))$, then $S((\mathbf{y}, \mathbf{w}))$ *PH*-majorizes $S(\mathbf{y})$. If $\mathbf{x}$ is a vector with elements $\{y_1, y_2, \ldots, y_m, w_1, w_2, \ldots, w_n\}$ in descending order, then $S(\mathbf{x})$ *PH*-majorizes $S((\mathbf{y}, \mathbf{w}))$, which implies that $S(\mathbf{x})$ *PH*-majorizes $S(\mathbf{y})$.

Recall that $T = (t_{i,j})$. We want to show that T is *PH*-majorized by $S(\mathbf{x})$, where $\mathbf{x}$ is the vector with elements $\{-t_{1,1}, -t_{2,2}, \ldots, -t_{m,m}\}$ in descending order. To do so, we only need to show that the *PH*-representation $(\mathbf{e}(j), T)$ has an equivalent Coxian representation of the form $(\boldsymbol{\beta}(j), S(\mathbf{x}))$ for $j = 1, \ldots, m$, where $\mathbf{e}(j)$ is a row vector with all elements being zero except for the j-th element being one, and $\boldsymbol{\beta}(j)$ is substochastic. Since T is a lower triangular *PH*-generator, the proof can be done inductively. It is clear that $(\mathbf{e}(1), T)$ is equivalent to *PH*-representation $((1), (t_{1,1}))$. It is easy to see that the *PH*-generator $(t_{1,1})$ is *PH*-majorized by $S(\mathbf{y})$, where $\mathbf{y}$ is the vector with elements $\{-t_{1,1}, -t_{2,2}, \ldots, -t_{m-1,m-1}\}$ in descending order. Then $(\mathbf{e}(1), T)$ has a Coxian representation $(\boldsymbol{\beta}(1), S(\mathbf{y}))$. By induction, suppose that $(\mathbf{e}(j), T)$ has an equivalent Coxian representation of the form $(\boldsymbol{\beta}(j), S(\mathbf{y}))$, for $j = 1, \ldots, m-1$. Since $S(\mathbf{y})$ is *PH*-majorized by $S(\mathbf{x})$, $(\mathbf{e}(j), T)$ has a Coxian representation with Coxian generator $S(\mathbf{x})$, $j = 1, \ldots, m-1$. Next, we show that $(\mathbf{e}(m), T)$ has an equivalent Coxian representation of the form $(\boldsymbol{\beta}(m), S(\mathbf{x}))$.

Recall that $\mathbf{T}^0 = -T\mathbf{e}$. Denote by t_j^0 the j-th element of $\mathbf{T}^0$, $j = 1, \ldots, m$. Denote by $f_j^*(s)$ the LST of the *PH*-distribution $(\mathbf{e}(j), T)$, for $j = 1, \ldots, m$. Then we have

$$\begin{aligned} f_m^*(s) &= \left(\frac{-t_{m,m}}{s - t_{m,m}}\right)\left(\frac{t_m^0}{(-t_{m,m})} + \sum_{j=1}^{m-1} \frac{t_{m,j}}{(-t_{m,m})} f_j^*(s)\right) \\ &= \frac{t_m^0}{(-t_{m,m})}\left(\frac{-t_{m,m}}{s - t_{m,m}}\right) + \sum_{j=1}^{m-1} \frac{t_{m,j}}{(-t_{m,m})} f_j^*(s)\left(\frac{-t_{m,m}}{s - t_{m,m}}\right), \end{aligned} \tag{1.79}$$

which indicates that $(\mathbf{e}(m), T)$ has Coxian representation

$$\left(\left(\frac{t_m^0}{(-t_{m,m})},(0,\cdots,0)\right)+\sum_{j=1}^{m-1}\frac{t_{m,j}}{(-t_{m,m})}(0,\,\boldsymbol{\beta}(j)),\ \begin{pmatrix} t_{m,m} & 0 \\ \begin{pmatrix} y_1 \\ 0 \end{pmatrix} & S(y) \end{pmatrix}\right). \tag{1.80}$$

It is readily seen that the Coxian generator in Eq. (1.80) is PH-majorized by $S(\mathbf{x})$. Therefore, $(\mathbf{e}(m),\, T)$ has an equivalent Coxian representation with the ordered Coxian generator $S(\mathbf{x})$. This completes the proof of Theorem 1.4.3.

Many PH-representations without the triangular structure also have an equivalent Coxian representation of the same order since those PH-representations actually represent Coxian distributions. The following proposition presents an example.

Proposition 1.4.2 *Assume that T is a PH-generator of order m that satisfies $DT = T'D$ for some diagonal matrix D with positive diagonal elements, where T' is the transpose of T. Then any PH-representation $(\boldsymbol{\alpha},\, T)$ has a Coxian representation $(\boldsymbol{\beta},\, S(\boldsymbol{\lambda}))$, where $\boldsymbol{\lambda} = (\lambda_1,\, \lambda_2,\, \ldots,\, \lambda_m)$ and, $\{-\lambda_1, -\lambda_2, \ldots, -\lambda_m\}$ are all the eigenvalues of T in descending order of their modulus (counting multiplicities).*

Proof. This proposition is an immediate consequence of Theorem 1.4.6.

Exercise 1.4.7* Show that, if T is a symmetric PH-generator, then any PH-representation $(\boldsymbol{\alpha},\, T)$ has an equivalent Coxian representation of the same order. (Hint: Use results in Micchelli and Willoughby (1979) and Proposition 1.4.2.)

Exercise 1.4.8* Show that, if T is a tri-diagonal PH-generator, then any PH-representation $(\boldsymbol{\alpha},\, T)$ has an equivalent Coxian representation of the same order. (Hint: Verify the condition $DT = T'D$.)

More generally, the next result shows that all PH-representations $(\boldsymbol{\alpha},\, T)$ for which all eigenvalues of T are real represent Coxian distributions.

Theorem 1.4.4 (O'Cinneide (1993)) *For any PH-generator T with only real eigenvalues, any PH-representation $(\boldsymbol{\alpha},\, T)$ represents a Coxian distribution.*

Theorem 1.4.4 guarantees that any $(\boldsymbol{\alpha},\, T)$ satisfying the corresponding condition has a Coxian representation. Given a PH-representation with a triangular PH-generator or a PH-generator that satisfies the conditions in Theorems 1.4.3 and 1.4.4, and Proposition 1.4.2, how do we find the corresponding Coxian representation? More generally, given a PH-representation, how do we find an equivalent bi-diagonal representation, which is not necessarily a PH-representation, for it? The *spectral polynomial algorithm* (SPA) to be introduced next is a simple and efficient tool in doing just that. For given PH-generator T with eigenvalues $\{-\lambda_1, -\lambda_2, \ldots, -\lambda_m\}$ (counting multiplicities), define

$$\begin{aligned} \mathbf{p}_1 &= -T\mathbf{e}/\lambda_1; \\ \mathbf{p}_n &= (\lambda_{n-1}I + T)\mathbf{p}_{n-1}/\lambda_n, \quad n = 2,\, 3, \ldots,\, m. \end{aligned} \tag{1.81}$$

Let $P = (\mathbf{p}_1,\, \mathbf{p}_2, \ldots,\, \mathbf{p}_m)$.

Theorem 1.4.5 *Matrix representations* $(\boldsymbol{\alpha}, T)$ *and* $(\boldsymbol{\alpha}P, S(\boldsymbol{\lambda}))$ *are equivalent, i.e.,* $1-\boldsymbol{\alpha}\exp(Tt)\mathbf{e} = 1-\boldsymbol{\alpha}P\exp(S(\boldsymbol{\lambda})t)\mathbf{e}$*, for* $t \geq 0$*. In addition, we have* $TP = PS(\boldsymbol{\lambda})$ *and* $P\mathbf{e} = \mathbf{e}$.

Proof. First, we show that P has unit row sums. By the well-known Caley-Hamilton theorem (Lancaster and Tismenetsky (1985)), we know $(\lambda_1 I + T)\cdots(\lambda_m I + T) = 0$, which leads to

$$\begin{aligned}
&(\lambda_1 I + T)\cdots(\lambda_{m-1} I + T)\lambda_m + \left(\prod_{k=1}^{m-1}(\lambda_k I + T)\right)T \\
&= (\lambda_1 I + T)\cdots(\lambda_{m-2} I + T)\lambda_{m-1}\lambda_m + \left(\prod_{k=1}^{m-2}(\lambda_k I + T)\right)T\lambda_m + \left(\prod_{k=1}^{m-1}(\lambda_k I + T)\right)T \\
&\vdots \\
&= \lambda_1\cdots\lambda_m I + T\lambda_2\cdots\lambda_m + \sum_{n=1}^{m-1}\left(\prod_{k=1}^{n}(\lambda_k I + T)\right)T\left(\prod_{k=n+2}^{m}\lambda_k\right) \\
&= 0.
\end{aligned} \tag{1.82}$$

Dividing by $\lambda_1\lambda_2\ldots\lambda_m$ and postmultiplying by $\mathbf{e}$ on both sides of Eq. (1.82), we obtain $\mathbf{p}_1 + \mathbf{p}_2 + \ldots + \mathbf{p}_m = \mathbf{e}$. Thus, the matrix P has unit row sums.

Rewriting Eq. (1.81) in matrix form, we obtain $TP = PS(\boldsymbol{\lambda})$, which leads to $T^n P = PS^n(\boldsymbol{\lambda})$, for all integer n. Then we have, for $t \geq 0$,

$$\begin{aligned}
\exp(Tt)\mathbf{e} &= \exp(Tt)P\mathbf{e} = \sum_{n=0}^{\infty}\frac{t^n}{n!}T^n P\mathbf{e} \\
&= \sum_{n=0}^{\infty}\frac{t^n}{n!}PS^n(\boldsymbol{\lambda})\mathbf{e} = P\exp(S(\boldsymbol{\lambda})t)\mathbf{e}.
\end{aligned} \tag{1.83}$$

Premultiplying by any substochastic vector $\boldsymbol{\alpha}$ on both sides of Eq. (1.83), we obtain $\boldsymbol{\alpha}\exp(Tt)\mathbf{e} = \boldsymbol{\alpha}P\exp(S(\boldsymbol{\lambda})t)\mathbf{e}$, which is equivalent to the expected result $1-\boldsymbol{\alpha}\exp(Tt)\mathbf{e} = 1-\boldsymbol{\alpha}P\exp(S(\boldsymbol{\lambda})t)\mathbf{e}$. This completes the proof of Theorem 1.4.5.

Exercise 1.4.9 Consider a lower triangular *PH*-generator $T = (t_{i,j})$ of order $m = 10$, defined as $t_{i,i} = -i$, $i = 1, \ldots, m$, $t_{i,j} = 1$, for $j < i$, and $i, j = 1, \ldots, m$, and $t_{i,j} = 0$, for $i < j$ and $i, j = 1, \ldots, m$. Find matrix P and $S(\boldsymbol{\lambda})$, where $\boldsymbol{\lambda} = (m, m-1, \ldots, 1)$.

Exercise 1.4.10 (Exercise 1.2.21 continued) Consider the following *PH*-representation:

$$\boldsymbol{\alpha} - (0.6, 0.2), \quad T = \begin{pmatrix} -4.5 & 2 \\ 0.625 & -2.5 \end{pmatrix}. \tag{1.84}$$

The eigenvalues of T are $\{-2, -5\}$. Find its equivalent matrix-representation by setting $\lambda_1 = 5$ and $\lambda_2 = 2$. Find its equivalent matrix-representation by setting $\lambda_1 = 2$ and $\lambda_2 = 5$. (Note: Compare the solutions to the matrix-representation in Exercise 1.2.21.)

The spectral polynomial algorithm defined in Eq. (1.81) and Theorem 1.4.5 is simple and efficient in finding equivalent bi-diagonal matrix representations for *PH*-representations. However, the equivalent bi-diagonal matrix representation $(\boldsymbol{\alpha}P, S(\boldsymbol{\lambda}))$ may not be a *PH*-representation since (a) $\boldsymbol{\alpha}P$ may not be nonnegative, or (b) $S(\boldsymbol{\lambda})$ may not be a *PH*-generator. Apparently, the new matrix representation is a *PH*-representation if (i) all eigenvalues of T are real and (ii) the matrix P is nonnegative (a sufficient condition). A number of conditions have been identified for the two requirements.

Theorem 1.4.6 (He et al. (2011)) *Assume that PH-generator T is triangular or there exists a positive diagonal matrix D such that $DT = T'D$. Then all eigenvalues of T are real and the matrix P in the SPA is nonnegative, if, in the vector $\boldsymbol{\lambda}$, the eigenvalues $\{-\lambda_1, -\lambda_2, \ldots, -\lambda_m\}$ are arranged in descending order of their modulus.*

Note that Theorem 1.4.6 gives a proof to Proposition 1.4.2.

By Theorem 1.4.5, $(\boldsymbol{\alpha}, T)$ represents a Coxian distribution if $(\boldsymbol{\alpha}P, S(\boldsymbol{\lambda}))$ represents a Coxian distribution. On the other hand, $(\boldsymbol{\alpha}P, S(\boldsymbol{\lambda}))$ represents a Coxian distribution if $\boldsymbol{\alpha}P$ is nonnegative and all elements in $\boldsymbol{\lambda}$ are real.

Exercise 1.4.11 Consider the following *PH*-generator of order 3:

$$T = \begin{pmatrix} -5 & 0 & 1.5 \\ 1 & -2 & 0 \\ 0 & 2 & -2 \end{pmatrix}. \tag{1.85}$$

Find an equivalent matrix representation for $((0, 1, 0), T)$ and $((1, 0, 0), T)$.

Exercise 1.4.11 demonstrates that $\boldsymbol{\alpha}P$ may not be nonnegative. Thus, $(\boldsymbol{\alpha}P, S(\boldsymbol{\lambda}))$ may not be a Coxian representation. When can $(\boldsymbol{\alpha}P, S(\boldsymbol{\lambda}))$ be a Coxian representation? Let $\boldsymbol{\beta} = \boldsymbol{\alpha}P$. If P is invertible, then $\boldsymbol{\alpha} = \boldsymbol{\beta}P^{-1}$, which implies that $\boldsymbol{\beta}$ is nonnegative if $\boldsymbol{\alpha}$ is a convex combination of the rows in matrix P^{-1}. For Exercise 1.4.11, we obtain

$$P = \begin{pmatrix} 0.77 & -0.11 & 0.33 \\ 0.22 & 0.39 & 0.38 \\ 0 & 0.13 & 0.86 \end{pmatrix} \quad \text{and} \quad P^{-1} = \begin{pmatrix} 1.14 & 0.53 & -0.67 \\ -0.74 & 2.60 & -0.85 \\ 0.11 & -0.40 & 1.28 \end{pmatrix}. \tag{1.86}$$

The rows of P^{-1} are m vectors and they can form a *convex hull* (Rockafellar (1970)). Thus, if P is invertible, $(\boldsymbol{\alpha}P, S(\boldsymbol{\lambda}))$ is a Coxian representation if and only if

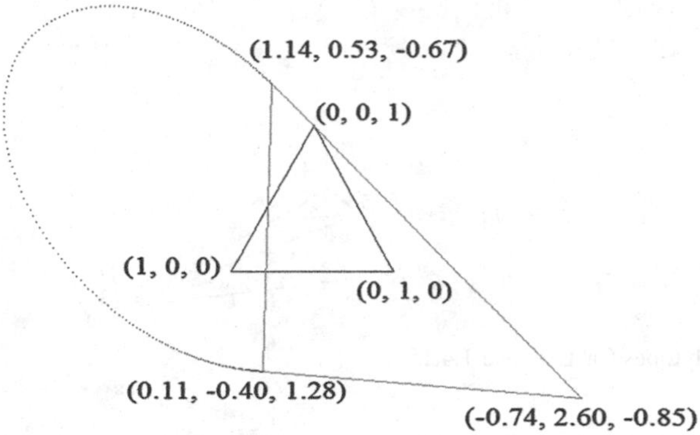

Fig. 1.11 The convex set of feasible $\boldsymbol{\alpha}$s for T in Eq. (1.85)

$\boldsymbol{\alpha}$ is in the convex hull with the rows of P^{-1} as its extreme points (see Fig. 1.11, where the convex hull of interest has three extreme points: (1.14, 0.53, −0.67), (0.11, −0.40, 1.28), and (−0.74, 2.60, −0.85)).

Exercise 1.4.12 For the *PH*-generator defined in Eq. (1.85), (i) compute matrix P^{-1}; (ii) plot the triangle formed by the three rows of P^{-1}; (iii) plot the triangle formed by $\{(1, 0, 0), (0, 1, 0), (0, 0, 1)\}$; and (iv) identify all $\boldsymbol{\alpha}$s such that $(\boldsymbol{\alpha}P, S(\boldsymbol{\lambda}))$ is a Coxian representation. (Hint: Since all vectors involved have a unit sum, they can be plotted in a two-dimensional space. See Fig. 1.11.)

The ice-cream cone in Fig. 1.11 covers all $\boldsymbol{\alpha}$ such that $(\boldsymbol{\alpha}, T)$ is a probability distribution, which is produced based on a theorem in Dehen and Latouche (1982). It is clear that (i), for any $\boldsymbol{\alpha}$ in the larger triangle area, $(\boldsymbol{\alpha}, T)$ has an equivalent Coxian representation $(\boldsymbol{\alpha}P, S(\boldsymbol{\lambda}))$; and (ii), for any $\boldsymbol{\alpha}$ in the intersection of the larger and smaller triangles, $(\boldsymbol{\alpha}, T)$ is a *PH*-representation and $(\boldsymbol{\alpha}P, S(\boldsymbol{\lambda}))$ is a Coxian representation. For other $\boldsymbol{\alpha}$ in the interior of the ice-cream cone, $(\boldsymbol{\alpha}, T)$ represents a *PH*-distribution, but it may not be a *PH*-representation since $\boldsymbol{\alpha}$ may not be nonnegative. Nonetheless, for such an $\boldsymbol{\alpha}$, an equivalent *PH*-representation, most likely with a greater order, can be found for $(\boldsymbol{\alpha}, T)$.

Exercise 1.4.13 Calculate the matrix P in the SPA for the following *PH*-generators:

$$T_1 = \begin{pmatrix} -5 & 0 & 1 \\ 0 & -3 & 1.5 \\ 1 & 1.5 & -4 \end{pmatrix}, \quad T_2 = \begin{pmatrix} -5 & 0 & 0.1 \\ 4 & -4 & 0 \\ 0 & 1 & -1 \end{pmatrix}, \quad T_3 = \begin{pmatrix} -5 & 0 & 2 \\ 3 & -3 & 0 \\ 0 & 1 & -1 \end{pmatrix}. \tag{1.87}$$

Comment on the equivalent bi-diagonal representations of $(\boldsymbol{\alpha}, T)$ for any stochastic vector $\boldsymbol{\alpha}$.

Commentary In He and Zhang (2008), an algorithm is developed for computing a minimal Coxian representation for a Coxian distribution. In He et al. (2011), an

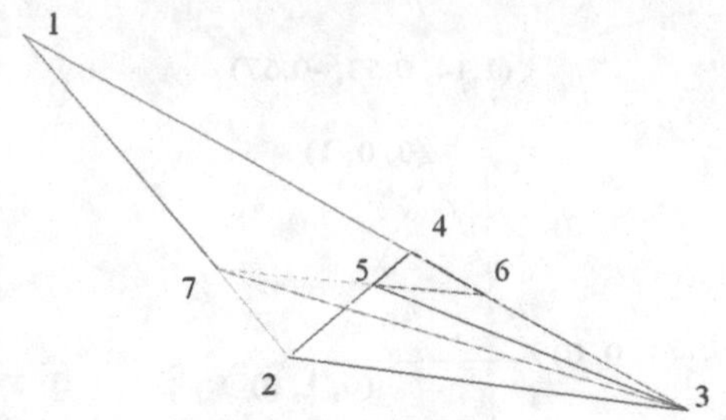

Fig. 1.12 Six polytopes for Exercise 1.4.15

algorithm is developed to find a Coxian representation for any *PH*-representation with only real eigenvalues. More discussion on *PH*-representations can be found in Sect. 1.7.5.

Additional Exercises and Extensions

Exercise 1.4.14 (He and Zhang (2006)) Assume that T is a *PH*-generator of order m.

(1) Suppose that all eigenvalues of T are real and satisfy $\lambda_1 \geq \lambda_2 \geq \ldots \geq \lambda_m$. Show that the first and last columns in P are nonnegative.
(2) For $m = 2$, show that P is nonnegative.

Exercise 1.4.14 says that every *PH*-distribution with a *PH*-representation of order two is a Coxian distribution and has a Coxian representation of order two.

Exercise 1.4.15 Assume that the three eigenvalues of T_1 given in Eq. (1.87) are $-\lambda_1$,$-\lambda_2$, and $-\lambda_3$. Assume that $\lambda_1 \geq \lambda_2 \geq \lambda_3$. Set the vector $\boldsymbol{\lambda}$ in the SPA in Eq. (1.81) as follows:

(i) $\boldsymbol{\lambda} = (\lambda_1, \lambda_2, \lambda_3)$; (Triangle (1, 2, 3))
(ii) $\boldsymbol{\lambda} = (\lambda_1, \lambda_3, \lambda_2)$; (Triangle (2, 3, 4))
(iii) $\boldsymbol{\lambda} = (\lambda_2, \lambda_1, \lambda_3)$; (Triangle (1, 3, 7))
(iv) $\boldsymbol{\lambda} = (\lambda_2, \lambda_3, \lambda_1)$; (Triangle (3, 6, 7))
(v) $\boldsymbol{\lambda} = (\lambda_3, \lambda_1, \lambda_2)$; (Triangle (3, 4, 5))
(vi) $\boldsymbol{\lambda} = (\lambda_3, \lambda_2, \lambda_1)$. (Triangle (3, 5, 6))

Compute the corresponding matrix P for all six cases. Plot the two triangles mentioned in Exercise 1.4.12 for all the six cases. Comment on their relationships with *PH*-representations. Comment on why elements in $\boldsymbol{\lambda}$ should be required in descending order in the SPA. (He and Zhang (2006))

Figure 1.12 shows that all other five polytopes are included in the largest one, which corresponds to case (i) with $\boldsymbol{\lambda} = (\lambda_1, \lambda_2, \lambda_3)$. This explains why it is required to have $\lambda_1 \geq \lambda_2 \geq \lambda_3$ in the SPA.

Exercise 1.4.16 For $\mathbf{x} = (x_1, \ldots, x_j, x_{j+1}, \ldots, x_m)$, let $\mathbf{y} = (x_1, \ldots, x_{j+1}, x_j, \ldots, x_m)$. We say that $\mathbf{y}$ is obtained by an *increasing simple permutation* if $x_{j+1} \geq x_j$. Suppose that **y** is a permutation of **x**, and **y** can be obtained from **x** by a serial of increasing simple permutations. Use the proof of Theorem 1.4.3 to show that $S(\mathbf{x})$ is *PH*-majorized by $S(\mathbf{y})$. For Exercise 1.4.15, explain why the triangle corresponding to the permutations of $(\lambda_1, \lambda_2, \lambda_3)$ are all covered by that of $(\lambda_1, \lambda_2, \lambda_3)$. Use the result to explain why the triangle corresponding to $(\lambda_3, \lambda_1, \lambda_2)$ is covered by that of $(\lambda_1,\lambda_3,\lambda_2)$. Use the result to explain why the triangle corresponding to $(\lambda_3, \lambda_2, \lambda_1)$ is covered by all others.

Exercise 1.4.17 Construct *PH*-representations of order 50, 100, and 150. Find their equivalent bi-diagonal representations by using the SPA. Comment on the effectiveness of the SPA.

Exercise 1.4.18* (Open problem) Generate *PH*-generators with only real eigenvalues. Use the SPA to compute matrix P for them. Prove or disprove the conjecture that at least two rows of P are nonnegative for $m \geq 4$. Explain the conjecture geometrically. (Note: The case with $m = 2$ has been solved in Exercise 1.4.14, and the case with $m = 3$ has been solved in He and Zhang (2006). The conjecture provides a theoretical basis for a Coxianization algorithm of *PH*-generators given in He et al. (2011).)

Exercise 1.4.19* (Open problem) For a given *PH*-representation, find an equivalent *PH*-representation of the smallest order. (Note: See Example 1.4.1 for some examples. This exercise has been an open problem about *PH*-representations since 1975. See Sect. 1.7.5 for more information and discussion.)

Theorem 1.4.5 indicates that, if all eigenvalues of *PH*-generator T are real, then every *PH*-representation $(\boldsymbol{\alpha}, T)$ has an equivalent Coxian representation. An interesting question is how to find such an equivalent Coxian representation. Denote by $\{-\lambda_1,-\lambda_2, \ldots,-\lambda_m\}$ (counting multiplicities) all the eigenvalues of T. Define P by Eq. (1.81) with $\mathbf{x} = \boldsymbol{\lambda} = (\lambda_1, \lambda_2, \ldots, \lambda_m)$. Suppose that there are m_1 rows of P that are not nonnegative. By *permutations* of the rows and columns of the matrix T, the first $m - m_1$ rows of the matrix P can be made nonnegative and each of the last m_1 rows has at least one negative element. We partition matrices T and P as follows:

$$T = \begin{pmatrix} T_{1,1} & T_{1,2} \\ T_{2,1} & T_{2,2} \end{pmatrix} \quad \text{and} \quad P = \begin{pmatrix} P_1 \\ P_2 \end{pmatrix}, \tag{1.88}$$

where $T_{2,2}$ is an $m_1 \times m_1$ matrix and P_2 is an $m_1 \times m$ matrix. The matrix P_1 is nonnegative. Define

$$T_1 = \begin{pmatrix} S(\boldsymbol{\lambda}) & 0 \\ T_{2,1}P_1 & T_{2,2} \end{pmatrix}. \tag{1.89}$$

Theorem 1.4.7 (He et al. (2011)) *Assume that all eigenvalues of a PH-generator T are real, and $\lambda_1 \geq \lambda_2 \geq \ldots \geq \lambda_m$. Then the matrix T_1 defined in Eq.* (1.89) *is a PH-generator, and PH-majorizes T.*

Based on Theorem 1.4.7, the following *Coxianization algorithm* can be developed.

Step 1 Use the SPA to compute $\{P, S(\boldsymbol{\lambda})\}$, where $\boldsymbol{\lambda} = (\lambda_1, \lambda_2, \ldots, \lambda_m)$, and $\{-\lambda_1, -\lambda_2, \ldots, -\lambda_m\}$ are all the eigenvalues of the current *PH*-generator T in descending order of modulus. If P is nonnegative, go to Step 4. Otherwise, go to Step 2.

Step 2 Use Eqs. (1.88) and (1.89) to find $\{m_1, T_1\}$. If all eigenvalues of $T_{2,2}$ are real, set $m =: m + m_1$ and $T =: T_1$, then go back to Step 1. Otherwise, go to Step 3.

Step 3 The algorithm cannot continue and is terminated unsuccessfully.

Step 4 The Coxian generator $S(\boldsymbol{\lambda})$ *PH*-majorizes the original T. For the current $\mathbf{x} = \boldsymbol{\lambda}$ and the original matrix T, use the SPA to find P, which is nonnegative.

If the algorithm is terminated successfully, the number of nonnegative rows in the matrix P increases by at least one in each iteration. Then the algorithm would terminate after at most $m - 1$ iterations. Thus, the order of the final ordered Coxian generator would be at most $m(m + 1)/2$. Numerical results show that the order of the final solution is significantly smaller than $m(m + 1)/2$. Although it has not been proved theoretically that the algorithm always terminates successfully, it does so for all examples we have tested.

Exercise 1.4.20 (Coxianization of *PH*-generators with only real eigenvalues) Use the above algorithm to find an equivalent Coxian representation of the following *PH*-distribution:

$$\boldsymbol{\alpha} = (0.2,\ 0.0,\ 0.8), \quad T = \begin{pmatrix} -5 & 0 & 0.05 \\ 2 & -2 & 0 \\ 0 & 1 & -1 \end{pmatrix}. \tag{1.90}$$

If some of the eigenvalues of T are complex, the above Coxianization algorithm will not find a *PH*-representation. Then the following methods can be used to find structured equivalent matrix representations (Mocanu and Commault (1999)). Define a special *unicyclic generator*

$$T(\lambda, z) = \begin{pmatrix} -\lambda & \lambda & & \\ & \ddots & \ddots & \\ & & -\lambda & \lambda \\ z\lambda & & & -\lambda \end{pmatrix}_{n\times n}, \tag{1.91}$$

where $\lambda > 0$ and $0 \le z < 1$. Then define

$$M(\lambda, n, z, m) = \begin{pmatrix} T(\lambda, z) & -T(\lambda, z)\mathbf{e}(1, 0, \ldots, 0) & & \\ & \ddots & \ddots & \\ & & T(\lambda, z) & -T(\lambda, z)\mathbf{e}(1, 0, \ldots, 0) \\ & & & T(\lambda, z) \end{pmatrix}_{(nm)\times(nm)}. \tag{1.92}$$

Given a *PH*-representation $(\boldsymbol{\alpha}, T)$, its distinct eigenvalues are λ_j with multiplicity m_j, $j = 1, \ldots, n$. For complex eigenvalue λ_j, denote by m_j its total multiplicity (the sum of the multiplicities of λ_j and its conjugate). We construct the following *PH*-generators for each eigenvalue (or pair of eigenvalues):

(i) If λ_j is real, define $M_j = M(\lambda_j, 1, 0, m_j)$.
(ii) If λ_j is complex, assume $\lambda_j = a_j + b_j\sqrt{-1}$. Let

- n_j be the smallest integer for which $a_j/b_j > \tan(\pi/n_j)$;
- $\sigma_j = \frac{1}{2}\left(2a_j - b_j \tan\left(\frac{\pi}{n_j}\right) + b_j \cot\left(\frac{\pi}{n_j}\right)\right)$;
- $z_j = \left(1 - \left(a_j - b_j \tan\left(\frac{\pi}{n_j}\right)\right)\frac{1}{\sigma_j}\right)^{n_j}$.

Then we define $M_j = M(\sigma_j, n_j, z_j, m_j)$.
Now, we introduce *PH*-generator S (*Hyper-Feedback-Erlang*):

$$S = \begin{pmatrix} M_1 & & & \\ & M_2 & & \\ & & \ddots & \\ & & & M_n \end{pmatrix}. \tag{1.93}$$

Similar to the SPA, we compute matrix P such that $TP = PS$ and $P\mathbf{e} = \mathbf{e}$. Let $\boldsymbol{\beta} = \boldsymbol{\alpha}P$. Then $(\boldsymbol{\alpha}, T)$ and $(\boldsymbol{\beta}, S)$ are equivalent matrix-representations. If $\boldsymbol{\beta}$ is nonnegative, then $(\boldsymbol{\beta}, S)$ is an equivalent *PH*-representation. If $\boldsymbol{\beta}$ is not nonnegative, then $(\boldsymbol{\beta}, S)$ is a special matrix-exponential (*ME*) representation.
Note: The linear system $TP = PS$ and $P\mathbf{e} = \mathbf{e}$ can be solved as follows. Denote by $\mathbf{x}$ the *direct-sum* of P (i.e., row vector $\mathbf{x}$ is obtained by stringing out the rows of P starting from the first row). Then we obtain $\mathbf{x}(T' \otimes I - I \otimes S) = 0$ from $TP = PS$ and $\mathbf{x}(I \otimes \mathbf{e}) = \mathbf{e}'$ from $P\mathbf{e} = \mathbf{e}$. Replace the first columns of $T' \otimes I - I \otimes S$ with $I \otimes \mathbf{e}$, denoted by A, and the first columns of the right hand side 0 with $\mathbf{e}'$. Then we solve $\mathbf{x}A = (\mathbf{e}', 0, \ldots, 0)$, which leads to $\mathbf{x} = (\mathbf{e}', 0, \ldots, 0)A^{-1}$. The invertibility of A is guaranteed theoretically.
Next, we introduce *PH*-generator S (*Hypo-Feedback-Erlang*):

$$S = \begin{pmatrix} M_1 & -M_1\mathbf{e}(1,0,\ldots,0) & & \\ & \ddots & \ddots & \\ & & M_{n-1} & -M_{n-1}\mathbf{e}(1,0,\ldots,0) \\ & & & M_n \end{pmatrix}. \tag{1.94}$$

Similar to the SPA, we compute matrix P such that $TP = PS$ and $P\mathbf{e} = \mathbf{e}$. Let $\boldsymbol{\beta} = \boldsymbol{\alpha}P$. Then $(\boldsymbol{\alpha}, T)$ and $(\boldsymbol{\beta}, S)$ are equivalent matrix-representations. If $\boldsymbol{\beta}$ is nonnegative, then $(\boldsymbol{\beta}, S)$ is an equivalent *PH*-representation.

We remark that the above method has been used in Horvath and Telek (2012) to generate *FM*-random variates (see Sect. 1.7.3).

Exercise 1.4.21 For the following *PH*-representation, find its equivalent Hyper-feedback-Erlang and Hypo-feedback-Erlang representations.

$$\boldsymbol{\alpha} = (0.2,\ 0.0,\ 0.8), \quad T = \begin{pmatrix} -5 & 0 & 1 \\ 2 & -2 & 0 \\ 0 & 1 & -1 \end{pmatrix}. \tag{1.95}$$

Plot the density functions of $(\boldsymbol{\alpha}, T)$ and $(\boldsymbol{\beta}, S)$ to verify that the representations are correct.

Exercise 1.4.22 (Time reversed (dual) *PH*-representation) For *PH*-representation $(\boldsymbol{\alpha}, T)$, let $\mathbf{u} = -\boldsymbol{\alpha} T^{-1}$ and $\Delta(\mathbf{u}) = \mathrm{diag}(\mathbf{u})$ (i.e., a matrix with diagonal elements from $\mathbf{u}$ and all others equal zero). Define $\boldsymbol{\beta} = (-T\mathbf{e})'\Delta(\mathbf{u})$ and $S = (\Delta(\mathbf{u}))^{-1} T' \Delta(\mathbf{u})$. Show that $(\boldsymbol{\beta}, S)$ is an equivalent *PH*-representation of $(\boldsymbol{\alpha}, T)$. (Hint: It is necessary to verify that $(\boldsymbol{\beta}, S)$ is a *PH*-representation first.)

Exercise 1.4.23* (Unicyclic representation) (O'Cinneide (1999) and He and Zhang (2005)) The following type of *PH*-representation is called a unicyclic representation:

$$\left(\boldsymbol{\alpha} = (\alpha_1, \ldots \alpha_m), \quad T = \begin{pmatrix} -\lambda_1 & & & p\lambda_1 \\ \lambda_2 & -\lambda_2 & & \\ & \ddots & \ddots & \\ & & \lambda_m & -\lambda_m \end{pmatrix}_{m \times m} \right), \tag{1.96}$$

where $0 \le p < 1$. Show that every *PH*-representation of order 3 has a unicyclic representation of order 3. Show, by giving a counterexample, that not every *PH*-representation has an equivalent unicyclic representation of the same order.

Exercise 1.4.24* (O'Cinneide (1999) and Yao (2002)) For *PH*-distribution $F(t)$ with a *PH*-representation of order m, show that $F^{(m)}(t) > 0$ for $t > 0$. (Note: Recall that $F^{(m)}(t)$ is the m-th derivative of $F(t)$.)

For vector $\mathbf{x} = (x_1, x_2, \ldots, x_m)$, rearrange the elements of $\mathbf{x}$ in ascending order and denote the elements as $x_{[1]} \le x_{[2]} \le \ldots \le x_{[m]}$, where $([1], [2], \ldots, [m])$ is a permutation of $(1, 2, \ldots, m)$. Vector $\mathbf{x}$ is *weakly supermajorized* by vector $\mathbf{y}$, denoted as $\mathbf{x} \prec^{w} \mathbf{y}$, if $x_{[1]} + x_{[2]} + \ldots + x_{[k]} \ge y_{[1]} + y_{[2]} + \ldots + y_{[k]}$, for $k = 1, \ldots, m$. For *PH*-generator T, define $\mathbf{r} = (r_1, r_2, \ldots, r_m) = -\mathbf{e}'T$ and

$$\mathbf{b}^* = \left(\sum_{i=m}^{m} \left(\sum_{j=m-i+1}^{m} r_{[j]} \right)^{-1}, \ldots, \sum_{i=k}^{m} \left(\sum_{j=m-i+1}^{m} r_{[j]} \right)^{-1}, \ldots, \sum_{i=1}^{m} \left(\sum_{j=m-i+1}^{m} r_{[j]} \right)^{-1} \right)'. \tag{1.97}$$

Theorem 1.4.8 (He et al. (2012a)) *Assume that T is a PH-generator of order m. Then $-T^{-1}\mathbf{e}$ is weakly supermajorized by $\mathbf{b}^*$ defined in Eq.* (1.97).

Exercise 1.4.25* (He et al. (2012a)) For *PH*-random variable X with *PH*-representation ($\boldsymbol{\alpha}$, T), show $E[X] \geq -(\mathbf{e}'T\mathbf{e})^{-1}$. In addition, for $k = 2, 3, \ldots$, show $E[X^k] \geq k!(-\mathbf{e}'T\mathbf{e})^{-k}$.

1.5 Multivariate Phase-Type Distributions

Consider a continuous time Markov chain $\{Y(t),\ t \geq 0\}$ with finite state space $\{1, 2, \ldots, m\}$, infinitesimal generator Q, and initial distribution $\boldsymbol{\alpha}$. Let $\mathcal{A}_k$ be a subset of $\{1, 2, \ldots, m\}$, for $k = 1, \ldots, K$. Assume that the intersection of $\{\mathcal{A}_k, k = 1, \ldots, K\}$ is nonempty and the Markov chain will be absorbed into the intersection with probability one.

Definition 1.5.1 Define

$$X_k = \min\{t: \quad Y(t) \in \mathcal{A}_k, \quad t \geq 0\}, \quad k = 1, , \ldots, K. \tag{1.98}$$

Random vector $\mathbf{X} = (X_1,\ X_2,\ \ldots,\ X_K)$ is said to have a *multivariate PH-distribution* (*MPH*) and a representation ($\boldsymbol{\alpha}$, Q, $\mathcal{A}_k$, $k = 1, \ldots, K$).

It is clear that the *PH*-distribution is a special case of the multivariate *PH*-distributions with $K = 1$, i.e., a multivariate *PH*-distribution with a single absorption subset.

Example 1.5.1 A four-state continuous time Markov chain $\{Y(t), t \geq 0\}$ is defined as follows: $Y(0)$ has distribution $\boldsymbol{\alpha} = (0.2, 0.5, 0.15, 0.15)$, and infinitesimal generator Q is given by

$$Q = \begin{matrix} 1 \\ 2 \\ 3 \\ 4 \end{matrix}\begin{pmatrix} -15 & 2 & 1 & 12 \\ 0 & -2 & 2 & 0 \\ 2 & 1 & -5 & 2 \\ 1 & 1.5 & 0 & -2.5 \end{pmatrix}. \tag{1.99}$$

Define $\mathcal{A}_1 = \{2, 4\}$ and $\mathcal{A}_2 = \{3, 4\}$. Given $\boldsymbol{\alpha}$, X_1 and X_2 are the times until the Markov chain enters $\mathcal{A}_1$ and $\mathcal{A}_2$ for the first time, respectively. Then we have defined $\mathbf{X} = (X_1, X_2)$, a multivariate *PH*-distribution.

Exercise 1.5.1 In Example 1.5.1, show that X_1 has a *PH*-distribution. Find a *PH*-representation of X_1. What is the distribution of X_2?

First, some basic properties on *MPH*s are presented.

Proposition 1.5.1 (Assaf and Levikson (1982) and Assaf et al. (1984))

(1) *If* $\mathbf{X}$ *and* $\mathbf{Y}$ *are independent MPHs,* ($\mathbf{X}$, $\mathbf{Y}$) *is also an MPH.*

(2) *If* $X_1, X_2, \ldots, X_K$ *are independent PH-distributions, then* $(X_1, X_2, \ldots, X_K)$ *is a multivariate PH-distribution.*
(3) *If* $\mathbf{X} = (X_1, X_2, \ldots, X_K)$ *is an MPH, then* X_k *is a PH-distribution, for* $k = 1, \ldots, K$.
(4) *If* $\mathbf{X} = (X_1, X_2, \ldots, X_K)$ *is an MPH,* $X_{\min} = \min\{X_1, X_2, \ldots, X_K\}$ *is a PH-distribution.*
(5) *If* $\mathbf{X} = (X_1, X_2, \ldots, X_K)$ *is an MPH,* $X_{\text{sum}} = X_1 + X_2 + \ldots + X_K$ *is a PH-distribution.*

Proof. Part (1) can be proved by constructing an underlying Markov chain for $(\mathbf{X}, \mathbf{Y})$ from the two independent underlying Markov chains for $\mathbf{X}$ and $\mathbf{Y}$. Absorption subsets for individual random variables in $\mathbf{X}$ and $\mathbf{Y}$ can be defined. Part (2) is obtained from part (1).

Since X_k is the absorption time into the set $\mathcal{A}_k$, by definition, X_k has a *PH*-distribution, for $k = 1, \ldots, K$. This proves part (3).

Define $\mathcal{A}$ as the union of $\{\mathcal{A}_k, k = 1, \ldots, K\}$. Then $X_{\min}$ is the absorption time into the set $\mathcal{A}$. Consequently, $X_{\min}$ has a *PH*-distribution. Then part (4) is proved.

We prove part (5) only for $K = 2$. Using $\mathcal{A}_1$ and $\mathcal{A}_2$, the state space of the underlying Markov chain can be divided into four subsets $\mathcal{B}_0 = \mathcal{A}^c{}_1 \cap \mathcal{A}^c{}_2$, $\mathcal{B}_1 = \mathcal{A}_1 \cap \mathcal{A}_2^c$, $\mathcal{B}_2 = \mathcal{A}^c{}_1 \cap \mathcal{A}_2$, and $\mathcal{B}_3 = \mathcal{A}_1 \cap \mathcal{A}_2$, where $\mathcal{A}^c{}_i$ is the complement set of $\mathcal{A}_i$, for $i = 1, 2$. Suppose that the infinitesimal generator of the underlying Markov chain has the form

$$Q = \begin{matrix} \mathcal{B}_0 \\ \mathcal{B}_1 \\ \mathcal{B}_2 \\ \mathcal{B}_3 \end{matrix} \begin{pmatrix} T_{0,0} & T_{0,1} & T_{0,2} & T_{0,3} \\ T_{1,0} & T_{1,1} & T_{1,2} & T_{1,3} \\ T_{2,0} & T_{2,1} & T_{2,2} & T_{2,3} \\ T_{3,0} & T_{3,1} & T_{3,2} & T_{3,3} \end{pmatrix}. \tag{1.100}$$

We define a new Markov chain with infinitesimal generator

$$\hat{Q} = \begin{matrix} \mathcal{B}_0 \\ \mathcal{B}_0 \\ \mathcal{B}_1 \\ \mathcal{B}_0 \\ \mathcal{B}_2 \\ \Delta \end{matrix} \begin{pmatrix} 0.5T_{0,0} & (0, \quad T_{0,1}) & (0, \quad T_{0,2}) & T_{0,3}\mathbf{e} \\ 0 & \begin{pmatrix} T_{0,0} & T_{0,1} \\ T_{1,0} & T_{1,1} \end{pmatrix} & \begin{matrix} 0 & 0 \\ 0 & 0 \end{matrix} & \begin{matrix} T_{0,2}\mathbf{e} + T_{0,3}\mathbf{e} \\ T_{1,2}\mathbf{e} + T_{1,3}\mathbf{e} \end{matrix} \\ 0 & \begin{matrix} 0 & 0 \\ 0 & 0 \end{matrix} & \begin{pmatrix} T_{0,0} & T_{0,2} \\ T_{2,0} & T_{2,2} \end{pmatrix} & \begin{matrix} T_{0,1}\mathbf{e} + T_{0,3}\mathbf{e} \\ T_{2,1}\mathbf{e} + T_{2,3}\mathbf{e} \end{matrix} \\ 0 & 0 \quad 0 & 0 \quad 0 & 0 \end{pmatrix}. \tag{1.101}$$

Note that the time spent in phases in the set $\mathcal{B}_0$ is counted in both X_1 and X_2 before the absorption into $\mathcal{A}_1$ or $\mathcal{A}_2$. Thus, the transition rates are given by matrix $0.5T_{0,0}$. Then the absorption time of phase Δ of the above Markov chain is $X_{\text{sum}} = X_1 + X_2$. The general case can be shown similarly and is left as an exercise (Exercise 1.5.7). This completes the proof of Proposition 1.5.1.

Theorem 1.5.1 (Assaf et al. (1984)) *The set of multivariate PH-distributions of order K is dense in the set of all multivariate distributions on* $[0, \infty]^K$.

Next, we discuss the joint probability distribution of a multivariate *PH*-distribution **X**.

Example 1.5.2 Consider the Markov chain $\{I(t), t \geq 0\}$ defined in Example 1.3.1 (see Eq. (1.29)). Let $\mathcal{A}_1 = \{(2, 1), (2, 2)\}$ and $\mathcal{A}_2 = \{(1, 2), (2, 2)\}$ be absorption sets. Then $\mathbf{X} = (X_1, X_2)$, corresponding to $\mathcal{A}_1$ and $\mathcal{A}_2$, has a multivariate *PH*-distribution. According to Example 1.3.1, X_1 and X_2 are independent exponential random variables with parameters λ_1 and λ_2, respectively. On the other hand, the joint distribution of X_1 and X_2 can also be obtained by, for $t_1 < t_2$,

$$\begin{aligned}
&P\{X_1 > t_1, X_2 > t_2\} \\
&= (1,0,0)\exp\left\{\begin{pmatrix} -\lambda_1-\lambda_2 & \lambda_1 & \lambda_2 \\ 0 & -\lambda_2 & 0 \\ 0 & 0 & -\lambda_1 \end{pmatrix} t_1\right\}\begin{pmatrix} 1 & 0 & 0 \\ 0 & 1 & 0 \\ 0 & 0 & 0 \end{pmatrix} \\
&\quad \cdot \exp\left\{\begin{pmatrix} -\lambda_1-\lambda_2 & \lambda_1 & \lambda_2 \\ 0 & -\lambda_2 & 0 \\ 0 & 0 & -\lambda_1 \end{pmatrix} (t_2 - t_1)\right\}\begin{pmatrix} 1 & 0 & 0 \\ 0 & 0 & 0 \\ 0 & 0 & 1 \end{pmatrix}\begin{pmatrix} 1 \\ 1 \\ 1 \end{pmatrix} \\
&= \exp(-\lambda_1 t_1)\exp(-\lambda_2 t_2).
\end{aligned} \tag{1.102}$$

Equation (1.102) is justified as follows. At time t_1, if the underlying Markov chain associated with X_1 and X_2 has been in phase (2, 1) or (2, 2), then we must have $X_1 < t_1$. Thus, to find $P\{X_1 > t_1, X_2 > t_2\}$, we only need to add probabilities for phases (1, 1) and (1, 2) at time t_1. Similarly, at time t_2, we only add probabilities for phases (1, 1) and (2, 1).

Similarly, the probability for $t_1 > t_2$ can be obtained as $\exp(-\lambda_1 t_1 - \lambda_2 t_2)$. Together, the probability $P\{X_1 > t_1, X_2 > t_2\}$ is found.

Example 1.5.3 (Marshall and Olkin (1967) multivariate exponential distribution) Random vector $\mathbf{X} = (X_1, \ldots, X_n)$ is said to have the *Marshall-Olkin multivariate exponential distribution* if there exist independent exponential random variables $Y_1, \ldots, Y_k$ such that for $i = 1, \ldots, n$, $X_i = \min\{Y_j : j \in J_i\}$, where $J_i \subset \{1, \ldots, k\}$. According to Example 1.3.1, independent underlying Markov chains can be introduced for $Y_1, \ldots, Y_k$. Thus, a underlying Markov chain $\{I(t), t \geq 0\}$ can be defined for $(Y_1, \ldots, Y_k)$. Consequently, X_j can be defined as the absorption time of a subset of states of the Markov chain $\{I(t), t \geq 0\}$. Then **X** is also a multivariate *PH*-distribution.

Let Y_1, Y_2, and Y_3 be independent exponential random variables with parameters λ_1, λ_2, and λ_3, respectively. Define $X_1 = \min\{Y_1, Y_3\}$ and $X_2 = \min\{Y_2, Y_3\}$. Then (X_1, X_2) has a Marshall-Olkin exponential distribution and a multivariate *PH*-distribution with matrix representation: $n = 2$, $K = 3$, $\boldsymbol{\alpha} = (1, 0, 0, 0)$,

$$Q=\begin{matrix}(1,1,1)\\(2,1,1)\\(1,2,1)\\\Delta\end{matrix}\begin{pmatrix}-\lambda_1-\lambda_2-\lambda_3 & \lambda_1 & \lambda_2 & \lambda_3\\ 0 & -\lambda_2-\lambda_3 & 0 & \lambda_2+\lambda_3\\ 0 & 0 & -\lambda_1-\lambda_3 & \lambda_1+\lambda_3\\ 0 & 0 & 0 & 0\end{pmatrix}, \tag{1.103}$$

and $J_1 = \{(2, 1, 1), \Delta\}$ and $J_2 = \{(1, 2, 1), \Delta\}$. The joint distribution function of (X_1, X_2) can be found by, for $t_1 \leq t_2$,

$$\begin{aligned}&P\{X_1 > t_1, X_2 > t_2\}\\ &= (1,0,0,0)\exp(Qt_1)\begin{pmatrix}1&0&0&0\\0&0&0&0\\0&0&1&0\\0&0&0&0\end{pmatrix}\exp(Q(t_2-t_1))\begin{pmatrix}1&0&0&0\\0&1&0&0\\0&0&0&0\\0&0&0&0\end{pmatrix}\begin{pmatrix}1\\1\\1\\1\end{pmatrix}\\ &= \exp(-(\lambda_1+\lambda_2+\lambda_3)t_2)+\exp(-(\lambda_2+\lambda_3)t_2)(\exp(-\lambda_1t_1)-\exp(-\lambda_1t_2)).\end{aligned} \tag{1.104}$$

The joint probability distribution for $t_1 > t_2$ can be found similarly. Random vector $\mathbf{X}$ has a *singular part* given by $P\{X_1 = X_2 > t\} = P\{Y_3 = \min\{Y_1, Y_2, Y_3\} > t\} = \exp(-(\lambda_1 + \lambda_2 + \lambda_3)t)\lambda_3/(\lambda_1 + \lambda_2 + \lambda_3)$.

Exercise 1.5.2 Find the joint density function of $\mathbf{X}$ defined in Example 1.5.3. Plot the joint density function for cases with $\lambda_1 = 1$, $\lambda_2 = 2$, and $\lambda_3 = 5$.

Exercise 1.5.3 In Example 1.5.3, assume that

$$Q=\begin{matrix}(1,1,1)\\(2,1,1)\\(1,2,1)\\\Delta\end{matrix}\begin{pmatrix}-\lambda_1-\lambda_2-\lambda_3 & \lambda_1 & \lambda_2 & \lambda_3\\ 0 & -\lambda_4 & 0 & \lambda_4\\ 0 & 0 & -\lambda_5 & \lambda_5\\ 0 & 0 & 0 & 0\end{pmatrix}, \tag{1.105}$$

where $\lambda_1 = 1, \lambda_2 = 2, \lambda_3 = 5, \lambda_4 = 2$, and $\lambda_5 = 10$. Find and plot the joint density function of $\mathbf{X}$. (Hint: Use the matrix expression in Eq. (1.104).)

For the general case, the joint probability distribution of a multivariate *PH*-distribution can be found as follows.

Theorem 1.5.2 (Assaf et al. (1984)) *For multivariate PH-distribution* $\mathbf{X}$*, we have, for* $t_1 \leq t_2 \leq \ldots \leq t_K$*,*

$$\begin{aligned}&P\{X_1 > t_1, \ldots, X_K > t_K\}\\ &\quad = \boldsymbol{\alpha}\exp(Tt_1)P_1\exp(T(t_2-t_1))P_2\cdots\exp(T(t_K-t_{K-1}))P_K\mathbf{e},\end{aligned} \tag{1.106}$$

where P_k *is a diagonal matrix whose i-th diagonal element is* 1*, if* $i \notin \mathcal{A}_k$*; or* 0*, if* $i \in \mathcal{A}_k$*. Probabilities for all other cases can be obtained by permutation. The absolutely continuous part of the joint distribution is given by, for* $t_1 \leq t_2 \leq \ldots \leq t_K$*,*

$$\begin{aligned} f(t_1,\ldots,t_K) &= (-1)^K \boldsymbol{\alpha} \exp(Tt_1)(TP_1 - P_1T)\exp(T(t_2-t_1))(TP_2-P_2T) \\ &\quad \cdots \exp(T(t_K - t_{K-1}))TP_K\mathbf{e}. \end{aligned} \tag{1.107}$$

Example 1.5.4 Consider a Markov chain with infinitesimal generator

$$Q = \begin{pmatrix} T_1 & T_{1,2} & \mathbf{T}_{1,3} \\ 0 & T_2 & \mathbf{T}_{2,3} \\ 0 & 0 & 0 \end{pmatrix}, \tag{1.108}$$

where T_1 and T_2 are PH-generators of orders m_1 and m_2, respectively. Define $\mathcal{A}_1 = \{m_1 + 1,\ m_1 + 2, \ldots,\ m_1 + m_2,\ m_1 + m_2 + 1\}$ and $\mathcal{A}_2 = \{m_1 + m_2 + 1\}$. Then we can interpret X_1 as the total time spent in states $\{1, 2, \ldots, m_1\}$ and X_2 the total time spent in states $\{1, 2, \ldots, m_1, m_1 + 1, m_1 + 2, \ldots, m_1 + m_2\}$. The joint probability distribution can be obtained as, for $t_1 \le t_2$,

$$P\{X_1 > t_1, X_2 > t_2\} = \boldsymbol{\alpha}\exp(Qt_1)\begin{pmatrix} I & 0 & 0 \\ 0 & 0 & 0 \\ 0 & 0 & 0 \end{pmatrix}\exp(Q(t_2-t_1))\begin{pmatrix} I & 0 & 0 \\ 0 & I & 0 \\ 0 & 0 & 0 \end{pmatrix}\mathbf{e}, \tag{1.109}$$

and, for $t_1 > t_2$,

$$\begin{aligned} &P\{X_1 > t_1, X_2 > t_2\} \\ &= \boldsymbol{\alpha}\exp(Qt_2)\begin{pmatrix} I & 0 & 0 \\ 0 & I & 0 \\ 0 & 0 & 0 \end{pmatrix}\exp(Q(t_1-t_2))\begin{pmatrix} I & 0 & 0 \\ 0 & 0 & 0 \\ 0 & 0 & 0 \end{pmatrix}\mathbf{e}. \end{aligned} \tag{1.110}$$

Exercise 1.5.4 (Example 1.5.4 continued) In Example 1.5.4, plot the joint density of random vector $\mathbf{X}$ for $\boldsymbol{\alpha} = (0.4, 0.6, 0, 0, 0, 0)$,

$$T_1 = \begin{pmatrix} -5 & 1 \\ 2 & -4 \end{pmatrix}, \quad T_{1,2} = \begin{pmatrix} 1 & 1 & 1 \\ 0 & 1 & 0 \end{pmatrix}, \quad T_2 = \begin{pmatrix} -2 & 2 & 0 \\ 0 & -2 & 2 \\ 0 & 0 & -2 \end{pmatrix} \tag{1.111}$$

Commentary The materials in this section come mainly from Assaf et al. (1984), including Definition (1.5.1). More results on *MPH*s can be found in Kulkarni (1989) and O'Cinneide (1990b). There are several ways to define *MPH*s with minor differences (see Kulkarni (1989) and Bladt and Nielsen (2010)). For specific applications, one of the definitions might be more suitable. *MPH*s have found applications in many areas such risk and insurance analysis (e.g., Asmussen (2000), Li (2003), and Cai and Li (2005)).

Additional Exercises and Extensions

Kulkarni (1989) offers an alternate definition of multivariate *PH*-distribution, which is more amenable for computation. Define continuous time Markov chain $\{I(t), t \geq 0\}$ with a single absorption state, state space $\{1, 2, \ldots, m + 1\}$, an infinitesimal generator given in Eq. (1.14), and an initial distribution $(\boldsymbol{\alpha}, 1 - \boldsymbol{\alpha}\mathbf{e})$. We still denote X as the absorption time to state $m + 1$. Then X has a *PH*-distribution $(\boldsymbol{\alpha}, T)$. Let $r_k(.)$ be a nonnegative reward function defined on $\{1, 2, \ldots, m\}$, for $k = 1, 2, \ldots, K$.

Definition 1.5.2 Define

$$X_k = \int_0^X r_k(I(t))dt, \quad k = 1, 2, \ldots, K. \tag{1.112}$$

Then we call the random vector $\mathbf{X} = (X_1, X_2, \ldots, X_K)$ a multivariate *PH*-distribution. A matrix representation of $\mathbf{X}$ is $(\boldsymbol{\alpha}, T, R)$, where R is a $K \times m$ matrix with the (k, j)-th element $r_k(j)$. For example, if $r_k(.) \equiv 1$, then $X_k = X$.

Exercise 1.5.5 (Exercise 1.5.4 continued) Define X_1 as the total time spent in states $\{1, 2, \ldots, m_1\}$ and X_2 the total time spent in states $\{m_1 + 1, m_1 + 2, \ldots, m_1 + m_2\}$. Show that the joint LST of X_1 and X_2 is given by

$$E[e^{-(s_1X_1+s_2X_2)}] = \boldsymbol{\alpha}(s_1I - T_1)^{-1}T_{1,2}(s_2I - T_2)^{-1}(-T_2)\mathbf{e}. \tag{1.113}$$

Derive formulas for the means and variances of X_1 and X_2.

Exercise 1.5.6 (Erlang type multivariate phase-type distribution) Define a continuous time Markov chain with an absorption state for which the initial probability distribution is $(\boldsymbol{\alpha}, 0, \ldots, 0)$ and the transitions on transient states are governed by

$$T = \begin{pmatrix} T_1 & T_{1,2} & & & \\ & T_2 & T_{2,3} & & \\ & & \ddots & \ddots & \\ & & & T_{K-1} & T_{K-1,K} \\ & & & & T_K \end{pmatrix}. \tag{1.114}$$

The states can be divided into K levels according to the natural structure of the above transition matrix, where level k corresponds to the transition block T_k. Define X_k as the total time the Markov chain spent in level k. Then $(X_1, \ldots, X_K)$ is a multivariate *PH*-random variable. What is the function $r_k(.)$, for $k = 1, \ldots, K$? Show that

$$E[e^{-(s_1X_1+s_2X_2+\ldots+s_KX_K)}] \\ = \boldsymbol{\alpha}(s_1I - T_1)^{-1}T_{1,2}(s_2I - T_2)^{-1}T_{2,3}\cdots(s_KI - T_K)^{-1}(-T_K)\mathbf{e}. \tag{1.115}$$

Show that X_k is a PH-distribution and find a PH-representation for it. Find the correlation between X_i and X_j, for $i < j$ and $i, j = 1, \ldots, m$. Define $c_k(.) = kr_k(.)$, for $k = 1, \ldots, K$. Use $\{c_k(.), k = 1, \ldots, K\}$ to define a new MPH $(X_1, \ldots, X_K)$ and find its joint LST.

Exercise 1.5.7* For multivariate PH-distribution $\mathbf{X}$, show that $X_{\text{sum}} = X_1 + X_2 + \ldots + X_K$ is a PH-distribution, for $K = 3, 4, \ldots$. (See the proof of Proposition 1.5.1 for the case with $K = 2$.)

Exercise 1.5.8 Consider the Markov chain defined in Example 1.2.2. Assume $\boldsymbol{\alpha} = (0.1, 0.9, 0, 0, 0)$. Define X_1 as the total time the Markov chain spent in phases 1 and 3, and X_2 the total time the Markov chain spent in phases 2 and 4.

(i) Explain that (X_1, X_2) is a multivariate PH-distribution. Find the joint LST of (X_1, X_2).
(ii) Let $Y_1 = 2X_1 + 0.5X_2$ and $Y_2 = 5X_1 + 10X_2$. Argue that (Y_1, Y_2) is a multivariate PH-distribution. Find the joint LST of (Y_1, Y_2).

Exercise 1.5.9* (*Bivariate exponential distribution*, Bladt and Nielsen (2010)) Consider a bivariate PH-distribution (Y_1, Y_2) with matrix representation

$$\boldsymbol{\alpha} = ((0, \cdots, 0, 1)_{1\times m}, 0_{1\times m}), \quad T = \begin{pmatrix} S^* & D \\ 0 & T^* \end{pmatrix}, \tag{1.116}$$

where

$$S^* = \begin{pmatrix} -1 & & & & \\ 1 & -2 & & & \\ & \ddots & \ddots & & \\ & & m-2 & -(m-1) & \\ & & & m-1 & -m \end{pmatrix}, \tag{1.117}$$

$$T^* = \begin{pmatrix} -m & & & & \\ m-1 & -(m-1) & & & \\ & \ddots & \ddots & & \\ & & 2 & -2 & \\ & & & 1 & -1 \end{pmatrix}.$$

Define Y_1 as the total time spent in phases $\{1, 2, \ldots, m\}$ and Y_2 as the total time spent in phases $\{m + 1, \ldots, 2\ m\}$. For

$$D = D_{\min} = I \quad \text{and} \quad D = D_{\max} = \begin{pmatrix} & & 1 \\ & \cdot^{\cdot^{\cdot}} & \\ 1 & & \end{pmatrix}, \tag{1.118}$$

show that the random variables Y_1 and Y_2 are both exponential with parameter 1. (Note: For both cases, Y_1 and Y_2 are not independent.) Denote by $Corr_{\min}(m, m)$ and $Corr_{\max}(m, m)$ the correlation between Y_1 and Y_2 associated with $(S^*, T^*, D_{\min})$ and $(S^*, T^*, D_{\max})$, respectively. Show

$$\begin{aligned} Corr_{\min}(m,m) &= 1 - \sum_{i=1}^{m} \frac{1}{i^2} \xrightarrow{m\to\infty} 1 - \frac{\pi^2}{6}; \\ Corr_{\max}(m,m) &= 1 - \frac{1}{m}\sum_{i=1}^{m} \frac{1}{i} \xrightarrow{m\to\infty} 1. \end{aligned} \tag{1.119}$$

Exercise 1.5.10* (Bivariate exponential distribution, He et al. (2012b)) Consider a bivariate *PH*-distribution (Y_1, Y_2) with matrix representation

$$\left((\boldsymbol{\alpha}, 0),\ T = \begin{pmatrix} S & D \\ 0 & T \end{pmatrix}\right), \tag{1.120}$$

where $\boldsymbol{\alpha} = -\mathbf{e}'S/m$, S and T are *PH*-generators of order m, D is a nonnegative matrix of order m, $S\mathbf{e} + D\mathbf{e} = 0$, and the matrices S, D, and T satisfy

(i) $S\mathbf{e} = -\mathbf{e}$;
(ii) $\mathbf{e}'T = -\mathbf{e}'$;
(iii) $-\mathbf{e}'S \geq 0$; and
(iv) $\mathbf{e}'D = \mathbf{e}'$,

where $\mathbf{e}'$ is the transpose of $\mathbf{e}$. (Note: The bivariate (Y_1, Y_2) is defined in Exercise 1.5.9.)

(1) Show that the marginal distributions of (Y_1, Y_2) are the exponential distribution with parameter 1.
(2) Show that the correlation between Y_1 and Y_2 is given by

$$Corr(Y_1,\ Y_2) = \frac{1}{m}\mathbf{e}'(-S)^{-1}D(-T)^{-1}\mathbf{e} - 1. \tag{1.121}$$

(3) Show that

$$1 - \sum_{i=1}^{m} \frac{1}{i^2} = Corr_{\min}(m,m) \leq Corr(Y_1,Y_2) \leq Corr_{\max}(m,m) = 1 - \frac{1}{m}\sum_{i=1}^{m} \frac{1}{i}. \tag{1.122}$$

1.6 Phase-Type Distributions: Parameter Estimation

Parameter estimation of probability distributions is a fundamental issue in probability theory, stochastic modeling, and statistics. The literature on parameter estimation is enormous. In this section, we focus on parameter estimation of *PH*-distributions.

Parameter estimation of *PH*-distributions refers to finding a *PH*-representation $(\boldsymbol{\alpha}, T)$ of order m for *PH*-random variable Y from *sample* $\{y_1, y_2, \ldots, y_n\}$. For the sample, y_k is called a *sample point* and n is called the *sample size*. We assume that the sample points are collected independently. Then we define $\{Y_1, Y_2, \ldots, Y_n\}$ as independent and identically distributed random variables (i.i.d.r.v.s), which have the same distribution as Y. Sample point y_k is a realization of Y_k. As usual, we define $\{I(t), t \geq 0\}$ as the underlying Markov chain associated with Y.

Example 1.6.1 A random sample of size 150 is generated from *PH*-distribution $(\boldsymbol{\alpha}, T)$ by using the play method (see Exercise 1.3.16), where

$$\boldsymbol{\alpha} = (0.1,\ 0,\ 0.9), \quad T = \begin{pmatrix} -15 & 0 & 10 \\ 0 & -2 & 0 \\ 0 & 2 & -2 \end{pmatrix}. \tag{1.123}$$

Let N_k be the number of sample points in the interval $[k \times 0.4, (k+1) \times 0.4]$. Ratios $\{N_k/(150 \times 0.4), k = 0, 1, \ldots\}$ approximate the original density function of $(\boldsymbol{\alpha}, T)$, which form an *empirical distribution* and are plotted in Fig. 1.13. The solid line in Fig. 1.13 is the density function of the original *PH*-distribution.

The *EM approach* has been widely used in parameter estimation of probability distributions. It consists of an *expectation step* and a *maximization step*. Thus, the EM algorithm is iterative in nature. This section presents an EM algorithm that is developed in Asmussen and Nerman (1991) and Asmussen et al. (1996)

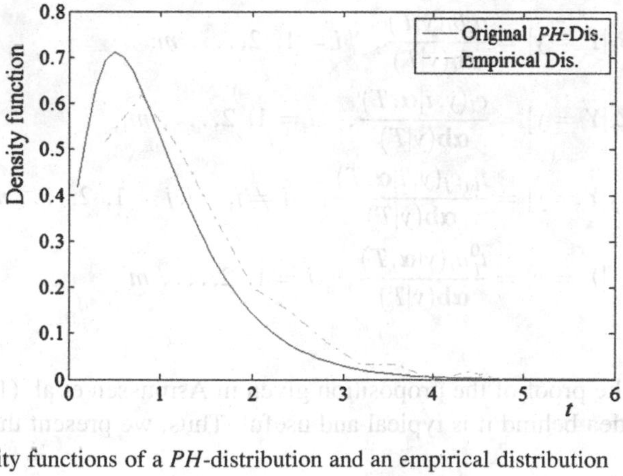

Fig. 1.13 Density functions of a *PH*-distribution and an empirical distribution

for *PH*-distributions. The EM algorithm is based on (i) the maximum likelihood approach and (ii) some properties of *PH*-distributions.

We begin with some properties of *PH*-distributions related to the expectation step of the EM algorithm. In this step, we assume that the distribution of Y is known and we want to characterize the underlying Markov chain associated with Y's *PH*-representation. Define, for the underlying Markov chain $\{I(t), t \geq 0\}$,

(i) B_i: 1, if $I(0) = i$; 0, otherwise, for $i = 1, 2, \ldots, m$.
(ii) Z_i: the total time spent in state i, for $i = 1, 2, \ldots, m$.
(iii) $N_{i,j}$: the total number of transitions from i to j, for $i \neq j$, $i = 1, 2, \ldots, m$, and $j = 1, 2, \ldots, m, m + 1$.

By the absorption nature of the underlying Markov chain, it is clear that Z_i and $N_{i,j}$ are finite with probability one, for $i = 1, 2, \ldots, m$, and $j = 1, 2, \ldots, m, m + 1$.

For $y > 0$, define the following vectors associated with *PH*-representation $(\boldsymbol{\alpha}, T)$:

$$\begin{aligned} \mathbf{a}(y|\boldsymbol{\alpha}, T) &= \boldsymbol{\alpha} \exp(Ty); \\ \mathbf{b}(y|T) &= \exp(Ty)\mathbf{T}^0; \\ \mathbf{c}(y, i|\boldsymbol{\alpha}, T) &= \int_0^y \boldsymbol{\alpha} \exp(Tu)\mathbf{e}_i \exp(T(y-u))\mathbf{T}^0 du, \end{aligned} \tag{1.124}$$

where $\mathbf{e}_i$ is the column vector with the i-th element equal to one and all other elements equal to zero. By Eq. (1.24), $\mathbf{a}(y|\boldsymbol{\alpha}, T)$ contains the probabilities that the underlying Markov chain is in states $\{1, 2, \ldots, m\}$ at time y. The vector $\mathbf{b}(y|T)$ contains the rates of absorption at time y. Then the conditional expectations of B_i, Z_i, and $N_{i,j}$ can be found as follows.

Proposition 1.6.1 (Asmussen et al. (1996)) *For $y > 0$, we have*

$$\begin{aligned} E_{(\boldsymbol{\alpha},T)}[B_i|Y=y] &= \frac{\alpha_i b_i(y|T)}{\boldsymbol{\alpha}\mathbf{b}(y|T)}, \quad i = 1, 2, \ldots, m; \\ E_{(\boldsymbol{\alpha},T)}[Z_i|Y=y] &= \frac{c_i(y, i|\boldsymbol{\alpha}, T)}{\boldsymbol{\alpha}\mathbf{b}(y|T)}, \quad i = 1, 2, \ldots, m; \\ E_{(\boldsymbol{\alpha},T)}[N_{i,j}|Y=y] &= \frac{t_{i,j} c_j(y, i|\boldsymbol{\alpha}, T)}{\boldsymbol{\alpha}\mathbf{b}(y|T)}, \quad i \neq j, \quad i,j = 1, 2, \ldots, m; \\ E_{(\boldsymbol{\alpha},T)}[N_{i,m+1}|Y=y] &= \frac{t_i^0 a_i(y|\boldsymbol{\alpha}, T)}{\boldsymbol{\alpha}\mathbf{b}(y|T)}, \quad i = 1, 2, \ldots, m. \end{aligned} \tag{1.125}$$

Proof. While the proof of the proposition given in Asmussen et al. (1996) is quite technical, the idea behind it is typical and useful. Thus, we present the proof in its entirety.

First note $P\{Y \in (y, y+\delta y)\} = \int_y^{y+\delta y} \boldsymbol{\alpha} \exp(Tt)\mathbf{T}^0 \mathrm{d}t \approx \boldsymbol{\alpha} \exp(Ty)\mathbf{T}^0 \delta y$ for small positive δy. The conditional expectation of B_i can be obtained as

$$
\begin{aligned}
E_{(\boldsymbol{\alpha},T)}[B_i|Y=y] &\approx \frac{P\{I(0)=i,\ Y\in(y,y+\delta y)\}}{P\{Y\in(y,y+\delta y)\}} \\
&= \frac{P\{I(0)=i\}P\{Y\in(y,y+\delta y)|I(0)=i\}}{P\{Y\in(y,y+\delta y)\}} \\
&\approx \frac{\alpha_i \mathbf{e}'_i \exp(Ty)\mathbf{T}^0\delta y}{\boldsymbol{\alpha}\exp(Ty)\mathbf{T}^0\delta y} \\
&\approx \frac{\alpha_i b_i(y|T)}{\boldsymbol{\alpha}\mathbf{b}(y|T)}.
\end{aligned}
\tag{1.126}
$$

Note that δy in Eq. (1.126) is a small positive number and the equalities in Eq. (1.126) hold when δy approaches zero. The conditional expectation of Z_i can be obtained as

$$
\begin{aligned}
E_{(\boldsymbol{\alpha},T)}[Z_i|Y=y] &= E\left[\int_0^\infty I_{\{I(t)=i\}}\mathrm{d}t|Y=y\right] \\
&= \int_0^\infty P\{I(t)=i|Y=y\}\mathrm{d}t \\
&\approx \int_0^\infty \frac{P\{I(t)=i,Y\in(y,y+\delta y)\}}{P\{Y\in(y,y+\delta y)\}}\mathrm{d}t \\
&= \int_0^y \frac{P\{I(t)=i\}P\{Y\in(y,y+\delta y)|I(t)=i\}}{P\{Y\in(y,y+\delta y)\}}\mathrm{d}t \\
&\approx \int_0^y \frac{\boldsymbol{\alpha}\exp(Tt)\mathbf{e}_i\mathbf{e}'_i\exp(T(y-t))\mathbf{T}^0\delta y}{\boldsymbol{\alpha}\exp(Ty)\mathbf{T}^0\delta y}\mathrm{d}t \\
&\approx \frac{c_i(y,i|\boldsymbol{\alpha},T)}{\boldsymbol{\alpha}\mathbf{b}(y|T)}.
\end{aligned}
\tag{1.127}
$$

Again, the equalities in (1.127) holds if δy approaches zero. In Eq. (1.127), we use $P\{I(t)=i\} = \boldsymbol{\alpha}\exp(Tt)\mathbf{e}_i$ and, for $\delta y > 0$, $P\{Y \in (y, y+\delta y)|I(t)=i\} \approx \mathbf{e}'_i \exp(T(y-t))\mathbf{T}^0\delta y$.

To find the expectation of $N_{i,j}$, we define $N^\varepsilon_{i,j} = \sum_{k=0}^{\infty} I_{\{I(k\varepsilon)=i,\ I((k+1)\varepsilon)=j\}}$, $\varepsilon > 0$, $i \neq j$. It is clear that $N^\varepsilon{}_{i,j}$ approximates $N_{i,j}$ (with reasonable mathematical rigor). Also note that the expected total number of transitions before absorption is finite (see Eq. (1.128)), i.e., $E[\sum_{i\neq j} N_{i,j}]$ is finite. It is easy to see that $\sum_{i\neq j} N^\varepsilon_{i,j}$ converges to $\sum_{i\neq j} N_{i,j}$. Let $[y/\varepsilon]$ be the smallest integer that is greater than or equal to y/ε. We further have, for $i \neq j$, $i, j = 1, 2, \ldots, m$,

$$\begin{aligned}
&E_{(\boldsymbol{\alpha},T)}[N^{\varepsilon}_{i,j}|Y=y]\\
&\approx \sum_{k=0}^{\lfloor y/\varepsilon\rfloor-1}\frac{P\{I(k\varepsilon)=i,\ I((k+1)\varepsilon)=j, y\in(y,y+dy)\}}{P\{y\in(y,y+dy)\}}\\
&\approx \sum_{k=0}^{\lfloor y/\varepsilon\rfloor-1}\frac{P\{I(k\varepsilon)=i\}P\{\ I((k+1)\varepsilon)=j|I(k\varepsilon)=i\}P\{y\in(y,y+dy)|I((k+1)\varepsilon)=j\}}{P\{y\in(y,y+dy)\}}\\
&\approx \sum_{k=0}^{\lfloor y/\varepsilon\rfloor-1}\frac{(\boldsymbol{\alpha}\exp(Tk\varepsilon)\mathbf{e}_i)(\mathbf{e}'_i\exp(T\varepsilon)\mathbf{e}_j)(\mathbf{e}'_j\exp(T(y-(k+1)\varepsilon))\mathbf{T}^0)}{\boldsymbol{\alpha}\exp(Ty)\mathbf{T}^0}\\
&\xrightarrow{\varepsilon\to 0}\ \frac{\int_0^y(\boldsymbol{\alpha}\exp(Tt)\mathbf{e}_i)t_{i,j}(\mathbf{e}'_j\exp(T(y-t))\mathbf{T}^0)dt}{\boldsymbol{\alpha}\exp(Ty)\mathbf{T}^0},
\end{aligned} \tag{1.128}$$

where $\exp(T\varepsilon)\approx I+\varepsilon T$ and $\mathbf{e}_i'\mathbf{e}_j=0$, for $i\neq j$, are utilized. The *dominated convergence theorem* yields the expected result for $E[N_{i,j}|Y=y]$. For $N_{i,m+1}$, we have, for $i=1, 2, \ldots, m$,

$$\begin{aligned}
&E_{(\boldsymbol{\alpha},T)}[N_{i,m+1}|Y=y]\\
&=\lim_{\varepsilon\to 0+}P\{I(Y-\varepsilon)=i|Y=y\}\\
&\approx \lim_{\varepsilon\to 0+}\frac{P\{I(y-\varepsilon)=i\}P\{y\in(y,y+dy)|I(y-\varepsilon)=i\}}{P\{y\in(y,y+dy)\}}\\
&=\lim_{\varepsilon\to 0+}\frac{(\boldsymbol{\alpha}\exp(T(y-\varepsilon))\mathbf{e}_i)(\mathbf{e}'_i\exp(T\varepsilon)\mathbf{T}^0)}{\boldsymbol{\alpha}\exp(Ty)\mathbf{T}^0}\\
&=\frac{\boldsymbol{\alpha}\exp(Ty)\mathbf{e}_i t_i^0}{\boldsymbol{\alpha}\exp(Ty)\mathbf{T}^0}.
\end{aligned} \tag{1.129}$$

This completes the proof of Proposition 1.6.1.

Exercise 1.6.1 For the exponential distribution with parameter λ, find the four expectations in Eq. (1.125). Explain the results intuitively. For Erlang distribution with parameter (m, λ), find the four expectations in Eq. (1.125).

Next, we present some properties related to the maximization step. While sample point $Y = y$ provides information about Y, it does not provide information about the underlying Markov chains directly. We further decompose Y in order to characterize the underlying Markov chain. To describe the underlying Markov chain $\{I(t), t \geq 0\}$, define

(1) W: the total number of transitions of the absorption process.
(2) I_n: the phase of the underlying Markov chain after the n-th transition, for $n = 0, 1, 2, \ldots$.
(3) S_n: the sojourn time of the underlying Markov chain in state I_n, after the n-th transition, for $n = 0, 1, 2, \ldots$.

Variables $\{I_0, I_1, I_2, \ldots, I_{W-1}, S_0, S_1, S_2, \ldots, S_{W-1}\}$ provide complete information about the absorption process (note: $I_W = m + 1$ and $S_W = \infty$). It is easy to see that $Y = S_0 + S_1 + S_2 + \ldots + S_{W-1}$.

Exercise 1.6.2 Show that $\{I_n, n = 0, 1, 2, \ldots\}$ is a discrete time Markov chain with transition probability matrix

$$P_T = (p_{i,j})_{(m+1)\times(m+1)} = I - \begin{pmatrix} \mathrm{diag}(t_{1,1}^{-1}, t_{2,2}^{-1}, \cdots, t_{m,m}^{-1}) & 0 \\ 0 & 0 \end{pmatrix} \begin{pmatrix} T & \mathbf{T}^0 \\ 0 & 0 \end{pmatrix}. \quad (1.130)$$

Matrix P_T is the transition probability matrix of the embedded Markov chain at transitions for the underlying Markov chain. For the Erlang distribution defined in Eq. (1.19), the matrix P_T is given by

$$P_T = \begin{pmatrix} 0 & & & 1 \\ 1 & 0 & & 0 \\ & \ddots & \ddots & \vdots \\ & & 1 & 0 & 0 \\ 0 & \cdots & 0 & 0 & 1 \end{pmatrix}. \quad (1.131)$$

Exercise 1.6.3 Explain that S_n has an exponential distribution with parameter $-t_{I_n,I_n}$.

Proposition 1.6.2 *Given $W = w$, the joint density function of $\{I_0, I_1, I_2, \ldots, I_{W-1}, S_0, S_1, S_2, \ldots, S_{W-1}\}$ is given by, for sample point $\mathbf{x} = \{i_0, i_1, i_2, \ldots, i_{w-1}, s_0, s_1, s_2, \ldots, s_{w-1}\}$,*

$$f(\mathbf{x};\ \boldsymbol{\alpha}, T) = \alpha_{i_0}\left(\prod_{i=1}^{m} \exp(t_{i,i}z_i)\right)\left(\prod_{i=1}^{m}\left(\prod_{j=1:\, j\neq i}^{m+1} (t_{i,j})^{n_{i,j}}\right)\right), \quad (1.132)$$

where z_i is the total time spent in state i, for $i = 1, 2, \ldots, m$, and $n_{i,j}$ is the total number of transitions from i to j, for $i \neq j$, $i = 1, 2, \ldots, m$, and $j = 1, 2, \ldots, m, m + 1$, for the sample point x. Recall $t_{i,m+1} = t_i^0$, for $i = 1, 2, \ldots, m$.

Proof. First note that the elements of P_T are given by $p_{i,j} = t_{i,j}/(-t_{i,i})$, for $i \neq j$, $i, j = 1, 2, \ldots, m$; $p_{i,i} = 0$, for $i = 1, 2, \ldots, m$; and $p_{i,m+1} = t^0{}_i/(-t_{i,i})$, for $i = 1, 2, \ldots, m$. Using the density function of an exponential distribution and conditional probabilities, we obtain

$$\begin{aligned} f(\mathbf{x};\ \boldsymbol{\alpha}, T) &= \alpha_{i_0}(-t_{i_0,i_0})\exp(t_{i_0,i_0}s_0)p_{i_0,i_1}\cdots(-t_{i_{w-1},i_{w-1}})\exp(t_{i_{w-1},i_{w-1}}s_{w-1})p_{i_{w-1},m+1} \\ &= \alpha_{i_0}\exp(t_{i_0,i_0}s_0)t_{i_0,i_1}\cdots t_{i_{w-2},i_{w-1}}\exp(t_{i_{w-1},i_{w-1}}s_{w-1})t^0_{i_{w-1},m+1}, \end{aligned} \quad (1.133)$$

which leads to Eq. (1.132). This completes the proof of Proposition 1.6.2.

Now, we are ready to present the EM algorithm. The EM algorithm is an iterative one and is based on the well-known *maximum likelihood method*. The idea of the maximum likelihood method is to choose parameters so that the likelihood of the given sample is maximized. However, the relationship between the given sample and the likelihood function is too complicated for *PH*-distributions, because information about the underlying Markov chain is missing. To resolve the issue, the maximum likelihood method is divided into two steps: an expectation step and a maximization step. This method constructs a likelihood function with some measures associated with the distribution. The expectation step estimates these measures by expectation. Then we replace the measures with their expectations in the likelihood function. In this step, the given sample is utilized. Once the likelihood function is determined, then a set of parameters is chosen so that the likelihood function is maximized. In general, the EM algorithm can be described as follows:

Step 0: Choose an initial *PH*-representation $(\boldsymbol{\alpha}, T)$.
E-step: Use the current $(\boldsymbol{\alpha}, T)$ and the given sample $\{y_1, y_2, \ldots, y_n\}$ to estimate B_i, Z_i, and $N_{i,j}$ (to be defined again in Eq. (1.134)).
M-step: Use the estimates of B_i, Z_i, and $N_{i,j}$ obtained in the E-step to generate a new set of parameters of the *PH*-distribution (i.e., $(\boldsymbol{\alpha}, T)$) that maximizes a likelihood function (to be defined in Eq. (1.135)).

More specifically, given a sample $\{y_1, y_2, \ldots, y_n\}$ and the current $(\boldsymbol{\alpha}^{(k)}, T^{(k)})$, we can find conditional expectations of B_i, Z_i, and $N_{i,j}$ for each sample point. In the E-step, we calculate, for $i \neq j$, $i, j = 1, 2, \ldots, m$,

$$\begin{aligned}
B_i^{(k+1)} &= \sum_{v=1}^{n} E_{(\boldsymbol{\alpha},T)^{(k)}}[B_i^{[v]} | Y = y_v] = \sum_{v=1}^{n} \frac{\alpha_i^{(k)} b_i(y_v | T^{(k)})}{\boldsymbol{\alpha}^{(k)} \mathbf{b}(y_v | T^{(k)})}; \\
Z_i^{(k+1)} &= \sum_{v=1}^{n} E_{(\boldsymbol{\alpha},T)^{(k)}}[Z_i^{[v]} | Y = y_v] = \sum_{v=1}^{n} \frac{c_i(y_v, i | \boldsymbol{\alpha}^{(k)}, T^{(k)})}{\boldsymbol{\alpha}^{(k)} \mathbf{b}(y_v | T^{(k)})}; \\
N_{i,j}^{(k+1)} &= \sum_{v=1}^{n} E_{(\boldsymbol{\alpha},T)^{(k)}}[N_{i,j}^{[v]} | Y = y_v] = \sum_{v=1}^{n} \frac{t_{i,j}^{(k)} c_j(y_v, i | \boldsymbol{\alpha}^{(k)}, T^{(k)})}{\boldsymbol{\alpha}^{(k)} \mathbf{b}(y_v | T^{(k)})}; \\
N_{i,m+1}^{(k+1)} &= \sum_{v=1}^{n} E_{(\boldsymbol{\alpha},T)^{(k)}}[N_{i,m+1}^{[v]} | Y = y_v] = \sum_{v=1}^{n} \frac{t_i^{0(k)} a_i(y_v | \boldsymbol{\alpha}^{(k)}, T^{(k)})}{\boldsymbol{\alpha}^{(k)} \mathbf{b}(y_v | T^{(k)})}.
\end{aligned} \tag{1.134}$$

Note that we use "(k)" as a superscript for parameters and measures of the k-th iteration, and we use "$[v]$" as a superscript for parameters and measures associated with the sample point y_v, $v = 1, 2, \ldots, n$.

By Proposition 1.6.1, for a given sample, $B_i^{(k+1)}$ is the expected total number of times that the underlying Markov chain starts in phase i for all n replications; $Z_i^{(k+1)}$ is the expected total time spent in phase i; and $N_{i,j}^{(k+1)}$ is the expected total number of transitions from phase i to j.

Exercise 1.6.4 Show that (i) $\sum_{i=1}^{m} B_i^{(k+1)} = n$; (ii) $\sum_{i=1}^{m} Z_i^{(k+1)} = \sum_{v=1}^{n} y_v$; and (iii) $\sum_{i=1}^{m} N_{i,m+1}^{(k+1)} = n$. Explain the equalities intuitively.

Next, we use B_i, Z_i, and $N_{i,j}$ to construct a likelihood function based on the fictitious information on the n sample points. Denote by $\{b_i, z_i, n_{i,j}\}$ the realizations of random variables $\{B_i, Z_i, N_{i,j}\}$, respectively.

Let $\mathbf{x}^{[v]} = \{i_0^{[v]}, i_1^{[v]}, \ldots, i_{w^{[v]}-1}^{[v]}, s_0^{[v]}, s_1^{[v]}, \ldots, s_{w^{[v]}-1}^{[v]}\}$, which corresponds to sample point y_v. We call $\{\mathbf{x}^{[1]}, \mathbf{x}^{[2]}, \ldots, \mathbf{x}^{[n]}\}$ a *fictitious sample* corresponding to $\{y_1, y_2, \ldots, y_n\}$. By Proposition 1.6.2, we construct the following likelihood function

$$f(\mathbf{x}^{[1]}, \cdots, \mathbf{x}^{[n]}; \boldsymbol{\alpha}, T) = \left(\prod_{i=1}^{m} \alpha_i^{b_i}\right)\left(\prod_{i=1}^{m} \exp(t_{i,i} z_i)\right)\left(\prod_{i=1}^{m}\prod_{j=1:\, j\neq i}^{m+1} (t_{i,j})^{n_{i,j}}\right). \quad (1.135)$$

where z_i and $n_{i,j}$ are defined the same as in Proposition 1.6.2, but for all sample points, and b_i is the total number of times that the underlying Markov chain starts in phase i.

Example 1.6.2 Assume random variable X has distribution function $F(x)$ and density function $f(x) = F'(x)$. It is easy to see $F(x + \delta x) - F(x) \approx \delta x f(x)$. For fixed δx, approximately, maximizing $f(x)$ is equivalent to maximizing the probability $P\{X \in (x, x + \delta x)\}$.

The basic idea of the maximum likelihood method is to choose parameters such that the likelihood function is maximized, as demonstrated in Example 1.6.2 for the simplest case. Thus, our goal is to choose $(\boldsymbol{\alpha}, T)$ to maximize the likelihood function given in Eq. (1.135).

Following the general theory for exponential families (Albert 1962) or using explicit calculations based on Eq. (1.135), the maximum likelihood estimates, based on the fictitious sample $\{\mathbf{x}^{[1]}, \mathbf{x}^{[2]}, \ldots, \mathbf{x}^{[n]}\}$, are

$$\alpha_i^* = \frac{b_i}{n}, \quad t_{i,j}^* = \frac{n_{i,j}}{z_i}, \quad t_{i,m+1}^* = \frac{n_{i,m+1}}{z_i}, \quad t_{i,i}^* = -\left(t_{i,m+1}^* + \sum_{j\neq i:\, 1}^{m} t_{i,j}^*\right). \quad (1.136)$$

Unfortunately, sample $\{\mathbf{x}^{[1]}, \mathbf{x}^{[2]}, \ldots, \mathbf{x}^{[n]}\}$ is fictitious and $\{b_i, z_i, n_{i,j}, i = 1, 2, \ldots, m, j = 1, 2, \ldots, m, m+1\}$ cannot be calculated directly. Thus, we use the estimates obtained in the E-step to replace $\{b_i, z_i, n_{i,j}, i = 1, 2, \ldots, m, j = 1, 2, \ldots, m, m+1\}$ in Eq. (1.136). Consequently, the M-step yields:

$$\begin{aligned} \alpha_i^{(k+1)} &= \frac{B_i^{(k+1)}}{n}, \quad t_{i,j}^{(k+1)} = \frac{N_{i,j}^{(k+1)}}{Z_i^{(k+1)}}, \quad i \neq j, \\ t_{i,m+1}^{(k+1)} &= \frac{N_{i,m+1}^{(k+1)}}{Z_i^{(k+1)}}, \quad t_{i,i}^{(k+1)} = -\left(t_{i,m+1}^{(k+1)} + \sum_{j\neq i:\, 1}^{m} t_{i,j}^{(k+1)}\right). \end{aligned} \quad (1.137)$$

Now, we summarize the details of the *EM algorithm* as follows.

EM-Step 0: Given a sample $\{y_1, y_2, \ldots, y_n\}$, choose m and initial PH-representation $(\boldsymbol{\alpha}^{(0)}, T^{(0)})$ of order m for the underlying Markov chain.

EM-E-step: For the current solution $(\boldsymbol{\alpha}^{(k)}, T^{(k)})$, use Eq. (1.134) to calculate $\{B_i^{(k+1)}, Z_i^{(k+1)}, N_{i,j}^{(k+1)}, i = 1, 2, \ldots, m, j = 1, 2, \ldots, m, m+1, i \neq j\}$. In this step, use Eq. (1.124) to calculate three vectors, whose computation is discussed in Exercise 1.6.7.

EM-M-step: Use Eq. (1.137) to calculate $(\boldsymbol{\alpha}^{(k+1)}, T^{(k+1)})$. Set $k =: k + 1$, go back to the E-step.

The following properties hold for the solutions obtained.

Exercise 1.6.5 In the EM algorithm, show that $(\boldsymbol{\alpha}^{(k)}, T^{(k)})$ keeps the structure of $(\boldsymbol{\alpha}^{(0)}, T^{(0)})$, i.e., if an element in $\boldsymbol{\alpha}^{(0)}$ or $T^{(0)}$ is zero, then the element is zero in $\boldsymbol{\alpha}^{(k)}$ or $T^{(k)}$ for all k.

Exercise 1.6.5 indicates that the structure of the final PH-representation is determined by the structure of the initial PH-representation.

Exercise 1.6.6 In the EM algorithm, show $y_v = \sum_{i=1}^{m} E_{(\boldsymbol{\alpha},T)}[Z_i^{[v]} | Y = y_v]$, $v = 1, 2, \ldots, n$.

Equation (1.134) and Exercise 1.6.6 lead to

$$\begin{aligned} \sum_{v=1}^{n} y_v &= \sum_{v=1}^{n} \sum_{i=1}^{m} E_{(\boldsymbol{\alpha}^{(k)},T^{(k)})}[Z_i^{[v]} | Y = y_v] \\ &= \sum_{v=1}^{n} \sum_{i=1}^{m} E_{(\boldsymbol{\alpha}^{(k+1)},T^{(k+1)})}[Z_i^{[v]}] \\ &= nE_{(\boldsymbol{\alpha}^{(k+1)},T^{(k+1)})}[Y], \end{aligned} \tag{1.138}$$

which indicates that the means of the fitted PH-distributions $(\boldsymbol{\alpha}^{(k)}, T^{(k)})$ are the same as the sample mean. Thus, for the EM algorithm, the mean of the fitted PH-distribution is the sample mean.

Exercise 1.6.7 Show that the derivatives of the vectors defined in Eq. (1.124), with respect to y, are given by

$$\begin{aligned} \frac{\mathrm{d}\mathbf{a}(y|\boldsymbol{\alpha}, T)}{\mathrm{d}y} &= \mathbf{a}(y|\boldsymbol{\alpha}, T)T; \\ \frac{\mathrm{d}\mathbf{b}(y|T)}{\mathrm{d}y} &= T\mathbf{b}(y|T); \\ \frac{\mathrm{d}\mathbf{c}(y, i|\boldsymbol{\alpha}, T)}{\mathrm{d}y} &= T\mathbf{c}(y, i|\boldsymbol{\alpha}, T) + a_i(y|\boldsymbol{\alpha}, T)\mathbf{T}^0, \quad i = 1, 2, \ldots, m. \end{aligned} \tag{1.139}$$

In the EM algorithm, the vectors $\mathbf{a}(y|\boldsymbol{\alpha}, T)$, $\mathbf{b}(y|T)$, and $\mathbf{c}(y, i|\boldsymbol{\alpha}, T)$ have to be calculated in each iteration. The vectors can be evaluated using matrix exponential

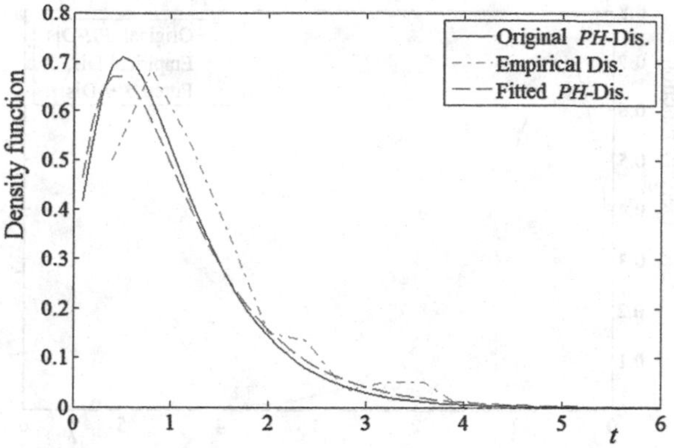

Fig. 1.14 Fitted Distribution with $k = 100$

functions or by solving the system of homogeneous differential equations given in Eq. (1.139). Standard methods in the literature can be used to solve the system numerically with high precision (e.g., Moler and Van Loan (1978)).

Example 1.6.3 (Example 1.6.1 continued) We choose $(\boldsymbol{\alpha}^{(0)}, T^{(0)})$ as follows:

$$\boldsymbol{\alpha}^{(0)} = (0.6,\ 0,\ 0.4), \quad T^{(0)} = \begin{pmatrix} -5 & 1 & 1 \\ 1 & -5 & 1 \\ 1 & 1 & -5 \end{pmatrix}. \tag{1.140}$$

Using the EM algorithm, we find $(\boldsymbol{\alpha}^{(k)}, T^{(k)})$ for $k = 100$, which is

$$\begin{aligned} \boldsymbol{\alpha}^{(100)} &= (0.7902,\ 0,\ 0.2098), \\ T^{(100)} &= \begin{pmatrix} -2.5027 & 1.0212 & 1.2682 \\ 0.1128 & -2.3084 & 0.1900 \\ 0.6538 & 1.9523 & -3.1426 \end{pmatrix}. \end{aligned} \tag{1.141}$$

The densities of the PH-distribution $(\boldsymbol{\alpha}^{(100)}, T^{(100)})$ and the original PH-distribution $(\boldsymbol{\alpha}, T)$ are plotted in Fig. 1.14. As is shown in Fig. 1.14, the fitted PH-distribution seems to approximate well the original distribution.

In the EM-algorithm, an important step is to choose the initial PH-representation. By Exercise 1.4.25, we know that for any PH-distribution, $\mathbf{e}'(-T)\mathbf{e} \geq 1/E[X]$ holds. Since $(\boldsymbol{\alpha}^{(k)}, T^{(k)})$ has the same mean, $T^{(0)}$ should be chosen so that $\mathbf{e}'(-T^{(0)})\mathbf{e}$ is greater than or equal to the reciprocal of the sample mean. For Example 1.6.3, we choose $(\boldsymbol{\alpha}^{(0)}, T^{(0)})$ as follows:

$$\boldsymbol{\alpha}^{(0)} = (0.6,\ 0,\ 0.4), \quad T^{(0)} = \begin{pmatrix} -5 & 0 & 3 \\ 1 & -6 & 0 \\ 0 & 1 & -6 \end{pmatrix}, \tag{1.142}$$

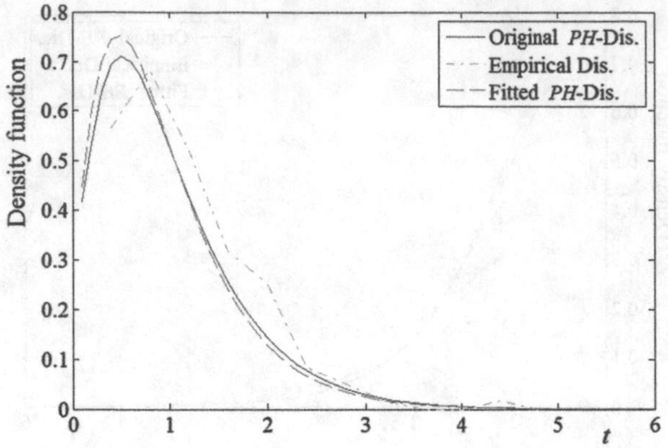

Fig. 1.15 Fitted *PH*-Distribution with $k = 50$

which has the same structure as the original *PH*-representation given in Eq. (1.123) and satisfies $\mathbf{e}'(-T^{(0)})\mathbf{e} \geq 1/E[X]$. Then we obtain

$$\boldsymbol{\alpha}^{(50)} = (0.9348,\ 0.0,\ 0.0652),$$
$$T^{(50)} = \begin{pmatrix} -2.2227 & 0 & 2.0815 \\ 0.3800 & -4.1799 & 0 \\ 0 & 1.0151 & -2.4497 \end{pmatrix}, \tag{1.143}$$

which is close to the original *PH*-representation shown in Fig. 1.15.

Commentary The content of this section comes from Asmussen et al. (1996). Additional references include Johnson and Taaffe (1990a, b), Horváth and Telek (2002), and Riska et al. (2004). The convergence of the EM algorithm has been addressed in papers such as Dempster et al. (1977) and Wu (1983). Software packages for parameter estimation of *PH*-distributions, based on the EM algorithm, have been developed and some are available online: http://home.imf.au.dk/asmus/pspapers.html; and http://webspn.hit.bme.hu/~telek/tools.htm.

Additional Exercises and Extensions

Exercise 1.6.8 A sample $\{y_1, y_2, \ldots, y_n\}$ is collected. Fit an exponential distribution with parameter λ using the following methods.

1. Use the EM algorithm to find λ.
2. Using the density function of the exponential distribution, we construct a likelihood function as $f(y_1, \ldots, y_n;\ \lambda) = \lambda^n \exp(-(y_1 + \ldots + y_n)\lambda)$. Find λ that maximizes the likelihood function $f(y_1, \ldots, y_n;\ \lambda)$.

Similar to parameter estimation, fitting of *PH*-distributions to a probability distribution function can be done by using the EM algorithm (Asmussen et al.

(1996)). For distribution function $F(x)$ and positive integer n, define $\{y_1, y_2, \ldots, y_n\}$ by $F(y_k) = k/(n+1)$, $k = 1, 2, \ldots, n$. Then consider $\{y_1, y_2, \ldots, y_n\}$ as a sample and use the EM algorithm to fit a *PH*-distribution to it.

Example 1.6.4 Consider a *Weibull distribution* with scale parameter $A = 1.2$ and shape parameter $B = 1.8$. The distribution function is given by

$$F_{A,B}(t) = 1 - \exp\left(-\left(\frac{t}{1.2}\right)^{1.8}\right), \quad t \geq 0. \tag{1.144}$$

For $n = 100$, use the above procedure to compute $\{y_1, y_2, \ldots, y_n\}$ and use the EM algorithm to fit a *PH*-distribution of order 5 by choosing

$$\begin{aligned} \boldsymbol{\alpha}^{(0)} &= (0.2, 0.1, 0.2, 0.2, 0.3), \\ T^{(0)} &= \begin{pmatrix} -5 & 1 & 1 & 0.5 & 0.5 \\ 1 & -5 & 1 & 0.5 & 0.5 \\ 1 & 1 & -5 & 0.5 & 0.5 \\ 1 & 2 & 1 & -6 & 0.5 \\ 0.5 & 1 & 1 & 0.5 & -4 \end{pmatrix}. \end{aligned} \tag{1.145}$$

Then we obtain

$$\begin{aligned} \boldsymbol{\alpha}^{(150)} &= (0, 0, 0, 0.0078, 0.9922), \\ T^{(150)} &= \begin{pmatrix} -4.5097 & 1.7464 & 0.2276 & 0.0021 & 0 \\ 0.5465 & -3.5799 & 0.1019 & 0.0007 & 0 \\ 1.5395 & 2.1751 & -4.1638 & 0.0453 & 0 \\ 0.6616 & 1.2305 & 1.2402 & -3.1341 & 0.0018 \\ 0.1009 & 0.1421 & 0.5765 & 1.8858 & -2.7053 \end{pmatrix}. \end{aligned} \tag{1.146}$$

The Weibull distribution, the empirical distribution, and the fitted *PH*-distribution $(\boldsymbol{\alpha}^{(150)}, T^{(150)})$ are plotted in Fig. 1.16.

Exercise 1.6.9 Consider an Erlang distribution with parameters $\lambda = 1$ and $m = 5$. For $n = 50$, use the above procedure to compute $\{y_1, y_2, \ldots, y_n\}$ and use the EM algorithm to fit a *PH*-distribution of order 5 by choosing (i) $T^{(0)}$ with a bi-diagonal structure; and (ii) $T^{(0)} = \mathbf{e}\mathbf{e}'/(m+1) - I$. Choose different types of $\boldsymbol{\alpha}$ for your fitting.

Exercise 1.6.10 Consider a *PH*-distribution with *PH*-representation $(\boldsymbol{\alpha}, T)$. Show that

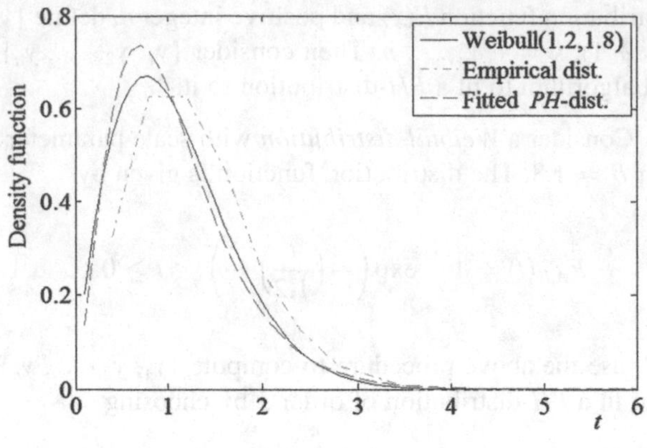

Fig. 1.16 Weibull(1.2, 1.8), an empirical distribution, and the fitted PH-distribution

$$\begin{aligned} E[B_i] &= \alpha_i, \quad i = 1, 2, \ldots, m; \\ E[Z_i] &= \boldsymbol{\alpha}(-T)^{-1}\mathbf{e}_i, \quad i = 1, 2, \ldots, m; \\ E_{(\boldsymbol{\alpha},T)}[N_{i,j}] &= \boldsymbol{\alpha}(-T)^{-1}\mathbf{e}_i t_{i,j}, \quad i \neq j, \quad i,j = 1, 2, \ldots, m; \\ E_{(\boldsymbol{\alpha},T)}[N_{i,m+1}] &= \boldsymbol{\alpha}(-T)^{-1}\mathbf{e}_i t_i^0, \quad i = 1, 2, \ldots, m. \end{aligned} \tag{1.147}$$

Then show that the expected total number of transitions until absorption is given by $\boldsymbol{\alpha}T^{-1}\text{diag}(t_{1,1}, \cdots, t_{m,m})\mathbf{e}$. Let $P = I - (\text{diag}(t_{1,1}, \ldots, t_{2,2}))^{-1}\, T$. Show that

$$\boldsymbol{\alpha}T^{-1}\text{diag}(t_{1,1}, \cdots, t_{m,m})\mathbf{e} = \boldsymbol{\alpha}\left(\sum_{n=0}^{\infty} P^n\right)\mathbf{e}. \tag{1.148}$$

Use Eq. (1.148) to interpret $\boldsymbol{\alpha}T^{-1}\text{diag}(t_{1,1}, \cdots, t_{m,m})\mathbf{e}$ probabilistically.

We remark that the likelihood function may be constructed differently. Usually, the likelihood function, for sample $\{y_1, y_2, \ldots, y_n\}$, is introduced as follows:

$$f(y_1, y_2, \ldots, y_n;\, \boldsymbol{\alpha}, T) = \prod_{v=1}^{n} \left(\boldsymbol{\alpha}\exp(Ty_v)\mathbf{T}^0\right). \tag{1.149}$$

It is not clear how to choose $(\boldsymbol{\alpha}, T)$ to maximize the above likelihood function. Compared to Eq. (1.149), the function in Eq. (1.135) takes more details on the underlying Markov chain into consideration. However, the fictitious sample used in Eq. (1.135) is not observable. The observed sample $\{y_1, y_2, \ldots, y_n\}$ only provides partial information about the absorption process. In the EM algorithm, the expectation step is introduced to generate such information for the maximization step, in which the likelihood function (1.135) is utilized.

1.7 Additional Topics

Sections 1.2, 1.3, 1.4, 1.5, and 1.6 cover the basic theory on *PH*-distributions and multivariate *PH*-distributions. In this section, we discuss generalizations and some theoretical problems of *PH*-distributions.

1.7.1 Discrete PH-distributions

Consider a discrete time Markov chain with $m + 1$ states and transition probability matrix

$$P = \begin{pmatrix} T & \mathbf{T}^0 \\ 0 & 1 \end{pmatrix}. \tag{1.150}$$

The time until absorption into state $m + 1$, denoted as X, is defined as a discrete time *PH*-random variable. Given an initial probability vector $(\boldsymbol{\alpha}, 1 - \boldsymbol{\alpha}\mathbf{e})$, it is easy to see that

$$P\{X = n\} = \begin{cases} 1 - \boldsymbol{\alpha}\mathbf{e}, & \text{if } n = 0; \\ \boldsymbol{\alpha}T^{n-1}\mathbf{T}^0, & \text{if } n = 1, 2, \ldots. \end{cases} \tag{1.151}$$

Exercise 1.7.1 Interpret the elements of T^{n-1} probabilistically. Interpret the elements of $(I - T)^{-1}$ probabilistically. Show that $E[X] = \boldsymbol{\alpha}(I - T)^{-1}\mathbf{e}$ by (i) using the results in Eq. (1.151); and (ii) applying the probabilistic interpretation of $(I - T)^{-1}$.

Exercise 1.7.2 Show that $E[z^X] = 1 - \boldsymbol{\alpha}\mathbf{e} + \boldsymbol{\alpha}z(I - zT)^{-1}\mathbf{T}^0$, $0 < z < 1$. Find the moments of X.

It is readily seen that the discrete *PH*-distribution is a generalization of the geometric distribution. We call $(\boldsymbol{\alpha}, T)$ a discrete *PH*-representation. The theory of discrete *PH*-distributions is almost parallel to that of continuous *PH*-distributions and yet differences exist (see Neuts (1981), O'Cinneide (1990a), and Latouche and Ramaswami (1999)). It is worth noting that any discrete probability distribution with a finite support is a discrete *PH*-distribution. Telek (2000) has shown a result similar to Theorem 1.4.1 for discrete *PH*-distributions. For additional coverage of discrete *PH*-distributions, see Latouche and Ramaswami (1999).

Exercise 1.7.3 Assume that $\{X_n, n = 1, 2, \ldots\}$ are independent and identically distributed random variables with a common LST $f^*(s)$. Assume that N has a discrete *PH*-distribution with *PH*-representation $(\boldsymbol{\alpha}, T)$. All random variables are independent. Define $Y = \sum_{n=1}^{N} X_n$. Show

$$E[e^{-sY}] = 1 - \boldsymbol{\alpha}\mathbf{e} + f^*(s)\boldsymbol{\alpha}(I - f^*(s)T)^{-1}(I - T)\mathbf{e}. \tag{1.152}$$

Exercise 1.7.4 Show that any discrete probability distribution with a finite support is a discrete *PH*-distribution, and find a *PH*-representation for it.

1.7.2 Matrix-Exponential (ME) Distributions

(Lipsky and Ramaswami (1985), Lipsky (1992), Asmussen and Bladt (1996), and Fackrell (2003)). *ME*-distributions are generalizations of *PH*-distributions by removing all constraints on the elements of ($\boldsymbol{\alpha}$, T, $\mathbf{e}$), as long as the expression in Eq. (1.9) is a probability distribution. The set of *ME*-distributions is larger than that of *PH*-distributions. For instance, the density function of an *ME*-distribution may be zero at some positive points. The minimal equivalent *ME*-representation of a *PH*-distribution usually has a smaller order, which can be significantly smaller than that of its minimal *PH*-representation. On the other hand, it is not convenient to use *ME*-representations in stochastic modeling because the probabilistic interpretation enjoyed by *PH*-distributions is lost.

Example 1.7.1 (Cox (1955a, b) and Asmussen and Bladt (1996)) Consider an *ME*-distribution with an *ME*-representation

$$\boldsymbol{\alpha} = (1,\, 0,\, 0), \quad T = \begin{pmatrix} 0 & -1-4\pi^2 & 1+4\pi^2 \\ 3 & 2 & -6 \\ 2 & 2 & -5 \end{pmatrix}, \quad \mathbf{e} = \begin{pmatrix} 1 \\ 1 \\ 1 \end{pmatrix}. \tag{1.153}$$

Then the density function is given by

$$\begin{aligned} f(t) &= \boldsymbol{\alpha}\exp(Tt)(-T\mathbf{e}) \\ &= \left(1+\frac{1}{4\pi^2}\right)(1-\cos(2\pi t))e^{-t}, \quad t \geq 0. \end{aligned} \tag{1.154}$$

The probability distribution is not a phase-type distribution since $f(t) = 0$ if t is a positive integer (See Fig. 1.17)

Compared to the examples in Sect. 1.2, it is clear that *ME*-distributions can approximate more complicated distribution functions at a significantly lower order.

Example 1.7.2 (Bladt and Nielsen (2011)) Consider *PH*-random variable X with *PH*-representation ($\boldsymbol{\alpha}$, T). The *first order moment distribution* is defined as $f_1(t) = tf(t)/E[X]$, $t \geq 0$, where $f_1(t)$ and $f(t)$ are the density functions of the first order moment distribution and X, respectively. It is shown that the first order moment distribution of X is also a *PH*-distribution with two equivalent matrix representations:

(i) *ME*-representation: $\left(\boldsymbol{\alpha}_1 = \left(\frac{\boldsymbol{\alpha}T^{-1}}{\boldsymbol{\alpha}T^{-1}\mathbf{e}},\, 0\right),\quad T_1 = \begin{pmatrix} T & -T \\ 0 & T \end{pmatrix},\quad \mathbf{e}\right)$;

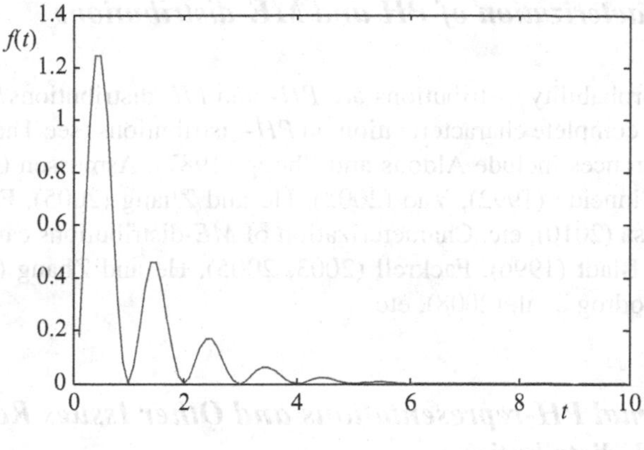

Fig. 1.17 The density function for Example 1.7.1

(ii) *PH*-representation:

$$\boldsymbol{\alpha}_1 = \left(\frac{\rho_2}{\rho_1}\boldsymbol{\pi}_1 \bullet (-T\mathbf{e}), 0\right),$$
$$T_1 = \begin{pmatrix} \Delta^{-1}(\boldsymbol{\pi}_2)T'\Delta(\boldsymbol{\pi}_2) & \frac{\rho_1}{\rho_2}\Delta^{-1}(\boldsymbol{\pi}_2)\Delta(\boldsymbol{\pi}_1) \\ 0 & \Delta^{-1}(\boldsymbol{\pi}_1)T'\Delta(\boldsymbol{\pi}_1) \end{pmatrix}, \tag{1.155}$$

where $\boldsymbol{\pi}_i = \boldsymbol{\alpha}T^{-i}(\boldsymbol{\alpha}T^{-i}\mathbf{e})^{-1}$, $\rho_i = \boldsymbol{\alpha}(-T)^{-i}\mathbf{e}$, $i = 1, 2$, and $\mathbf{a}\bullet\mathbf{b} = (a_1b_1, a_2b_2, \ldots, a_nb_n)$.

Exercise 1.7.5 For the *PH*-distributions in Example 1.2.1, find the matrix representations for their first order moment distributions. Plot the density functions using the three representations.

Exercise 1.7.6 Show that the two matrix-representations in Example 1.7.2 are equivalent. (Hint: Calculate their LSTs.)

1.7.3 Further Generalizations of PH-distributions

(i) *Bilateral PH- and bilateral ME- distributions* Both *PH* and *ME*- distributions are defined on the nonnegative half of the real line. One generalization of *PH* and *ME*-distributions is to include the negative half of the real line. Such generalizations are called bilateral distributions. See Shanthikumar (1985), Ahn and Ramaswami (2005), and Bladt et al. (2013) for more details.
(ii) *Semi-Markovian PH-distributions* (Shi et al. (1996)).
(iii) *Multivariate ME-distributions* (Bladt and Nielsen (2010)).

1.7.4 Characterization of PH and ME-distributions

What kind of probability distributions are *PH*- and *ME*-distributions? O'Cinneide (1990a) gives a complete characterization on *PH*-distributions (see Theorem 1.2.2). Additional references include Aldous and Shepp (1987), Asmussen (1989, 2000), Maier and O'Cinneide (1992), Yao (2002), He and Zhang (2005), Fackrell et al. (2010), Sanguesa (2010), etc. Characterization of *ME*-distributions can be found in Asmussen and Bladt (1996), Fackrell (2003, 2005), He and Zhang (2007), Bean et al. (2008), Bodrog et al. (2008), etc.

1.7.5 Minimal PH-representations and Other Issues Related to PH-distributions

A *PH*-representation with the smallest order among all equivalent *PH*-representations is called a *minimal PH-representation*. Finding a minimal *PH*-representation for a *PH*-distribution has been an open problem since 1975. As shown in Commault and Mocanu (2003), the problem is equivalent to a well-known minimal representation problem in *control theory*. Note that, He and Zhang (2008) develop an algorithm to find an equivalent Coxian representation of the smallest order. Pulungan and Hermanns (2009) consider a special structure for a minimal representation. Both only address the minimal representation within a subset of *PH*-distributions. In general, finding the minimal *PH*-representation is a fundamental issue with great impact on both theoretical studies and applications of *PH*-distributions, and it is a challenging problem.

1.7.6 Additional Exercises

Neuts (1995) provides a large number of exercises related to *PH*-distributions and their applications. Exercises in Chaps. 4 and 5 in Neuts (1995) are particularly useful to understand *PH*-distributions and their application.

References

Ahn S, Ramaswami V (2005) Bilateral phase-type distributions. Stoch Models 21:239–259

Albert A (1962) Estimating the infinitesimal generator of a continuous time, finite state Markov process. Ann Math Stat 33:727–753

Aldous D, Shepp L (1987) The least variable phase type distribution is Erlang. Stoch Models 3:467–473

Asmussen S (1989) Exponential families generated by phase-type distributions and other Markov lifetimes. Scand J Stat 16:319–334

Asmussen S (1995) Stationary distributions for fluid flow models with or without Brownian noise. Stoch Models 11:21–49

Asmussen S (2000) Ruin probabilities. World Scientific, Hong Kong

Asmussen S (2003) Applied probability and queues, 2nd edn. Springer, New York

Asmussen S, Nerman O (1991) Fitting phase-type distribution via the EM algorithm. In: Vest Nielsen K (ed) Symposium i Anvendt Statistik, Copenhagen, UNI-C, Copenhagen, 21–23 Jan 1991, pp 335–346

Asmussen S, O'Cinneide C (1998) Representation for matrix-geometric and matrix-exponential steady-state distributions with applications to many-server queues. Stoch Models 14:369–387

Asmussen S, Bladt M (1996) Renewal theory and queueing algorithms for matrix-exponential distributions. In: Alfa AS, Chakravarthy S (eds) Proceedings of the first international conference on matrix analytic methods in stochastic models. Marcel Dekker, New York

Asmussen S, Nerman O, Olsson M (1996) Fitting phase-type distributions via the EM algorithm. Scand J Stat 23:419–441

Assaf D, Levikson B (1982) Closure of phase type distributions under operations arising in reliability theory. Ann Prob 10:265–269

Assaf D, Langberg NA, Savits TH, Shaked M (1984) Multivariate phase-type distributions. Oper Res 32:688–702

Bean NG, Fackrell M, Taylor P (2008) Characterization of matrix-exponential distributions. Stoch Models 24:339–363

Berman A, Plemmons RJ (1979) Nonnegative matrices in the mathematical sciences. Academic, New York

Bladt M, Nielsen BF (2010) Multivariate matrix-exponential distributions. Stoch Models 26:1–26

Bladt M, Nielsen BF (2011) Moment distributions of phase type. Stoch Models 27:651–663

Bladt M, Esparza LJR, Nielsen BF (2013) Bilateral matrix-exponential distributions. In: Latouche G, Ramaswami V, Sethuraman J, Sigman K, Squillante MS, Yao DD (eds) Matrix-analytic Methods in stochastic models, Springer proceedings in mathematics & statistics, Springer, New York, pp 41–56

Bodrog L, Hovarth A, Telek M (2008) Moment characterization of matrix-exponential and Markovian arrival processes. Ann Oper Res 160:51–68

Cai J, Li HJ (2005) Multivariate risk model of phase type. Insur Math Econ 36:137–152

Commault C, Mocanu S (2003) Phase-type distributions and representations: some results and open problems for system theory. Int J Control 76:566–580

Cox DR (1955a) On the use of complex probabilities in the theory of stochastic processes. Math Proc Camb Philos Soc 51:313–319

Cox DR (1955b) The analysis of non-Markovian stochastic processes by the inclusion of supplementary variables. Math Proc Camb Philos Soc 51:433–441

Cumani A (1982) On the canonical representation of Markov processes modeling failure time distributions. Microelectron Reliab 22:583–602

Dehen M, Latouche G (1982) A geometric interpretation of the relations between the exponential and the generalized Erlang distributions. Adv Appl Prob 14:885–897

Dempster AP, Laird NM, Rubin DB (1977) Maximum likelihood from incomplete data via the EM algorithm. J Roy Stat Soc, Ser B 39:1–38

Erlang AK (1917) Solution of some problems in the theory of probabilities of significance in automatic telephone exchange. Post Office Elec Eng J 10:189–197

Fackrell M (2003) Characterization of matrix-exponential distributions, Ph.D. thesis, School of Applied Mathematics, University of Adelaide, South Australia

Fackrell M (2005) Fitting with matrix-exponential distributions. Stoch Models 21:377–400

Fackrell M, He QM, Taylor P, Zhang HQ (2010) The algebraic degree of phase-type distributions. J Appl Prob 47:611–629

He QM, Zhang HQ (2005) A note on unicyclic representation of *PH*-distributions. Stoch Models 21:465–483

He QM, Zhang HQ (2006) Spectral polynomial algorithms for computing bi-diagonal representations for matrix-exponential distributions and phase-type distributions. Stoch Models 22:289–317

He QM, Zhang HQ (2007) On matrix exponential distributions. Adv Appl Prob 39:271–292

He QM, Zhang HQ (2008) An algorithm for computing the minimal Coxian representation. INFORMS J Comput 20:179–190

He QM, Zhang HQ, Xue JG (2011) Algorithms for Coxianization of *PH*-generators. INFORMS J Comput 23:153–164

He QM, Zhang HQ, Vera J (2012a) Majorization and extremal *PH*-distributions. In: Latouche G, Ramaswami V, Sethuraman J, Sigman K, Squillante MS, Yao DD (eds) Matrix-analytic methods in stochastic models. Springer proceedings in mathematics & statistics, Springer, New York, pp 107–122

He QM, Zhang HQ, Vera J (2012b) On some properties of bivariate exponential distributions. Stoch Models 28:187–206

Horváth A, Telek M (2002) PhFit: a general purpose phase type fitting tool. In: Tools 2002. Lecture notes in computer science, vol 2324. Springer, London, pp 82–91

Horváth G, Telek M (2012) Acceptance-rejection methods for generating random variates from matrix exponential distribution and rational arrival processes. In: Latouche G, Ramaswami V, Sethuraman J, Sigman K, Squillante MS, Yao DD (eds) Matrix-Analytic Methods in Stochastic Models. Springer proceedings in mathematics & statistics, Springer, New York, pp 123–144

Johnson MA, Taaffe MR (1990a) Matching moments to phase distributions: nonlinear programming approaches. Stoch Models 6:259–281

Johnson MA, Taaffe MR (1990b) Matching moments to phase distributions: density function shapes. Stoch Models 6:283–306

Kulkarni VG (1989) A new class of multivariate phase type distributions. Oper Res 37:151–158

Lancaster P, Tismenetsky M (1985) The theory of matrices, 2nd edn. Academic, New York

Latouche G, Ramaswami V (1999) Introduction to matrix analytic methods in stochastic modeling. ASA & SIAM, Philadelphia

Law AM, Kelton WD (2000) Simulation modeling and analysis. McGraw Hill, New York

Li HJ (2003) Association of multivariate phase-type distributions with applications to shock models. Stat Prob Lett 64:1043–1059

Lipsky L (1992) Queueing theory: a linear algebraic approach. Macmillan, New York

Lipsky L, Ramaswami V (1985) A unique minimal representation of Coxian service centres, Technical report, Department of Computer Science and Engineering, University of Nebraska, Lincoln

Maier RS, O'Cinneide CA (1992) A closure characterization of phase-type distributions. J Appl Prob 29:92–103

Marshall AW, Olkin I (1967) A multivariate exponential distribution. J Am Stat Asso 62: 30–44

Micchelli CA, Willoughby RA (1979) On functions which preserve the class of Stieltjes matrices. Linear Algebra Appl 23:141–156

Minc H (1988) Non-negative matrices. Wiley, New York

Mocanu S, Commault C (1999) Sparse representations of phase type distributions. Stoch Models 15:759–778

Moler C, Van Loan C (1978) Nineteen dubious ways to compute the exponential of a matrix. SIAM Rev 20:801–836

Neuts MF (1981) Matrix-geometric solutions in stochastic models – an algorithmic approach. The Johns Hopkins University Press, Baltimore

Neuts MF (1992) Two further closure-properties of *PH*-distributions. Asia-Pac J Oper Res 9:77–85

Neuts MF (1995) Algorithmic probability: a collection of problems. Chapman & Hall, London

Neuts MF (1975) Probability distributions of phase type. In: Liber Amicorum Prof. Emeritus H. Florin, University of Louvain, Belgium. pp 173–206

O'Cinneide CA (1989) On non-uniqueness of representations of phase-type distributions. Stoch Models 5:247–259

O'Cinneide CA (1990a) Characterization of phase-type distributions. Stoch Models 6:1–57

O'Cinneide CA (1990b) On the limitations of multivariate phase-type families. Oper Res 38:519–526

O'Cinneide CA (1991a) Phase-type distributions and invariant polytope. Adv Appl Prob 23:515–535

O'Cinneide CA (1991b) Phase-type distributions and majorization. Ann Appl Prob 1:219–227

O'Cinneide CA (1993) Triangular order of triangular phase-type distributions. Stoch Models 9:507–529

O'Cinneide CA (1999) Phase-type distributions: open problems and a few properties. Stoch Models 15:731–757

Pulungan R, Hermanns H (2009) Acyclic minimality by construction – almost. Quant Eval Syst, QEST09:63–72

Riska A, Diev V, Smirni E (2004) An EM-based technique for approximating long-tailed data sets with PH distributions. Perform Eval 55:147–164

Rockafellar RT (1970) Convex analysis. Princeton University Press, Princeton

Ross SM (2010) Introduction to probability models, 10th edn. Academic, New York

Sanguesa C (2010) On the minimal value in Maier's property concerning phase-type distributions. Stoch Models 26:124–140

Sengupta B (1989) Markov processes whose steady state distribution is matrix-exponential with an application to the *GI/PH/*1 queue. Adv Appl Prob 21:159–180

Shaked M, Shanthikumar JG (2006) Stochastic orders. Springer, New York

Shanthikumar JG (1985) Bilateral phase-type distributions. Nav Res Log 32:119–136

Shi DH, Guo J, Liu L (1996) *SPH*-distributions and rectangle-iterative algorithm. In: Chakravarthy S, Alfa AS (eds) Matrix-analytic methods in stochastic models. Marcel Dekker, New York, pp 207–224

Telek M (2000) The minimal coefficient of variation of discrete phase type distributions. In: Latouche G, Taylor PG (eds) Advances in algorithmic methods for stochastic models, Proceedings of the 3rd international conference on matrix analytic methods, Notable Publications, New Jersey, pp 391–400

Wu CFJ (1983) On the convergence properties of the EM algorithm. Ann Stat 11:95–103

Yao RH (2002) A proof of the steepest increase conjecture of a phase-type density. Stoch Models 18:1–6

Chapter 2
From the Poisson Process to Markovian Arrival Processes

Abstract This chapter introduces Markovian arrival processes. Topics covered include: (i) the Poisson process; (ii) definitions of Markovian arrival processes; (iii) performance measures; (iv) batch Markovian arrival processes; and (v) Markovian arrival processes with marked arrivals.

Many stochastic systems have input to the system over time. Examples include: queueing systems for which there is a stream of customer arrivals; queueing networks for which there are multiple streams of customer arrivals; inventory systems for which both demands and replenishments form streams of inputs; insurance systems for which claims that are random in size arrive randomly, and electrical systems for which electric shock waves arrive randomly. Analysis of these stochastic systems requires a mathematical tool that can describe the input process(es) analytically. A useful tool should provide a good approximation of real time arrival processes and lead to analytically and/or numerically tractable models.

The *Poisson process* is one of the simplest counting processes. It possesses the memoryless property, which often leads to Markovian models for stochastic systems. *Markovian arrival process* (*MAP*) is a generalization of the Poisson process, which keeps many useful properties of the Poisson process. For instance, the memoryless property of the Poisson process is partially preserved by the Markovian arrival process by conditioning on the phase of an underlying Markov chain. The set of Markovian arrival processes is versatile. In fact, any stochastic counting process can be approximated arbitrarily closely by a sequence of Markovian arrival processes. The Poisson process and Markovian arrival processes have proven to be indispensable stochastic modeling tools.

The objective of this chapter is to introduce the Poisson process and Markovian arrival processes, and to present a number of their basic properties. In particular, we give (i) four definitions for the Poisson process and (ii) four definitions for the Markovian arrival processes. These definitions provide insight into, and

Q.-M. He, *Fundamentals of Matrix-Analytic Methods*,
DOI 10.1007/978-1-4614-7330-5_2,

consequently enhance our understanding of, the stochastic processes from different perspectives. Performance measures related to these processes are defined and analyzed. Computational methods are developed for the performance measures.

In addition to the Poisson process and Markovian arrival processes, a number of well-known stochastic processes have been developed to model input processes, including *renewal process*, *regenerative process*, *Lévy process*, *stationary process*, *Brownian motion*, *Martingale*, *Gaussian process*, etc. (see Sigman (1995), Asmussen (2003), and Ross (2010)) The studies and applications of those processes usually require advanced mathematics and are not covered in this book.

We refer to Ross (2010) for basic concepts on counting processes and renewal processes: *interarrival time*, *renewal process*, *renewal point*, *independent increment*, *stationary increment*, *arrival rate*, *superposition*, *decomposition*, etc. We also refer to Sigman (1995) and Asmussen (2003) for advanced treatment of counting processes.

2.1 The Poisson Process

Assume that $\{N(t), t \geq 0\}$ is a *counting process*, where $N(t)$ is the number of *events* (*arrivals*) occurring in $[0, t]$. An event can be the arrival of a customer, a demand, or a group of customers. In fact, an event can be defined in any way or as anything that is necessary and meaningful. Next, we give three equivalent definitions of the Poisson process. We begin with a constructive definition.

Definition 2.1.1 The counting process $\{N(t), t \geq 0\}$ is called a Poisson process if $\{N(t) \leq n\} = \{X_1 + X_2 + \ldots + X_n + X_{n+1} > t\}$, for $n = 0, 1, 2, \ldots$, and $t \geq 0$, where $\{X_1, X_2, \ldots, X_n, \ldots\}$ are independent and identically distributed random variables with a common exponential distribution with parameter λ (see Fig. 2.1).

Definition 2.1.1 can be viewed as a "constructive" definition since it describes the arrivals of events by specifying the interarrival times between events or the arrival times of events sequentially. In fact, Definition 2.1.1 says that the time between two consecutive events has an exponential distribution.

Alternatively, Poisson processes can be defined as follows. Let $N(s, t) = N(t) - N(s)$ for $t > s$. The counting process $\{N(t), t \geq 0\}$ has *independent increments* if $N(s, t)$ and $N(u, v)$ are independent for all non-overlapping intervals (s, t) and (u, v). The counting process $\{N(t), t \geq 0\}$ has *stationary increments* if the distribution of $N(s, s + t)$ is independent of s.

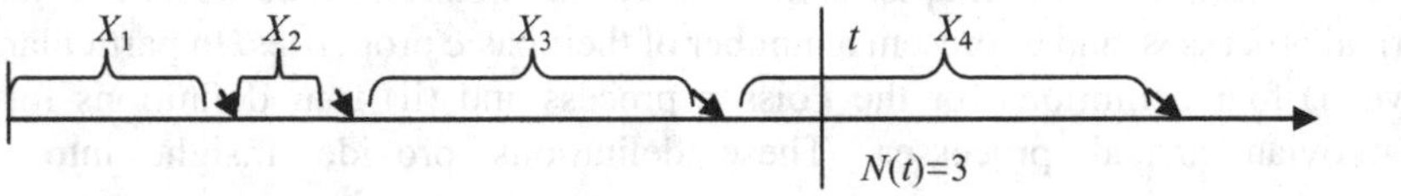

Fig. 2.1 A Poisson process

Definition 2.1.2 The process $\{N(t), t \geq 0\}$ is called a Poisson process if

(i) $N(0) = 0$;
(ii) $\{N(t), t \geq 0\}$ possesses the independent increment property; and
(iii) $P\{N(t+s) - N(s) = n\} = e^{-\lambda t}(\lambda t)^n/n!$, $n = 0, 1, 2, \ldots$, i.e., a Poisson distribution with parameter λt.

Definition 2.1.3 The process $\{N(t), t \geq 0\}$ is called a Poisson process if

(i) $N(0) = 0$;
(ii) $\{N(t), t \geq 0\}$ possesses the independent increment property and the stationary increment property; and
(iii) $P\{N(h) = 1\} = \lambda h + o(h)$ and $P\{N(h) \geq 2\} = o(h)$. (Note: Recall that $o(h)/h$ converges to zero, if $h \to 0$.)

Each of the three definitions reveals useful properties of the Poisson process from a unique perspective. Definition 2.1.2 can be viewed as a "macro" definition since it describes what may occur in visible intervals. In fact, Definition 2.1.2 says that the number of events in any interval has a Poisson distribution. Definition 2.1.3 can be viewed as a "micro" definition since it describes what may occur in infinitesimally small intervals. In fact, Definition 2.1.3 says that the probability an event occurs in a small interval is proportional to the length of the interval.

Theorem 2.1.1 *Definitions* 2.1.1, 2.1.2, *and* 2.1.3 *are equivalent*. (*See Chap*. 5 *in Ross* (2010) *for a proof*.)

We remark that a thorough and in-depth understanding of the Poisson process helps us greatly in learning about more complicated arrival processes such as Markovian arrival processes. Next, we use a number of exercises and properties to enhance our understanding of the Poisson process and to learn when and how to use the three definitions.

Exercise 2.1.1 Assume that X_λ and Y_μ are independent Poisson random variables with parameters λ and μ, respectively. Show that the means of X_λ and Y_μ are λ and μ, respectively. Show that $X_\lambda + Y_\mu$ has a Poisson distribution with parameter $\lambda + \mu$.

Exercise 2.1.1 implies that, for a Poisson process with parameter λ, the total number of events occurring in two non-overlapping intervals has a Poisson distribution with parameter $\lambda(t + s)$, where t and s are the lengths of the two intervals.

Exercise 2.1.2 Assume that $\{N(t), t \geq 0\}$ is a Poisson process with parameter $\lambda = 0.4$. Let $X = N(1, 1.5) + N(4, 5.5)$. What is the distribution of X? Interpret X intuitively.

Exercise 2.1.3 For exponential random variable X with parameter λ, prove that $P\{X > h\} = 1 - \lambda h + o(h)$.

In Definition 2.1.1, note that $\{X_1 > h\}$ is equivalent to $\{N(h) = 0\}$. Then Exercise 2.1.3 implies that $P\{N(h) = 0\} = 1 - \lambda h + o(h)$, which is consistent with Definition 2.1.3. Let $P_n(t) = P\{N(t + s) - N(s) = n\}$, for $s, t \geq 0$,

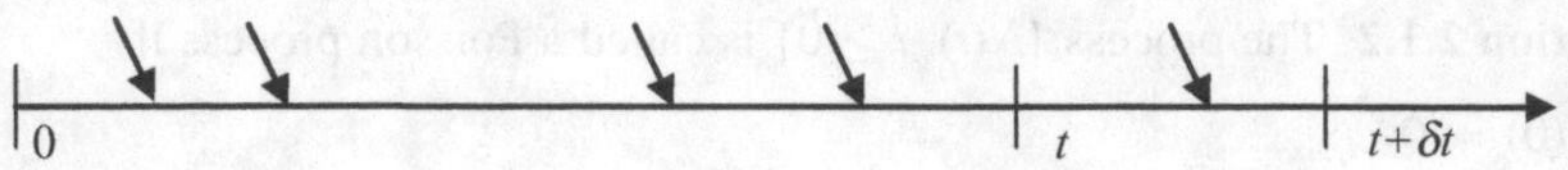

Fig. 2.2 Events in [0, t] and (t, $t + \delta t$)

and $n = 0, 1, 2, \ldots$. Using Exercise 2.1.3 and conditioning on the number of events in (t, $t+\delta t$), Definition 2.1.3 leads to (see Fig. 2.2), for $n \geq 1$,

$$\begin{aligned}
&P_n(t+\delta t)\\
&\quad = P\{N(t+\delta t) = n\}\\
&\quad = \sum_{k=0}^{\infty} P\{N(t,\ t+\delta t) = k\}P\{N(t+\delta t) = n|N(t,\ t+\delta t) = k\}\\
&\quad = \sum_{k=0}^{n} P\{N(t,t+\delta t) = k\}P\{N(t) = n-k,\ N(t,t+\delta t) = k|N(t,t+\delta t) = k\}\\
&\quad = P_n(t)(1-\lambda\delta t) + P_{n-1}(t)\lambda\delta t + o(\delta t).
\end{aligned} \tag{2.1}$$

For $n = 0$, Eq. (2.1) becomes $P_0(t+\delta t) = P_0(t)(1-\lambda\delta t) + o(\delta t)$, which can be rewritten as

$$\frac{P_0(t+\delta t) - P_0(t)}{\delta t} = -\lambda P_0(t) + \frac{o(\delta t)}{\delta t}. \tag{2.2}$$

Letting δt go down to zero in Eq. (2.2), we obtain the differential equation $\mathrm{d}P_0(t)/\mathrm{d}t = -\lambda P_0(t)$ for $t > 0$. With initial condition $P_0(0) = 1$, solving the differential equation leads to $P_0(t) = \exp(-\lambda t)$. Intuitively, $P_0(t)$ is the probability that there is no arrival in [0, t], which is also the probability that the first arrival takes place after t, i.e., $P_0(t) = P\{X_1 > t\} = \exp(-\lambda t)$.

For $n \geq 1$, similarly, Eq. (2.1) leads to a differential equation and a solution for $P_n(t)$, which are given in Exercise 2.1.4.

Exercise 2.1.4 For a Poisson process with parameter λ, use the induction method and Eq. (2.1) to show that $\mathrm{d}P_n(t)/\mathrm{d}t = -\lambda P_n(t) + \lambda P_{n-1}(t)$ and $P_n(0) = 0$, for $n = 1, 2, \ldots$. Solve the differential equation to find $P_n(t)$.

Exercise 2.1.4 indicates that the properties in Definition 2.1.2 can be derived from Definition 2.1.3. On the other hand, it is easy to show the properties in Definition 2.1.3 from Definition 2.1.2. Thus, the two definitions are equivalent.

The Poisson process has a single parameter λ, which has an explicit interpretation.

Exercise 2.1.5 Use Definitions 2.1.1, 2.1.2, and 2.1.3 to show $E[N(t)] = \lambda t$. What is $E[N(t+s)-N(s)]$?

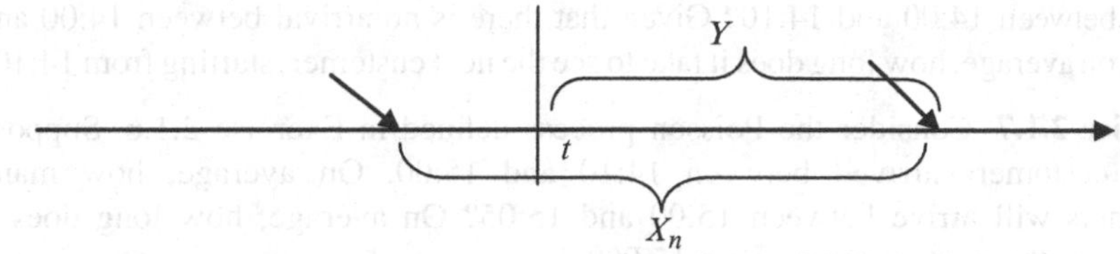

Fig. 2.3 Memoryless property of Poisson processes

By Exercise 2.1.5, we obtain $E[N(1)] = \lambda$, which says that the parameter λ is the expected number of events per unit time. Further, using Definition 2.1.1 and the *law of large numbers* in probability theory (or renewal theory), it can be shown that

$$P\left\{\lim_{t\to\infty}\frac{N(t)}{t} = \lambda\right\} = 1. \tag{2.3}$$

Equation (2.3) indicates that $N(t)/t$ converges *almost surely* to λ.

The parameter λ is called the *arrival rate* and the events of a Poisson process are called *arrivals*. Since λ is also the only parameter of the exponential distribution of the interarrival times, Eq. (2.3) gives it an interpretation. Such a connection between the Poisson process and the exponential distribution enhances our understanding on both of them. We remark that the understanding of the parameter λ as the arrival rate is crucial to learning Markovian arrival processes and, more generally, continuous time Markov chains.

Proposition 2.1.1 *In a Poisson process with parameter λ, let Y be the time elapsed until the first event after time t (see Fig. 2.3). Then Y has an exponential distribution with parameter λ.*

Proof. By Definition 2.1.1, the interarrival time (i.e., X_n) has an exponential distribution. Then Y, the *residual interarrival time*, is a conditional random variable of X_n. By the memoryless property of X_n, Y has an exponential distribution with the same parameter as X_n. This completes the proof of Proposition 2.1.1.

Recall that one of the most important properties of the exponential distribution is its memoryless property (Theorem 1.1.1). Proposition 2.1.1 indicates that the Poisson process also possesses the memoryless property in the sense that what is going to occur in the future (after time t) is independent of what has happened before or at time t. This memoryless property plays a vital role in stochastic modeling with the Poisson process. The fact that the residual time Y has an exponential distribution is used every time a continuous time Markov chain is constructed for a stochastic model.

Exercise 2.1.6 Customers arrive to a grocery store according to a Poisson process with $\lambda = 5$ per minute. Suppose that the current time is two o'clock. On average, how long does it take to see the next customer? On average, how many customers will arrive between 14:00 and 14:10? Given that there is no arrival between 14:00 and 14:10, on average, how long does it take to see the next customer, starting from 14:10?

Exercise 2.1.7 Consider the Poisson process defined in Exercise 2.1.6. Suppose 1,000 customers arrived between 14:10 and 15:00. On average, how many customers will arrive between 15:00 and 15:05? On average, how long does it take to see the next customer after 15:00?

Like LST, the *probability generating function* is another popular tool in the study of probability distributions and stochastic processes.

Exercise 2.1.8 Show that the probability generating function of $N(t)$ is given by

$$E[z^{N(t)}] \equiv \sum_{n=0}^{\infty} z^n P\{N(t) = n\} = \exp(-\lambda t(1 - z)), \quad \text{for } 0 \le z \le 1. \qquad (2.4)$$

Use this result to give another proof of Exercise 2.1.5. Using this result, higher moments of $N(t)$ can be found. Show that $E[N^2(t)] = (\lambda t)^2 + \lambda t$ and find the variance of $N(t)$. (Hint: $N^2(t) = N(t)(N(t) - 1) + N(t)$.)

Several stochastic processes can be combined to form a single process, which is called the *superposition* of stochastic processes.

Example 2.1.1 *Assume that* $\{N_1(t), t \ge 0\}$ *and* $\{N_2(t), t \ge 0\}$ *are two independent Poisson processes with parameters* λ_1 *and* λ_2*, respectively. Define* $N(t) = N_1(t) + N_2(t)$, $t \ge 0$. *Then* $\{N(t), t \ge 0\}$ *is a Poisson process with parameter* $\lambda_1 + \lambda_2$.

Proof. The result can be proved by verifying Definition 2.1.2. It is clear that (i) $N(0) = N_1(0) + N_2(0) = 0$; (ii) the independent increment property of $\{N(t), t \ge 0\}$ is inherited from that of $\{N_1(t), t \ge 0\}$ and $\{N_2(t), t \ge 0\}$; and (iii) by Exercise 2.1.1, $N(t)$ has a Poisson distribution with parameter $\lambda_1 t + \lambda_2 t = (\lambda_1 + \lambda_2)t$. This completes the proof.

Example 2.1.1 indicates that the time until the next arrival (from either Poisson process) has an exponential distribution with parameter $\lambda_1 + \lambda_2$.

Exercise 2.1.9 For Example 2.1.1, show that the probability that the next arrival comes from Poisson process i (i.e., $\{N_i(t), t \ge 0\}$) is $\lambda_i/(\lambda_1 + \lambda_2)$, for $i = 1, 2$. (Hint: Use Proposition 1.1.3.)

Exercise 2.1.10 Prove Example 2.1.1 by verifying Definition 2.1.1 (Hint: Use the memoryless property to show that the interarrival times are independent and identically distributed exponential distributions.)

Proposition 2.1.2 (Superposition of Poisson processes) *Assume that* $\{N_1(t), t \ge 0\}$, ..., *and* $\{N_n(t), t \ge 0\}$ *are independent Poisson processes with arrival rates* λ_1, ..., *and* λ_n*, respectively. Then* $\{N(t) = N_1(t) + \ldots + N_n(t), t \ge 0\}$ *is a Poisson process with arrival rate* $\lambda_1 + \ldots + \lambda_n$. *The probability that the next arrival comes from Poisson process* i *is* $\lambda_i/(\lambda_1 + \ldots + \lambda_n)$, for $i = 1, \ldots, n$.

Proof. The proposition is proved using Example 2.1.1, Exercise 2.1.9, and the induction method. By Example 2.1.1, Proposition 2.1.2 holds for $n = 2$. We rewrite

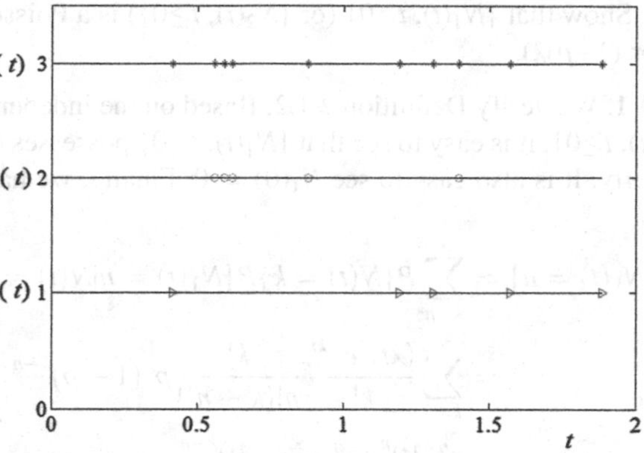

Fig. 2.4 Decomposition of a Poisson process with $\lambda = 5$ for $p = 0.3$

$N_1(t) + \ldots + N_n(t)$ as $(N_1(t) + \ldots + N_{n-1}(t)) + N_n(t)$. By induction, $\{N_1(t) + \ldots + N_{n-1}(t), t \geq 0\}$ is a Poisson process that is independent of $\{N_n(t), t{\geq}0\}$. Applying Example 2.1.1 again, the expected result is obtained. This completes the proof of Proposition 2.1.2.

A stochastic process can be decomposed into several stochastic processes, which is called the *decomposition* of stochastic processes. Consider Poisson process $\{N(t), t \geq 0\}$ with parameter λ. When an event occurs, a *Bernoulli trial* is conducted. With probability p (or $1-p$) the event is marked as type 1 (or 2), independent of all other events. Denote by $N_1(t)$ (or $N_2(t)$) the number of events marked as type 1 (or 2) in $[0, t]$. In this way, two stochastic processes, $\{N_1(t), t \geq 0\}$ and $\{N_2(t), t \geq 0\}$, are defined. Clearly, we have $N(t) = N_1(t) + N_2(t)$. The arrival epochs of the three processes are plotted in Fig. 2.4: dots on line 3 are for the arrivals of $\{N(t), t \geq 0\}$, line 2 for $\{N_1(t), t \geq 0\}$, and line 1 for $\{N_2(t), t \geq 0\}$.

By conditioning on $N(t)$, we have, for any fixed $t > 0$,

$$
\begin{aligned}
&E[z_1^{N_1(t)} z_2^{N_2(t)}] \\
&= \sum_{n=0}^{\infty} P\{N(t) = n\} E[z_1^{N_1(t)} z_2^{N_2(t)} | N(t) = n] \\
&= \sum_{n=0}^{\infty} \frac{\exp(-\lambda t)}{n!} (\lambda t)^n \left(\sum_{k=0}^{n} \binom{n}{k} p^k (1-p)^{n-k} z_1^k z_2^{n-k} \right) \qquad (2.5) \\
&= \sum_{n=0}^{\infty} \frac{\exp(-\lambda t)}{n!} (\lambda t)^n (z_1 p + z_2(1-p))^n \\
&= \exp(-\lambda p t(1 - z_1)) \exp(-\lambda(1 - p)t(1 - z_2)) \\
&= E[z_1^{N_1(t)}] E[z_2^{N_2(t)}].
\end{aligned}
$$

Equation (2.5) implies that, for fixed t, $N_1(t)$ and $N_2(t)$ are two independent Poisson random variables with parameters $p\lambda$ and $(1-p)\lambda$, respectively.

Example 2.1.2 Show that $\{N_1(t), t \geq 0\}$ (or $\{N_2(t), t \geq 0\}$) is a Poisson process with parameter $p\lambda$ (or $(1-p)\lambda$).

Proof. **Method 1**. We verify Definition 2.1.2. Based on the independent increment property of $\{N(t), t \geq 0\}$, it is easy to see that $\{N_1(t), t \geq 0\}$ possesses the independent increment property. It is also easy to see $N_1(0) = 0$. Finally, we have, for $n \geq 0$,

$$\begin{aligned} P\{N_1(t) = n\} &= \sum_{k=n}^{\infty} P\{N(t) = k\} P\{N_1(t) = n | N(t) = k\} \\ &= \sum_{k=n}^{\infty} \frac{(\lambda t)^k e^{-\lambda t}}{k!} \frac{k!}{n!(k-n)!} p^n (1-p)^{k-n} \\ &= \frac{p^n (\lambda t)^n e^{-\lambda t}}{n!} \sum_{k=n}^{\infty} \frac{(\lambda t)^{k-n}}{(k-n)!} (1-p)^{k-n} \\ &= \frac{(\lambda p t)^n e^{-\lambda t + (1-p)\lambda t}}{n!} \\ &= \frac{(\lambda p t)^n e^{-\lambda p t}}{n!}. \end{aligned} \tag{2.6}$$

Thus, $N_1(t)$ has a Poisson distribution with parameter $\lambda p t$. By Definition 2.1.2, $\{N_1(t), t \geq 0\}$ is a Poisson process with parameter λp.

Method 2. We verify Definition 2.1.1. The time between two consecutive arrivals of $\{N_1(t), t \geq 0\}$ can be expressed as $Y = \sum_{n=1}^{N} X_n$, where N is a random variable and has a geometric distribution with parameter $1-p$. By Proposition 1.1.4, Y has an exponential distribution with parameter $p\lambda$. Because of the memoryless property, the interarrival times are independent. Thus, by Definition 2.1.1, $\{N_1(t), t \geq 0\}$ is a Poisson process with parameter $p\lambda$. This completes the proof of Example 2.1.2.

Equation (2.5) and Example 2.1.2 are not sufficient to show that the Poisson processes $\{N_1(t), t \geq 0\}$ and $\{N_2(t), t \geq 0\}$ are independent. In fact, to show that the two processes are independent, we must prove that random vectors $(N_1(t_1), N_1(t_2), \ldots, N_1(t_k))$ and $(N_2(t_1), N_2(t_2), \ldots, N_2(t_k))$ are independent for all possible $\{k, t_1, t_2, \ldots, t_k\}$. A complete proof for that result and Proposition 2.1.3 can be found in Chap. 5, Ross (2010).

Proposition 2.1.3 (Decomposition of the Poisson process) *Assume that the events of a Poisson process with parameter λ are marked independently into type* $1, 2, \ldots,$ *and n events with probability distribution* $\{p_1, p_2, \ldots, p_n\}$. *Denote by* $\{N_1(t), t \geq 0\}$, $\ldots$, *and* $\{N_n(t), t \geq 0\}$ *the counting processes of individual types of events. Then* $\{N_1(t), t \geq 0\}$, $\ldots$, *and* $\{N_n(t), t \geq 0\}$ *are n independent Poisson processes with parameters* $p_1\lambda, p_2\lambda, \ldots,$ and $p_n\lambda$, *respectively.*

Proof. The result is obtained by using the result for $n = 2$ (see Proposition 5.2 in Ross (2010)) and the induction method. This completes the proof of Proposition 2.1.3.

The decomposition method described in Proposition 2.1.3 gives the Poisson process the flexibility in modeling real stochastic processes. For instance, $N_1(t)$ can be used for counting the arrivals of men, $N_2(t)$ for the arrivals of women, $N_3(t)$ for the arrivals of couples (a man and a woman together), etc. The decomposition results are useful for interpreting events in Markovian arrival processes as well (see Sects. 2.2, 2.4, and 2.5).

Exercise 2.1.11 Families arrive at an amusement park according to a Poisson process with rate $\lambda = 25$ per hour. A family may consist of a mother and a child, a father and a child, or a mother, a father, and a child. The percentages of families of the three types are 40 %, 20 %, and 40 %. What is the arrival process of families with a mother, a father, and a child?

In a Poisson process, the arrival time of the n-th event is $S_n = X_1 + X_2 + \ldots + X_n$. According to Exercise 1.1.9, S_n has an Erlang distribution with parameters (n, λ). An interesting and useful result is the conditional distribution of S_n. More specifically, we want to find the joint conditional distribution of $\{S_1, S_2, \ldots, S_n \mid N(t) = n\}$. Choose small positive numbers $\{\delta_1, \delta_2, \ldots, \delta_n\}$. First, we have, for $0 \le t_1 < t_1 + \delta_1 < t_2 < \ldots < t_n < t_n + \delta_n \le t$,

$$\begin{aligned} &P\{S_i \in (t_i, t_i + \delta_i),\ i = 1, \ldots, n|N(t) = n\} \\ &\quad = \frac{P\{S_i \in (t_i, t_i + \delta_i),\ i = 1, \ldots, n,\ N(t) = n\}}{P\{N(t) = n\}}. \end{aligned} \tag{2.7}$$

The event considered in Eq. (2.7) is equivalent to (i) there is no arrival in $(0, t_1)$, $(t_1+\delta_1, t_2)$, …, $(t_{n-1}+\delta_{n-1}, t_n)$, and $(t_n+\delta_n, t)$; and (ii) there is exactly one arrival in each of the intervals $(t_1, t_1+\delta_1)$, …, and $(t_n, t_n+\delta_n)$. Recall that $P\{N(t_{j-1}+\delta_{j-1}, t_j) = 0\} = \exp(-\lambda(t_j - t_{j-1} - \delta_{j-1}))$, and $P\{N(t_j, t_j+\delta_j) = 1\} = \lambda\delta_j + o(\delta_j)$, for $j = 1, \ldots, n$. By Definitions 2.1.2 and 2.1.3, we have

$$\begin{aligned} &P\{S_i \in (t_i, t_i + \delta_i],\ i = 1, \ldots, n|N(t) = n\} \\ &\quad \approx \frac{e^{-\lambda t_1}\lambda\delta_1 e^{-\lambda(t_2-t_1-\delta_1)}\lambda\delta_2 \cdots e^{-\lambda(t_n-t_{n-1}-\delta_{n-1})}\lambda\delta_n e^{-\lambda(t-t_n-\delta_n)}}{e^{-\lambda t}(\lambda t)^n/n!} \\ &\quad \approx n!(\delta_1 \cdots \delta_n)/t^n. \end{aligned} \tag{2.8}$$

Letting δ_1, …, and δ_n go down to zero, the joint density function of $\{S_1, S_2, \ldots, S_n \mid N(t) = n\}$, which is equal to the probability in Eq. (2.8) divided by $\delta_1 \ldots \delta_n$, is obtained as $n!/t^n$, for fixed t.

Denote by $\{U_1, \ldots, U_n\}$ n independent random variables with a common *continuous uniform distribution* on $[0, t]$. Denote by $U_{(1)}$, …, and $U_{(n)}$ the *order statistics* of U_1, …, and U_n, i.e., $U_{(1)} \le U_{(2)} \cdots \le U_{(n)}$. For example, $U_{(1)} = \min\{U_1, \ldots, U_n\}$ and $U_{(n)} = \max\{U_1, \ldots, U_n\}$. The density function of U_j is $1/t$ in the interval $[0, t]$. The joint density function of $\{U_1, \ldots, U_n\}$ is $1/t^n$ in the set $[0, t]^n = [0, t] \times \ldots \times [0, t]$. Also note that the total number of permutations of n different numbers is $n!$. Then it is easy to show that the joint density function of $(U_{(1)}, \ldots, U_{(n)})$ is $n!/t^n$ on the set $\{(t_1, \ldots, t_n), 0 \le t_1 \le \ldots \le t_n \le t\}$. Then we have shown the following useful result.

Proposition 2.1.4 *For fixed t, random vectors* $\{S_1, S_2, \ldots, S_n \mid N(t) = n\}$ *and* $(U_{(1)}, U_{(2)}, \ldots, U_{(n)})$ *have the same joint density distribution* $n!/t^n$, *on the set* $\{(t_1, \ldots, t_n), 0 \le t_1 \le \ldots \le t_n \le t\}$, *and* 0, *otherwise.*

By Proposition 2.1.4, it is easy to find

$$\begin{aligned} &E[S_1 + S_2 + \ldots + S_n | N(t) = n] \\ &\quad = E[U_{(1)} + U_{(2)} + \ldots + U_{(n)}] = E[U_1 + U_2 + \ldots + U_n] = \frac{nt}{2}. \end{aligned} \tag{2.9}$$

Equation (2.9) shows that the average arrival time of n events is $t/2$, given that exactly n events occur in $[0, t]$. That average arrival time is independent of arrival rate λ. On the other hand, without condition $N(t) = n$, we have $E[S_1 + S_2 + \ldots + S_n] = 1/\lambda + 2/\lambda + \ldots + n/\lambda = n(n + 1)/(2\lambda)$, which depends on arrival rate λ.

Proposition 2.1.4 is handy in many cases. As an example, we use it to solve the following problem.

Example 2.1.3 Consider Poisson process $\{N(t), t \ge 0\}$. Suppose that an event occurs at time s. With probability $p(s)$, the event is marked as type 1. With probability $1-p(s)$, the event is marked as type 2. Let $N_1(t)$ be the total number of type 1 events in $[0, t]$. Let $N_2(t)$ be the total number of type 2 events in $[0, t]$. It is easy to see that $N(t) = N_1(t) + N_2(t)$ holds. Show

$$\begin{aligned} E[N_1(t)] &= \lambda \int_0^t p(s)ds; \\ E[N_2(t)] &= \lambda \int_0^t (1 - p(s))ds. \end{aligned} \tag{2.10}$$

Proposition 2.1.4 can be applied to prove Eq. (2.10). Let $I(t) = 1$, w.p. $p(t)$; 0, w.p. $1-p(t)$. Then $E[I(t)] = p(t)$. The first equality in Eq. (2.10) is obtained as follows

$$\begin{aligned} E[N_1(t)] &= \sum_{n=0}^{\infty} e^{-\lambda t} \frac{(\lambda t)^n}{n!} E[I(S_1) + \ldots + I(S_n) | N(t) = n] \\ &= \sum_{n=1}^{\infty} e^{-\lambda t} \frac{(\lambda t)^n}{n!} E[I(U_{(1)}) + \ldots + I(U_{(n)})] \\ &= \sum_{n=1}^{\infty} e^{-\lambda t} \frac{(\lambda t)^n}{n!} E[I(U_1) + \ldots + I(U_n)] \\ &= \sum_{n=1}^{\infty} e^{-\lambda t} \frac{(\lambda t)^n}{n!} n E[I(U_1)] \\ &= \left(\sum_{n=1}^{\infty} e^{-\lambda t} \frac{(\lambda t)^n}{n!} n \right) \int_0^t \frac{p(s)}{t} \mathrm{ds} \\ &= \lambda t \int_0^t \frac{p(s)}{t} \mathrm{ds}. \end{aligned} \tag{2.11}$$

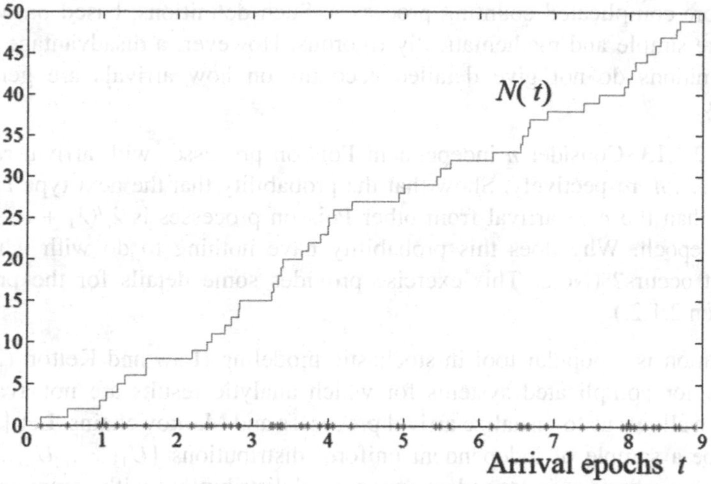

Fig. 2.5 A sample path of Poisson process $\{N(t), t \geq 0\}$ with $\lambda = 5$

The second equality in Eq. (2.10) can be proved similarly. Alternatively, the second equality in Eq. (2.10) can be obtained from the first one using $E[N(t)] = E[N_1(t)] + E[N_2(t)] = \lambda t$.

Commentary Like the exponential distribution, the Poisson process is introduced and studied in many textbooks. We refer to Ross (2010) for more about the Poisson process.

Additional Exercises and Extensions

Exercise 2.1.12 (Definition 2.1.4) Show that Poisson process $\{N(t), t \geq 0\}$ is a continuous time Markov chain. Draw a sample path of $N(t)$, given that $N(0) = 0$ (see Fig. 2.5). Explain why $\{N(t), t \geq 0\}$ is also called a *pure birth* process (a special continuous time Markov chain) with state space $\{0, 1, 2, \ldots\}$. The number of events of the Poisson process in $[s, t]$ is the number of jumps in the Markov chain in $[s, t]$. Show that the infinitesimal generator of the Markov chain is given by

$$Q = \begin{matrix} 0 \\ 1 \\ \vdots \\ \vdots \end{matrix} \begin{pmatrix} -\lambda & \lambda & & \\ & -\lambda & \lambda & \\ & & \ddots & \ddots \\ & & & \ddots \end{pmatrix}. \tag{2.12}$$

In Fig. 2.5, the horizontal axis represents the time t, and the dots on the horizontal axis indicate the arrival epochs. The vertical axis is for the number of arrivals.

Exercise 2.1.12 can also serve as a definition of the Poisson process. As will be shown in the next few sections, pure birth type Markov processes can be used to define more complicated counting processes. Such definitions, based on Markov chains, are simple and mathematically rigorous. However, a disadvantage is that such definitions do not give detailed accounts on how arrivals are generated intuitively.

Exercise 2.1.13 Consider n independent Poisson processes with arrival rates λ_i, $i = 1, 2, \ldots, n$, respectively. Show that the probability that the next type i arrival is earlier than the next arrival from other Poisson processes is $\lambda_i/(\lambda_1 + \ldots + \lambda_n)$ at any time epoch. Why does this probability have nothing to do with when the last event occurs? (Note: This exercise provides some details for the proof of Proposition 2.1.2.)

Simulation is a popular tool in stochastic modeling (Law and Kelton (2000)), especially for complicated systems for which analytic results are not available. Simulation allows us to visualize arrival processes and Markov chains. Let $\{u_1, \ldots, u_n, \ldots\}$ be a sample of independent uniform distributions $\{U_1, \ldots, U_n, \ldots\}$ on [0, 1]. Then a sample of independent exponential distributions with parameter λ can be obtained by $x_n = -\log(1-u_n)/\lambda$, $n = 1, 2, \ldots$. A sample of arrival times can be obtained as $\{x_1 + \ldots + x_n, n = 1, 2, \ldots\}$. In practice, if a sample of arrival times $\{x_1 + \ldots + x_n, n = 1, 2, \ldots, N\}$ is collected, then an *estimate* of the arrival rate λ can be calculated by $N/(x_1 + \ldots + x_N)$.

Exercise 2.1.14 Write a simulation program to visualize the arrivals of Poisson processes with $\lambda = 0.2$, 1, 2, and 5. (i) Compare the given arrival rate λ to the estimate of the arrival rate. (ii) For the sample generated by simulation, calculate $N(n, n + 1) = N(n + 1) - N(n)$, for $n = 0, 1, 2, \ldots$. Compare $N(n, n + 1)$ to λ.

We remark that, while the average arrival rate in $[0, n]$ converges to λ, the number of arrivals in any unit interval $[n, n + 1]$ can be very different from λ. This indicates that the transient (local) behavior of a stochastic process can be quite different from its steady state behavior. Thus, in system design, it is important to distinguish between long-term and short-term objectives. Throughout this book, we have a number of exercises that calculate long-term performance measures as well as short-term (local) ones. Those exercises demonstrate the difference between transient and steady-state behaviors of stochastic processes.

Exercise 2.1.15 (*Compound Poisson process*: discrete case) Buses arrive to a bus station according to a Poisson process with arrival rate λ. Each bus takes Y passengers to the station. The random variable Y has a distribution $\{p_1, p_2, \ldots, p_N\}$, where N is a positive integer. Let $Z(t)$ be the total number of passengers arrived in the bus station in $[0, t]$. Show that the probability generating function of $Z(t)$ is given by

$$E[z^{Z(t)}] = \exp\left(-\lambda t\left(1 - \sum_{n=1}^{N} p_n z^n\right)\right). \tag{2.13}$$

Show that $E[Z(t)] = \lambda t E[Y]$. Explain the result intuitively. (Hint: To prove Eq. (2.13), condition on the number of buses arrived in $[0, t]$.)

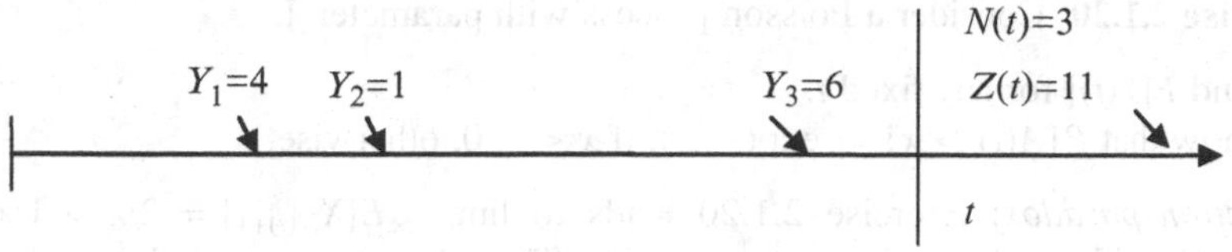

Fig. 2.6 A compound Poisson process $Z(t)$

Intuitively, a compound Poisson process can be obtained by marking a Poisson process (see Fig. 2.6). We use probability distribution $\{p_1, p_2, \ldots, p_N\}$ to determine the number of customers arrived at the arrival epochs of the Poisson process.

Exercise 2.1.16 (Compound Poisson process: continuous case) Trucks arrive to a warehouse according to a Poisson process with arrival rate λ. Each truck carries Y tons of products to the warehouse. The LST of random variable Y is $f^*(s)$. Let $Z(t)$ be the total amount of products carried to the warehouse in $[0, t]$. Show that the LST of $Z(t)$ is given by, for $s > 0$,

$$E[\exp(-sZ(t))] = \exp(-\lambda t(1 - f^*(s))). \tag{2.14}$$

Show that $E[Z(t)] = \lambda t E[Y]$. Explain the result intuitively.

Exercise 2.1.17 A *renewal process* is a counting process such that the interarrival times between consecutive customers are i.i.d.r.v.s. Show that the Poisson process is a special renewal process and find the distribution of its interarrival times. (Hint: Recall Definition 2.1.1.)

Exercise 2.1.18 (The mean queue length at time t in the $M/G/\infty$ queue) Consider an $M/G/\infty$ queue, i.e., a queueing system with an infinite number of identical servers. Customers arrive according to a Poisson process with parameter λ. As soon as a customer arrives, the customer enters a server to receive service. Service times have a common general distribution, denoted by $G(t)$. As soon as the service is completed, the customer leaves the system. Let $q(t)$ be the queue length at time t (i.e., the total number of customers in the system).

(i) Show that $E[q(t)] = \lambda \int_0^t (1 - G(s))ds$, given that the system was empty at time zero. (Hint: Mark the customer arrived at time s by its status at time $t > s$, and use Eq. (2.10).)
(ii) Show that $q(t)$ has a Poisson distribution with parameter $\lambda \int_0^t (1 - G(s))ds$, given that the system was empty at time zero.
(iii) Assume that $q(0) = k$. Find the mean and distribution of $q(t)$.

Exercise 2.1.19 Prove Example 2.1.1 by verifying Definition 2.1.3.

Define $A(t) = t - S_{N(t)}$ and $Y(t) = S_{N(t)+1} - t$. Function $A(t)$ is called *the age* of the Poisson process at time t, and $Y(t)$ is called *the excess* (*residual/remaining*) at time t. Then the interval that covers t is of the length $X_{N(t)+1} = A(t) + Y(t)$.

Exercise 2.1.20 Consider a Poisson process with parameter λ.

(1) Find $E[Y(t)]$ for any fixed t.
(2) Show that $P\{A(t) > x\} = \exp(-\lambda x)$, if $x < t$; 0, otherwise.

Inspection paradox: Exercise 2.1.20 leads to $\lim_{t\to\infty} E[X_{N(t)+1}] = 2/\lambda > 1/\lambda = E[X_1]$. Consider t as an inspection epoch. The inspection epoch has a larger probability to fall into a longer interval than a shorter one. Thus, the inspector is expected to see a longer interval. Also note that X_n in Definition 2.1.1 is exponential and has the memoryless property for any fixed n. However, $X_{N(t)+1}$ is no longer exponential for $t > 0$ (see Bladt and Nielsen (2010)).

2.2 Markovian Arrival Processes: Definitions and Examples

In this section, we take a step-by-step approach to constructing *Markovian arrival processes* (*MAPs*), which are introduced in Neuts (1979). Then we provide four definitions of Markovian arrival processes.

In general, a (continuous time) Markovian arrival process is a counting process that is defined on top of a finite state continuous time Markov chain (called the *underlying Markov chain*). Recall from Sect. 1.2 that underlying Markov chains are also used in the definition of *PH*-distributions. However, different from that for a *PH*-distribution, an underlying Markov chain for a Markovian arrival process has no absorption state (phase). A Markovian arrival process counts the number of arrivals (e.g., customers), which can be associated with changes of state (i.e., transitions) in the underlying Markov chain. The arrivals can also occur during the stay in each state of the underlying Markov chain. For an *MAP*, the transitions of state with arrival, transitions of state without arrival, and arrivals without a transition of state, are all referred to as *events*. Arrival rates of events can be customized for different states, demonstrating the versatility inherent to *MAP*s.

First, we define a continuous time Markov chain for the construction of Markovian arrival processes.

Example 2.2.1 Consider a two state continuous time Markov chain $\{I(t), t \geq 0\}$ with infinitesimal generator

$$D = \frac{1}{2}\begin{pmatrix} -2.5 & 2.5 \\ 10 & -10 \end{pmatrix}. \tag{2.15}$$

The sojourn time in state 1 has an exponential distribution with parameter 2.5 and the sojourn time in state 2 has an exponential distribution with parameter 10. After its stay in state 1 (or 2), the Markov chain transits to state 2 (or 1) with probability 1. A sample path of the Markov chain is plotted in Fig. 2.7. The stationary distribution of $\{I(t), t \geq 0\}$ can be obtained by solving the linear system $\boldsymbol{\theta} D = 0$ and $\boldsymbol{\theta}\mathbf{e} = 1$. It is easy to obtain $\boldsymbol{\theta} = (4/5, 1/5)$.

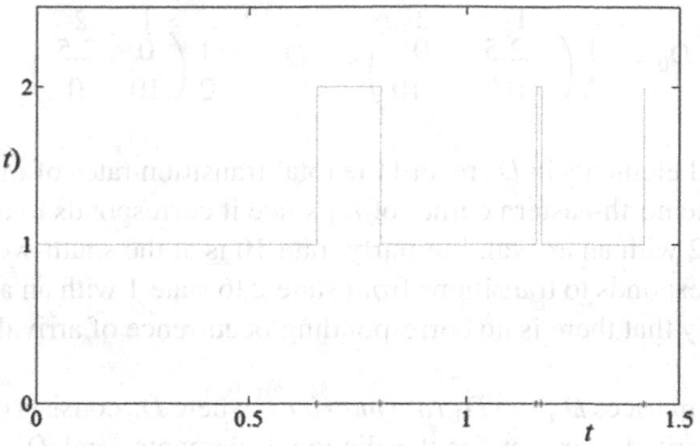

Fig. 2.7 A sample path of *MAP* $\{(N_1(t), I(t)), t \geq 0\}$

Next, we construct several Markovian arrival processes on top of the Markov chain $\{I(t), t \geq 0\}$ defined in Example 2.2.1.

Markovian arrival process 1: Whenever the Markov chain $\{I(t), t \geq 0\}$ jumps from one state to another, an arrival (e.g., the arrival of a customer) occurs. That is, an arrival occurs at every transition epoch. Let $N_1(t)$ be the number of arrivals in $[0, t]$. Then we use $\{(N_1(t), I(t)), t \geq 0\}$ to denote the *MAP*, where $N_1(t)$ records the number of arrivals and $I(t)$ keeps track of the state of the underlying Markov chain. If the initial state of $\{I(t), t \geq 0\}$ (i.e., $I(0)$) is determined (e.g., $I(0)$ has probability distribution $\{\alpha_i, i = 1, 2\}$), the *MAP* is well-defined. Figure 2.7 shows a sample path of the *MAP*, where the dots on the horizontal axis represent arrival epochs, the vertical axis is for the phases of the underlying Markov chain, and the graph presents a sample of the underlying Markov chain.

To represent the arrival process, we need (and only need) the arrival rates (recall the Poisson process). For $\{(N_1(t), I(t)), t \geq 0\}$, the arrival rates are exactly the transition rates for individual states. Let us consider what happens when the process is in state 1 by *censoring* state 2 (i.e., removing all the time periods in which the Markov chain is in state 2). It appears that arrivals occur according to a Poisson process with parameter 2.5 since the sojourn time in state 1 is exponentially distributed with parameter 2.5. Therefore, if the underlying process is in state 1, we say that the arrival rate is 2.5, which is exactly the transition rate of state 1. If the underlying Markov chain is in state 1, the time until the next arrival is exponentially distributed with parameter 2.5. Similarly, the arrival rate is 10 if the underlying process is in state 2. If the underlying Markov chain is in state 2, the time until the next arrival is exponentially distributed with parameter 10. Also note that after each arrival, the state changes and the arrival rate changes accordingly.

There are two numbers involved in our description of the arrival process: 2.5 and 10. Rate 2.5 is associated with arrivals and transitions from state 1 to state 2. Rate 10 is associated with arrivals and transitions from state 2 to state 1. Placing the rates into two matrices to reflect both the occurrence of arrival and the change of state, we obtain

$$D_0 = \begin{matrix} 1 \\ 2 \end{matrix} \overset{\begin{matrix} 1 & 2 \end{matrix}}{\begin{pmatrix} -2.5 & 0 \\ 0 & -10 \end{pmatrix}}, \quad D_1 = \begin{matrix} 1 \\ 2 \end{matrix} \overset{\begin{matrix} 1 & 2 \end{matrix}}{\begin{pmatrix} 0 & 2.5 \\ 10 & 0 \end{pmatrix}}. \tag{2.16}$$

The diagonal elements in D_0 record the total transition rates of individual states. Rate 2.5 is at the north-eastern corner of D_1 since it corresponds to transitions from state 1 to state 2 with an arrival. Similarly, rate 10 is at the south-western corner of D_1 since it corresponds to transitions from state 2 to state 1 with an arrival. Zeros in D_0 and D_1 imply that there is no corresponding occurrence of arrival and/or change of state.

We call the matrices D_0 and D_1 *rate matrices*, where D_0 consists of the transition rates without arrivals (except for the diagonal elements) and D_1 consists of the transition rates with arrivals. The 2-tuple (D_0, D_1) represents the *MAP*. It is easy to see that $D = D_0 + D_1$, which describes the change of state in the underlying Markov chain, if the arrivals are not considered.

We remark that there are a number of reasons for organizing the transition rates of *MAP* $\{(N_1(t), I(t)), t \geq 0\}$ into two matrices $\{D_0, D_1\}$. The above-mentioned reason reflects whether or not there is an arrival or a transition of state. The block representation of matrix Q in Eq. (2.17) below offers a second reason, while another reason for the computation of performance measures associated with *MAP*s will be given in Sect. 2.3. It is more important to understand the meaning of the rates than placing the rates in any particular manner.

Proposition 2.2.1 *The stochastic process* $\{(N_1(t), I(t)), t \geq 0\}$ *is a continuous time Markov chain with state space* $\{0, 1, 2, \ldots\} \times \{1, 2\}$ *and infinitesimal generator*

$$Q = \begin{matrix} (0,1) \\ (0,2) \\ (1,1) \\ (1,2) \\ \vdots \\ \vdots \end{matrix} \overset{\begin{matrix} (0,1) & (0,2) & (1,1) & (1,2) & (2,1) & (2,2) & \cdots \end{matrix}}{\begin{pmatrix} -2.5 & 0 & 0 & 2.5 & & & \\ 0 & -10 & 10 & 0 & & & \\ & & -2.5 & 0 & 0 & 2.5 & \\ & & 0 & -10 & 10 & 0 & \ddots \\ & & & & \ddots & & \ddots \\ & & & & & \ddots & \ddots \end{pmatrix}} = \begin{pmatrix} D_0 & D_1 & & \\ & D_0 & D_1 & \\ & & \ddots & \ddots \end{pmatrix}. \tag{2.17}$$

Proof. First note that the sojourn time in state $(n, 1)$, for $n \geq 0$, is exponentially distributed with parameter 2.5, and the sojourn time in state $(n, 2)$ is exponentially distributed with parameter 10. At the end of its stay in state $(n, 1)$, the Markov chain transits to $(n + 1, 2)$ with probability one. At the end of its stay in state $(n, 2)$,

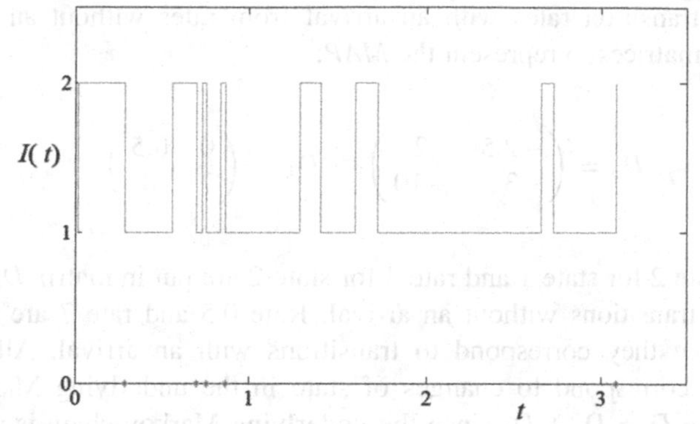

Fig. 2.8 A sample path of *MAP* $\{(N_2(t), I(t)), t \geq 0\}$

the Markov chain transits to $(n + 1, 1)$ with probability one. By definition, the stochastic process is a continuous time Markov chain. The transition rate from $(n, 1)$ to $(n + 1, 2)$ is 2.5, and 10 from $(n, 2)$ to $(n + 1, 1)$, as shown in Eq. (2.17). This completes the proof of Proposition 2.2.1.

Exercise 2.2.1 Use Proposition 2.2.1 to find $P\{N_1(t) \leq 2 \mid (N_1(0), I(0)) = (0, 1)\}$. (Note: Use knowledge about continuous time Markov chains only.) Find $P\{N_1(t) = 2 \mid (N_1(0), I(0)) = (0, 1)\}$.

Markovian arrival process 2: When the continuous time Markov chain $\{I(t), t \geq 0\}$ transits from state 1 to state 2, an arrival occurs with probability 0.2; and no arrival occurs with probability 0.8. When the Markov chain transits from state 2 to state 1, an arrival occurs with probability 0.7; and no arrival occurs with probability 0.3. We use $\{(N_2(t), I(t)), t \geq 0\}$ to represent this *MAP*, where $N_2(t)$ records the number of arrivals in $[0, t]$. A sample path of the *MAP* is shown in Fig. 2.8.

To represent this *MAP*, we have a look at the arrival rates for individual states.

- When the underlying Markov chain is in state 1, the total transition rate is 2.5; the rate of transitions with an arrival is $0.2 \times 2.5 = 0.5$; and the rate of transitions without an arrival is $0.8 \times 2.5 = 2$. The arrival rate can also be obtained as follows. First we censor the time spent in state 2. The time spent in state 1 is divided into exponential times with parameter 2.5. At the end of each exponential time, a Bernoulli trial is conducted. With probability 0.2, there is an arrival. According to Proposition 1.1.4, the time until a customer arrives has an exponential distribution with parameter $0.2 \times 2.5 = 0.5$. Alternatively, by Proposition 2.1.3, transitions from 1 to 2 can be marked into 2 types: transitions with an arrival and transitions without an arrival. By censoring state 2, arrivals form a Poisson process with arrival rate $0.2 \times 2.5 = 0.5$. Independently, transitions without an arrival form a Poisson process with rate $0.8 \times 2.5 = 2$.
- When the underlying Markov chain is in state 2, the total rate of transitions is 10; the rate of transitions with an arrival is $0.7 \times 10 = 7$; and the rate of transitions without an arrival is $0.3 \times 10 = 3$.

Separating transition rates with an arrival from rates without an arrival, we introduce two matrices to represent the *MAP*:

$$D_0 = \begin{pmatrix} -2.5 & 2 \\ 3 & -10 \end{pmatrix}, \quad D_1 = \begin{pmatrix} 0 & 0.5 \\ 7 & 0 \end{pmatrix}. \tag{2.18}$$

Note that rate 2 for state 1 and rate 3 for state 2 are put in matrix D_0 since they correspond to transitions without an arrival. Rate 0.5 and rate 7 are included in matrix D_1 since they correspond to transitions with an arrival. All four rates {2, 3, 0.5, 7} correspond to changes of state in the underlying Markov chain. Again, we have $D = D_0 + D_1$ since the underlying Markov chain is not affected by the arrivals. Similar to $\{(N_1(t), I(t)), t \geq 0\}$, $\{(N_2(t), I(t)), t \geq 0\}$ is also a continuous time Markov chain with state space $\{0, 1, 2, \ldots\} \times \{1, 2\}$ and infinitesimal generator

$$Q = \begin{array}{c} (0,1) \\ (0,2) \\ (1,1) \\ (1,2) \\ \vdots \end{array} \begin{array}{c} \begin{array}{cccccccc} (0,1) & (0,2) & (1,1) & (1,2) & (2,1) & (2,2) & \cdots \end{array} \\ \begin{pmatrix} -2.5 & 2 & 0 & 0.5 & & & \\ 3 & -10 & 7 & 0 & & & \\ & & -2.5 & 2 & 0 & 0.5 & \\ & & 3 & -10 & 7 & 0 & \ddots \\ & & & & \ddots & \ddots & \ddots \\ & & & & & \ddots & \ddots \end{pmatrix} \end{array}$$
$$= \begin{pmatrix} D_0 & D_1 & & \\ & D_0 & D_1 & \\ & & \ddots & \ddots \end{pmatrix}. \tag{2.19}$$

Since there may not be an arrival at a transition of state, in the above Markov chain, there are transitions from $(n, 1)$ to $(n, 2)$ and from $(n, 2)$ to $(n, 1)$.

Markovian arrival process 3: During each period that the Markov chain $\{I(t), t \geq 0\}$ is in state 1, we define a Poisson process with rate 4 for arrivals. Such arrivals are not accompanied by a change of state in the underlying Markov chain. When the underlying Markov chain transits to state 2, the Poisson process defined on state 1 is turned off. It is turned on whenever the underlying Markov chain returns to state 1. Arrivals do not occur in state 2 nor are there any arrivals associated with transitions from state 2 to state 1. We use $\{(N_3(t), I(t)), t \geq 0\}$ to represent this *MAP*, where $N_3(t)$ records the number of arrivals in $[0, t]$. A sample path of $\{(N_3(t), I(t)), t \geq 0\}$ is plotted in Fig. 2.9.

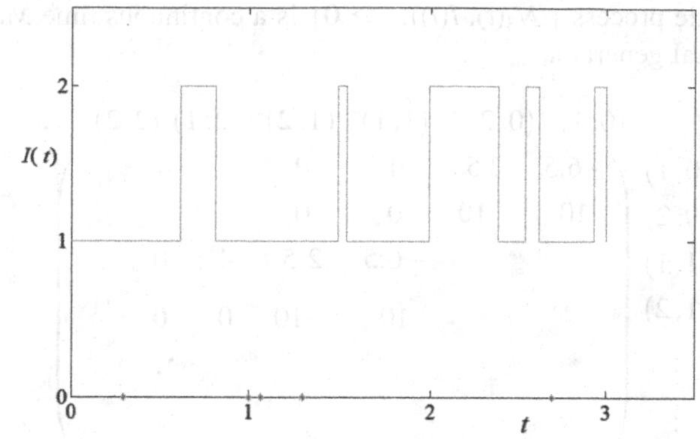

Fig. 2.9 A sample path of *MAP* $\{(N_3(t), I(t)), t \geq 0\}$

In state 1, the time until the next transition of the underlying Markov chain, denoted by X_1, is exponentially distributed with parameter 2.5, and the time until the next arrival of the Poisson process, denoted by Z_1, is exponentially distributed with parameter 4. Then the time until the next event (an arrival or a transition) is $\min(X_1, Z_1)$. By Proposition 1.1.2, $\min(X_1, Z_1)$ is exponentially distributed with parameter $2.5 + 4 = 6.5$, which is the total rate of events in state 1. By Proposition 1.1.3, the next *event* is an arrival with probability 4/6.5, and the next event is a transition of state with probability 2.5/6.5. Consequently, the rate for an arrival is $6.5 \times (4/6.5) = 4$ and the rate for a transition is $6.5 \times (2.5/6.5) = 2.5$. In state 2, there is no imposed Poisson arrival. Thus, the total rate of events is 10, which is the rate for transitions from state 2 to state 1 without an arrival.

Putting all the rates into two matrices, we obtain

$$D_0 = \begin{pmatrix} -2.5-4 & 2.5 \\ 10 & -10 \end{pmatrix}, \quad D_1 = \begin{pmatrix} 4 & 0 \\ 0 & 0 \end{pmatrix}. \tag{2.20}$$

Rate 4 is at the north-western corner of matrix D_1 since it corresponds to the events of arrivals from the imposed Poisson process without a change of state. Rate 4 is also added to the transition rate of state 1 at the north-western corner of D_0 since the total rate of events in state 1 is $2.5 + 4$. This implies that the sojourn time in state 1 has an exponential distribution with parameter 6.5. Note that the actual sojourn time in state 1 for the underlying Markov chain $\{I(t), t \geq 0\}$ is still exponentially distributed with parameter 2.5 since the underlying Markov chain is not affected by the imposed Poisson process.

The arrival event in state 1 can be observed in the Markovian arrival process, but cannot be observed in the underlying Markov chain. In fact, by definition, such events occur within a state and have no effect on the transitions of the underlying Markov chain. Therefore, we still have $D = D_1 + D_2$.

The stochastic process $\{(N_3(t), I(t)), t \geq 0\}$ is a continuous time Markov chain with infinitesimal generator

$$Q = \begin{array}{c} (0,1) \\ (0,2) \\ (1,1) \\ (1,2) \\ \vdots \\ \vdots \end{array} \begin{pmatrix} -6.5 & 2.5 & 4 & 0 & & & \\ 10 & -10 & 0 & 0 & & & \\ & & -6.5 & 2.5 & 4 & 0 & \\ & & 10 & -10 & 0 & 0 & \ddots \\ & & & & \ddots & \ddots & \ddots \\ & & & & & \ddots & \ddots \end{pmatrix} \quad \text{columns: } (0,1)\ (0,2)\ (1,1)\ (1,2)\ (2,1)\ (2,2) \ldots$$

$$= \begin{pmatrix} D_0 & D_1 & & \\ & D_0 & D_1 & \\ & & \ddots & \ddots \end{pmatrix}. \tag{2.21}$$

Exercise 2.2.2 Show that a Poisson process with parameter λ is an *MAP* defined on a single-state Markov chain and infinitesimal generator $D = (0)$. Consequently, the matrix representation of the Poisson process is $(D_0 = (-\lambda), D_1 = (\lambda))$.

Markovian arrival process 4: Combining Markovian arrival processes $\{(N_2(t), I(t)), t \geq 0\}$ and $\{(N_3(t), I(t)), t \geq 0\}$, we obtain a new *MAP* $\{(N_4(t), I(t)), t \geq 0\}$, where $N_4(t) = N_2(t) + N_3(t)$. Putting all the rates without arrivals (where there must be a change of state) into D_0 and all the rates with arrivals (where there could be no change of state) into D_1, we obtain

$$D_0 = \begin{pmatrix} -2.5-4 & 2 \\ 3 & -10 \end{pmatrix}, \quad D_1 = \begin{pmatrix} 4 & 0.5 \\ 7 & 0 \end{pmatrix}. \tag{2.22}$$

For this new *MAP*, there are arrivals in state 1 according to a Poisson process with parameter 4. There are also arrivals at state transition epochs. If there is a transition from state 1 to state 2, there is no arrival with probability 2/2.5, and there is an arrival with probability 0.5/2.5. There is no arrival during each stay in state 2. If there is a transition from state 2 to state 1, there is no arrival with probability 3/10, and there is an arrival with probability 7/10. A sample path of $\{(N_4(t), I(t)), t \geq 0\}$ can be found in Fig. 2.10.

The *MAP* $\{(N_4(t), I(t)), t \geq 0\}$ can also be interpreted in the following ways.

- When the Markov chain is in state 1, the total rate of events is 6.5. Censoring out state 2, we obtain a Poisson process with arrival rate 6.5. Mark the Poisson process into three types of events: an arrival with probability 4/6.5; a transition from state 1 to 2 with an arrival with probability $0.2 \times 2.5/6.5$; and a transition from state 1 to 2 without an arrival with probability $0.8 \times 2.5/6.5$. Consequently, the arrival rates of the three types of events are 4, 0.5, and 2, respectively.

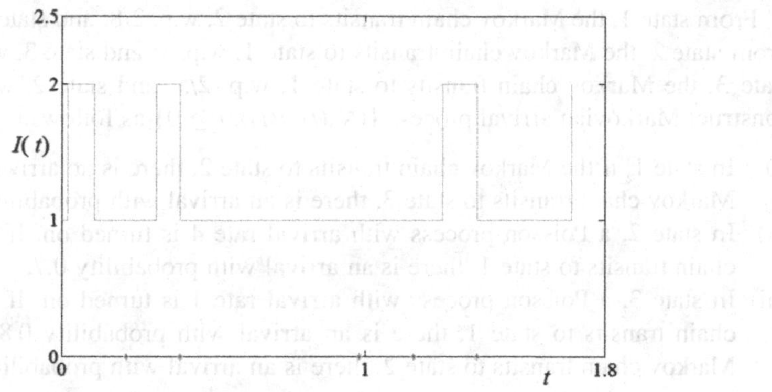

Fig. 2.10 A sample path of *MAP* $\{(N_4(t), I(t)), t \geq 0\}$

- When the Markov chain is in state 2, the total rate of events is 10. Censoring out state 1, we obtain a Poisson process with arrival rate 10. When there is an event, the probability that there is an arrival and the state remains in 2 is 0/10 (the rate for such events is 0); the probability that there is an arrival and the state changes to 1 is 7/10 (the rate for such events is 7); and the probability that there is no arrival and the state changes to 1 is 3/10 (the rate for such events is 3).

Similar to the previous three cases, the stochastic process $\{(N_4(t), I(t)), t \geq 0\}$ is a continuous time Markov chain with infinitesimal generator

$$
\begin{array}{c} \\ (0,1)\\(0,2)\\(1,1)\\Q=\quad(1,2)\\(2,1)\\(2,2)\end{array}
\begin{array}{c}
\begin{array}{ccccccc} (0,1) & (0,2) & (1,1) & (1,2) & (2,1) & (2,2) & \cdots \end{array}\\
\begin{pmatrix}
-6.5 & 2 & 4 & 0.5 & & & \\
3 & -10 & 7 & 0 & & & \\
 & & -6.5 & 2 & 4 & 0.5 & \\
 & & 3 & -10 & 7 & 0 & \ddots \\
 & & & & \ddots & \ddots & \ddots \\
 & & & & & \ddots & \ddots
\end{pmatrix}
\end{array}
\tag{2.23}
$$

$$
= \begin{pmatrix} D_0 & D_1 & \\ & D_0 & D_1 \\ & & \ddots & \ddots \end{pmatrix}
$$

Now, we consider a more complicated case.

Example 2.2.2 Consider a three-state continuous time Markov chain $\{I(t), t \geq 0\}$ with

$$
D = \begin{pmatrix} -8 & 2 & 6 \\ 10 & -10 & 0 \\ 2 & 3 & -5 \end{pmatrix}. \tag{2.24}
$$

From state 1, the Markov chain transits to state 2, w.p. 2/8; and state 3, w.p. 6/8. From state 2, the Markov chain transits to state 1, w.p. 1; and state 3, w.p. 0. From state 3, the Markov chain transits to state 1, w.p. 2/5; and state 2, w.p. 3/5. We construct Markovian arrival process $\{(N_5(t), I(t)), t \geq 0\}$ as follows.

(i) In state 1, if the Markov chain transits to state 2, there is no arrival; and if the Markov chain transits to state 3, there is an arrival with probability 0.4.
(ii) In state 2, a Poisson process with arrival rate 4 is turned on. If the Markov chain transits to state 1, there is an arrival with probability 0.7.
(iii) In state 3, a Poisson process with arrival rate 1 is turned on. If the Markov chain transits to state 1, there is an arrival with probability 0.8; and if the Markov chain transits to state 2, there is an arrival with probability 0.5.

Based on the above construction, the total rates of events in states 1, 2, and 3 are 8 + 0, 10 + 4, and 5 + 1, respectively. The transition and arrival rates are given as follows.

1. In state 1, the transition rate to state 2 without an arrival is 2, to state 3 without an arrival is $0.6 \times 6 = 3.6$, to state 1 with an arrival is 0, to state 2 with an arrival is 0, and to state 3 with an arrival is $0.4 \times 6 = 2.4$.
2. In state 2, the transition rate to state 1 without an arrival is $0.3 \times 10 = 3$, to state 3 without an arrival is 0, to state 1 with an arrival is $0.7 \times 10 = 7$, to state 2 with an arrival is 4, and to state 3 with an arrival is 0.
3. In state 3, the transition rate to state 1 without an arrival is $0.2 \times 2 = 0.4$, to state 2 without an arrival is $0.5 \times 3 = 1.5$, to state 1 with an arrival is $0.8 \times 2 = 1.6$, to state 2 with an arrival is $0.5 \times 3 = 1.5$, and to state 3 with an arrival is 1.

Summarizing the transition rates into two matrices, we obtain a matrix representation for the Markovian arrival process:

$$D_0 = \begin{pmatrix} -8 & 2 & 3.6 \\ 3 & -10-4 & 0 \\ 0.4 & 1.5 & -5-1 \end{pmatrix}, \quad D_1 = \begin{pmatrix} 0 & 0 & 2.4 \\ 7 & 4 & 0 \\ 1.6 & 1.5 & 1 \end{pmatrix}. \qquad (2.25)$$

Again, the arrivals within each state have no effect on the underlying Markov chain. Therefore, we have $D = D_1 + D_2$.

MAPs $\{(N_1(t), I(t)), t \geq 0\}$ to $\{(N_5(t), I(t)), t \geq 0\}$ demonstrate that the events with arrivals can occur

(i) In individual states according to Poisson processes; and
(ii) When the state of the underlying Markov chain changes.

In fact, these are the only two ways to generate arrivals in all *MAP*s. As stated in the second paragraph of this section, in an *MAP*, an event can be (i) a transition of state only; (ii) a transition with an arrival; or (iii) an arrival without a change of state. The first formal definition of *MAP*s is based on this observation.

Definition 2.2.1 Given matrices $\{D_0, D_1\}$ of order m, where m is a positive integer, all the elements of the two matrices are nonnegative except the diagonal elements of D_0, which are negative, and $D = D_0 + D_1$ is an infinitesimal generator. Let $D_0 = (d_{0,(i,j)})$ and $D_1 = (d_{1,(i,j)})$. Then (D_0, D_1) defines *MAP* $\{(N(t), I(t)), t \geq 0\}$ as follows.

1. Set $N(0) = 0$.
2. Define a continuous time Markov chain $\{I(t), t \geq 0\}$ by D.
3. For phase (state) i with $d_{1,(i,i)} > 0$, define a Poisson process with arrival rate $d_{1,(i,i)}$, for $i = 1, 2, \ldots, m$. The Poisson process is turned on, if $I(t) = i$; and is turned off, otherwise.
4. If $I(t) = i$ and an arrival from the imposed Poisson process occurs, $N(t)$ increases by one, for $i = 1, 2, \ldots, m$.
5. At the end of each stay in state i, with probability $d_{0,(i,j)}/(-d_{(i,i)})$ (note that $d_{(i,i)} = d_{0,(i,i)} + d_{1,(i,i)}$), $I(t)$ transits from phase i to j and $N(t)$ remains the same (i.e., without an arrival); and, with probability $d_{1,(i,j)}/(-d_{(i,i)})$, $I(t)$ transits from phase i to j and $N(t)$ increases by one (i.e., with an arrival), for $i \neq j$, and $i, j = 1, \ldots, m$.

We call (D_0, D_1) a matrix representation and $\{I(t), t \geq 0\}$ the underlying Markov chain. We shall also call the state of the underlying Markov chain the phase. If the initial phase of $\{I(t), t \geq 0\}$ (i.e., $I(0)$) is determined (e.g., use a probability distribution $\{\alpha_i, i = 1, \ldots, m\}$), the *MAP* is well-defined.

The above definition of *MAP*s can be restated in terms of transition and arrival rates as follows: $d_{0,(i,j)}$ is the rate of transitions from phase i to j without an arrival, for $i \neq j$; $d_{1,(i,j)}$ is the rate of transitions from phase i to j with an arrival; and $-d_{0,(i,i)}$ is the total rate of events in phase i. The use of transition rates in defining *MAP*s becomes handy when introducing Markov chains for stochastic systems.

Example 2.2.3 The following are some well-known *MAP*s.

(1) The Poisson process (see Exercise 2.2.2) with parameter λ: $D_0 = (-\lambda)$, $D_1 = (\lambda)$, $D = (0)$, and $m = 1$. Note that $d_{(1,1)} = 0$ since there is only one phase and there is no phase transition.

(2) An *Erlang renewal process* with m phases is defined as a (standard) *renewal process* for which interarrival times are independent and identically distributed Erlang random variables with parameters m and λ. Using the *PH*-representation of the Erlang distribution, a matrix representation of this arrival process can be given as $(D_0 = \lambda J(m), D_1 = \lambda K(m))$, where

$$J(m) = \begin{pmatrix} -1 & 1 & & \\ & -1 & \ddots & \\ & & \ddots & 1 \\ & & & -1 \end{pmatrix}_{m\times m}, \quad K(m) = \begin{pmatrix} 0 & 0 & & \\ & 0 & \ddots & \\ & & \ddots & 0 \\ 1 & & & 0 \end{pmatrix}_{m\times m}. \tag{2.26}$$

A sample path of the Erlang renewal process is plotted in Fig. 2.11.

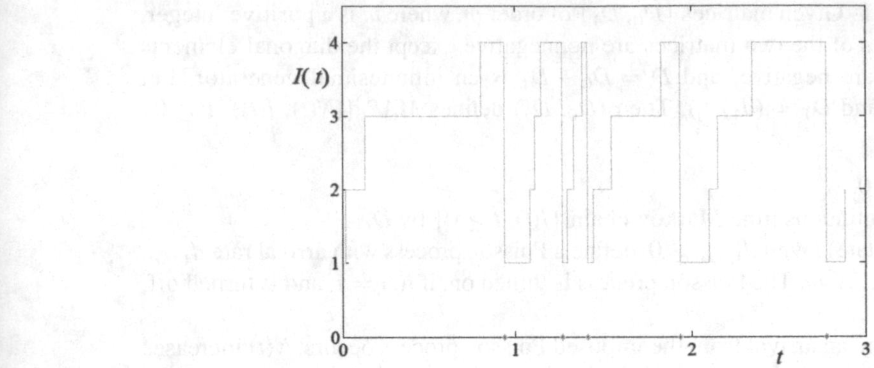

Fig. 2.11 A sample path of an Erlang renewal process with $m = 4$

(3) *PH*-renewal process: A (standard) *renewal process* whose interarrival times are independent and have a common *PH*-distribution ($\boldsymbol{\alpha}$, T) with $\boldsymbol{\alpha}\mathbf{e} = 1$. A matrix representation of the *MAP* is

$$D_0 = T, \quad D_1 = \mathbf{T}^0\boldsymbol{\alpha}. \tag{2.27}$$

The infinitesimal generator of its underlying Markov chain is $T + \mathbf{T}^0\boldsymbol{\alpha}$. Transitions associated with T have no arrival, while transitions associated with $\mathbf{T}^0\boldsymbol{\alpha}$ have an arrival. Read the discussion before Lemma 1.3.1 for an intuitive explanation of the matrix representation.

(4) A bursty arrival process:

$$D_0 = \begin{pmatrix} -10 & 0 \\ 1 & -1 \end{pmatrix}, \quad D_1 = \begin{pmatrix} 9 & 1 \\ 0 & 0 \end{pmatrix}. \tag{2.28}$$

When the underlying Markov chain is in state 1, a number of arrivals can occur in a short period of time.

(5) A *Markov switched Poisson process* (*MSPP*):

$$D_0 = \begin{pmatrix} -\lambda_1 & & & \\ & -\lambda_2 & & \\ & & \ddots & \\ & & & -\lambda_m \end{pmatrix}, \quad D_1 = (d_{i,j}). \tag{2.29}$$

By definition, $D_0 + D_1$ must be an infinitesimal generator. In this *MAP*, arrivals switch between m Poisson processes with arrival rates $\{\lambda_1, \lambda_2, \ldots, \lambda_m\}$. After each arrival, the *MAP* may switch to another Poisson process (i.e., the phase is changed) or stay in the same Poisson process (i.e., the phase remains the same). For example,

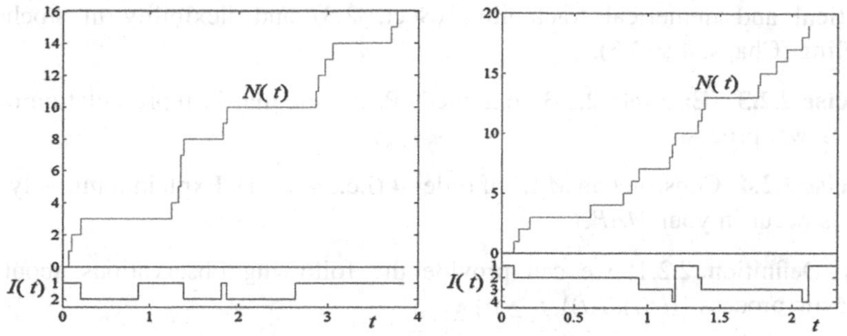

Fig. 2.12 Sample paths of $N(t)$ and $I(t)$ for (4) and (5) of Example 2.2.3

$$D_0 = \begin{pmatrix} -10 & 0 & 0 & 0 \\ 0 & -8 & 0 & 0 \\ 0 & 0 & -10 & 0 \\ 0 & 0 & 0 & -8 \end{pmatrix}, \quad D_1 = \begin{pmatrix} 0 & 10 & 0 & 0 \\ 0 & 7 & 1 & 0 \\ 0 & 0 & 9 & 1 \\ 8 & 0 & 0 & 0 \end{pmatrix}. \tag{2.30}$$

Figure 2.12 presents sample paths of the underlying Markov chain (lower portion of the figures) and the counting process (upper portion of the figures).

(6) A *Markov modulated Poisson process* (*MMPP*): $D_0 = D - \Lambda$, $D_1 = \Lambda$, where Λ is a diagonal matrix with positive diagonal elements. In this *MAP*, arrivals occur within individual phases. No arrival is associated with the change of phase. Thus, the arrival rates are modulated by continuous time Markov chain D. For example,

$$D = \begin{pmatrix} -10 & 10 \\ 2 & -2 \end{pmatrix}, \quad \Lambda = \begin{pmatrix} 4 & 0 \\ 0 & 7 \end{pmatrix}. \tag{2.31}$$

For this example, the Poisson arrival rates are 4 and 7 for phases 1 and 2, respectively. *MMPP*s are *MAP*s with only one type of arrivals. That is, arrivals come only from imposed Poisson processes.

Compared to Poisson processes (see Fig. 2.5), the *MAP* defined in (4) of Example 2.2.3 is significantly more bursty. That is, the process has some long silent periods without any arrivals and some short periods with a large number of arrivals. This can be explained using the transition rates in matrices D_0 and D_1. In phase 1, the process stays for a very short period before the next event. However, the next event is most likely an arrival and the process stays in phase 1. Therefore, in a very short period of time, it is possible to have many arrivals until the process transits to phase 2. In phase 2, the process stays there (on average) for a long time before going back to phase 1. There is no arrival during that long period.

The flexibility to generate arrival processes with different features is one of the main reasons that *MAP*s can be used widely in practice. Other reasons include analytical and numerical tractability (Sect. 2.3) and flexibility in stochastic modeling (Chaps. 4 and 5).

Exercise 2.2.3 (Example 2.2.3 continued) Prove the matrix representation of the *PH*-renewal process.

Exercise 2.2.4 Construct an *MAP* of order 4 (i.e., $m = 4$). Explain intuitively how arrivals occur in your *MAP*.

By Definition 2.2.1, we can provide the following observations about the stochastic process $\{(N(t), I(t)), t \geq 0\}$

(i) In state (n, i), the next transition occurs after an exponential time and the next arrival from the imposed Poisson process in phase i also comes after an exponential time. According to Proposition 1.1.2, the sojourn time in state (n, i) (i.e., the time until the next event: a transition only, an arrival, or a transition plus an arrival) is exponentially distributed with parameter $-d_{0,(i,i)}$, for $n \geq 0$ and $i = 1, \ldots, m$.
(ii) At the end of its stay in state (n, i), the process $\{(N(t), I(t)), t \geq 0\}$ transits to state (n, j) with probability $d_{0,(i,j)}/(-d_{0,(i,i)})$, for $i \neq j$, and $i, j = 1, \ldots, m$; and to state $(n + 1, j)$ with probability $d_{1,(i,j)}/(-d_{0,(i,i)})$, for $i, j = 1, \ldots, m$.

By the definition of continuous time Markov chains and the above observations, it is easy to see that $\{(N(t), I(t)), t \geq 0\}$ is a continuous time Markov chain, which leads to the second definition of *MAP*s.

Definition 2.2.2 Define $\{(N(t), I(t)), t \geq 0\}$ as a continuous time Markov chain with state space $\{0, 1, 2, \ldots\}\times\{1, 2, \ldots, m\}$, where m is a positive integer, and infinitesimal generator

$$Q = \begin{pmatrix} D_0 & D_1 & & \\ & D_0 & D_1 & \\ & & \ddots & \ddots \\ & & & \ddots \end{pmatrix}. \tag{2.32}$$

Assume that $N(0) = 0$. Then $\{(N(t), I(t)), t \geq 0\}$ is called an *MAP* with matrix representation (D_0, D_1).

According to the definition of continuous time Markov chains, D_0 and D_1 in matrix Q must satisfy the conditions given in Definition 2.2.1. It is also clear that $N(t)$ is nondecreasing and forms a counting process. Definition 2.2.2 gives a reason to call the process $\{(N(t), I(t)), t \geq 0\}$ a *Markov modulated counting process*.

Exercise 2.2.5 Show that, for the Markov chain defined in Definition 2.2.2, (i) the sojourn time in phase i is exponentially distributed with parameter $d_{(i,i)}$, for $i = 1, 2, \ldots, m$, where $D = (d_{(i,i)}) = D_0 + D_1$ (Hint: Use Proposition 1.1.4); and (ii) $\{I(t), t \geq 0\}$ is a continuous time Markov chain with infinitesimal generator D.

Let τ_n be the interarrival time between the $(n-1)$-st and n-th arrivals of an *MAP* with matrix representation (D_0, D_1). Let I_n be the phase right after the n-th arrival, i.e., $I_n = I(\tau_1+\tau_2 + \ldots+\tau_n + 0)$. Since there can be events without arrivals between consecutive arrivals, in general, τ_n may not be exponentially distributed. Next, we find the joint distribution of τ_n and I_n.

If $N(t) = 0$, the transitions of Markov chain Q is governed by D_0 in $(0, t)$ (see Eq. (2.32)). Thus, we have $P\{I(t) = j, N(t) = 0 \mid I(0) = i\} = (\exp(D_0t))_{i,j}$. Consequently, we obtain $P\{I(t) = j, \tau_1 > t \mid I(0) = i\} = P\{I(t) = j, N(t) = 0 \mid I(0) = i\} = (\exp(D_0t))_{i,j}$. For the Poisson process, in an infinitesimally small interval, it is well-known that the probability of having one event is proportional to the length of the interval, say δt, and the probability of having two or more events is $o(\delta t)$. The event $\{I_1 = j, \tau_1 < \delta t \mid I(0) = i\}$ implies that at least one arrival event takes place in $(0, \delta t)$, which is accompanied by a transition from phase i to j. Then we must have $P\{I_1 = j,\ \tau_1 < \delta t \mid I(0) = i\} = d_{1,(i,j)}\delta t + o(\delta t)$, for $i, j = 1, 2, \ldots, m$. By the Markovian property, the event $\{I_1 = j, \tau_1 < \delta t \mid I(0) = i\}$ is equivalent to $\{I_n = j, \tau_n < \delta t \mid I(\tau_1+\tau_2 + \ldots+\tau_{n-1} + 0) = i\}$. The phase transitions between two consecutive arrivals are governed by D_0 as well.

By conditioning on the phase at time $\tau_1+\tau_2 + \ldots+\tau_{n-1} + t$, the following equation can be established, for $i, j = 1, 2, \ldots, m$,

$$\begin{aligned}
&P\{I_n = j,\ t < \tau_n \le t+\delta t\ |I_{n-1} = i\} \\
&\quad = P\{I(\tau_1 + \ldots + \tau_{n-1} + \tau_n + 0) = j,\ t < \tau_n \le t+\delta t\ |I(\tau_1 + \ldots + \tau_{n-1} + 0) = i\} \\
&\quad = P\{I(\tau_n + 0) = j,\ t < \tau_n \le t+\delta t\ |I(0) = i\} \\
&\quad = \sum_{k=1}^{m} P\{I(t) = k,\ \tau_n > t|I(0) = i\}P\{I(\tau_n + 0) = j,\ t < \tau_n \le t+\delta t\ |I(t) = k\} \\
&\quad = \sum_{k=1}^{m} (\exp(D_0t))_{i,k}(d_{1,(k,j)}\delta t + o(\delta t)) \\
&\quad = (\exp(D_0t)D_1)_{i,j}\delta t + o(\delta t).
\end{aligned} \tag{2.33}$$

By integration over t, Eq. (2.33) leads to

$$P\{I_n = j,\ \tau_n \le t\ |I_{n-1} = i\} = \left(\int_0^t \exp(D_0x)\mathrm{d}xD_1\right)_{i,j}. \tag{2.34}$$

By definition (C'inlar (1969)), the stochastic process $\{(I_n, \tau_n), n = 0, 1, 2, \ldots\}$ is a *Markov renewal process*, which leads to the third definition of *MAP*s.

Definition 2.2.3 Assume that (D_0, D_1) satisfies conditions in Definition 2.2.1. Define a Markov renewal process $\{(I_n, \tau_n), n = 0, 1, 2, \ldots\}$ by Eq. (2.34) and the corresponding continuous time Markov chain $\{I(t), t \ge 0\}$ with infinitesimal generator $D = D_0 + D_1$. Define $N(t)$ as $N(0) = 0$, and $\{N(t) \leq n\} = \{\tau_1+\tau_2+\ldots+\tau_{n+1} > t\}$, for $t > 0$. The stochastic process $\{(N(t), I(t)), t \geq 0\}$ is called a Markovian arrival process.

Exercise 2.2.6 Assume that D is irreducible and D_1 is nonzero. Show that the matrix $-D_0$ is an M-matrix (see Exercise 1.2.24) and D_0 is invertible. Find $P\{I_n = j, \tau_n < \infty \mid I_{n-1} = i\} = \left(\int_0^\infty \exp(D_0 t)\mathrm{d}tD_1\right)_{i,j}$ explicitly. Explain $-D_0^{-1}D_1$ and $-D_0^{-1}D_1\mathbf{e} = \mathbf{e}$ probabilistically.

Exercise 2.2.7 For the Erlang renewal process defined in Eq. (2.26), assume $m = 5$. If $I(0) = 3$, find the distribution of the time until the first arrival. If $I(0)$ has an initial distribution $\boldsymbol{\alpha}$, find the distribution of the time until the first arrival.

Exercise 2.2.8 Show that, in an *MAP*, given the current phase, the time until the next arrival has a *PH*-distribution. Find a *PH*-representation for that *PH*-distribution. If $I(t)$ has distribution $\boldsymbol{\alpha}$, what is the distribution of the time until the first arrival after t?

Similar to the Poisson process, *MAP*s can be superposed or decomposed.

Superposition of independent Markovian arrival processes Assume that $\{(N_1(t), I_1(t)), t \geq 0\}$ and $\{(N_2(t), I_2(t)), t \geq 0\}$ are independent Markovian arrival processes with matrix representations (D_0, D_1) and (C_0, C_1), respectively. Define $I(t) = (I_1(t), I_2(t))$. Then $\{(N_1(t) + N_2(t), I(t)), t \geq 0\}$ is a Markovian arrival process with matrix representation

$$(D_0 \otimes I + I \otimes C_0, D_1 \otimes I + I \otimes C_1). \tag{2.35}$$

Proposition 2.2.2 *The superposition of two independent Markovian arrival processes is a Markovian arrival process whose matrix representation can be constructed using the Kronecker product given in* Eq. (2.35).

Exercise 2.2.9 Prove the matrix representation in Eq. (2.35) probabilistically, and show $(D_0\otimes I + I\otimes C_0 + D_1\otimes I + I\otimes C_1)\mathbf{e} = 0$. (Hint: Use the memoryless property in each phase.)

Example 2.2.4 Here is an example for the superposition of *MAP*s, for which

$$\begin{aligned} D_0 &= \begin{pmatrix} -10 & 0 \\ 1 & -1 \end{pmatrix}, \quad D_1 = \begin{pmatrix} 9 & 1 \\ 0 & 0 \end{pmatrix}, \\ C_0 &= \begin{pmatrix} -5 & 1 \\ 2 & -7 \end{pmatrix}, \quad C_1 = \begin{pmatrix} 0 & 4 \\ 2 & 3 \end{pmatrix}. \end{aligned} \tag{2.36}$$

Then we have

$$D_0 \otimes I + I \otimes C_0$$
$$= \begin{pmatrix} -10 & 0 & 0 & 0 \\ 0 & -10 & 0 & 0 \\ 1 & 0 & -1 & 0 \\ 0 & 1 & 0 & -1 \end{pmatrix} + \begin{pmatrix} -5 & 1 & 0 & 0 \\ 2 & -7 & 0 & 0 \\ 0 & 0 & -5 & 1 \\ 0 & 0 & 2 & -7 \end{pmatrix}$$
$$= \begin{pmatrix} -15 & 1 & 0 & 0 \\ 2 & -17 & 0 & 0 \\ 1 & 0 & -6 & 1 \\ 0 & 1 & 2 & -8 \end{pmatrix},$$
$$D_1 \otimes I + I \otimes C_1$$
$$= \begin{pmatrix} 9 & 0 & 1 & 0 \\ 0 & 9 & 0 & 1 \\ 0 & 0 & 0 & 0 \\ 0 & 0 & 0 & 0 \end{pmatrix} + \begin{pmatrix} 0 & 4 & 0 & 0 \\ 2 & 3 & 0 & 0 \\ 0 & 0 & 0 & 4 \\ 0 & 0 & 2 & 3 \end{pmatrix} = \begin{pmatrix} 9 & 4 & 1 & 0 \\ 2 & 12 & 0 & 1 \\ 0 & 0 & 0 & 4 \\ 0 & 0 & 2 & 3 \end{pmatrix}.$$

Exercise 2.2.10 Outline a recursive method to find a matrix representation of the superposition process of n independent *MAP*s.

Decomposition (marking or thinning) of Markovian arrival processes Assume that $\{(N(t), I(t)), t \geq 0\}$ is a Markovian arrival process with matrix representation (D_0, D_1). We mark arrivals into type 1 and type 2 independently by probabilities $\{p, 1-p\}$. We obtain a Markovian arrival process with marked arrivals (see Sect. 2.5) $\{(N_1(t), N_2(t), I(t)), t \geq 0\}$ with matrix representation $(D_0, pD_1, (1-p)D_1)$. Process $\{(N_1(t), I(t)), t \geq 0\}$ (or $\{(N_2(t), I(t)), t \geq 0\}$) is a Markovian arrival process with matrix representation $(D_0 + (1-p)D_1, pD_1)$ (or $(D_0 + pD_1, (1-p)D_1)$). The reason is that, if there is an arrival associated with rate $(D_1)_{i,j}$, with probability p, this arrival is marked as type 1 and the corresponding rate is $p(D_1)_{i,j}$; and with probability $1-p$, the arrival is marked as type 2 and the corresponding rate is $(1-p)(D_1)_{i,j}$. For the first case, the rate $(1-p)D_1$ is added to D_0 since it does not bring a type 1 arrival. Note that the two Markovian arrival processes $\{(N_1(t), I(t)), t \geq 0\}$ and $\{(N_2(t), I(t)), t \geq 0\}$ may not be independent. The reason is that, for both marked processes, future arrival rates depend on the phase of the (same) underlying Markov chain. Recall that for the Poisson process, however, such a decomposition leads to two independent Poisson processes (see Proposition 2.1.3).

Proposition 2.2.3 *Assume that* $\{(N(t), I(t)), t \geq 0\}$ *is a Markovian arrival process with matrix representation* (D_0, D_1)*. We decompose the MAP into n types of arrivals using independent markings with probabilities* $\{p_1, p_2, \ldots, p_n\}$*. Then process* $\{(N_j(t), I(t)), t \geq 0\}$ *is a Markovian arrival process with matrix representation* $(D_0 + (1-p_j)D_1, p_jD_1)$*, for* $j = 1, \ldots, n$.

Proposition 2.2.3 gives a special example of the Marked *MAP* introduced in Sect. 2.5.

Commentary We refer to Rudemo (1973), Neuts (1979), Lucantoni et al. (1990), and Lucantoni (1991), for more details on *MAP*s. Several survey papers summarize the development and applications of *MAP*s since 1979 (e.g., Chakravarthy (2001, 2010), Artalejo et al. (2010), and Ramaswami (2010)). *MAP*s are similar to the *world driven* demand processes in the literature of inventory theory (Song and Zipkin (1993)).

Additional Exercises and Extensions

Simulation has been used in Chap. 1 to visualize the absorption process of *PH*-distributions (see Exercise 1.3.16) and in Sect. 2.1 to visualize the Poisson process. To simulate a Markovian arrival process, we need to keep track of the phase, sojourn time, and type of events. Generate sojourn times by using parameters $\{-d_{0,(i,i)},\ i = 1, \ldots, m\}$ for exponential distributions and the relationship $X = -\log(U)/(-d_{0,(i,i)})$, where U is the uniform random variable on [0, 1]. Determine the type of an event and the phase right after the event by using the matrix

$$P = [(\mathrm{diag}(-D_0))^{-1}(D_0, D_1) + (I,\ 0)]J, \tag{2.37}$$

where $\mathrm{diag}(-D_0)$ is obtained by setting all off-diagonal elements of $-D_0$ to zero, J is a matrix with all elements in the upper triangular part and on the diagonal being one, and elements in the lower triangular part being zero. The matrix P is an $m\times(2m)$ matrix. Each row in P gives a cumulative probability distribution for states $\{(i, j)_0, j = 1, \ldots, m, i \neq j;\ (i, j)_1, j = 1, \ldots, m\}$, for $i = 1, 2, \ldots, m$. For fixed i, state $(i, j)_0$ represents a phase transition from i to j without an arrival; and state $(i, j)_1$ represents a phase transition from i to j with an arrival.

Exercise 2.2.11 (Example 2.2.3 continued) Write a simulation program to visualize the *MAP*s defined in Example 2.2.3. Plot the phase process $\{I(t), t \geq 0\}$ and the arrival process $\{N(t), t \geq 0\}$, and estimate the arrival rate for each *MAP*.

Exercise 2.2.12 Consider an *MAP* with matrix representation (D_0, D_1). Define $P = (p_{i,j}) = -D_0^{-1}D_1$, where $p_{i,j}$ is the probability that, given that the phase process $\{I(t), t\geq 0\}$ is in i initially, the phase becomes j right after the next arrival. Let $\boldsymbol{\pi}$ satisfy $\boldsymbol{\pi}P = \boldsymbol{\pi}$ and $\boldsymbol{\pi}\mathbf{e} = 1$. The vector $\boldsymbol{\pi}$ is the stationary distribution of the phase right after arrivals. In steady state, the interarrival times have a *PH*-distribution $(\boldsymbol{\pi}, D_0)$. (i) In steady state, find the distribution of the interarrival times for an Erlang renewal process with m phases; (ii) In steady state, find the distribution of the interarrival times for a *PH*-renewal process with *PH*-representation $(\boldsymbol{\alpha}, T)$. For both cases, find P and $\boldsymbol{\pi}$. Explain your results intuitively.

The next example demonstrates that, by introducing phases properly, a wide range of *MAP*s can be constructed.

Exercise 2.2.13 (Stochastic modeling) Consider a stochastic process $\{(N(t), I(t)), t \geq 0\}$ (which may not be Markovian) with two states $\{1, 2\}$ for $I(t)$ such that the sojourn time in state j has general distribution $F_j(t)$. Arrivals are generated in state 1 with a constant interarrival time c. Note: whenever the underlying process enters state 1, the arrival process is re-initialized. Apparently, the process is not an *MAP*. First, by Theorem 1.2.1, we can find *PH*-distributions $(\boldsymbol{\alpha}_1, T_1)$ and $(\boldsymbol{\alpha}_2, T_2)$ to approximate $F_1(t)$ and $F_2(t)$, respectively. Second, we use Erlang distribution Erlang($n = 5$, $\lambda = 5/c$) to approximate constant c (Note: This method is called the *Erlangization method* in stochastic modeling (e.g., Asmussen et al. (2002), Ramaswami et al. (2008), and Stanford et al. (2011))). As the order n increases, the Erlang distribution becomes closer to the constant c probabilistically. Denote by $I_1(t)$ and $I_2(t)$ the phases of the underlying Markov chains for the two phase-type distributions, and $I_3(t)$ the phase of the Erlang distribution if $I(t) = 1$. Then we introduce *MAP* $\{(N(t), I(t), I_1(t), I_2(t), I_3(t)), t \geq 0\}$. Show that a matrix representation of the *MAP* is given by

$$D_0 = \begin{pmatrix} T_1 \otimes I + I \otimes \left(\frac{5}{c} J(5)\right) & (\mathbf{T}_1^0 \otimes \mathbf{e})\boldsymbol{\alpha}_1 \\ \mathbf{T}_2^0(\boldsymbol{\alpha}_2 \otimes (1, 0, \ldots, 0)) & T_2 \end{pmatrix},$$
$$D_1 = \begin{pmatrix} I \otimes \left(\frac{5}{c} K(5)\right) & 0 \\ 0 & 0 \end{pmatrix}, \tag{2.38}$$

where matrices $J(5)$ and $K(5)$ are defined in Eq. (2.26). The *MAP* approximates the original stochastic process. As the order of the Erlang distribution increases, the approximation becomes more accurate.

Exercise 2.2.13 gives a concrete example to demonstrate that Markovian arrival processes can approximate non-Markovian arrival processes. Exercise 2.2.13 also indicates that the set of *MAP*s is versatile, which will be confirmed by Theorem 2.4.3.

Exercise 2.2.14 (Non-uniqueness of the matrix representation) Show that the following two *PH*-renewal processes represent the same Poisson process with arrival rate 1:

$$\text{(i) } D_0 = \begin{pmatrix} -2 & 1 \\ 2 & -3 \end{pmatrix}, \quad D_1 = \begin{pmatrix} 0.2 & 0.8 \\ 0.2 & 0.8 \end{pmatrix}; \text{ and}$$
$$\text{(ii) } D_0 = \begin{pmatrix} -2 & 2 \\ 2 & -5 \end{pmatrix}, \quad D_1 = \begin{pmatrix} 0 & 0 \\ 2 & 1 \end{pmatrix}. \tag{2.39}$$

(Hint: First, show that the two Markovian arrival processes are *PH*-renewal processes. Then show that the interarrival times have an exponential distribution (see Example 1.4.1). For case (ii), ignore the first arrival.)

Exercise 2.2.15 (Reducibility of the underlying Markov chain) Consider an *MAP* with matrix representation

$$D_0 = \begin{pmatrix} -2 & 0 \\ 2 & -5 \end{pmatrix} \text{ and } D_1 = \begin{pmatrix} 2 & 0 \\ 2 & 1 \end{pmatrix}. \tag{2.40}$$

What is special about phase 1? Is the underlying Markov chain *D* irreducible?

Exercise 2.2.16 Consider an *MAP* with matrix representation

$$D_0 = \begin{pmatrix} -2 & 2 & 0 \\ 1 & -1 & 0 \\ 0 & 1 & -4 \end{pmatrix} \text{ and } D_1 = \begin{pmatrix} 0 & 0 & 0 \\ 0 & 0 & 0 \\ 1 & 1 & 1 \end{pmatrix}. \tag{2.41}$$

What is special about phases {1, 2}? Is the underlying Markov chain *D* irreducible?

Continuous time *MAP*s can be defined (purely) based on Poisson processes as follows.

Exercise 2.2.17 (Definition 2.2.4) (He (2010)) Let $\{\alpha_i, 1 \le i \le m\}$ be nonnegative numbers with a unit sum, $\{d_{0,(i,j)}, i \neq j \text{ and } i, j = 1, \ldots, m\}$ and $\{d_{1,(i,j)}, i, j = 1, \ldots, m\}$ be nonnegative numbers, and *m* be a finite positive integer. Assume $-d_{0,(i,i)} \equiv d_{0,(i,1)} + \ldots + d_{0,(i,\ i-1)} + d_{0,(i,i+1)} + \ldots + d_{0,(i,m)} + d_{1,(i,1)} + \ldots + d_{1,(i,i)} + \ldots + d_{1,(i,m)} > 0$, for $i = 1, \ldots, m$. We define stochastic process $\{(N(t), I(t)), t{\ge}0\}$ as follows.

1. Define $m(2m-1)$ independent Poisson processes with parameters $\{d_{0,(i,j)}, i \neq j$ and $i, j = 1, \ldots, m\}$ and $\{d_{1,(i,j)}, i, j = 1, \ldots, m\}$. If $d_{0,(i,j)} = 0$ or $d_{1,(i,j)} = 0$, the corresponding Poisson process has no event.
2. Determine $I(0)$ by the probability distribution $\{\alpha_i, i = 1, \ldots, m\}$. Set $N(0) = 0$.
3. If $I(t) = i$, for $1 \le i \le m$, $I(t)$ and $N(t)$ remain the same until the first event occurs in the $2m-1$ Poisson processes corresponding to $\{d_{0,(i,j)}, j = 1, \ldots, m, j \neq i\}$ and $\{d_{1,(i,j)}, j = 1, \ldots, m\}$. If the next event comes from the Poisson process corresponding to $d_{0,(i,j)}$, the phase variable $I(t)$ transits from phase *i* to phase *j* and $N(t)$ does not change at the epoch, for $j = 1, \ldots, m$, and $j \neq i$. If the next event comes from the Poisson process corresponding to $d_{1,(i,j)}$, the phase variable $I(t)$ transits from phase *i* to phase *j* and $N(t)$ increases by one at the epoch, i.e., an arrival occurs, for $j = 1, \ldots, m$.

For the process $\{(N(t), I(t)), t{\ge}0\}$, show the following results.

(i) Prove that the sojourn time of $\{(N(t), I(t)), t{\ge}0\}$ in state (n, i) has an exponential distribution with parameter $-d_{0,(i,i)}$, for $i = 1, \ldots, m$.

(ii) Prove that the probability that the next phase is *j* and no arrival occurs at the transition epoch is given by $d_{0,(i,j)}/(-d_{0,(i,i)})$, given that the current state is (n, i), for $i \neq j$ and $i, j = 1, \ldots, m$, and $n = 0, 1, 2, \ldots$.

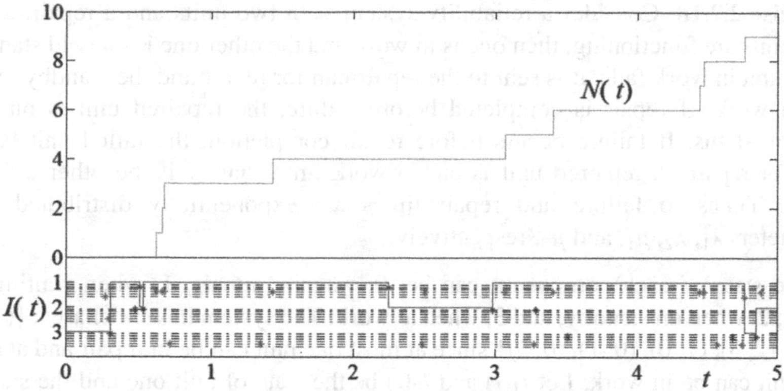

Fig. 2.13 Sample paths of underlying Poisson processes, $I(t)$, and $N(t)$ of an *MAP*

(iii) Prove that the probability that the next phase is j and an arrival occurs at the transition epoch is given by $d_{1,(i,j)}/(-d_{0,(i,i)})$, given that the current state is (n, i), for $i, j = 1, \ldots, m$ and $n = 0, 1, 2, \ldots$.
(iv) Prove that the sojourn time of $\{I(t), t\geq 0\}$ in phase i has an exponential distribution with parameter $-d_{0,(i,i)} + (-d_{1,(i,i)})$, for $i = 1, \ldots, m$. (Hint: Use Proposition 1.1.2.)
(v) Let $D_0 = (d_{0,(i,j)})$, $D_1 = (d_{1,(i,j)})$. Show that $\{I(t), t\geq 0\}$ is a continuous time Markov chain with infinitesimal generator $D = D_0 + D_1$.
(vi) Show that $\{(N(t), I(t)), t\geq 0\}$ is a continuous time Markov chain with the infinitesimal generator given in Eq. (2.32).

Consequently, $\{(N(t), I(t)), t\geq 0\}$ is an *MAP* with matrix-representation (D_0, D_1). Definition 2.2.4 provides an intuitive interpretation to the parameters of *MAP*s and is handy in stochastic modeling. Definition 2.2.4 also provides a straightforward method to simulate *MAP*s.

In Fig. 2.13, $N(t)$ and $I(t)$ are simulated for a process with $m = 3$. The function $N(t)$ increases by one if and only if (i) one of the 15 Poisson processes generates an arrival; and (ii) the Poisson process is indexed by $I(t)$. For each phase, there should be $2m - 1 = 5$ Poisson processes. Two of the blue colored Poisson processes for each phase are for transitions without an arrival. The three red colored Poisson processes for each phase are for arrivals. Again, arrivals of the Poisson processes are counted in $N(t)$ only if it is indexed as $I(t)$ at its arrival time epoch.

We remark that we do not formally prove that Definitions 2.2.1, 2.2.2, 2.2.3, and 2.2.4 are equivalent definitions of *MAP*s. On the other hand, some of the exercises in this section are sufficient for readers to see that the four definitions are indeed equivalent. Among the definitions, 2.2.1 is a constructive definition based on a Markov chain and Poisson processes; 2.2.2 is a "macro" definition based on a Markov chain; 2.2.3 is a "micro" definition based on a Markov renewal process; and 2.2.4 is a constructive definition based on Poisson processes.

We use the following examples to demonstrate the use of Definition 2.2.4.

Exercise 2.2.18 Consider a reliability system with two units and a repairman. If both units are functioning, then one is in work and the other one is on cold standby. If the unit in work fails, it is sent to the repairman for repair and the standby unit is put in work. If repair is completed before failure, the repaired unit is on cold standby status. If failure occurs before repair completion, the failed unit has to wait for repair. A repaired unit is put in work immediately if the other unit has failed. Times to failure and repair times are exponentially distributed with parameters λ_1, λ_2, μ_1, and μ_2, respectively.

The state of each component can be: 0, when in repair; 1, when waiting for repair; 2, when on standby; and 3, when at work. The system has six phases: {(3, 2), (3, 0), (2, 3), (1, 0), (0, 1), (0, 3)} since at most one unit can be in repair and at most one unit can be in work. Let $I_1(t)$ and $I_2(t)$ be the state of unit one and the state of unit two at time t, respectively. Process $\{I(t) = (I_1(t), I_2(t)), t \geq 0\}$ can be defined by the following underlying Poisson processes.

We are interested in the number of repairs in [0, t]. We define the events of Poisson processes associated with repair rates as arrivals (corresponding to D_1). Then the rest of the Poisson processes correspond to D_0. Let $N(t)$ be the number of such arrivals associated with continuous time Markov chain $\{I(t), t \geq 0\}$. Then it is easy to see that $\{(N(t), I(t)), t \geq 0\}$ is an *MAP* with matrix representation

$$D_0 = \begin{matrix} (3,2) \\ (2,3) \\ (3,0) \\ (0,3) \\ (1,0) \\ (0,1) \end{matrix} \begin{pmatrix} -\lambda_1 & 0 & 0 & \lambda_1 & 0 & 0 \\ 0 & -\lambda_2 & \lambda_2 & 0 & 0 & 0 \\ 0 & 0 & -\lambda_1-\mu_2 & 0 & \lambda_1 & 0 \\ 0 & 0 & 0 & -\lambda_2-\mu_1 & 0 & \lambda_2 \\ 0 & 0 & 0 & 0 & -\mu_2 & 0 \\ 0 & 0 & 0 & 0 & 0 & -\mu_1 \end{pmatrix}, \qquad (2.42)$$

$$D_1 = \begin{matrix} (3,2) \\ (2,3) \\ (3,0) \\ (0,3) \\ (1,0) \\ (0,1) \end{matrix} \begin{pmatrix} 0 & 0 & 0 & 0 & 0 & 0 \\ 0 & 0 & 0 & 0 & 0 & 0 \\ \mu_2 & 0 & 0 & 0 & 0 & 0 \\ 0 & \mu_1 & 0 & 0 & 0 & 0 \\ 0 & 0 & 0 & \mu_2 & 0 & 0 \\ 0 & 0 & \mu_1 & 0 & 0 & 0 \end{pmatrix}.$$

Exercise 2.2.19 (Exercise 2.2.18 continued) For the model introduced in Exercise 2.2.18, assume that the two units are identical (i.e., the same distribution for the repair time and the same distribution for the time to failure). We are interested in the number of repairs in [0, t]. We define the events of Poisson processes associated with repair rates as arrivals (corresponding to D_1). Then the rest of the Poisson processes correspond to D_0. Let $N(t)$ be the number of such arrivals associated with continuous time Markov chain $\{I(t), t{\geq}0\}$. Show that $\{(N(t), I(t)), t{\geq}0\}$ is an *MAP* with matrix representation

$$D_0 = \begin{array}{c} \\ (3,2) \\ (3,0) \\ (1,0) \end{array} \begin{array}{c} \begin{array}{ccc} (3,2) & (3,0) & (1,0) \end{array} \\ \begin{pmatrix} -\lambda & \lambda & 0 \\ 0 & -\lambda-\mu & \lambda \\ 0 & 0 & -\mu \end{pmatrix} \end{array}, \quad D_1 = \begin{array}{c} \\ (3,2) \\ (3,0) \\ (1,0) \end{array} \begin{array}{c} \begin{array}{ccc} (3,2) & (3,0) & (1,0) \end{array} \\ \begin{pmatrix} 0 & 0 & 0 \\ \mu & 0 & 0 \\ 0 & \mu & 0 \end{pmatrix} \end{array}. \tag{2.43}$$

Exercise 2.2.20* Generalize Exercises 2.2.18 and 2.2.19 to systems, for which the times to failure and repair times are *PH*-distributed.

Exercise 2.2.21 (A shoe-shine shop) Customers arrive to a shoe-shine shop according to a Poisson process with parameter λ. The shoe-shine shop has one worker. The shoe-shine process consists of two stages. Each stage has a chair. If an arriving customer finds an empty chair at the first stage, the customer will take the seat. If the worker is available, the worker will start the first stage service. The first stage service takes an exponential time with parameter μ_1 to complete. When the first stage service is completed, the customer moves into another chair for the second stage service. The service time for the second stage has an exponential time with parameter μ_2. There is no other waiting seat in the shop. If both seats are occupied, an arriving customer leaves the system without service immediately. (1) Introduce an *MAP* to count the number of customers who have their shoes shined. (2) Introduce an *MAP* to count the number of customers who leave the system without having their shoes shined.

Exercise 2.2.22 Generalize Exercise 2.2.21 to a system with *MAP* arrivals and *PH*-service times.

Exercise 2.2.23 (Periodic arrival process) Arrivals of patients to a hospital demonstrate periodicity on weekdays. We construct the following arrival process, which offers different arrival rates and patterns for 5 days in a week.

$$D_0 = \begin{pmatrix} nJ(n)+D_{(1,0)} & nK(n) & & & \\ & nJ(n)+D_{(2,0)} & nK(n) & & \\ & & nJ(n)+D_{(3,0)} & nK(n) & \\ & & & nJ(n)+D_{(4,0)} & nK(n) \\ nK(n) & & & & nJ(n)+D_{(5,0)} \end{pmatrix},$$

$$D_1 = \begin{pmatrix} D_{(1,1)} & & & & \\ & D_{(2,1)} & & & \\ & & D_{(3,1)} & & \\ & & & D_{(4,1)} & \\ & & & & D_{(5,1)} \end{pmatrix}, \tag{2.44}$$

where $J(n)$ and $K(n)$ are defined in Eq. (2.26), n is a positive integer, $(D_{(k,\,0)}, D_{(k,\,1)})$ is the matrix representation of an *MAP*, for $k = 1, 2, 3, 4$, and 5. Explain the matrix representation intuitively.

Exercise 2.2.24 For an *MAP* with (D_0, D_1), we assume that the initial distribution of the phase process is given by $\boldsymbol{\alpha}$ (i.e., the distribution of $I(0)$ is $\boldsymbol{\alpha}$). Show that the arrival time of the third customer has a *PH*-distribution by finding a *PH*-representation for it.

Exercise 2.2.25 For an *MAP* with (D_0, D_1), we assume that the initial distribution of the phase process is given by $\boldsymbol{\alpha}$. Show that the distribution of excess $Y(t)$ (see Exercise 2.1.20) is of phase-type with *PH*-representation $(\boldsymbol{\alpha}\exp((D_0 + D_1)t), D_0)$. Assume that the matrix representation (D_0, D_1) is irreducible. Show that the *PH*-representation of $\lim_{t\to\infty}Y(t)$ is $(\boldsymbol{\theta}, D_0)$, where $\boldsymbol{\theta}$ satisfies $\boldsymbol{\theta}(D_0 + D_1) = 0$ and $\boldsymbol{\theta}\mathbf{e} = 1$.

Exercise 2.2.26* Consider an *MAP* with (D_0, D_1). Assume that (D_0, D_1) is irreducible. Let $\boldsymbol{\theta}$ satisfy $\boldsymbol{\theta}(D_0 + D_1) = 0$ and $\boldsymbol{\theta}\mathbf{e} = 1$. In steady state, the *reversed underlying Markov chain* (see Andersen et al. (2004)) has an infinitesimal generator $\text{diag}^{-1}(\boldsymbol{\theta})(D_0 + D_1)'\text{diag}(\boldsymbol{\theta})$, where $(D_0 + D_1)'$ is the transpose of $D_0 + D_1$. Consider the age $A(t)$ and excess $Y(t)$ defined in Exercise 2.1.20. Show that, in steady state, $A(t)$ has a *PH*-distribution with *PH*-representation $(\boldsymbol{\theta}, \text{diag}^{-1}(\boldsymbol{\theta})D_0'\text{diag}(\boldsymbol{\theta}))$. Show that, in steady state, the mean of the interarrival time $A(t) + Y(t)$ covering time epoch t is given by $-2\boldsymbol{\theta}D_0^{-1}\mathbf{e}$.

2.3 Markovian Arrival Processes: Performance Measures

Since an *MAP* is a continuous time Markov chain, it can be analyzed using existing theories on Markov chains. On the other hand, *MAP*s, as Markov chains, have a special structure. Taking advantages of such a special structure makes the analysis much more efficient. Based on conditional probabilities and transform methods (the LST and the probability generating function), explicit expressions for some performance measures of *MAP*s are derived in this section. Our focus is on the number of arrivals $N(t)$. We find the (conditional) mean and variance of $N(t)$. As will be shown, the phase of the underlying process $I(t)$ plays a crucial role in the analysis.

For *MAP* $\{(N(t), I(t)), t \geq 0\}$ with matrix representation (D_0, D_1), define

$$\begin{aligned} p_{i,j}(n,t) &= P\{N(t) = n,\ I(t) = j | I(0) = i\}, \quad i,j = 1,\ldots,m. \\ P(n,t) &= (p_{i,j}(n,t))_{m\times m}. \end{aligned} \tag{2.45}$$

Note that $P(0, 0) = I$, and $P(n, 0) = 0$, for $n = 1, 2, \ldots$, and $p_{i,j}(n, t)$ is the probability that there are n arrivals in $[0, t]$ and the underlying Markov chain is in phase j at time t, given that the Markov chain is initially in phase i.

Recall that the sojourn time in state (n, i) is exponentially distributed. Conditioning on the number of arrivals in $(t, t+\delta t)$, we obtain, for small δt,

$$p_{i,j}(n,t+\delta t) = p_{i,j}(n,t)\big(1-(-d_{0,(j,j)})\delta t\big) + \sum_{k=1:k\neq j}^{m} p_{i,k}(n,t)d_{0,(k,j)}\delta t \\ + \sum_{k=1}^{m} p_{i,k}(n-1,t)d_{1,(k,j)}\delta t + o(\delta t), \tag{2.46}$$

where $p_{i,j}(n,t)\big(1-(-d_{0,(j,j)})\delta t\big)$ corresponds to the event that there are no arrivals and no changes of phase in the interval (t, $t+\delta t$); $\sum_{k=1:k\neq j}^{m} p_{i,k}(n,t)d_{0,(k,j)}\delta t$ corresponds to the event that there are no arrivals in the interval (t, $t+\delta t$) and the phase changes exactly once, which is from k to j; $\sum_{k=1}^{m} p_{i,k}(n-1,t)d_{1,(k,j)}\delta t$ corresponds to the event that an arrival occurs in the interval (t, $t+\delta t$) and the phase changes exactly once, which is from phase k to j (notice that $k=j$ is possible for this case since arrivals may occur within a phase), and $o(\delta t)$ contains the probabilities of all other events. Letting δt go down to zero, we obtain, for $i, j = 1, \ldots, m$,

$$\begin{aligned}\frac{\mathrm{d}p_{i,j}(n,t)}{\mathrm{d}t} &= \lim_{\delta t\to 0+} \frac{p_{i,j}(n,t+\delta t) - p_{i,j}(n,t)}{\delta t} \\ &= p_{i,j}(n,t)d_{0,(j,j)} + \sum_{k=1:k\neq j}^{m} p_{i,k}(n,t)d_{0,(k,j)} + \sum_{k=1}^{m} p_{i,k}(n-1,t)d_{1,(k,j)}.\end{aligned} \tag{2.47}$$

By writing Eq. (2.47) in a matrix form, we obtain

Lemma 2.3.1

$$\begin{aligned}\frac{\mathrm{d}P(0,t)}{\mathrm{d}t} &= P(0,t)D_0; \\ \frac{\mathrm{d}P(n,t)}{\mathrm{d}t} &= P(n,t)D_0 + P(n-1,t)D_1, \quad n = 1, 2, \ldots.\end{aligned} \tag{2.48}$$

Note that the elegant matrix form differential equation (2.48) is obtained from Eq. (2.47) only if the transition rates are properly organized in matrices D_0 and D_1. This is another reason why the transition rates are placed in D_0 and D_1 in a unique way.

Define $P^*(z,t) = \sum_{n=0}^{\infty} z^n P(n,t)$, the conditional probability generating function of $N(t)$, for $z \geq 0$. By Eq. (2.48) and routine calculations, we obtain the following result.

Theorem 2.3.1 *For MAP* (D_0, D_1), *we have, for* $z \geq 0$,

$$P^*(z,t) = \exp((D_0 + zD_1)t). \tag{2.49}$$

Proof. Multiplying by z^n on both sides of Eq. (2.48) and summing up both sides, yields

$$\begin{aligned}\sum_{n=0}^{\infty} z^n \frac{\mathrm{d}P(n,t)}{\mathrm{d}t} &= \sum_{n=0}^{\infty} z^n P(n,t)D_0 + z\sum_{n=1}^{\infty} z^{n-1}P(n-1,t)D_1 \\ \Rightarrow \quad \frac{\mathrm{d}P^*(z,t)}{\mathrm{d}t} &= P^*(z,t)D_0 + zP^*(z,t)D_1 \\ &= P^*(z,t)(D_0 + zD_1).\end{aligned} \tag{2.50}$$

Note that $P^*(z, 0) = I$ since $P(0, 0) = I$, and $P(n, 0) = 0$, for $n = 1, 2, \ldots$. Equation (2.49) is obtained directly from Eq. (2.50). This completes the proof of Theorem 2.3.1.

Exercise 2.3.1 Find $P^*(z, t)$ for (i) a Poisson process with parameter λ; and (ii) an Erlang arrival process with $m = 2$.

The number of arrivals in $[0, t]$ depends on the initial phase. By definition, the i-th element of $P^*(z, t)\mathbf{e}$ is the probability generating function of the number of arrivals in $[0, t]$, given $I(0) = i$. Let $\boldsymbol{\alpha}$ be the initial distribution of the underlying Markov chain (i.e., the distribution of $I(0)$). Then the probability generating function of $N(t)$ is given by $\boldsymbol{\alpha}P^*(z,t)\mathbf{e}$.

Exercise 2.3.2 Find $\boldsymbol{\alpha}P^*(z,t)\mathbf{e}$ for Erlang renewal processes.

Exercise 2.3.3 Let $D^*(z) = D_0 + zD_1$. Using the generating function $P^*(z, t)$ for the Poisson process (see Exercise 2.1.8), explain $P^*(z,t) = \exp(D^*(z)t)$ intuitively.

By Theorem 2.3.1, we can find moments of $N(t)$. For example, the mean of $N(t)$ can be found explicitly as follows. Define the *stationary arrival rate* as $\lambda = \boldsymbol{\theta}D_1\mathbf{e}$, where $\boldsymbol{\theta}$ is the stationary distribution vector of D, i.e., $\boldsymbol{\theta}D = 0$ and $\boldsymbol{\theta}\mathbf{e} = 1$. At an arbitrary time, if the phase is i, the arrival rate at the moment is the i-th element of $D_1\mathbf{e}$. By conditioning on the phase at that epoch, the average arrival rate at the epoch is $\boldsymbol{\theta}D_1\mathbf{e}$. This explains why λ is called the stationary arrival rate.

Note: If D is irreducible, then $\boldsymbol{\theta}$ is unique and can be calculated as follows. Replace the first column of D with $\mathbf{e}$ and denote the resultant matrix as C. Then $\boldsymbol{\theta}$ can be obtained by solving equation $\boldsymbol{\theta}C = (1,0,\ldots,0)$. Consequently, $\boldsymbol{\theta}$ is the first row of C^{-1}. That the matrix C is invertible can be shown as follows. If C is singular, there exists nonzero vector $\mathbf{u}$ such that $\mathbf{u}C = 0$. Postmultiplying by $\mathbf{e}$ on both sides of $\mathbf{u}C = 0$, we obtain $\mathbf{ue} - \mathbf{ud}(1) = 0$, where $\mathbf{d}(1)$ is the first column of D. From $\mathbf{u}C = 0$, we obtain $\mathbf{u}D + (\mathbf{ue}-\mathbf{ud}(1), 0, \ldots, 0) = \mathbf{u}D = 0$. This implies that all elements of $\mathbf{u}$ must have the same sign (since D is irreducible). Then $\mathbf{ue} \neq 0$. On the other hand, $\mathbf{u}D = 0$ implies $\mathbf{ud}(1) = 0$, which contradicts $\mathbf{ue} = \mathbf{ud}(1) \neq 0$. Therefore, C is invertible and $\boldsymbol{\theta}$ can be found uniquely.

Theorem 2.3.2 *Assume that the underlying Markov chain $\{I(t), t \geq 0\}$ is irreducible, i.e., D is irreducible. Given initial distribution $\boldsymbol{\alpha}$, we have*

$$E[N(t)] = \lambda t + \boldsymbol{\alpha}(\exp(Dt) - I)(D - \mathbf{e}\boldsymbol{\theta})^{-1}D_1\mathbf{e}, \quad t \geq 0. \tag{2.51}$$

Proof. We use the probability generating function $P^*(z, t)$ to find the mean. By Theorem 2.3.1, it is easy to obtain

$$\begin{aligned}
\left.\frac{\partial P^*(z,t)}{\partial z}\right|_{z=1}
&= \sum_{n=0}^{\infty}\frac{t^n}{n!}\left.\frac{\mathrm{d}(D_0+zD_1)^n}{\mathrm{d}z}\right|_{z=1} \\
&= \sum_{n=1}^{\infty}\frac{t^n}{n!}\sum_{k=0}^{n-1}(D_0+zD_1)^k\left.\frac{\mathrm{d}(D_0+zD_1)}{\mathrm{d}z}(D_0+zD_1)^{n-k-1}\right|_{z=1} \\
&= \sum_{n=1}^{\infty}\frac{t^n}{n!}\sum_{k=0}^{n-1}D^kD_1D^{n-1-k}.
\end{aligned} \tag{2.52}$$

(See Exercise 2.3.31 for derivatives of matrices.) Postmultiplying by $\mathbf{e}$ on both sides of Eq. (2.52) and using $D\mathbf{e} = 0$, yields

$$\left.\frac{\partial P^*(z,t)}{\partial z}\right|_{z=1}\mathbf{e} = \left(\sum_{n=1}^{\infty}\frac{t^n}{n!}D^{n-1}\right)D_1\mathbf{e}. \tag{2.53}$$

To simplify the right-hand side of Eq. (2.53), we postmultiply the first part of the right hand side of Eq. (2.53) by $D-\mathbf{e}\boldsymbol{\theta}$ to obtain

$$\sum_{n=1}^{\infty}\frac{t^n}{n!}D^{n-1}(D-\mathbf{e}\boldsymbol{\theta}) = \sum_{n=1}^{\infty}\frac{t^n}{n!}D^n - t\mathbf{e}\boldsymbol{\theta} = \exp(Dt) - I - t\mathbf{e}\boldsymbol{\theta}. \tag{2.54}$$

We show that the matrix $D - \mathbf{e}\boldsymbol{\theta}$ is invertible. If $D - \mathbf{e}\boldsymbol{\theta}$ is not invertible, there exists a nonzero vector $\mathbf{u}$ such that $\mathbf{u}(D - \mathbf{e}\boldsymbol{\theta}) = 0$. If $\mathbf{ue} = 0$, then $\mathbf{u}D = 0$. Since D is irreducible, then $\mathbf{u}$ must be nonnegative or nonpositive and nonzero, which contradicts $\mathbf{ue} = 0$. If $\mathbf{ue} \neq 0$, without loss of generality, we assume $\mathbf{ue} = 1$, which leads to $\mathbf{u}D = \boldsymbol{\theta}$. Postmultiplying by $\mathbf{e}$, we obtain $0 = \boldsymbol{\theta}\mathbf{e} = 1$, which is a contradiction. Therefore, $D - \mathbf{e}\boldsymbol{\theta}$ is invertible. Postmultiplying by the inverse of $D - \mathbf{e}\boldsymbol{\theta}$ on both sides of Eq. (2.54), yields

$$\sum_{n=1}^{\infty}\frac{t^n}{n!}D^{n-1} = (\exp(Dt) - I - t\mathbf{e}\boldsymbol{\theta})(D-\mathbf{e}\boldsymbol{\theta})^{-1}. \tag{2.55}$$

Combining with Eqs. (2.53) and (2.55) leads to

$$\begin{aligned}
\left.\frac{\partial P^*(z,t)}{\partial z}\right|_{z=1}\mathbf{e} &= \sum_{n=1}^{\infty}\frac{t^n}{n!}D^{n-1}D_1\mathbf{e} \\
&= (\exp(Dt) - I - t\mathbf{e}\boldsymbol{\theta})(D-\mathbf{e}\boldsymbol{\theta})^{-1}D_1\mathbf{e} \\
&= \lambda t\mathbf{e} + (\exp(Dt) - I)(D-\mathbf{e}\boldsymbol{\theta})^{-1}D_1\mathbf{e}.
\end{aligned} \tag{2.56}$$

Note that $\boldsymbol{\theta}(D-\mathbf{e}\boldsymbol{\theta})^{-1} = -\boldsymbol{\theta}$. Premultiplying by $\boldsymbol{\alpha}$ on both sides of Eq. (2.56), we obtain Eq. (2.51). By straightforward calculations, it can be shown that matrices $\exp(Dt)-I$ and $D-\mathbf{e}\boldsymbol{\theta}$ are commutable. Thus, in Eqs. (2.55) and (2.56), $(\exp(Dt) - I)(D - \mathbf{e}\boldsymbol{\theta})^{-1}$ can be written as $(D-\mathbf{e}\boldsymbol{\theta})^{-1}(\exp(Dt)-I)$. This completes the proof of Theorem 2.3.2.

Note: The proof of Theorem 2.3.2 is quite technical, but the techniques used in the proof are typical methods in dealing with *MAP*s.

Similar to the Poisson process, by Theorem 2.3.2, λ can be interpreted as the long-run average number of arrivals per unit time, i.e., $\lambda = \lim_{t\to\infty} N(t)/t$, almost surely. Particularly, for a stationary process, the initial distribution of the phase $\boldsymbol{\alpha}$ is replaced with $\boldsymbol{\theta}$. Using $D\mathbf{e} = 0$ and $\boldsymbol{\theta}D = 0$, we obtain

$$E_{\boldsymbol{\theta}}[N(t)] \equiv \boldsymbol{\theta} \left.\frac{\partial P^*(z,t)}{\partial z}\right|_{z=1} \mathbf{e} = \boldsymbol{\theta} \sum_{n=1}^{\infty} \frac{t^n}{n!} \sum_{k=0}^{n-1} D^k D_1 D^{n-1-k} \mathbf{e} = \boldsymbol{\theta} D_1 \mathbf{e} t = \lambda t. \quad (2.57)$$

Note: Assume that $I(0)$ has distribution $\boldsymbol{\pi}$, the distribution of the phase right after an arbitrary arrival. Then Eq. (2.51) may not be reduced to a simple form.

The arrival rate λ plays an important role in the application of *MAP*s.

Example 2.3.1 Consider an *MAP* with

$$D_0 = \begin{pmatrix} -\lambda_1 & \lambda_1 & & \\ & -\lambda_2 & \ddots & \\ & & \ddots & \lambda_{m-1} \\ & & & -\lambda_m \end{pmatrix}_{m\times m}, \quad D_1 = \begin{pmatrix} 0 & 0 & & \\ & 0 & \ddots & \\ & & \ddots & 0 \\ \lambda_m & & & 0 \end{pmatrix}_{m\times m}. \quad (2.58)$$

By routine calculations, we found $\boldsymbol{\theta} = \left(\sum_{i=1}^{m} 1/\lambda_i\right)^{-1} \left(\lambda_1^{-1}, \quad \cdots, \quad \lambda_m^{-1}\right)$ and arrival rate $\lambda = \left(\sum_{i=1}^{m} 1/\lambda_i\right)^{-1}$.

Exercise 2.3.4 Find the arrival rates for *MAP*s defined in Examples 2.2.2 and 2.2.3.

Exercise 2.3.5 Using Theorem 2.3.2 to prove Eq. (2.57).

Theorem 2.3.3 *Assume that the underlying Markov chain $\{I(t), t \geq 0\}$ is irreducible, i.e., D is irreducible. Given initial distribution $\boldsymbol{\theta}$, which is the stationary distribution of D, we have*

$$\begin{aligned} Var(N(t)) = &(\lambda - 2\lambda^2 - 2\boldsymbol{\theta}D_1(D - \mathbf{e}\boldsymbol{\theta})^{-1}D_1\mathbf{e})t \\ &+ 2\boldsymbol{\theta}D_1(D - \mathbf{e}\boldsymbol{\theta})^{-1}(\exp(Dt) - I)(D - \mathbf{e}\boldsymbol{\theta})^{-1}D_1\mathbf{e}. \end{aligned} \quad (2.59)$$

Proof. Similar to Eq. (2.52), we obtain

$$\left.\frac{\partial^2 P^*(z,t)}{\partial z^2}\right|_{z=1} = \sum_{n=2}^{\infty}\frac{t^n}{n!}\sum_{k=1}^{n-1}\left(\sum_{l=0}^{k-1} D^l D_1 D^{k-1-l}\right) D_1 D^{n-1-k} + \sum_{n=2}^{\infty}\frac{t^n}{n!}\sum_{k=0}^{n-2} D^k D_1\left(\sum_{l=0}^{n-2-k} D^l D_1 D^{n-2-k-l}\right), \tag{2.60}$$

which leads to

$$\left.\frac{\partial^2 P^*(z,t)}{\partial z^2}\right|_{z=1}\mathbf{e} = 2\sum_{n=2}^{\infty}\frac{t^n}{n!}\sum_{l=0}^{n-2} D^l D_1 D^{n-2-l} D_1 \mathbf{e}. \tag{2.61}$$

Premultiplying by $\boldsymbol{\theta}$ on both sides of Eq. (2.61), we obtain

$$\begin{aligned}\boldsymbol{\theta}\left.\frac{\partial^2 P^*(z,t)}{\partial z^2}\right|_{z=1}\mathbf{e} &= 2\boldsymbol{\theta} D_1\left(\sum_{n=2}^{\infty}\frac{t^n}{n!}D^{n-2}\right)D_1\mathbf{e}\\ &= 2\boldsymbol{\theta} D_1\left(\sum_{n=2}^{\infty}\frac{t^n}{n!}D^{n-2}\right)(D-\mathbf{e}\boldsymbol{\theta})(D-\mathbf{e}\boldsymbol{\theta})^{-1}D_1\mathbf{e}\\ &= 2\boldsymbol{\theta} D_1\left(\sum_{n=1}^{\infty}\frac{t^n}{n!}D^{n-1} - tI - \frac{t^2}{2}\mathbf{e}\boldsymbol{\theta}\right)(D-\mathbf{e}\boldsymbol{\theta})^{-1}D_1\mathbf{e}.\end{aligned} \tag{2.62}$$

Since $Var(N(t)) = E[N(t)(N(t)-1)] + E[N(t)] - (E[N(t)])^2$ and $E[N(t)] = \lambda t$, using Eq. (2.51), we obtain

$$Var(N(t)) = \boldsymbol{\theta}\left.\frac{\partial^2 P^*(z,t)}{\partial z^2}\right|_{z=1}\mathbf{e} + E[N(t)] - (E[N(t)])^2. \tag{2.63}$$

Then Eq. (2.59) is obtained by routine calculations. This completes the proof of Theorem 2.3.3.

Exercise 2.3.6 For a stationary Markovian arrival process, i.e., the distribution of $I(0)$ is $\boldsymbol{\theta}$, find the *coefficient of variation* of $N(t)$. Show that the distribution of $I(t)$ is $\boldsymbol{\theta}$, for all $t > 0$.

Exercise 2.3.7 The *index of dispersion* of $N(t)$ is defined as $Var(N(t))/E[N(t)]$. For a stationary Markovian arrival process, find an expression for the index of dispersion of $N(t)$.

For the Poisson process, the numbers of arrivals in non-overlapping intervals are independent. For *MAP*s, on one hand, the numbers of arrivals in non-overlapping intervals are correlated. On the other hand, the numbers are conditionally independent, i.e., if $I(t)$ is known, then the arrival process in $(t, t+s)$ is independent of that in $[0, t]$. Based on the conditional independence, the correlation can be obtained explicitly for the stationary case. Let $N(t, t + s) = N(t + s) - N(t)$. Let $p_{i,j}(n_1, t, n_2, s) = P\{N(t) = n_1, N(t,s + t) = n_2, I(t + s) = j \mid I(0) = i\}$, for $i, j = 1, \ldots, m$ and $n_1, n_2 = 0, 1, 2, \ldots$. Define $p^*_{i,j}(z_1, t, z_2, s)$ as the *joint probability generating function* of $\{p_{i,j}(n_1, t, n_2, s), n_1, n_2 = 0, 1, 2, \ldots\}$, for $i, j = 1, \ldots, m$; and $P^*(z_1, t, z_2, s) = (p^*_{i,j}(z_1, t, z_2, s))_{m\times m}$.

Theorem 2.3.4 $P^*(z_1, t, z_2, s) = \exp((D_0 + z_1D_1)t)\exp((D_0 + z_2D_1)s)$.

Proof. Conditioning on $I(t)$, by the Markovian property of $I(t)$, we obtain

$$
\begin{aligned}
p_{i,j}(n_1, t, n_2, s) &= \sum_{k=1}^{m} p_{i,k}(n_1, t) P\{N(t, t+s) = n_2, I(t+s) = j | I(0) = i, I(t) = k\} \\
&= \sum_{k=1}^{m} p_{i,k}(n_1, t) p_{k,j}(n_2, s),
\end{aligned}
\tag{2.64}
$$

which leads to $P^*(z_1, t, z_2, s) = P^*(z_1, t)P^*(z_2, s)$. The result is obtained by Theorem 2.3.1. This completes the proof of Theorem 2.3.4.

Exercise 2.3.8 In Theorem 2.3.4, simplify the joint probability generating function for $z_1 = z_2 = z$. Explain the simplified formula $P^*(z, t + s) = P^*(z, t)P^*(z, s)$ intuitively.

Corollary 2.3.1 *Assume that D is irreducible. Given that I*(0) *has distribution* $\boldsymbol{\theta}$, *we have* (*i*) $E[N(t, t + s)] = \lambda s$; *and* (*ii*)

$$
\begin{aligned}
&E[N(t)N(t, t+s)] \\
&= \lambda^2 ts + \boldsymbol{\theta} D_1 (D - \mathbf{e}\boldsymbol{\theta})^{-1} (\exp(Dt) - I)(\exp(Ds) - I)(D - \mathbf{e}\boldsymbol{\theta})^{-1} D_1 \mathbf{e}.
\end{aligned}
\tag{2.65}
$$

Proof. Note that $\boldsymbol{\theta} D = 0$ implies $\boldsymbol{\theta}\exp(Dt) = \boldsymbol{\theta}$. Then part (i) is obtained directly from Theorem 2.3.4. By Theorem 2.3.4, we obtain

$$
\boldsymbol{\theta} \left. \frac{\partial^2 P^*(z_1, t, z_2, s)}{\partial z_1 \partial z_2} \right|_{z_1 = z_2 = 1} \mathbf{e} = \boldsymbol{\theta} D_1 \left(\sum_{n=1}^{\infty} \frac{t^n}{n!} D^{n-1} \right) \left(\sum_{n=1}^{\infty} \frac{s^n}{n!} D^{n-1} \right) D_1 \mathbf{e}. \tag{2.66}
$$

Equation (2.65) is obtained using Eqs. (2.54) and (2.66). This completes the proof of Corollary 2.3.1.

Exercise 2.3.9 Assume that $I(0)$ has a general distribution. Discuss the computation of the correlation between $N(t)$ and $N(t, t + s)$.

Exercise 2.3.10 For $t_1 < t_2 < \ldots < t_n$, show that the (conditional) joint probability generating function of $N(t_1), N(t_1, t_2), \ldots$, and $N(t_{n-1}, t_n)$ is given by (Note: $t_0 = 0$)

Table 2.1 Comparison of the Poisson process and *MAP*s

	Poisson process	*MAP*
Parameters	λ	(D_0, D_1)
Interarrival time	Exponential (λ)	General distribution
Arrival rate	λ	$\boldsymbol{\theta} D_1 \mathbf{e}$
$E[N(t)]$	λt	$\boldsymbol{\theta} D_1 \mathbf{e}\, t + \boldsymbol{\alpha}(\exp(Dt) - I)(D - \mathbf{e}\boldsymbol{\theta})^{-1} D_1 \mathbf{e}$
Properties	Memoryless	(i) Partial memoryless (ii) Approximate any counting process
Closure property	Independent superposition	Independent superposition

$$
\begin{aligned}
P^*(z_1, z_2, \ldots, z_n) &= E[z_1^{N(t_1)} z_2^{N(t_1,t_2)} \cdots z_n^{N(t_{n-1},t_n)}] \\
&= \prod_{k=1}^{n} \exp((D_0 + z_k D_1)(t_k - t_{k-1})).
\end{aligned} \tag{2.67}
$$

Exercise 2.3.11 Find the joint probability generating function of $N(t_1)$ and $N(s + t_1, s + t_2)$ for $t_1 < t_2$. Explain the result intuitively.

Exercise 2.3.12 Assume that D is irreducible and $I(0)$ has distribution $\boldsymbol{\theta}$. Prove that, for $t, u, v, s \geq 0$,

$$
\begin{aligned}
&E[N(t, t+u)N(t+u+v, t+u+v+s)] \\
&\quad = \lambda^2 us + \boldsymbol{\theta} D_1 (D - \mathbf{e}\boldsymbol{\theta})^{-1}(\exp(Du) - I) \\
&\quad \cdot \exp(Dv)(\exp(Ds) - I)(D - \mathbf{e}\boldsymbol{\theta})^{-1} D_1 \mathbf{e}.
\end{aligned} \tag{2.68}
$$

Using Eq. (2.68), the correlation between $N(t, t + u)$ and $N(t + u + v, t + u + v + s)$ can be obtained, i.e., the correlation between the numbers of arrivals of any two non-overlapping intervals can be obtained explicitly for the stationary case.

Commentary We remark that higher moments of $N(t)$ can be found using Eq. (2.49), but the formulas are much more complex than those presented in this section (see Neuts et al. (1992)). More detailed work on measures of *MAP*s can be found in Narayana and Neuts (1992), Neuts and Li (1997), and Nielsen et al. (2007). The following table summarizes variables and performance measures of the Poisson process and *MAP*s (Table 2.1).

Additional Exercises and Extensions

Exercise 2.3.13 Prove $\lim_{t\to\infty} E[N(t)]/t = \lambda$ for any initial distribution $\boldsymbol{\alpha}$. Use the stationary distribution of the underlying Markov chain to explain the result intuitively. Use the simulation program developed for Exercise 2.2.11 to verify the result.

Exercise 2.3.14 Assume that $\{(N_1(t), I_1(t)), t \geq 0\}$ and $\{(N_2(t), I_2(t)), t \geq 0\}$ are independent Markovian arrival processes with arrival rates λ_1 and λ_2, respectively. What is the arrival rate of superposition process $\{(N_1(t) + N_2(t), I(t)), t \geq 0\}$?

Exercise 2.3.15 Assume that $\{(N(t), I(t)), t \geq 0\}$ is a Markovian arrival process with arrival rate λ. We mark arrivals independently by $(p, 1-p)$ into type 1 and type 2 arrivals. What are the arrival rates of type 1 and type 2 arrivals, respectively?

Exercise 2.3.16 Consider an *MAP* with matrix representation

$$D_0 = \begin{pmatrix} -5 & 0 & 0 \\ 0 & -2 & 1 \\ 0 & 0 & -1 \end{pmatrix}, \quad D_1 = \begin{pmatrix} 5 & 0 & 0 \\ 0 & 0 & 1 \\ 0 & 1 & 0 \end{pmatrix}. \tag{2.69}$$

Is the underlying Markov chain irreducible? How do we define and calculate the (stationary) arrival rate of the *MAP*? (Hint: The arrival rate depends on the initial phase. Note: This example shows why the irreducibility condition on D is assumed for many cases.)

Exercise 2.3.17 (Exercise 2.2.19 continued) In Exercise 2.2.19, if $\lambda = 0.01$ and $\mu = 0.5$, show that the stationary distribution of $\{I(t), t \geq 0\}$ is $\boldsymbol{\theta} = (0.9800, 0.0196, 0.0004)$ and the arrival rate is $\boldsymbol{\theta} D_1 \mathbf{e} = 0.009996$, i.e., repair completion occurs 0.009996 times per unit time.

Let $\mathbf{v}_n(\boldsymbol{\alpha}) = \boldsymbol{\alpha} \sum_{l=0}^{n} D^l D_1 D^{n-l}$, for $n = 0, 1, 2, \ldots$. Then

$$\boldsymbol{\alpha} \left. \frac{\partial^2 P^*(z,t)}{\partial z^2} \right|_{z=1} \mathbf{e} = 2 \left(\sum_{n=2}^{\infty} \frac{t^n \mathbf{v}_{n-2}(\boldsymbol{\alpha})}{n!} \right) D_1 \mathbf{e} \equiv 2\mathbf{u}(\boldsymbol{\alpha}, t) D_1 \mathbf{e}. \tag{2.70}$$

Exercise 2.3.18 (Generalization of Theorem 2.3.3 and Exercises 2.3.6 and 2.3.7) Show that the summation on the right hand side of Eq. (2.70) is finite for fixed t. Based on Eq. (2.70), outline an approximation method for computing $E[N(t)(N(t)-1)]$ for any *MAP*. Then outline methods for computing the variance, coefficient of variation, and index of dispersion of $N(t)$. (Note: Unlike Eq. (2.62), the summation in Eq. (2.70) cannot be simplified since the initial distribution of the phase has a general form. Thus, truncation has to be utilized in computing the second moment.)

For a general initial distribution $\boldsymbol{\alpha}$, similar to Eq. (2.66), we obtain

$$\begin{aligned} &\boldsymbol{\alpha} \left. \frac{\partial^2 P^*(z_1, t, z_2, s)}{\partial z_1 \partial z_2} \right|_{z_1 = z_2 = 1} \mathbf{e} \\ &= \boldsymbol{\alpha} \left(\sum_{n=1}^{\infty} \frac{t^n}{n!} \left(\sum_{k=0}^{n-1} D^k D_1 D^{n-1-k} \right) \right) \left(\sum_{n=1}^{\infty} \frac{s^n}{n!} D^{n-1} \right) D_1 \mathbf{e} \\ &\equiv \hat{\mathbf{u}}(\boldsymbol{\alpha}, t) \left(\sum_{n=1}^{\infty} \frac{s^n}{n!} D^{n-1} \right) D_1 \mathbf{e} \\ &= \hat{\mathbf{u}}(\boldsymbol{\alpha}, t) \Big((\exp(Ds) - I)(D - \mathbf{e}\boldsymbol{\theta})^{-1} D_1 \mathbf{e} - \lambda s \mathbf{e} \Big). \end{aligned} \tag{2.71}$$

Exercise 2.3.19 (Generalization of Corollary 2.3.1 and Exercise 2.3.10) Based on Eq. (2.71), outline an approximation method for computing $E[N(t, t+s)\, N(t+s+u, t+s+u+v)]$, i.e., the correlation between the numbers of arrivals in the intervals $[t, t+s]$ and $[t+s+u, t+s+u+v]$, for any *MAP*.

Assume that X is a *PH*-random variable with matrix representation $(\boldsymbol{\beta}, T)$ and $\{(N(t), I(t)), t\geq 0\}$ is an *MAP* with (D_0, D_1). Let N be the number of arrivals in $[0, X]$. Then the conditional moment generating function of N can be obtained as follows, for $0 \leq z \leq 1$,

$$\begin{aligned}
&\int_0^{\infty} \exp((D_0+zD_1)t)\boldsymbol{\beta}\exp(Tt)\mathbf{T}^0 dt \\
&= (I\otimes\boldsymbol{\beta})\int_0^{\infty} \exp((D_0\otimes I+I\otimes T+zD_1\otimes I)t)dt(I\otimes\mathbf{T}^0) \\
&= -(I\otimes\boldsymbol{\beta})(D_0\otimes I+I\otimes T+zD_1\otimes I)^{-1}(I\otimes\mathbf{T}^0) \\
&= -(I\otimes\boldsymbol{\beta})\left(I-z(-(D_0\otimes I+I\otimes T))^{-1}(D_1\otimes I)\right)^{-1}(-(D_0\otimes I+I\otimes T))^{-1}(I\otimes\mathbf{T}^0).
\end{aligned} \tag{2.72}$$

Exercise 2.3.20 (Terminating *MAP*s) Assume that $I(0)$ has a distribution $\boldsymbol{\alpha}$ and $\boldsymbol{\beta}\mathbf{e} = 1$. Show that $N+1$ has a discrete *PH*-distribution with matrix representation $(\boldsymbol{\alpha}\otimes\boldsymbol{\beta},\quad (-(D_0\otimes I+I\otimes T))^{-1}(D_1\otimes I))$.

For the Poisson process, the interarrival times are independent. For *MAP*s, on the one hand, the interarrival times are correlated. On the other hand, the interarrival times are conditionally independent. Based on the conditional independence, the correlation between interarrival times can be obtained explicitly for the stationary case as well. Recall that τ_n is the interarrival time between the $(n-1)$-st and the n-th arrivals of an *MAP* with matrix representation (D_0, D_1).

Exercise 2.3.21 Find the distributions of τ_1 and τ_2. Find the joint distribution of τ_1 and τ_2.

By conditioning on the phases right after the arrivals, the joint LST of $\{\tau_1,\tau_2,\ldots,\tau_n\}$ can be obtained as follows

$$\begin{aligned}
&(E[\exp(-(s_1\tau_1+s_2\tau_2+\cdots+s_n\tau_n)) : I(\tau_1+\tau_2+\cdots+\tau_n)=j|I(0)=i])_{m\times m} \\
&= \int_0^{\infty}\cdots\int_0^{\infty} \exp((-s_1I+D_0)t_1)D_1\mathrm{d}t_1\cdots\exp((-s_nI+D_0)t_n)D_1\mathrm{d}t_n \\
&= (s_1I-D_0)^{-1}D_1(s_2I-D_0)^{-1}D_1\cdots(s_nI-D_0)^{-1}D_1.
\end{aligned} \tag{2.73}$$

Assume that $I(0)$ has a distribution $\boldsymbol{\alpha}$. The joint LST of $\{\tau_1,\tau_2,\ldots,\tau_n\}$ is given as

$$\boldsymbol{\alpha}(s_1I-D_0)^{-1}D_1(s_2I-D_0)^{-1}D_1\cdots(s_nI-D_0)^{-1}D_1\mathbf{e}. \tag{2.74}$$

Exercise 2.3.22 Assume that $I(0)$ has the distribution $\boldsymbol{\pi}$ (see Exercise 2.2.12). Show that τ_n has the *PH*-distribution $(\boldsymbol{\pi}, D_0)$ for any n.

Exercise 2.3.23 Assume that $I(0)$ has a distribution $\boldsymbol{\alpha}$. Show that the covariance between τ_1 andτ_n is given by

$$\begin{aligned} &E[(\tau_1 - E[\tau_1])(\tau_n - E[\tau_n])] \\ &= \boldsymbol{\alpha}(-D_0)^{-1}\left((-D_0)^{-1}D_1\right)^{n-1}(-D_0)^{-1}\mathbf{e} \\ &\quad - \boldsymbol{\alpha}(-D_0)^{-1}\mathbf{e}\boldsymbol{\alpha}\left((-D_0)^{-1}D_1\right)^{n-1}(-D_0)^{-1}\mathbf{e}. \end{aligned} \tag{2.75}$$

Find an expression for the covariance between τ_k andτ_n. (Hint: Given $\boldsymbol{\alpha}$, find the phase distribution right after the arrival of the $(k-1)$-st arrival. Also note that $-D_0{}^{-1}D_1\mathbf{e} = \mathbf{e}$.)

Exercise 2.3.24 Assume that $I(0)$ has a distribution $\boldsymbol{\alpha}$. (i) Show that the distribution of the phase right after the k-th arrival is given by $\boldsymbol{\alpha}(-D_0{}^{-1}D_1)^k$. (ii) Show that the correlation between τ_k andτ_n is given by

$$\begin{aligned} &\rho(\tau_1, \tau_n) \\ &= \frac{\boldsymbol{\alpha}(-D_0)^{-1}\left((-D_0)^{-1}D_1\right)^{n-1}(-D_0)^{-1}\mathbf{e} - \boldsymbol{\alpha}(-D_0)^{-1}\mathbf{e}\boldsymbol{\alpha}\left((-D_0)^{-1}D_1\right)^{n-1}(-D_0)^{-1}\mathbf{e}}{\sqrt{2\boldsymbol{\alpha}D_0^{-2}\mathbf{e} - (\boldsymbol{\alpha}D_0^{-1}\mathbf{e})^2}\sqrt{2\boldsymbol{\alpha}\left((-D_0)^{-1}D_1\right)^{n-1}D_0^{-2}\mathbf{e} - \left(\boldsymbol{\alpha}\left((-D_0)^{-1}D_1\right)^{n-1}D_0^{-1}\mathbf{e}\right)^2}}. \end{aligned} \tag{2.76}$$

Hint: Use Exercise 2.2.6 and Proposition 2.2.1 to show that

$$\begin{aligned} E[\tau_k] &= \boldsymbol{\alpha}\left((-D_0)^{-1}D_1\right)^{k-1}(-D_0)^{-1}\mathbf{e}, \\ E[\tau_k^2] &= 2\boldsymbol{\alpha}\left((-D_0)^{-1}D_1\right)^{k-1}(-D_0)^{-2}\mathbf{e}, \\ \mathrm{var}(\tau_k) &= 2\boldsymbol{\alpha}\left((-D_0)^{-1}D_1\right)^{k-1}(-D_0)^{-2}\mathbf{e} - \left(\boldsymbol{\alpha}\left((-D_0)^{-1}D_1\right)^{k-1}(-D_0)^{-1}\mathbf{e}\right)^2. \end{aligned}$$

Exercise 2.3.25 Show that the random vector $(\tau_1, \tau_2, \ldots, \tau_n)$ has a multi-variate phase-type distribution. Assume that $I(0)$ has a distribution $\boldsymbol{\alpha}$. Find a matrix-representation for $(\tau_1, \tau_2, \ldots, \tau_n)$.

Exercise 2.3.26 (Random time transformation) Let $\mu(i)$ be a positive function on $\{1, 2, \ldots, m\}$. Let $x(t) = \int_0^t \mu(I(t))\mathrm{d}t$, $I_r(t) = I(x(t))$, and $N_r(t) = N(x(t))$, for $t \geq 0$. Show that $\{(N_r(t), I_r(t)), t \geq 0\}$ is an *MAP* with matrix representation $(M^{-1}D_0, M^{-1}D_1)$, where $M = \mathrm{diag}(\mu(1), \ldots, \mu(m))$. (Hint: Consider the differential equations for $P(n, t)$.)

Exercise 2.3.27 (Chapter 2, Neuts (1981)) For *PH*-renewal processes, show

$$
\begin{aligned}
E[N(t)] = & \frac{t}{E[w_1]} - \frac{1}{\boldsymbol{\alpha}\mathbf{e}} + \frac{E[w_1^2]}{2(E[w_1])^2} \\
& + \frac{\boldsymbol{\alpha}}{E[w_1]\boldsymbol{\alpha}\mathbf{e}}(\mathbf{e}\boldsymbol{\theta} - \exp((T + \mathbf{T}^0\boldsymbol{\alpha})t))T^{-1}\mathbf{e},
\end{aligned} \tag{2.77}
$$

where $E[w_1]$ is the mean of the *PH*-distribution (i.e., the mean of the interarrival time).

For special initial distribution $\boldsymbol{\alpha}$, the corresponding *MAP* may have some special properties. If $\boldsymbol{\alpha} = \boldsymbol{\theta}$, the arrival rate is always λ. If $\boldsymbol{\alpha} = \boldsymbol{\pi}$, the interarrival times have the same *PH*-distribution $(\boldsymbol{\pi}, D_0)$, although they do not have to be independent.

Exercise 2.3.28 For the *MAP* (D_0, D_1) given in Eq. (2.78), (i) find the distribution of the time until the next arrival from an arbitrary time epoch; (ii) in steady state, find the distribution of the time between two consecutive arrivals; and (iii) in steady state, find the distribution of the sum of two consecutive interarrival times. Plot the density functions.

$$
D_0 = \begin{pmatrix} -8 & 2 & 3.6 \\ 3 & -10-4 & 0 \\ 0.4 & 1.5 & -5-1 \end{pmatrix}, \quad D_1 = \begin{pmatrix} 0 & 0 & 2.4 \\ 7 & 4 & 0 \\ 1.6 & 1.5 & 1 \end{pmatrix}. \tag{2.78}
$$

Define $\lambda(t) = E[N(t)]/t$, for $t > 0$, and $\lambda(t, 1) = E[N(t + 1)] - E[N(t)]$.

Exercise 2.3.29 Consider an *MAP* with

$$
\left(\boldsymbol{\alpha} = (0.8, 0.2), \quad D_0 = \begin{pmatrix} -3 & 0 \\ 0.05 & -0.2 \end{pmatrix}, \quad D_1 = \begin{pmatrix} 2 & 1 \\ 0 & 0.15 \end{pmatrix} \right). \tag{2.79}
$$

Find the (average) arrival rate λ. Plot $\lambda(t)$ and $\lambda(t, 1)$.

We define $N(t, t + 1) = N(t + 1) - N(t)$.

Exercise 2.3.30 Simulate the *MAP* defined in Exercise 2.3.29. Plot $N(n, n+1)$, for $n = 0, 1, 2, \ldots$, up to 10 times.

Exercise 2.3.31 (Derivatives of matrices) Assume that $A(z) = (a_{i,j}(z)) = B(z)C(z)$, where the elements of matrices $B(z) = (b_{i,j}(z))$ and $C(z) = (c_{i,j}(z))$ are differentiable. Show that

$$
\frac{\mathrm{d}a_{i,j}(z)}{\mathrm{d}z} = \sum_{k=1}^{m} \left(\frac{\mathrm{d}b_{i,k}(z)}{\mathrm{d}z} \right) c_{k,j}(z) + \sum_{k=1}^{m} b_{i,k}(z) \left(\frac{\mathrm{d}c_{k,j}(z)}{\mathrm{d}z} \right). \tag{2.80}
$$

Then use Eq. (2.80) to prove

$$
\frac{\mathrm{d}A(z)}{\mathrm{d}z} = \frac{\mathrm{d}B(z)}{\mathrm{d}z} C(z) + B(z) \frac{\mathrm{d}C(z)}{\mathrm{d}z}, \tag{2.81}
$$

and, for $n = 1, 2, \ldots,$

$$\frac{\mathrm{d}(A(z))^n}{\mathrm{d}z} = \sum_{k=0}^{n-1} (A(z))^k \frac{\mathrm{d}A(z)}{\mathrm{d}z} (A(z))^{n-1-k}. \tag{2.82}$$

2.4 Batch Markovian Arrival Processes

Batch Markovian arrival processes (*BMAPs*) are direct generalizations of the Markovian arrival processes introduced in Sect. 2.2. In general, *BMAPs*, as well as *MMAPs* to be introduced in Sect. 2.5, are part of the family of Markovian arrival processes. Apart from the mathematical details, the most important aspect in these generalizations is the interpretation of the arrival rates. For *MAP* (D_0, D_1), elements of D_0 are interpreted as the (event) rates without an arrival and elements of D_1 are interpreted as the rates with an arrival. A natural way to define more complicated arrival processes is to divide the arrival rates in D_1 and assign different meanings to them.

Example 2.4.1 For the *MAP* (D_0, D_1) defined in Sect. 2.2, if an arrival event occurs, one customer arrives. Instead of interpreting D_1 as the rate of arrival events for one customer, it can be interpreted as follows.

(i) D_1 is the (matrix) rate of events for a batch of five customers.
(ii) If an arrival event occurs according to D_1, with probability 0.1, one customer arrives; with probability 0.2, two customers arrive; with probability 0.3, three customers arrive; and with probability 0.4, four customers arrive. For this case, we can write arrival rates in four matrices to reflect the number of customers in different arrival batches: $0.1D_1$, $0.2D_1$, $0.3D_1$, and $0.4D_1$.

Now, we give the first definition of *BMAPs*.

Definition 2.4.1 Assume that matrices $\{D_1, D_2, \ldots, D_N\}$ of order m are nonnegative, D_0 has nonnegative off-diagonal elements and negative diagonal elements, m is a positive integer, N is a positive integer (maybe infinite), and $D = D_0 + D_1 + \ldots + D_N$ is an infinitesimal generator. Let $D_0 = (d_{0,(i,j)})$, $D_n = (d_{n,(i,j)})$, for $n = 1, \ldots, N$. Then $(D_0, D_1, \ldots, D_N)$ defines *BMAP* $\{(N(t), I(t)), t \geq 0\}$ as follows.

1. Set $N(0) = 0$.
2. Define a continuous time Markov chain $\{I(t), t \geq 0\}$ by D.
3. For phase (state) i with $d_{n,(i,i)} > 0$, define a Poisson process with arrival rate $d_{n,(i,i)}$, for $n = 1, \ldots, N$ and $i = 1, \ldots, m$. The Poisson processes are turned on, if $I(t) = i$; otherwise, they are turned off.
4. For $i = 1, \ldots, m$, if $I(t) = i$ and an arrival from the imposed Poisson process $d_{n,(i,i)}$ occurs, $N(t)$ increases by n, for $1 \leq n \leq N$.
5. At the end of each stay in state i, with probability $d_{0,(i,j)}/(-d_{(i,i)})$ (note $d_{(i,i)} = d_{0,(i,i)} + d_{1,(i,i)} + \ldots + d_{N,(i,i)}$), $I(t)$ transits from phase i to j and $N(t)$ remains the same (i.e., without an arrival); and, with probability $d_{n,(i,j)}/(-d_{(i,i)})$, $I(t)$ transits

from phase i to j and $N(t)$ increases by n (i.e., with n arrivals), for $1 \le n \le N$ and $i \ne j$ and $i, j = 1, \ldots, m$.

We shall call $(D_0, D_1, \ldots, D_N)$ a matrix representation of *BMAP* $\{(N(t), I(t)), t \ge 0\}$. The Markov chain $\{I(t), t \ge 0\}$ is called the underlying Markov chain of $\{(N(t), I(t)), t \ge 0\}$. We remark that $d_{1,(i,j)} + \ldots + d_{N,(i,j)}$ is the rate for an arrival event of customers if the phase transits from i to j. If such an event occurs, with probability $d_{n,(i,j)}/(d_{1,(i,j)} + \ldots + d_{N,(i,j)})$, the batch size is n, for $1 \le n \le N$.

Example 2.4.2 The following are examples of *BMAP*s.

(1) *Compound Poisson* process: $D_0 = (-\lambda), D_1 = (\lambda_1), D_2 = (\lambda_2), \ldots, D_N = (\lambda_N)$, where $\lambda = \lambda_1 + \lambda_2 + \ldots + \lambda_N$. That is, for each arrival event, there are n customers with probability λ_n/λ. (Note: In the notation used in Sect. 2.1 (Exercise 2.1.15), $\lambda_n = p_n\lambda$, where $\{p_1, p_2, \ldots\}$ is a distribution of the batch size.)
(2) For *MAP*:

$$D_0 = \begin{pmatrix} -2-4 & 1.6 \\ 3 & -10 \end{pmatrix}, \quad D_1 = \begin{pmatrix} 4 & 0.4 \\ 7 & 0 \end{pmatrix}, \tag{2.83}$$

defined in Sect. 2.2, we split D_1 as follows to obtain *BMAP*

$$D_0 = \begin{pmatrix} -2-4 & 1.6 \\ 3 & -10 \end{pmatrix}, \quad \hat{D}_1 = \begin{pmatrix} 3 & 0.2 \\ 4 & 0 \end{pmatrix}, \quad \hat{D}_2 = \begin{pmatrix} 1 & 0.2 \\ 3 & 0 \end{pmatrix}. \tag{2.84}$$

Exercise 2.4.1 Explain the arrivals in the cases defined in Example 2.4.2 probabilistically.

Exercise 2.4.2 Assume that $(D_0, D_1, \ldots, D_N)$ represents a *BMAP*. Show that $(D_0, D_1 + \ldots + D_N)$ represents an *MAP*. How do you interpret this *MAP*?

Exercise 2.4.3 Assume that $(D_0, D_1, \ldots, D_N)$ represents a *BMAP*. Show that $(D - D_k, D_k)$ represents an *MAP*. How do you interpret this *MAP*?

Consider $\{(N(t), I(t)), t \ge 0\}$ as a stochastic process. The state space is $\{0, 1, 2, \ldots\} \times \{1, 2, \ldots, m\}$. The variable $N(t)$ is increasing in t. The value of $N(t)$ can increase by at most N at one jump. Similar to the *MAP* case, it can be shown that $\{(N(t), I(t)), t \ge 0\}$ is a continuous time Markov chain. Thus, similar to *MAP*s, *BMAP*s can be defined in the following ways.

Definition 2.4.2 Assume that $\{D_0, D_1, \ldots, D_N\}$ satisfy the conditions given in Definition 2.4.1. A *BMAP* is defined as a continuous time Markov chain $\{(N(t), I(t)), t \ge 0\}$ with $N(0) = 0$ and infinitesimal generator

$$Q = \begin{pmatrix} D_0 & D_1 & \cdots & D_N & & \\ & D_0 & D_1 & \cdots & D_N & \\ & & \ddots & \ddots & \ddots & \ddots \\ & & & \ddots & \ddots & \ddots \end{pmatrix}. \tag{2.85}$$

For a *BMAP*, define Markov renewal process $\{(I_n, J_n, \tau_n), n = 0, 1, 2, \ldots\}$, where I_n is the phase right after the n-th (batch) arrival, J_n is the size of the n-th batch, and τ_n is the interarrival time between the $(n-1)$-st and the n-th arrival events. Similar to Eq. (2.34), $\{(I_n, J_n, \tau_n), n = 0, 1, 2, \ldots\}$ satisfies, for $x \geq 0$ and $k = 1, \ldots, N$,

$$P\{I_n = j,\ J_n = k,\ \tau_n \leq x\,|I_{n-1} = i\} = \left(\int_0^x \exp(D_0 t)\mathrm{d}t D_k\right)_{i,j}. \qquad (2.86)$$

Definition 2.4.3 Assume that $\{D_0, D_1, \ldots, D_N\}$ satisfy the conditions given in Definition 2.4.1. Let $I(t)$ be the phase of a continuous time Markov chain with infinitesimal generator $D = D_0 + D_1 + \ldots + D_N$, at time t. Define a Markov renewal process $\{(I_n, J_n, \tau_n), n = 0, 1, 2, \ldots\}$ using Eq. (2.86). Define $\{N(t) < k\} = \cup_{n>0}\{\tau_1+\tau_2 + \ldots+\tau_n > t, J_1 + J_2 + \ldots + J_{n-1} < k \leq J_1 + J_2 + \ldots + J_n\}$. Then $\{(N(t), I(t)), t \geq 0\}$ is a *BMAP*.

Exercise 2.4.4 Show that $(-D_0)^{-1}(D_1 + \ldots + D_N)\mathbf{e} = \mathbf{e}$. Explain the results intuitively.

Exercise 2.4.5 Show that, in a *BMAP*, if $I(t)$ has a distribution $\boldsymbol{\alpha}$, the distribution of the time until the first arrival after t is *PH*-distributed with a matrix representation $(\boldsymbol{\alpha}, D_0)$. Show that the distribution of the time until the first arrival of a batch of size k after t is *PH*-distributed with matrix representation $(\boldsymbol{\alpha}, D - D_k)$.

To analyze the counting process $\{N(t), t \geq 0\}$ of a *BMAP*, define $\{p_{i,j}(n, t), i, j = 1, .., m\}$ and $P(n, t)$ as in Eq. (2.45). Similar to Eq. (2.46), conditioning on the number of arrivals in $(t, t+\delta t)$, we obtain

$$\begin{aligned} p_{i,j}(n, t+\delta t) = {} & p_{i,j}(n,t)(1-(-d_{0,(j,j)})\delta t) + \sum_{k=1:k\neq j}^{m} p_{i,k}(n,t)d_{0,(k,j)}\delta t \\ & + \sum_{l=1}^{\min\{N,n\}}\sum_{k=1}^{m} p_{i,k}(n-l,t)d_{l,(k,j)}\delta t + o(\delta t). \end{aligned} \qquad (2.87)$$

By Eq. (2.87), we obtain

$$\frac{\mathrm{d}P(n,t)}{\mathrm{d}t} = P(n,t)D_0 + P(n-1,t)D_1 + \cdots + P(n-\min\{n, N\}, t)D_{\min\{n,N\}}. \qquad (2.88)$$

Equation (2.88) can be interpreted intuitively as follows. On the right-hand side of Eq. (2.88), $P(n-i, t)D_i$ is the (matrix) rate of transitions that there are $n-i$ customers in the system at time t, and i customers arrive at time t. Consequently, there are $n - i + i = n$ customers right after time t.

By routine calculations, similar to the proof of Theorem 2.3.1, we obtain the following theorem.

Theorem 2.4.1

$$P^*(z,t) = \sum_{n=0}^{\infty} z^n P(n,t) = \exp\{(D_0 + zD_1 + z^2 D_2 + \cdots + z^N D_N)t\} \\ \equiv \exp(D^*(z)t). \tag{2.89}$$

The result in Theorem 2.4.1 is quite intuitive. In the expression, z^k is associated with D_k since D_k is the (matrix) arrival rate of batches of size k. $D^*(z)$ can be considered as a "rate generating function" of the number of customers arriving per unit time. Then $\exp(D^*(z)t)$ is the probability generating function of the number of customers arrived in $[0, t]$. This intuitive explanation is related to Exercise 2.3.3.

By routine calculations, it is easy to obtain:

$$\left.\frac{\partial P^*(z,t)}{\partial z}\right|_{z=1} = \sum_{n=1}^{\infty} \frac{t^n}{n!} \sum_{k=0}^{n-1} D^k \left(\sum_{j=1}^{N} jD_j\right) D^{n-1-k};$$
$$\left.\frac{\partial P^*(z,t)}{\partial z}\right|_{z=1} \mathbf{e} = \sum_{n=1}^{\infty} \frac{t^n}{n!} D^{n-1} \left(\sum_{j=1}^{N} jD_j\right) \mathbf{e}. \tag{2.90}$$

Note that $D\mathbf{e} = 0$ is used to obtain the second equality in Eq. (2.90). By Eq. (2.90), it is possible to obtain the expected number of arrivals in $[0, t]$ for any initial distribution $\boldsymbol{\alpha}$. Theorem 2.3.2 can be generalized as follows.

Theorem 2.4.2 *Assume that the underlying Markov chain $\{I(t), t \geq 0\}$ is irreducible, i.e., D is irreducible. Given initial distribution $\boldsymbol{\alpha}$, we have*

$$E[N(t)] = \lambda t + \boldsymbol{\alpha}(\exp(Dt) - I)(D - \mathbf{e}\boldsymbol{\theta})^{-1}\left(\sum_{j=1}^{N} jD_j \mathbf{e}\right), \quad t \geq 0, \tag{2.91}$$

where $\boldsymbol{\theta}$ is the stationary distribution vector of D, i.e., $\boldsymbol{\theta}D = 0$ and $\boldsymbol{\theta}\mathbf{e} = 1$, and $\lambda = \boldsymbol{\theta}\left(\sum_{j=1}^{N} jD_j\right)\mathbf{e}$.

In particular, for a stationary process, we have

$$\boldsymbol{\theta}\left.\frac{\partial P^*(z,t)}{\partial z}\right|_{z=1} \mathbf{e} = \boldsymbol{\theta} \sum_{n=1}^{\infty} \frac{t^n}{n!} \sum_{k=0}^{n-1} D^k \left(\sum_{j=1}^{N} jD_j\right) D^{n-1-k} \mathbf{e} = \boldsymbol{\theta}\left(\sum_{j=1}^{N} jD_j\right) \mathbf{e}t = \lambda t. \tag{2.92}$$

Equation (2.92) gives an interpretation to the (stationary) arrival rate λ.

Exercise 2.4.6 Find the arrival rates of the *BMAP*s defined in Example 2.4.2.

The *arrival rate of batches* (not the total number of arrivals) is given by $\hat{\lambda} = \boldsymbol{\theta}\left(\sum_{j=1}^{N} D_j\right)\mathbf{e}$. The *arrival rate of batches of size j* is $\lambda_j = \boldsymbol{\theta}D_j\mathbf{e}$.

Exercise 2.4.7 Find the batch arrival rates and the arrival rates of batches of size j for *BMAP*s defined in Example 2.4.2.

Superposition of independent batch Markovian arrival processes Consider independent *BMAP*s $(D_1, D_2, \ldots, D_N)$ of order m_1 and $(C_1, C_2, \ldots, C_M)$ of order m_2. The superposition process is a *BMAP* with matrix representation

$$D_0 \otimes I + I \otimes C_0, \quad D_n \otimes I + I \otimes C_n, \quad n = 1,\ 2,\ \ldots. \tag{2.93}$$

Exercise 2.4.8 In expression (2.93), why isn't $\sum_{k=0}^{n} (D_k \otimes I + I \otimes C_{n-k})$ the arrival rate of batches of size n, for $n = 1, 2, \ldots$? (Hint: For continuous time Markov chains, only one event can occur at a time.)

Decomposition (marking) of batch Markovian arrival processes Similar to the decomposition of the Poisson process and *MAP*s. There are various ways to mark arrivals: group marking, individual marking, etc. Explicit expressions can be obtained for the resulting Markovian arrival processes for various marking schemes. See Sect. 2.5 for details.

Example 2.4.3 For a *BMAP*, assume that a customer is marked as a type 1 customer with probability p. Then in any arrived batch of size k, the number of customers marked as type 1 has a binomial distribution with parameter k and p (recall Exercise 1.1.17). The resulting arrival process $\{(N_1(t), I(t)), t \geq 0\}$ of type 1 customers is a *BMAP* as well. A matrix representation of $\{(N_1(t), I(t)), t \geq 0\}$ is given by

$$D_0 + \sum_{k=0}^{N} (1-p)^k D_k, \quad \sum_{k=n}^{N} \frac{k!}{n!(k-n)!} p^n (1-p)^{k-n} D_k, \quad n = 1, 2, \ldots, N. \tag{2.94}$$

To end this section, a fundamental result about *BMAP*s given in Asmussen and Koole (1993) is presented without a proof. This result demonstrates the applicability of *BMAP*s in stochastic modeling.

Theorem 2.4.3 (Asmussen and Koole (1993)) *Any stochastic arrival process can be approximated by BMAPs.*

Commentary Almost all the results obtained for *MAP*s in Sect. 2.3 can be generalized to *BMAP*s. For example, the correlation between the numbers of arrivals in non-overlapping intervals can be found in a way similar to that in Sect. 2.3. By replacing D_1 with $\bar{D}_1 = \sum_{n=1}^{N} nD_n$ or $\hat{D}_1 = \sum_{n=1}^{N} D_n$ in the formulas in Sect. 2.3 properly, we can obtain results for the *BMAP*s (e.g., see Exercises 2.4.9 and 2.4.10).

Additional Exercises and Extensions

Exercise 2.4.9 (Extension of Theorem 2.3.3) Assume that the underlying Markov chain $\{I(t), t \geq 0\}$ is irreducible, i.e., D is irreducible. Given initial distribution $\boldsymbol{\theta}$, i.e., the stationary distribution of D, show that

$$Var(N(t)) = (\lambda_2 - 2\lambda^2 - 2\boldsymbol{\theta}\bar{D}_1(D - \mathbf{e}\boldsymbol{\theta})^{-1}\bar{D}_1\mathbf{e})t \\ + 2\boldsymbol{\theta}\bar{D}_1(D - \mathbf{e}\boldsymbol{\theta})^{-1}(\exp(Dt) - I)(D - \mathbf{e}\boldsymbol{\theta})^{-1}\bar{D}_1\mathbf{e}. \tag{2.95}$$

where $\lambda_2 = \boldsymbol{\theta}\sum_{n=1}^{N} n^2 D_n\mathbf{e}$. Note:

$$\begin{aligned}\left.\frac{\partial^2 P^*(z,t)}{\partial z^2}\right|_{z=1} &= \sum_{n=2}^{\infty}\frac{t^n}{n!}\sum_{k=1}^{n-1}\left(\sum_{l=0}^{k-1} D^l D^{*(1)}(1)D^{k-1-l}\right)D^{*(1)}(1)D^{n-1-k} \\ &\quad + \sum_{n=2}^{\infty}\frac{t^n}{n!}\sum_{k=0}^{n-1} D^k D^{*(2)}(1)D^{n-1-k} \\ &\quad + \sum_{n=2}^{\infty}\frac{t^n}{n!}\sum_{k=0}^{n-2} D^k D^{*(1)}(1)\left(\sum_{l=0}^{n-2-k} D^l D^{*(1)}(1)D^{n-2-k-l}\right).\end{aligned} \tag{2.96}$$

Exercise 2.4.10 (Extension of Exercise 2.3.12) Assume that the underlying Markov chain $\{I(t), t \geq 0\}$ is irreducible, i.e., D is irreducible. Assume that $I(0)$ has distribution $\boldsymbol{\theta}$. Prove that, for $t, u, v, s \geq 0$,

$$\begin{aligned} &E[N(t, t+u)N(t+u+v, t+u+v+s)] \\ &= \lambda^2 us + \boldsymbol{\theta}\bar{D}_1(D - \mathbf{e}\boldsymbol{\theta})^{-1}(\exp(Du) - I) \\ &\quad \cdot \exp(Dv)(\exp(Ds) - I)(D - \mathbf{e}\boldsymbol{\theta})^{-1}\bar{D}_1\mathbf{e}. \end{aligned} \tag{2.97}$$

Exercise 2.4.11 Prove $\lim_{t\to\infty} E[N(t)]/t = \lambda$ for any initial distribution $\boldsymbol{\alpha}$. Explain the result intuitively.

Exercise 2.4.12 For *BMAP* $(D_0, D_1, \ldots, D_N)$, denote by $N(j, t)$ the number of batches of size j or less arrived in $[0, t]$. Is $\{(N(j, t), I(t)), t \geq 0\}$ a Markovian arrival process? If the answer is yes, find a matrix representation for the Markovian arrival process.

Exercise 2.4.13 For a *BMAP* with the matrix representation given in Eq. (2.98), in steady state, (i) find the arrival rate, and arrival rates of individual batches; (ii) find the distribution of the interarrival time; and (iii) find the distribution of the time between two consecutive arrivals of batch size 2.

$$D_0 = \begin{pmatrix} -8 & 2 & 3.6 \\ 3 & -10-4 & 0 \\ 0.4 & 1.5 & -5-1 \end{pmatrix}, \quad D_1 = \begin{pmatrix} 0 & 0 & 0 \\ 7 & 2 & 0 \\ 1.6 & 0 & 1 \end{pmatrix}, \\ D_2 = \begin{pmatrix} 0 & 0 & 2.4 \\ 0 & 2 & 0 \\ 0 & 1.5 & 0 \end{pmatrix}. \tag{2.98}$$

Exercise 2.4.14 Answer questions without calculations. For a *BMAP* with the matrix representation given in Eq. (2.99), (i) find the arrival rate, and the arrival rates of individual batches; (ii) find the distribution of the interarrival time; and (iii) find the distribution of the time between two consecutive arrivals of batch size 2.

$$D_0 = \begin{pmatrix} -2 & 0 & 0 \\ 0 & -2 & 0 \\ 0 & 0 & -2 \end{pmatrix}, \quad D_1 = \begin{pmatrix} 0 & 2 & 0 \\ 0 & 0 & 2 \\ 0 & 0 & 0 \end{pmatrix}, \quad D_2 = \begin{pmatrix} 0 & 0 & 0 \\ 0 & 0 & 0 \\ 2 & 0 & 0 \end{pmatrix}. \quad (2.99)$$

Exercise 2.4.15 Buses arrive to a bus stop according to an *MAP* with

$$D_0 = \begin{pmatrix} -2 & 2 & 0 \\ 0 & -3 & 3 \\ 0 & 0 & -5 \end{pmatrix}, \quad D_1 = \begin{pmatrix} 0 & 0 & 0 \\ 0 & 0 & 0 \\ 5 & 0 & 0 \end{pmatrix}. \quad (2.100)$$

The number of passengers who get off a bus at the bus stop is random with distribution $p_n = c/n^2$, for $n = 1, 2, 3, \ldots$, where c is the normalization constant and $c = 1/(1 + 1/2^2 + \ldots + 1/n^2 + \ldots)$. Then passengers arrive at the bus stop according to a *BMAP*. (i) Find a matrix-representation for the *BMAP*; (ii) find the arrival rate of buses; and (iii) find the arrival rate of passengers.

Exercise 2.4.16 (Definition 2.4.4) Let $\{\alpha_i, i = 1, \ldots, m\}$ be nonnegative numbers with a unit sum, $\{d_{0,(i,j)}, i \neq j$, and $i, j = 1, \ldots, m\}$, $\{d_{n,(i,j)}, i, j = 1, \ldots, m\}$ be nonnegative numbers, for $n = 1, \ldots, N$ $(< \infty)$, and m be a finite positive integer. Assume $-d_{0,(i,i)} \equiv d_{0,(i,1)} + \ldots + d_{0,(i,i-1)} + d_{0,(i,i+1)} + \ldots + d_{0,(i,m)} + d_{1,(i,1)} + \ldots + d_{1,(i,i)} + \ldots + d_{1,(i,m)} + \ldots + d_{N,(i,1)} + \ldots + d_{N,(i,m)} > 0$, for $i = 1, \ldots, m$. We define stochastic process $\{(N(t), I(t)), t \geq 0\}$ as follows.

(1) Define $m((N + 1)m-1)$ independent Poisson processes with parameters $\{d_{0,(i,j)}, i \neq j$ and $i, j = 1, \ldots, m\}$ and $\{d_{n,(i,j)}, i, j = 1, \ldots, m, n = 1, \ldots, N\}$. If $d_{n,(i,j)} = 0$, the corresponding Poisson process has no event.
(2) Determine $I(0)$ by the probability distribution $\{\alpha_i, i = 1, \ldots, m\}$. Set $N(0) = 0$.
(3) If $I(t) = i$, for $i = 1, \ldots, m$, $I(t)$ and $N(t)$ remain the same until the first event occurs in the $(N + 1)m-1$ Poisson processes corresponding to $\{d_{0,(i,j)}, j = 1, \ldots, m$, and $j \neq i\}$ and $\{d_{n,(i,j)}, j = 1, \ldots, m, n = 1, \ldots, N\}$. If the next event comes from the Poisson process corresponding to $d_{0,(i,j)}$, the variable $I(t)$ changes from phase i to phase j and $N(t)$ does not change at the epoch, for $j = 1, \ldots, m$, and $j \neq i$; and if the next event comes from the Poisson process corresponding to $d_{n,(i,j)}$, the phase variable $I(t)$ transits from phase i to phase j and $N(t)$ increases by n at the epoch, i.e., that a batch of n arrivals is associated with the event, for $j = 1, \ldots, m$ and $1 \leq n \leq N$.

In Fig. 2.14, there are three batches of size 1 and two batches of size 2. For the process $\{(N(t), I(t)), t \geq 0\}$, show the following results.

(i) Prove that the sojourn time of $\{(N(t), I(t)), t \geq 0\}$ in state (n, i) has an exponential distribution with parameter $-d_{0,(i,i)}$, for $i = 1, \ldots, m$.

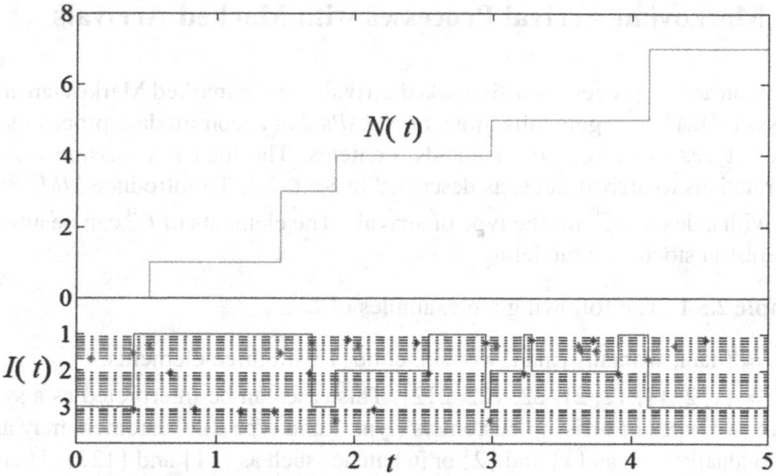

Fig. 2.14 Sample paths of underlying Poisson processes, $I(t)$ and $N(t)$ of a *BMAP*

(ii) Prove that the probability that the next phase is j and no arrival at the transition epoch is given by $d_{0,(i,j)}/(-d_{0,(i,i)})$, given that the current state is (n, i), for $i \neq j$ and $i, j = 1, \ldots, m$ and $n = 0, 1, 2, \ldots$.

(iii) Prove that the probability that the next phase is j and an arrival of size k occurs at the transition epoch is given by $d_{k,(i,j)}/(-d_{0,(i,i)})$, given that the current state is (n, i), for $i, j = 1, \ldots, m$, $k = 1, \ldots, N$, and $n = 0, 1, 2, \ldots$.

(iv) Prove that the sojourn time of $\{I(t), t\geq 0\}$ in phase i has an exponential distribution with parameter $-d_{0,(i,i)} - d_{1,(i,i)} - d_{2,(i,i)} - \ldots - d_{N,(i,i)}$, for $i = 1, \ldots, m$. (Hint: Use Proposition 1.1.4.)

(v) Let $D_0 = (d_{0,(i,j)})$, $D_k = (d_{k,(i,j)})$, $k = 1, \ldots, N$. Show that $\{I(t), t\geq 0\}$ is a continuous time Markov chain with infinitesimal generator $D = D_0 + D_1 + \ldots + D_K$.

(vi) Show that $\{(N(t), I(t)), t\geq 0\}$ is a continuous time Markov chain with the infinitesimal generator given in Eq. (2.85).

Exercise 2.4.17 (Terminating *BMAP*) Assume that X is a *PH*-random variable with matrix representation $(\boldsymbol{\beta}, T)$. For the *BMAP* defined in Definition 2.4.4, show that the number of arrivals in $[0, X]$ has a probability generating function

$$(\boldsymbol{\alpha}\otimes\boldsymbol{\beta})\left(I-(-(D_0\otimes I+I\otimes T))^{-1}\sum_{k=1}^{N}z^k(D_k\otimes I)\right)^{-1}(-(D_0\otimes I+I\otimes T))^{-1}(\mathbf{e}\otimes\mathbf{T}^0). \tag{2.101}$$

Exercise 2.4.18 Show that, for $s > 0$ and $0 \leq z \leq 1$,

$$\int_0^\infty e^{-st}E[z^{N(t)}]dt = (sI - D^*(z))^{-1} \tag{2.102}$$

2.5 Markovian Arrival Processes with Marked Arrivals

Markovian arrival processes with marked arrivals (or the marked Markovian arrival process) (*MMAP*) are generalizations of *BMAP*s that accommodate processes with different types of customers, demands, or items. The idea is to assign different interpretations to arrival rates, as described in Sect. 2.4. To introduce *MMAP*s, we begin with index set C^0 for the type of arrivals. The elements of C^0 can be anything of interest in stochastic modeling.

Example 2.5.1 The following are examples of C^0.

1. $C^0 = \{\text{man, woman, }\{\text{man, woman, child}\}\}$ with three elements.
2. $C^0 = \{1, 2, 11, 12, 21, 22, 122, 212\}$. This case can be interpreted as a system with two types of arriving customers (type 1 and type 2). Customers may arrive individually such as $\{1\}$ and $\{2\}$ or in batches such as $\{11\}$ and $\{122\}$. There are four types of batches $\{11, 12, 21, 22\}$ with two customers, in which the order of customers in a batch is considered. There are two types of batches with three customers $\{122, 212\}$.
3. $C^0 = \{1, 11, 111, \ldots, 1\ldots1\}$. There is only one type of customer, but customers can arrive in batches. It is easy to see that C^0 is the index set for *BMAP*s.

Next, we give three equivalent definitions of *MMAP*s.

Definition 2.5.1 Let C^0 be a set of indices. Assume that $\{D_h, h \in C^0\}$ are nonnegative matrices, D_0 is a matrix with nonnegative off-diagonal elements and negative diagonal elements, $D = D_0 + \sum_{h\in C^0} D_h$ is an infinitesimal generator of order m, and m is a positive integer. Let $D_0 = (d_{0,(i,j)})$, $D_h = (d_{h,(i,j)})$, for $h \in C^0$. Then $(D_0, D_h, h{\in}C^0)$ defines *MMAP* $\{(N_h(t), h{\in}C^0, I(t)), t \geq 0\}$.

1. Set $N_h(0) = 0, h \in C^0$.
2. Define a continuous time Markov chain $\{I(t), t \geq 0\}$ by D.
3. For phase i with $d_{h,(i,i)} > 0$, define a Poisson process with arrival rate $d_{h,(i,i)}$, for $h \in C^0$ and $1 \leq i \leq m$. The Poisson processes are turned on, if $I(t) = i$; otherwise, they are turned off,.
4. For $1 \leq i \leq m$ and $h \in C^0$, if $I(t) = i$ and an arrival from the imposed Poisson process $d_{h,(i,i)}$ occurs, $N_h(t)$ increases by one.
5. At the end of each stay in state i, with probability $d_{0,(i,j)}/(-d_{0,(i,i)})$, $I(t)$ transits from phase i to j and $N_h(t)$ remains the same (i.e., without an arrival) for $i \neq j$ and $h \in C^0$, and, with probability $d_{h,(i,j)}/(-d_{0,(i,i)})$, $I(t)$ transits from phase i to j, $N_h(t)$ increases by one, and other $N_u(t)$ remains the same, $u \in C^0$ and $u \neq h$, for $h \in C^0$ and $1 \leq i, j \leq m$.

We shall call $(D_0, D_h, h \in C^0)$ a matrix representation of *MMAP* $\{(N_h(t), h \in C^0, I(t)), t \geq 0\}$. The Markov chain $\{I(t), t \geq 0\}$ is called the underlying Markov chain of $\{(N_h(t), h \in C^0, I(t)), t \geq 0\}$.

Definition 2.5.2 Assume that $\{D_0, D_h, h \in C^0\}$ satisfy conditions in Definition 2.5.1. Define a continuous time Markov chain $\{(N_h(t), h \in C^0, I(t)), t \geq 0\}$ with transition rates given in $\{D_0, D_h, h \in C^0\}$ such that D_h corresponds to the increase of one in $N_h(t)$, for $h \in C^0$. Then $\{(N_h(t), h \in C^0, I(t)), t \geq 0\}$ is an *MMAP*.

For an *MMAP*, define a Markov renewal process $\{(I_n, J_n, \tau_n), n = 0, 1, 2, \ldots\}$, where I_n is the state right after the n-th arrival, J_n is the type of the n-th arrival, τ_n is the interarrival time between the $(n-1)$-st and the n-th arrivals, and

$$P\{I_n = j,\ J_n = h,\ \tau_n \leq x \,|I_{n-1} = i\} = \left(\int_0^x \exp(D_0 t)\mathrm{d}t D_h\right)_{i,j}. \tag{2.103}$$

Definition 2.5.3 Assume that $\{D_0, D_h, h \in C^0\}$ and D satisfy conditions in Definition 2.5.1. Let $I(t)$ be the phase of the continuous time Markov chain D at time t. Let

$$\begin{aligned}\{N_h(t) < n\} = \bigcup\nolimits_{k>0}\{&\tau_1 + \tau_2 + \ldots + \tau_k > t,\, I_{\{J_1=h\}} + I_{\{J_2=h\}} + \ldots \\ &+ I_{\{J_k=h\}} \geq n > I_{\{J_1=h\}} + I_{\{J_2=h\}} + \ldots + I_{\{J_{k-1}=h\}}\},\end{aligned} \tag{2.104}$$

where $I_{\{.\}}$ is the indicator function. Then $\{(N_h(t), h \in C^0, I(t)), t \geq 0\}$ is an *MMAP*.

The set of *MMAP*s is versatile. By properly adding phases and allocating transition rates, a broad range of arrival processes with special features can be constructed.

Example 2.5.2 *MMAP*s can be used to construct arrival processes with special features and special relationships between the arrivals.

(1) Cyclic arrivals:

$$D_0 = \begin{pmatrix} -2 & 0 \\ 0 & -3 \end{pmatrix},\ D_1 = \begin{pmatrix} 0 & 2 \\ 0 & 0 \end{pmatrix},\ D_2 = \begin{pmatrix} 0 & 0 \\ 3 & 0 \end{pmatrix}.$$

Type 1 and type 2 customers arrive cyclically. (See Fig. 2.15: Red "*" for type 1 arrivals and blue "o" for type 2 arrivals.)

(2) Bursty vs smooth:

$$D_0 = \begin{pmatrix} -1 & 0 \\ 0 & -100 \end{pmatrix},\ D_1 = \begin{pmatrix} 0 & 0 \\ 1 & 99 \end{pmatrix},\ D_2 = \begin{pmatrix} 0 & 1 \\ 0 & 0 \end{pmatrix}.$$

The type 1 process is bursty, while type 2 is smooth. (See Fig. 2.16)

(3) Individual vs group:

$$D_0 = \begin{pmatrix} -5 & 0 \\ 0 & -10 \end{pmatrix},\ D_1 = \begin{pmatrix} 4 & 0 \\ 0 & 9 \end{pmatrix},\ D_{2,1} = \begin{pmatrix} 0 & 1 \\ 1 & 0 \end{pmatrix}.$$

Every type 2 arrival is accompanied by a type 1 arrival.

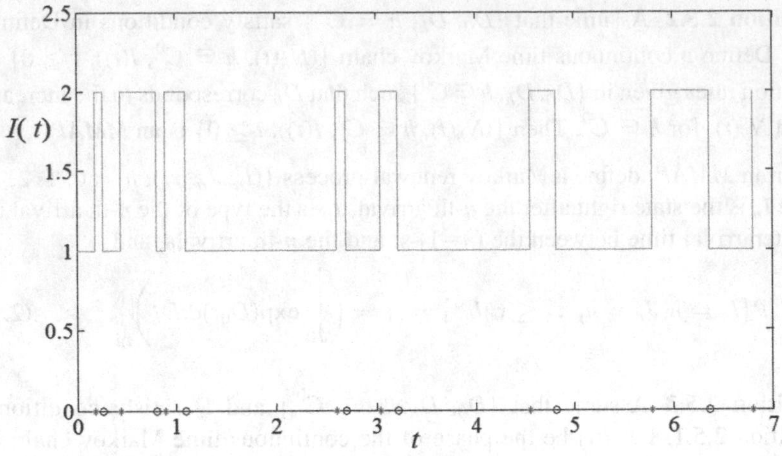

Fig. 2.15 A cyclic *MMAP*

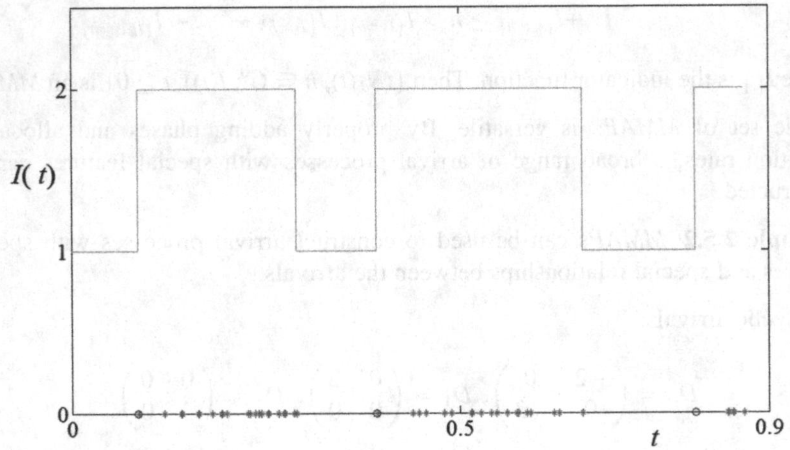

Fig. 2.16 A bursty *MMAP*

(4) Type 2 always follows type 1:

$$D_0 = \begin{pmatrix} -5 & 0 \\ 0 & -10 \end{pmatrix}, \; D_1 = \begin{pmatrix} 4 & 1 \\ 0 & 0 \end{pmatrix}, \quad D_{2,1} = \begin{pmatrix} 0 & 0 \\ 10 & 0 \end{pmatrix}.$$

(5) Orders in individual batches:

$$D_0 = \begin{pmatrix} -5 & 0 \\ 0 & -10 \end{pmatrix}, \quad D_{112} = \begin{pmatrix} 4 & 0 \\ 0 & 9 \end{pmatrix}, \quad D_{121} = \begin{pmatrix} 0 & 1 \\ 1 & 0 \end{pmatrix}; \text{ and}$$

$$D_0 = \begin{pmatrix} -5 & 0 \\ 0 & -10 \end{pmatrix}, \quad D_{311221} = \begin{pmatrix} 4 & 0 \\ 0 & 9 \end{pmatrix}, \quad D_{1213312} = \begin{pmatrix} 0 & 1 \\ 1 & 0 \end{pmatrix},$$

if the order within batches matters.

Following Example 2.5.2, more arrival processes with special features can be constructed. As demonstrated by Example 2.2.3 and Exercise 2.2.13, the interarrival times between events can be versatile. Thus, *MMAP*s can be used to model fairly general counting processes.

To analyze the numbers of arrivals in [0, *t*], we define several functions as follows:

$$\begin{aligned}
p_{i,j}(\{n_h,\ h \in C^0\}, t) &= P\{I(t) = j,\ N_h(t) = n_h,\ h \in C^0\ |I(0) = i\}; \\
P(\{n_h,\ h \in C^0\}, t) &= (p_{i,j}(\{n_h,\ h \in C^0\}, t)); \\
P^*(\{z_h,\ h \in C^0\}, t) &= \sum_{\{n_h \geq 0,\ h \in C^0\}} P(\{n_h,\ h \in C^0, t) \prod_{h \in C^0} z_h^{n_h}; \\
D^*(\{z_h,\ h \in C^0\}) &= D_0 + \sum_{h \in C^0} z_h D_h, \quad |z_h| \leq 1,\ h \in C^0.
\end{aligned} \tag{2.105}$$

By routine calculations, we obtain the following result.

Theorem 2.5.1 *For an MMAP, the joint probability generating function for the numbers of arrivals in* [0, *t*] *is given by*

$$P^*(\{z_h,\ h \in C^0\}, t) = \exp(D^*(\{z_h,\ h \in C^0\})\, t). \tag{2.106}$$

Example 2.5.3 Assume $C^0 = \{\text{man, woman, }\{\text{man, woman, child}\}\}$. The matrix representation of an *MMAP* is

$$\begin{aligned}
D_0 = \begin{pmatrix} -2-4 & 1.6 \\ 3 & -10 \end{pmatrix}, \quad D_M = \begin{pmatrix} 2 & 0.2 \\ 2 & 0 \end{pmatrix}, \\
D_W = \begin{pmatrix} 1 & 0.2 \\ 3 & 0 \end{pmatrix}, \quad D_{MWC} = \begin{pmatrix} 1 & 0 \\ 2 & 0 \end{pmatrix}.
\end{aligned} \tag{2.107}$$

Then we obtain

$$P^*(\{z_h,\ h \in C^0\}, t) = \exp((D_0 + z_M D_M + z_W D_W + z_{MWC} D_{MWC})\, t). \tag{2.108}$$

Equation (2.106) can be used to find various arrival rates. For example, taking partial derivative of both sides of Eq. (2.106) with respect to z_h, yields

$$\left.\frac{\partial P^*(\{z_c,\ c\in C^0\},t)}{\partial z_h}\right|_{\{z_c=1,\ c\in C^0\}}\mathbf{e}=\sum_{n=0}^{\infty}\frac{t^n}{n!}D^{n-1}D_h\mathbf{e}. \tag{2.109}$$

Given initial distribution $\boldsymbol{\alpha}$, we have

$$E[N_h(t)]=\lambda_h t+\boldsymbol{\alpha}(\exp(Dt)-I)(D-\mathbf{e}\boldsymbol{\theta})^{-1}D_h\mathbf{e},\quad t\geq 0, \tag{2.110}$$

where $\lambda_h=\boldsymbol{\theta}D_h\mathbf{e}$, which is the arrival rate of type h batches, where $\boldsymbol{\theta}$ is the stationary distribution vector of D, i.e., $\boldsymbol{\theta}D=0$ and $\boldsymbol{\theta}\mathbf{e}=1$.

$$\begin{aligned}Var(N_h(t))=&(\lambda_h-2\lambda_h^2-2\boldsymbol{\theta}D_h(D-\mathbf{e}\boldsymbol{\theta})^{-1}D_h\mathbf{e})t\\&+2\boldsymbol{\theta}D_h(D-\mathbf{e}\boldsymbol{\theta})^{-1}(\exp(Dt)-I)(D-\mathbf{e}\boldsymbol{\theta})^{-1}D_h\mathbf{e}.\end{aligned} \tag{2.111}$$

Exercise 2.5.1 (Example 2.5.3 continued) For the *MMAP* defined in Example 2.5.3, calculate the arrival rates for (i) batches with a man only; (ii) batches with a woman only; and (iii) batches with a man, a woman, and a child. Calculate the arrival rate of men, arrival rate of women, and arrival rate of children.

For the *MMAP* defined in Example 2.5.3, let $N_{\{M\}}(t)$ be the total number of men arrived in [0, t], $N_{\{W\}}(t)$ the total number of women arrived in [0, t], and $N_{\{C\}}(t)$ the total number of children arrived in [0, t]. It is easy to see that $N_{\{M\}}(t)=N_M(t)+N_{MWC}(t)$, $N_{\{W\}}(t)=N_W(t)+N_{MWC}(t)$, and $N_{\{C\}}(t)=N_{MWC}(t)$. Define

$$\begin{aligned}p_{i,j}(n_M,n_W,n_C,t)&=P\{N_{\{M\}}(t)=n_M,\ N_{\{W\}}(t)=n_W,N_{\{C\}}(t)\\&\qquad=n_C,I(t)=j|I(0)=i\},\quad 1\leq i,j\leq m;\\P(n_M,n_W,n_C,t)&=(p_{i,j}(n_M,n_W,n_C,t));\\P^*(z_{\{M\}},z_{\{W\}},z_{\{C\}},t)&=\sum_{n_M,n_W,n_C\geq 0}z_{\{M\}}^{n_M}z_{\{W\}}^{n_W}z_{\{C\}}^{n_C}P(n_M,n_W,n_C,t).\end{aligned} \tag{2.112}$$

Exercise 2.5.2 Show that

$$P^*(z_{\{M\}},z_{\{W\}},z_{\{C\}},t)=\exp\left(\left(D_0+z_{\{M\}}D_M+z_{\{W\}}D_W+z_{\{M\}}z_{\{W\}}z_{\{C\}}D_{MWC}\right)t\right). \tag{2.113}$$

Find formulas for the (stationary) arrival rates of men, women, and children, respectively. (Hint: To find the arrival rate of men, you can set $z_{\{W\}}=z_{\{C\}}=1$ in Eq. (2.113).)

Exercise 2.5.3 In Example 2.5.3, show that $\{(N_M(t), I(t)), t\geq 0\}$ is an *MAP*. Find a matrix representation for that *MAP*. (Note the difference between the arrival of all the men and the arrival of the men alone.)

Exercise 2.5.4 Find the arrival rates of individual items and individual batches for *MMAP*s defined in Example 2.5.2.

Exercise 2.5.5 (*BMAP*s) Assume $K = 1$ and $C^0 = \{1, 11, 111\}$. Define $\{N_h(t), h \in C^0, I(t), t \geq 0\}$ as an *MMAP* with matrix representation ($D_0, D_h, h \in C^0$). Define $N(t) = N_1(t) + 2N_{11}(t) + 3N_{111}(t)$. Show that $\{(N(t), I(t)), t \geq 0\}$ is a *BMAP* with matrix representation ($C_0 = D_0, C_1 = D_1, C_2 = D_{11}, C_3 = D_{111}$). Find the probability generating function of the *BMAP*.

Exercise 2.5.6 Assume $K = 2$ and $C^0 = \{1, 2, 11, 12, 21, 22\}$. Define $\{N_h(t), h \in C^0, I(t), t \geq 0\}$ as an *MMAP* with matrix representation ($D_0, D_h, h \in C^0$). Define $N_{\{1\}}(t) = N_1(t) + 2N_{11}(t) + N_{12}(t) + N_{21}(t)$, which is the total number of type 1 customers in $[0, t]$. Show that $\{(N_{\{1\}}(t), I(t)), t \geq 0\}$ is a *BMAP* with matrix representation ($C_0 = D_0 + D_2 + D_{22}, C_1 = D_1 + D_{12} + D_{21}, C_2 = D_{11}$). Find the probability generating function of the *BMAP*.

The covariances and correlations between $\{N_h(t), h \in C^0\}$ can be found explicitly. Without loss of generality, we assume $C^0 = \{1, 2\}$.

Theorem 2.5.2 *For the stationary version of the MMAP with* (D_0, D_1, D_2), *the covariance between* $N_1(t)$ *and* $N_2(t)$ *is given by*

$$\begin{aligned} Cov(N_1(t), N_2(t)) = &-\left(2\lambda_1\lambda_2 + \boldsymbol{\theta}\left(\sum_{k=1}^{2} D_k(D - \mathbf{e}\boldsymbol{\theta})^{-1} D_{3-k}\right)\mathbf{e}\right)t \\ &+ \boldsymbol{\theta}\left(\sum_{k=1}^{2} D_k(D - \mathbf{e}\boldsymbol{\theta})^{-1}(\exp(Dt) - I)(D - \mathbf{e}\boldsymbol{\theta})^{-1} D_{3-k}\right)\mathbf{e}. \end{aligned} \tag{2.114}$$

Proof. Note that

$$Cov(N_1(t), N_2(t)) = \frac{1}{2}\{Var(N_1(t) + N_2(t)) - Var(N_1(t)) - Var(N_2(t))\}. \tag{2.115}$$

By Theorem 2.3.3, we have, for $k = 1, 2$,

$$\begin{aligned} Var(N_k(t)) = &(\lambda_k - 2\lambda_k^2 - 2\boldsymbol{\theta} D_k(D - \mathbf{e}\boldsymbol{\theta})^{-1} D_k\mathbf{e})t \\ &+ 2\boldsymbol{\theta} D_k(D - \mathbf{e}\boldsymbol{\theta})^{-1}(\exp(Dt) - I)(D - \mathbf{e}\boldsymbol{\theta})^{-1} D_k\mathbf{e}, \end{aligned} \tag{2.116}$$

and

$$\begin{aligned} &Var(N_1(t) + N_2(t)) \\ &= (\lambda_1 + \lambda_2 - 2(\lambda_1 + \lambda_2)^2 - 2\boldsymbol{\theta}(D_1 + D_2)(D - \mathbf{e}\boldsymbol{\theta})^{-1}(D_1 + D_2)\mathbf{e})t \\ &\quad + 2\boldsymbol{\theta}(D_1 + D_2)(D - \mathbf{e}\boldsymbol{\theta})^{-1}(\exp(Dt) - I)(D - \mathbf{e}\boldsymbol{\theta})^{-1}(D_1 + D_2)\mathbf{e}. \end{aligned} \tag{2.117}$$

Then Eq. (2.114) is obtained by routine calculations. This completes the proof of Theorem 2.5.2.

To find the correlation between any two types of arrivals in an *MMAP*, we ignore all other types of arrivals and apply Theorem 2.5.2.

Exercise 2.5.7 (Exercise 2.5.6 continued) If $N_{\{1\}}(t) = N_1(t) + N_{21}(t)$ and $N_{\{2\}}(t) = N_2(t) + N_{21}(t)$, find the covariance between $N_{\{1\}}(t)$ and $N_{\{2\}}(t)$ in terms of the covariance between $N_1(t)$, $N_2(t)$, and $N_{21}(t)$.

Commentary He (1996, 2001), He and Neuts (1998), and Latouche et al. (2003) provide more details on the analysis of *MMAP*s and their application in queueing theory.

Additional Exercises and Extensions

Exercise 2.5.8 Show that the superposition process of two independent *MMAP*s is an *MMAP*. Find a matrix representation of the superposition process.

Exercise 2.5.9 Show that, if arrivals of an *MMAP* are marked by independent Burnoulli trials, the resulting process is again an *MMAP*.

Exercise 2.5.10 For the following *MMAP*, the two types of events arrive cyclically.

$$D_0 = \begin{pmatrix} D_{0,(1,1)} & 0 \\ 0 & D_{0,(2,2)} \end{pmatrix}, \quad D_1 = \begin{pmatrix} 0 & D_{1,(1,2)} \\ 0 & 0 \end{pmatrix}, \quad D_2 = \begin{pmatrix} 0 & 0 \\ D_{2,(2,1)} & 0 \end{pmatrix}. \tag{2.118}$$

Find the stationary arrival rates.

Exercise 2.5.11 Follow the approach used in Exercise 2.5.10 to construct a cyclic *MMAP* for which the interarrival times are *PH*-distributed with *PH*-representations $(\boldsymbol{\alpha}, T)$ and $(\boldsymbol{\beta}, S)$. Find the arrival rates.

Exercise 2.5.12 Construct an *MMAP* with three types of arrivals that arrive cyclically. You can assume that batch sizes are one and the interarrival times have *PH*-distributions.

Exercise 2.5.13 An *MMAP* has two types of arrivals, type 1 and type 2, that arrive individually (Batch size is one). The interarrival times are *PH*-distributed. After each arrival, the type of the next arrival is determined immediately. The time for a type 1 arrival has a *PH*-distribution with matrix representation

$$\boldsymbol{\alpha}_1 = (0.2,\ 0.8), \quad T_1 = \begin{pmatrix} -5 & 3 \\ 0.5 & -2 \end{pmatrix}. \tag{2.119}$$

With probability 0.6 (or 0.4), the next arrival is of type 1 (or 2). The time for a type 2 arrival has a *PH*-distribution with matrix representation

$$\boldsymbol{\alpha}_2 = (0.5, \ 0.5), \quad T_2 = \begin{pmatrix} -1 & 1 \\ 0 & -0.5 \end{pmatrix}. \tag{2.120}$$

With probability 0.2 (or 0.8), the next arrival is of type 2 (or 1). Find a matrix representation for the *MMAP*. Compute the arrival rates of type 1 and type 2 arrivals, respectively. (Note: Alternatively, the type of an arrival is determined at its arrival epoch. The next arrival time depends on the type of the arrival.)

Exercise 2.5.14 (Exercise 2.2.18 and 2.2.19 continued) For the model introduced in Exercise 2.2.19, we are interested in the number of repairs and the number of failures in $[0, \ t]$. We define the events of the Poisson process associated with repair rates as r arrivals (corresponding to D_r) and the events of the Poisson process associated with failure rates as f arrivals (corresponding to D_f). The rest of the Poisson processes correspond to D_0. Set $C^0 = \{\text{failure, repair}\}$. Let $N_r(t)$ be the number of repairs completed in $[0, \ t]$ and $N_f(t)$ the number of failures occurring in $[0, \ t]$. Then it can be shown that $\{(N_r(t), \ N_f(t), \ I(t)), \ t \geq 0\}$ is an *MMAP* with a matrix representation

$$
\begin{array}{c}
\begin{array}{ccc} (3,2) & (3,0) & (1,0) \end{array} \\
D_0 = \begin{array}{c} (3,2) \\ (3,0) \\ (1,0) \end{array} \begin{pmatrix} -\lambda & 0 & 0 \\ 0 & -\lambda-\mu & 0 \\ 0 & 0 & -\mu \end{pmatrix},
\end{array}
\qquad
\begin{array}{c}
\begin{array}{ccc} (3,2) & (3,0) & (1,0) \end{array} \\
D_r = \begin{array}{c} (3,2) \\ (3,0) \\ (1,0) \end{array} \begin{pmatrix} 0 & 0 & 0 \\ \mu & 0 & 0 \\ 0 & \mu & 0 \end{pmatrix},
\end{array}
$$

$$
\begin{array}{c}
\begin{array}{ccc} (3,2) & (3,0) & (1,0) \end{array} \\
D_f = \begin{array}{c} (3,2) \\ (3,0) \\ (1,0) \end{array} \begin{pmatrix} 0 & \lambda & 0 \\ 0 & 0 & \lambda \\ 0 & 0 & 0 \end{pmatrix}.
\end{array}
$$

$$\tag{2.121}$$

If $\lambda = 0.01$ and $\mu = 0.5$, the stationary distribution of $\{I(t), \ t \geq 0\}$ is $\boldsymbol{\theta} = (0.9800, \ 0.0196, \ 0.0004)$, the arrival rate of failure is $\boldsymbol{\theta} D_f \mathbf{e} = 0.009996$, and the arrival rate of repair is $\boldsymbol{\theta} D_r \mathbf{e} = 0.009996$.

Exercise 2.5.15 (Exercise 2.5.14 continued) For the model introduced in Exercise 2.5.14, assume that the times to failure have a common *PH*-distribution with matrix representation $(\boldsymbol{\alpha}, \ T)$ and the repair times have a common *PH*-distribution with matrix representation $(\boldsymbol{\beta}, \ S)$. Then an *MMAP* $\{(N_r(t), \ N_f(t), \ I(t)), \ t \geq 0\}$ can be constructed for the numbers of failures and repairs, which has a matrix representation

$$D_0 = \begin{array}{c} (3,2) \\ (3,0) \\ (1,0) \end{array} \overset{\begin{array}{ccc} (3,2) & (3,0) & (1,0) \end{array}}{\begin{pmatrix} T & 0 & 0 \\ 0 & T\otimes I + I\otimes S & 0 \\ 0 & 0 & S \end{pmatrix}}, \quad D_{\mathrm{r}} = \begin{array}{c} (3,2) \\ (3,0) \\ (1,0) \end{array} \overset{\begin{array}{ccc} (3,2) & (3,0) & (1,0) \end{array}}{\begin{pmatrix} 0 & 0 & 0 \\ I\otimes \mathbf{S}^0 & 0 & 0 \\ 0 & \boldsymbol{\alpha}\otimes(\mathbf{S}^0\boldsymbol{\beta}) & 0 \end{pmatrix}},$$

$$D_{\mathrm{f}} = \begin{array}{c} (3,2) \\ (3,0) \\ (1,0) \end{array} \overset{\begin{array}{ccc} (3,2) & (3,0) & (1,0) \end{array}}{\begin{pmatrix} 0 & (\mathbf{T}^0\boldsymbol{\alpha})\otimes\boldsymbol{\beta} & 0 \\ 0 & 0 & \mathbf{T}^0\otimes I \\ 0 & 0 & 0 \end{pmatrix}}. \tag{2.122}$$

Exercise 2.5.16 (Definition 2.5.4) Let $\{\alpha_i, i = 1, \ldots, m\}$ be nonnegative numbers with a unit sum, $\{d_{0,(i,j)}, i \neq j, \text{and } i, j = 1, \ldots, m\}$, $\{d_{h,(i,j)}, i, j = 1, \ldots, m, h \in C^0\}$ be nonnegative numbers, and m be a finite positive integer. Assume $-d_{0,(i,i)} \equiv \sum_{j=1, j\neq i}^{m} d_{0,(i,j)} + \sum_{h\in C^0} \sum_{j=1}^{m} d_{h,(i,j)} > 0$, for $i = 1, \ldots, m$. We define a stochastic process $\{(N_h(t), h \in C^0, I(t)), t \geq 0\}$ as follows.

(1) Define independent Poisson processes with parameters $\{d_{0,(i,j)}, 1 \leq i \neq j \leq m\}$ and $\{d_{h,(i,j)}, 1 \leq i, j \leq m, h \in C^0\}$. If $d_{0,(i,j)} = 0$ or $d_{h,(i,j)} = 0$, the corresponding Poisson process has no event.
(2) Determine $I(0)$ by the probability distribution $\{\alpha_i, 1 \leq i \leq m\}$. Set $N_h(0) = 0$, for $h \in C^0$.
(3) If $I(t) = i$, for $1 \leq i \leq m$, $I(t)$ and $\{N_h(t), h \in C^0\}$ remain the same until the first event occurs in the Poisson processes corresponding to $\{d_{0,(i,j)}, j = 1, \ldots, m, j \neq i\}$ and $\{d_{h,(i,j)}, j = 1, \ldots, m, h \in C^0\}$. If the next event comes from the Poisson process corresponding to $d_{0,(i,j)}$, the variable $I(t)$ changes from phase i to phase j and $\{N_h(t), h \in C^0\}$ do not change at the epoch, for $j = 1, \ldots, m, j \neq i$; If the next event comes from the Poisson process corresponding to $d_{h,(i,j)}$, the phase variable $I(t)$ changes from phase i to phase j, $N_h(t)$ increases by one (or by a pre-specified number, such as the batch size) at the epoch, and $N_l(t)$ remains the same for $l \neq h$ and $l \in C^0$, for $i, j = 1, \ldots, m, h \in C^0$.

For the process $\{(N_h(t), h \in C^0, I(t)), t \geq 0\}$, prove the following results.

(i) The process $\{I(t), t \geq 0\}$ is a continuous time Markov chain with infinitesimal generator $D = D_0 + \sum_{h\in C^0} D_h$.
(ii) The process $\{N_h(t), h \in C^0, I(t), t \geq 0\}$ is a marked Markovian arrival process with transition rates given by $(D_0, D_h, h \in C^0)$.

Exercise 2.5.17 (Terminating *MMAP*) Assume that X is a *PH*-random variable with matrix representation $(\boldsymbol{\beta}, T)$. For the *MMAP* defined in Definition 2.5.4, show that the conditional joint probability generating function of the numbers of arrivals in $[0, X]$ is given by

$$(\boldsymbol{\alpha}\otimes\boldsymbol{\beta})\left(I-(-(D_0\otimes I+I\otimes T))^{-1}\sum_{h\in C^0} z_h(D_h\otimes I)\right)^{-1}(-(D_0\otimes I+I\otimes T))^{-1}(\mathbf{e}\otimes\mathbf{T}^0). \tag{2.123}$$

(Note: The multivariate random vector with the probability generating function given in Eq. (2.123) is a discrete multivariate phase-type distribution.)

Similar to Bernoulli trials, thinning is another way to single out or remove some arrivals from the rest of the arrivals.

Exercise 2.5.18 (Thinning of *MAP*s) Consider an *MAP* (D_0, D_1). After an exponential time with parameter λ, the next arrival is marked as type 2. The arrivals during the exponential time are marked as type 1. Immediately after one arrival is marked as type 2, the exponential clock is restarted. Show that the resulting arrival process is an *MMAP* with matrix representation

$$C_0=\begin{pmatrix} D_0-\lambda I & \lambda I \\ 0 & D_0\end{pmatrix},\quad C_1=\begin{pmatrix} D_1 & 0 \\ 0 & 0\end{pmatrix},\quad C_2=\begin{pmatrix} 0 & 0 \\ D_1 & 0\end{pmatrix}. \tag{2.124}$$

Exercise 2.5.19 (Thinning of *MAP*s) Consider an *MAP* (D_0, D_1). After a phase-type time $(\boldsymbol{\beta}, T)$, the next arrival is thinned out. Immediately after one arrival is removed, the phase-type clock is restarted. Show that the resulting arrival process is an *MAP* with matrix representation

$$C_0=\begin{pmatrix} D\otimes I+I\otimes T & I\otimes\mathbf{T}^0 \\ 0 & D_0\end{pmatrix},\quad C_1=\begin{pmatrix} 0 & 0 \\ D_1\otimes\boldsymbol{\beta} & 0\end{pmatrix}. \tag{2.125}$$

Note that the same thinning operations can be applied to *BMAP*s and *MMAP*s and the resulting arrival processes are again *BMAP*s and *MMAP*s.

Exercise 2.5.20 (Example 2.5.3 continued) In steady state, find the distributions of the interarrival times of men, women, and families, respectively. Plot their density functions.

Exercise 2.5.21* Define superposition, decomposition and thinning for *MMAP*s.

Exercise 2.5.22 (Exercise 2.2.21 continued) For the shoe-shine shop problem, introduce an *MMAP* for the departure process of all customers (with or without service). Compute the rates of customers who receive and do not receive service. (Note: The departure processes of customers who receive and do not receive service are not independent. The *MMAP* can be used to study the correlation between the two.)

Exercise 2.5.23 (Exercise 2.2.23 continued) Generalize the *MAP* defined in Exercise 2.2.23 to an *MMAP* with five types of arrivals, one type for each day. Outline the steps for computing the arrival rate of individual types of arrivals.

Exercise 2.5.24 (Random time transformation of *MMAP*) Let $\mu(i)$ be a positive function on $\{1, 2, \ldots, m\}$. Let $x(t) = \int_0^t \mu(I(t))dt$, $I_r(t) = I(x(t))$, and $N_{r,h}(t) = N_h(x(t))$, for $h \in C^0$, for $t \geq 0$. Show that $\{(N_{r,h}(t), h \in C^0, I_r(t)), t \geq 0\}$ is an *MMAP* with matrix representation $(M^{-1}D_0, M^{-1}D_h, h \in C^0)$, where $M = \text{diag}(\mu(1), \ldots, \mu(m))$.

2.6 Additional Topics

*MAP*s have been used extensively in stochastic modeling. The main advantage of *MAP*s is that, if an *MAP* is utilized, a Markov chain can usually be introduced for the system. The Markov chain is usually analytically and numerically tractable. See Neuts (1992), Alfa and Neuts (1995), Asmussen (2000), Chakravarthy (2001, 2010), Alfa (2010), Artalejo et al. (2010), and Ramaswami (2010), for applications of and surveys on *MAP*s.

2.6.1 Discrete Time Markovian Arrival Processes (DMAP)

The theory on discrete time Markovian arrival processes is parallel to that of the continuous case, but some technical details are different, due to different probabilistic interpretations to arrivals rates and arrival probabilities. A discrete time batch Markovian arrival process is defined by matrices $\{D_n, n = 0, 1, 2, \ldots\}$, where all matrices are nonnegative, and $D = D_0 + D_1 + \ldots + D_n + \ldots$ is a stochastic matrix. The matrix D_n is the probability that an arrival is of size n. A major difference between the discrete time case and the continuous time case is that, for the discrete time case, two or more events can occur simultaneously, while there is at most one event occurring at a time for the continuous time case. See Blondia (1993), Alfa and Neuts (1995), Latouche and Ramaswami (1999), and Alfa (2010), for more about *DMAP*s.

Exercise 2.6.1 Assume that $(D_n, n = 0, 1, 2, \ldots)$ and $(C_n, n = 0, 1, 2, \ldots)$ are two *DMAP*s. Show that the superposition of the two arrival processes has a matrix representation $(D_0 \otimes C_0, \sum_{k=0}^{n} D_k \otimes C_{n-k}, n = 1, 2, \ldots)$.

Exercise 2.6.2 Assume $(D_n, n = 0, 1, 2, \ldots)$ represents a *DMAP*. Outline the steps to calculate the (average) arrival rate (the mean number of arrivals per unit time).

Exercise 2.6.3 (Probability generating function) Let $f_{i,j}^*(z,t)$ be the probability generating function of the total number of arrivals in $(0, t)$, and the phase of the underlying Markov chain be j, given that the underlying Markov chain is initially in state i. For $t = 1, 2, \ldots$, show that

$$\left(f_{i,j}^*(z,t)\right) = \left(D_0 + \sum_{n=1}^{\infty} z^n D_n\right)^t, \quad 0 \leq z \leq 1. \tag{2.126}$$

2.6.2 Identification and Characterization of MAPs

Identification is concerned with what kind of counting processes are *MAP*s. It also addresses the characterization of *MAP*s. For instance, the alternative representation of *MAP*s, the minimal order of *MAP*s, and a minimal representation of an *MAP* are some issues of interest. The results are useful in parameter estimation, computation, and stochastic modeling. References include Neuts et al. (1992), Neuts (1993), Ryden (1996), Heindl et al. (2006), Bodrog et al. (2008), Ramirez-Cobo et al. (2010), etc. Andersen et al. (2004) study the time reversal of Markovian arrival processes.

2.6.3 Parameter Estimation and Fitting of MAP, BMAP, and MMAP

For applications, we must be able to estimate parameters of arrival processes. On the other hand, parameter estimation and fitting of *MAP*s are difficult issues. In recent years, some progress has been made, especially for *MAP*s of lower orders (see Breuer (2002), Breuer and Alfa (2005), Horváth et al. (2005), Okamura et al. (2009), and Buchholz et al. (2010)). (http://webspn.hit.bme.hu/~telek/tools.htm)

2.6.4 Matrix Representations of MAPs

Similar to *PH*-representations of *PH*-distributions, matrix representations of *MAP*s are not unique. Consequently, selecting the proper matrix representation is an important issue. Usually, a matrix representation of a smaller order is desired. Studies on such issues can be found in Telek and Horváth (2007).

2.6.5 General Point Processes, Alternative Formalisms, and Variants of MAPs

Both the Poisson process and Markovian arrival processes are special point processes or counting processes. The general literature on point processes can be found in Sigman (1995) and Asmussen and Bladt (1999). *MAP*s are closely related to Markovian additive processes (C'inlar (1969, 1972), Pacheco and Prabhu (1995), and Miyazawa (2002)). In inventory theory, *MAP*s are called world driven demand processes (Song and Zipkin (1993) and Zipkin (2000)). Artalejo and Gomez-Corral (2010) introduces HDHR, which is a special but effective approach in stochastic modeling. The following are some variants of *MAP*s.

(i) Markovian arrival processes with time varying parameters. For many applications, time dependent arrival processes play an important role. For instance, arrivals of demand can be cyclic or time dependent. There are different ways to model time dependent arrival processes. By using underlying Markov chains, arrival rates can be assigned to individual phases to reflect the time dependent nature. Another approach is to define the arrival processes by using time dependent parameters.

(ii) Transient/terminating *MAP*s (He and Neuts (1998) and Latouche et al. (2003)).

2.6.6 Additional Exercises

Neuts (1995) provides a large number of exercises related to *MAPs*. Exercises in Chaps. 4 and 5 in Neuts (1995) are particularly useful to understand *MAPs* and their application.

References

Alfa AS (2010) Queueing theory for telecommunications: discrete time modelling of a single node system. Springer, New York

Alfa AS, Neuts MF (1995) Modelling vehicular traffic using the discrete time Markovian arrival process. Transport Sci 29:109–117

Andersen AT, Neuts MF, Nielsen BF (2004) On the time reversal of Markovian arrival processes. Stoch Models 20:237–260

Artalejo JR, Gómez-Corral A (2010) A state-dependent Markov-modulated mechanism for generating events and stochastic models. Math Methods Appl Sci 33:1342–1349

Artalejo J, Gomez-Corral A, He QM (2010) Markovian arrivals in stochastic modelling: a survey and some new results. Stat Oper Res Trans (SORT) 34:101–144

Asmussen S (2000) Matrix-analytic models and their analysis. Scand J Stat 27:193–226

Asmussen S (2003) Applied probability and queues, 2nd edn. Springer, New York

Asmussen S, Bladt M (1999) Point processes with finite-dimensional conditional probabilities. Stoch Processes Appl 82:127–142

Asmussen S, Koole G (1993) Marked point processes as limits of Markovian arrival streams. J Appl Prob 30:365–372

Asmussen S, Avram F, Usabel M (2002) Erlangian approximations for finite-horizon ruin probabilities. ASTIN Bull 32:267–281

Bladt M, Nielsen BF (2010) Multivariate matrix-exponential distributions. Stoch Models 26:1–26

Blondia C (1993) A discrete-time batch Markovian arrival process as B-ISDN traffic model. Belg J Oper Res Stat Comput Sci 32:3–23

Bodrog L, Horváth A, Telek M (2008) Moment characterization of matrix exponential and Markovian arrival processes. Ann Oper Res 160:51–68

Breuer L (2002) An EM algorithm for batch Markovian arrival processes and its comparison to a simpler estimation procedure. Ann Oper Res 112:123–138

Breuer L, Alfa AS (2005) An EM algorithm for platoon arrival processes in discrete time. Oper Res Lett 33:535–543

Buchholz P, Kemper P, Kriege J (2010) Multi-class Markovian arrival processes and their parameter fitting. Perform Eval 67:1092–1106

C'inlar E (1969) Markov renewal theory. Adv Appl Prob 1:123–187

C'inlar E (1972) Markov additive processes. I, II. *Z. Wahrscheinlichkeitsth*, verw. Geb. 24:85–93, 95–121

Chakravarthy SR (2001) The batch Markovian arrival process: a review and future work. In: Krishnamoorthy A, Raju N, Ramaswami V (eds) Advances in probability & stochastic processes. Notable Publications, pp 21–49

Chakravarthy SR (2010) Markovian arrival process. In: Cochran JJ (ed) Wiley encyclopedia of operations research and management science. Wiley, Published Online: 15 June, 2010

He QM (1996) Queues with marked customers. Adv Appl Prob 28:567–587

He QM (2001) The versatility of *MMAP*[*K*] and the *MMAP*[*K*]/*G*[*K*]/1 queue. Queueing Syst 38:397–418

He QM (2010) Construction of continuous time Markov arrival processes. J Syst Sci Syst Eng 19:351–366

He QM, Neuts MF (1998) Markov chains with marked transitions. Stoch Processes Appl 74:37–52

Heindl A, Mitchell K, van de Liefvoort A (2006) Correlation bounds for second-order *MAP*s with application to queueing network decomposition. Perform Eval 63:553–577

Horváth G, Buchholz P, Telek M (2005) A *MAP* fitting approach with independent approximation of the inter-arrival time distribution and the lag correlation. In: Proceedings of the 2nd international conference on quantitative evaluation of systems, Torino, pp 124–133

Latouche G, Ramaswami V (1999) Introduction to matrix analytic methods in stochastic modeling. ASA & SIAM, Philadelphia

Latouche G, Remiche M-A, Taylor P (2003) Transient Markovian arrival processes. Ann Appl Prob 13:628–640

Law AM, Kelton WD (2000) Simulation modeling and analysis. McGraw Hill, New York

Lucantoni DM (1991) New results on the single server queue with a batch Markovian arrival process. Stoch Models 7:1–46

Lucantoni DM, Meier-Hellstern KS, Neuts MF (1990) A single server queue with server vacations and a class of non-renewal arrival processes. Adv Appl Prob 22:676–705

Miyazawa M (2002) A paradigm of Markov additive processes for queues and their networks. In: Latouche G, Taylor PG (eds) Matrix analytic methods theory and applications. World Scientific, New Jersey, pp 265–289

Narayana S, Neuts MF (1992) The first two moment matrices of the counts for the Markovian arrival process. Stoch Models 8:459–477

Neuts MF (1979) A versatile Markovian point process. J Appl Prob 16:764–779

Neuts MF (1981) Matrix-geometric solutions in stochastic models – an algorithmic approach. The Johns Hopkins University Press, Baltimore

Neuts MF (1992) Models based on the Markovian arrival process. IEICE Trans Commun E75-B:1255–1265

Neuts MF (1993) The burstiness of point processes. Stoch Models 9:445–466

Neuts MF (1995) Algorithmic probability: a collection of problems. Chapman & Hall, London

Neuts MF, Li J-M (1997) An algorithm for the $P(n, t)$ matrices of a continuous *BMAP*. In: Chakravarthy SR, Alfa AS (eds) Matrix-analytic methods in stochastic models. Lecture notes in pure and applied mathematics, vol 183. Marcel Dekker, pp 7–19

Neuts MF, Liu D, Surya N (1992) Local poissonification of the Markovian arrival process. Stoch Models 8:87–129

Nielsen BF, Nilsson LAF, Thygesen UH, Beyer JE (2007) Higher order moments and conditional asymptotics of the batch Markovian arrival process. Stoch Models 23:1–26

Okamura H, Dohi T, Trivedi KS (2009) Markovian arrival process parameter estimation with group data. IEEE/ACM Trans Netw 17:1326–1339

Pacheco A, Prabhu NU (1995) Markov-additive processes of arrivals. In: Dshalalow JH (ed) Advances in queuing: theory, methods and open problems. CRC Press, Florida, USA, pp 167–194

Ramaswami V (2010) Poisson process and its generalizations. In: Cochran JJ (ed) Wiley encyclopedia of operations research and management science. Wiley

Ramaswami V, Woolford DG, Stanford DA (2008) The erlangization method for Markovian fluid flows. Ann Oper Res 160:215–225

Ramirez-Cobo P, Lillo RE, Wiper MP (2010) Nonidentifiability of the two-state Markovian arrival process. J Appl Prob 47:630–649
Ross SM (2010) Introduction to probability models, 10th edn. Academic, New York
Rudemo M (1973) Point processes generated by transitions of Markov chains. Adv Appl Prob 5:262–286
Ryden T (1996) On identifiability and order of continuous-time aggregated Markov chains, Markov-modulated Poisson processes, and phase-type distributions. J Appl Prob 33:640–653
Sigman K (1995) Stationary marked point processes: an intuitive approach. Chapman and Hall, New York
Song JS, Zipkin P (1993) Inventory control in a fluctuating demand environment. Oper Res 41:351–370
Stanford DA, Yu KQ, Ren JD (2011) Erlangian approximation to finite time ruin probabilities in perturbed risk models. Scand Actuarial J 2011:38–58
Telek M, Horvath G (2007) A minimal representation of Markovian arrival processes and a moments matching method. Perform Eval 64:1153–1168
Zipkin PH (2000) Foundations of inventory management. McGraw Hill, Boston

Chapter 3
From the Birth-and-Death Process to Structured Markov Chains

Abstract This chapter introduces Markov chains of QBD, *M/G/*1, and *GI/M/*1 types. Matrix-geometric solutions for the stationary distributions are presented with probabilistic interpretations. Algorithms are developed for computing performance measures. Markov chains of QBD, *M/G/*1, and *GI/M/*1 types with a tree structure are also introduced and analyzed. Some results on tail asymptotics of Markov chains with infinitely many background phases are given as well.

In this chapter, we examine a number of specially structured Markov chains. In general, Markov chains can be categorized according to the types of time and state space. The state of a process can be viewed at discrete time epochs or continuously. The state space can be countable or uncountable (e.g., a general topological space). Consequently, a Markov chain can be

(1) Discrete in time with a countable state space;
(2) Continuous in time with a countable state space;
(3) Discrete in time with a general state space; or
(4) Continuous in time with a general state space.

All four cases have been investigated extensively and basic results have been collected in many classical books on Markov chains (e.g., Karlin and Taylor (1975, 1981), Revuz (1975), Kemeny et al. (1976), Neuts (1981, 1989), Resnick (1992), Grassmann (2000), Asmussen (2003), Seneta (2006), Meyn and Tweedie (2009), and references therein).

In this chapter, we concentrate on Markov chains with a countable state space (Cases (1) and (2)). The other two cases are generally more difficult to deal with and require advanced mathematical tools to understand, research, and apply. Although we only focus on Markov chains with a countable state space, the material to be covered is still enormous. Thus, we further limit our attention to a few structured Markov chains with a countable state space, including QBD, *GI/M/*1, and *M/G/*1

Q.-M. He, *Fundamentals of Matrix-Analytic Methods*,
DOI 10.1007/978-1-4614-7330-5_3, © Springer Science+Business Media New York 2014

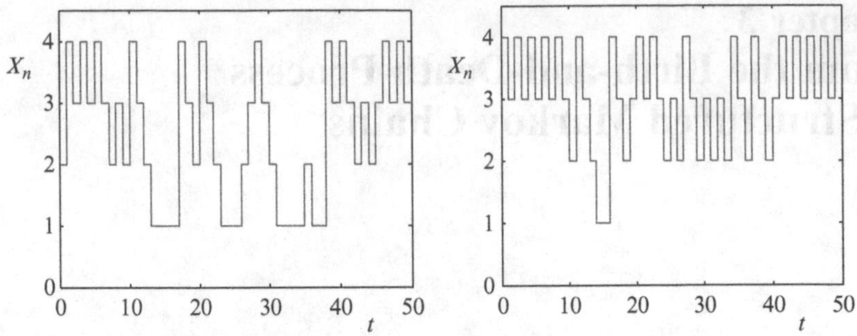

Fig. 3.1 Two sample paths of a Markov chain with four states

type Markov chains, and those types of Markov chains with a tree structure. We shall concentrate on the limiting probabilities and first passage times of such Markov chains. We refer readers to Neuts (1981, 1989), Latouche and Ramaswami (1999), and Li (2010) for further reading on structured Markov chains.

It is well-known that, under certain conditions, a continuous time Markov chain can be transformed into a discrete time Markov chain using the so-called *uniformization* technique. Thus, for many cases, the theory on continuous time Markov chains with a countable state space is similar to that of discrete time Markov chains. Consequently, while we shall focus on the discrete time case, we shall also make comments on extensions to the continuous time case.

A discrete time stochastic process $\{X_k, k = 0, 1, 2, \ldots\}$ with state space $\{0, 1, 2, \ldots\}$ is called a *time homogenous Markov chain* if, for all $k, i, j = 0, 1, 2, \ldots$,

$$\begin{aligned} &P\{X_{k+1} = j | X_k = i,\ X_u = i_u,\ u = 0, 1, \ldots, k-1\} \\ &= P\{X_{k+1} = j | X_k = i\} = p_{i,j}. \end{aligned} \tag{3.1}$$

Numbers $\{p_{i,j},\ i, j = 0, 1, 2, \ldots\}$ are called *transition probabilities*. Putting all the transition probabilities into a matrix, we obtain

$$P = \begin{array}{c} \\ 0 \\ 1 \\ 2 \\ \vdots \end{array} \begin{array}{c} \begin{array}{cccc} 0 & 1 & 2 & \cdots \end{array} \\ \begin{pmatrix} p_{0,0} & p_{0,1} & p_{0,2} & \cdots \\ p_{1,0} & p_{1,1} & p_{1,2} & \cdots \\ p_{2,0} & p_{2,1} & p_{2,2} & \cdots \\ \vdots & \vdots & \vdots & \vdots \end{pmatrix} \end{array}. \tag{3.2}$$

Matrix P is called a *transition probability matrix*, which is also a *stochastic matrix* (i.e., all elements of P are nonnegative and all row sums are one). Given the distribution of the initial state (i.e., the distribution of X_0), there is a one-to-one relationship between Markov chains and transition probability matrices. Consequently, we shall also call P a Markov chain. Figure 3.1 plots two sample paths of a Markov chain with four states.

We refer to Asmussen (2003) and Ross (2010) for basic concepts related to Markov chains with a countable state space: *irreducibility, aperiodicity, transience, recurrence, null recurrence, positive recurrence, ergodicity, absorption state, first passage time, taboo probability, steady state, steady state probability, limiting probability, stationary distribution*, etc.

3.1 The Birth-and-Death Process

A discrete time *birth-and-death* process $\{X_k, k = 0, 1, 2, \ldots\}$ is a Markov chain with transition probability matrix

$$P = \begin{matrix} 0 \\ 1 \\ 2 \\ \vdots \\ \vdots \end{matrix} \begin{pmatrix} p_{0,0} & p_{0,1} & & & \\ p_{1,0} & 0 & p_{1,2} & & \\ & p_{2,1} & 0 & p_{2,3} & \\ & & \ddots & \ddots & \ddots \\ & & & \ddots & \ddots & \ddots \end{pmatrix}, \tag{3.3}$$

where $p_{0,0} + p_{0,1} = 1$ and $p_{n,n-1} + p_{n,n+1} = 1$, for $n = 1, 2, \ldots$. As shown in Eq. (3.3), if $X_k = n$, then X_{k+1} is either $n + 1$ (birth) or $n-1$ (death), except for the boundary state $n = 0$, where only birth can occur. Assume that the transition probabilities $\{p_{0,1}, p_{n,n+1}, p_{n,n-1}, n = 1, 2, \ldots\}$ are positive. Then the Markov chain is *irreducible*, i.e., the probability of reaching any state j, in one or more transitions, from any state i is positive.

If they exist, the *limiting probabilities* of the Markov chain are defined as

$$\pi_n = \lim_{k\to\infty} P\{X_k = n|X_0 = j\} \equiv \lim_{k\to\infty} p_{j,n}^{(k)}, \quad n = 0, 1, 2, \ldots. \tag{3.4}$$

A useful property is that the limiting probabilities, if they exist, are unique and independent of the initial state. (Note: In general, the limit may depend on the initial state.) In addition, limiting probabilities can be explained from different perspectives. Four interpretations of limiting probabilities $\{\pi_n, n = 0, 1, 2, \ldots\}$ are provided below.

(a) Limiting probabilities as defined in Eq. (3.4).
(b) *Long-run average* of the time that the Markov chain spent in individual states, which can be expressed as follow:

$$\pi_n = \lim_{k\to\infty} \frac{\sum_{t=1}^{k} I_{\{X_t=n\}}}{k} = \lim_{k\to\infty} \frac{\sum_{t-1}^{k} p_{j,n}^{(t)}}{k}, \quad n = 0, 1, 2, \ldots, \tag{3.5}$$

where $I_{\{.\}}$ is the indicator function.
(c) *Stationary distribution*: If X_0 is distributed according to $\{\pi_n, n = 0, 1, 2, \ldots\}$, then X_k is also distributed according to $\{\pi_n, n = 0, 1, 2, \ldots\}$. Let $\boldsymbol{\pi} = (\pi_0, \pi_1, \ldots)$. It can be shown that $\boldsymbol{\pi}P = \boldsymbol{\pi}$ and $\boldsymbol{\pi}\mathbf{e} = 1$, which provide equations for

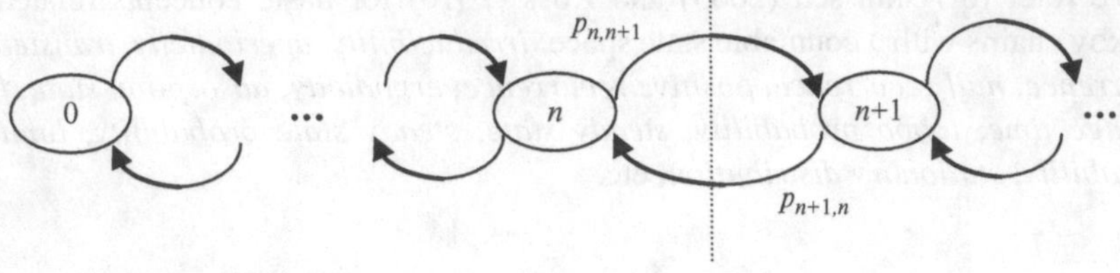

Fig. 3.2 Transitions of the birth-and-death process P

finding the limiting probabilities. (Note: We use the terms "stationary distribution" and "limiting probabilities" interchangeably in this book.)

(d) *Mean first return time* of any state: Denote by $\tau_{(n)}$ the *first return time* to state n. Then we have $\pi_n = 1/E[\tau_{(n)}]$, $n = 0, 1, 2, \ldots$.

The above four points of view on the limiting probabilities are true for all Markov chains with a countable state space, but may not be true for other types of Markov chains. There are conditions for the existence of the limiting probabilities. A sufficient condition is that the Markov chain is *ergodic* (i.e., irreducible and *positive recurrent*). We shall give conditions for ergodicity, but do not explore this issue in great length. For most of the cases, we shall assume that the limiting probabilities exist and we develop methods to find them. For general discrete time Markov chains, results for ergodicity can be found in, for example, Karlin and Taylor (1975) and Cohen (1982).

Exercise 3.1.1 (a) Let $P^{(k)} = \left(p_{i,j}^{(k)}\right)$ (defined by Eq. (3.4)). Show the *Chapman-Kolmogorov equation* $P^{(k)} = P^k$, for $k = 1, 2, \ldots$. (b) Assume that the limiting probabilities exist for a discrete time Markov chain P with a countable state space. Use the Chapman-Kolmogorov equation to prove $\boldsymbol{\pi}P = \boldsymbol{\pi}$ and $\boldsymbol{\pi}\mathbf{e} = 1$. (c) Define a cycle as the period between two consecutive visits to state n. Then $\tau_{(n)}$ is the cycle length. Explain $\pi_n = 1/E[\tau_{(n)}]$ intuitively.

We assume that the limiting probabilities defined by Eq. (3.4) exist for the birth-and-death process defined in Eq. (3.3). By Exercise 3.1.1, we must have $\boldsymbol{\pi}P = \boldsymbol{\pi}$ and $\boldsymbol{\pi}\mathbf{e} = 1$. Expanding $\boldsymbol{\pi}P = \boldsymbol{\pi}$ and reorganizing the linear equations, yields

$$\pi_n p_{n,n+1} = \pi_{n+1} p_{n+1,n}, \quad n = 0, 1, 2, \ldots. \tag{3.6}$$

Exercise 3.1.2 Show Eq. (3.6). Explain Eq. (3.6) intuitively (see Fig. 3.2).

By Eq. (3.6), if the limiting probabilities exist, we obtain,

$$\begin{aligned} \pi_n &= \pi_0 \frac{p_{0,1}p_{1,2}\cdots p_{n-1,n}}{p_{1,0}p_{2,1}\cdots p_{n,n-1}}, \quad n = 1, 2, \ldots; \\ \pi_0 &= \left(1 + \sum_{n=1}^{\infty} \frac{p_{0,1}p_{1,2}\cdots p_{n-1,n}}{p_{1,0}p_{2,1}\cdots p_{n,n-1}}\right)^{-1}. \end{aligned} \tag{3.7}$$

Theorem 3.1.1 *The Markov chain P defined in Eq.* (3.3) *is* ergodic (*i.e., irreducible and positive recurrent*) *if and only if* $p_{n,\,n+1} > 0$ *for all n, and*

$$\sum_{n=1}^{\infty} \frac{p_{0,1}p_{1,2}\cdots p_{n-1,n}}{p_{1,0}p_{2,1}\cdots p_{n,n-1}} < \infty. \tag{3.8}$$

Consequently, under these conditions, limiting probabilities exist, are positive, and are given by Eq. (3.7).

Exercise 3.1.3 Prove the necessity of condition (3.8) in Theorem 3.1.1. The conditions for ergodicity imply the irreducibility of the Markov chain. Explain the relationship between ergodicity and the conditions intuitively. (Note: Proofs of Theorem 3.1.1 can be found in many books on Markov chains and queueing theory (e.g., Cohen (1982)). Conditions for *null recurrence* and *transience* can be found in Cohen (1982) as well.)

Exercise 3.1.4 Assume $p_{0,1} = p_{1,2} = p_{2,3} = \ldots = p$ in Eq. (3.3). Show that, if $p < 0.5$,

$$\pi_n = \left(1 - \frac{p}{1-p}\right)\left(\frac{p}{1-p}\right)^n, \quad n = 0,\ 1,\ 2,\ \ldots. \tag{3.9}$$

In Exercise 3.1.4, what happens to the Markov chain if $p = 0.5$ or $p > 0.5$? Intuitively, the Markov chain may not return to state zero once it leaves state zero. Then the limiting probabilities are zero.

The probability distribution given in Eq. (3.9) is called a *geometric distribution*. This is the origin of the term *matrix-geometric solution* to be introduced in Sect. 3.2. In order to understand the idea behind the matrix-geometric solution, for this special case, an alternative method is used to find solution (3.9). The new method is suitable for finding the limiting probabilities of more general cases (e.g., QBD (Sect. 3.2) and *GI*/*M*/1 (Sect. 3.4)). First, we assume that the limiting probabilities exist and are positive. Expanding equation $\boldsymbol{\pi}P = \boldsymbol{\pi}$, we obtain

$$\begin{aligned} &\pi_0(1-p) + \pi_1(1-p) = \pi_0; \\ &\pi_{n-1}p + \pi_{n+1}(1-p) = \pi_n, \quad n = 1,\ 2,\ \ldots. \end{aligned} \tag{3.10}$$

We conjecture that the solution to Eq. (3.10) has the form $\pi_n = \pi_0 x^n$, for $n = 0$, 1, 2, ... Replacing π_n with $\pi_0 x^n$ in line 2 in Eq. (3.10) yields

$$p + x^2(1-p) = x. \tag{3.11}$$

Note that π_0 is positive since limiting probabilities exist and are positive. For the same reason, x must be positive if the limiting probabilities have the geometric form. Solving Eq. (3.11), we find $x = p/(1-p)$. By the normalization condition $\boldsymbol{\pi}\mathbf{e} = 1$, we find $\pi_0 = 1-x$. If $p < 0.5$, then $\{\pi_n = (1-x)x^n,\ n = 0,\ 1,\ \ldots\}$ is a solution to $\boldsymbol{\pi}P = \boldsymbol{\pi}$ and $\boldsymbol{\pi}\mathbf{e} = 1$. Since the limiting probabilities are unique, solution (3.9) gives the limiting probabilities of the Markov chain.

Exercise 3.1.5 Find all the solutions for the quadratic equation (3.11). Explain why solution $x = p/(1-p)$ has to be chosen for the limiting probabilities.

Exercise 3.1.6 Assume $p_{0,1} = p_{1,2} = p_{2,3} = \ldots = p$ in Eq. (3.3). Assume $p < 0.5$. Define $r_n^{(j)}$ as the expected number of visits to state $n + j$ before reaching any state in $\{0, 1, 2, \ldots, n\}$, given that the Markov chain starts in state n, for $j = 1, 2, \ldots$.

(i) Explain intuitively that $r_n^{(j)}$ is a constant in $n = 1, 2, \ldots$.

(ii) Define ${}_n p_{n,n+j}^{(k)} = P\{X_k = n+j,\ X_u,\ u = 1, 2, \ldots, k-1|X_0 = n\}$, the taboo probability of states $\{0, 1, \ldots, n\}$, for $n \geq 1, j \geq 1$, and $k \geq 0$. Show (a) $r_n^{(j)} = \sum_{k=1}^{\infty} {}_n p_{n,n+j}^{(k)}$, for $j = 1, 2, \ldots$; and (b) $r_n^{(2)} = \left(r_n^{(1)}\right)^2$.

(iii) Let $r_n = r_n^{(1)}$. Show that r_n satisfies equation

$$r_n = p + r_n^2(1-p), \quad n = 0,\ 1,\ 2,\ \ldots. \tag{3.12}$$

(Hint: Conditioning on the last state visited before reaching state $n + 1$, we obtain ${}_n p_{n,n+1}^{(k)} = {}_n p_{n,n+2}^{(k-1)}(1-p)$, for $k \geq 2$.)

(iv) Explain why Eq. (3.12) gives $r_n = p/(1-p)$. Also make a direct connection between solution r_n and the limiting probabilities given in Eq. (3.9). (Note: No proof or formal conclusion is necessary.)

(v) Define τ_n as the first passage time to reach state $n-1$, given that $X_0 = n > 0$. Explain intuitively that τ_n has a discrete *PH*-distribution with matrix representation

$$\boldsymbol{\alpha} = (1,\ 0,\ 0,\ \ldots), \quad T = \begin{pmatrix} 0 & p & & & \\ 1-p & 0 & p & & \\ & 1-p & 0 & p & \\ & & \ddots & \ddots & \ddots \end{pmatrix}. \tag{3.13}$$

(vi) Show that random variables $\{\tau_n,\ n = 1, 2, \ldots\}$ are independent and identically distributed.

(vii) Show that the first passage time from state n to state 0 is $\tau_n + \tau_{n-1} + \ldots + \tau_1$.

(viii) Let $f^*(z)$ be the probability generating function of τ_n. Show that $f^*(z)$ satisfies the equation, for $0 \leq z \leq 1$,

$$f^*(z) = z(1-p) + zp(f^*(z))^2. \tag{3.14}$$

(Hint: Condition on the first transition of the Markov chain.)

(ix) Show that $E[\tau_n] = 1/(1-2p)$, if $p < 0.5$. What is $E[\tau_n]$, if $p \geq 0.5$?

Commentary The birth-and-death process is introduced and studied in many textbooks in science and engineering. We recommend Cohen (1982) and Ross (2010) for further reading.

Additional Exercises and Extensions

Exercise 3.1.7 Consider a discrete time process $\{X_k, k = 0, 1, 2, \ldots\}$ with a finite number of states and transition probability matrix

$$P_N = \begin{pmatrix} p_{0,0} & p_{0,1} & & & \\ p_{1,0} & 0 & p_{1,2} & & \\ & \ddots & \ddots & \ddots & \\ & & p_{N-1,N-2} & 0 & p_{N-1,N} \\ & & & p_{N,N-1} & p_{N,N} \end{pmatrix}. \quad (3.15)$$

Assume that the limiting probabilities exist and are positive. Find the limiting probabilities. Discuss the relationship between the limiting probabilities of the Markov chain P, defined in Eq. (3.3), and that of the Markov chain P_N, defined in Eq. (3.15).

Exercise 3.1.8 Continuous time birth-and-death process $\{X(t), t \geq 0\}$ is defined as a continuous time Markov chain with infinitesimal generator Q given by

$$Q = \begin{pmatrix} -\lambda & \lambda & & \\ \mu & -(\lambda+\mu) & \lambda & \\ & \mu & -(\lambda+\mu) & \lambda \\ & & \ddots & \ddots & \ddots \end{pmatrix}. \quad (3.16)$$

If $\lambda < \mu$, then the limiting probabilities exist and satisfy $\boldsymbol{\pi} Q = 0$ and $\boldsymbol{\pi}\mathbf{e} = 1$, where $\boldsymbol{\pi} = (\pi_0, \pi_1, \ldots, \pi_n, \ldots)$. Show that the limiting probabilities are given by

$$\pi_n = \left(1 - \frac{\lambda}{\mu}\right)\left(\frac{\lambda}{\mu}\right)^n, \quad n = 0,\ 1,\ 2,\ \ldots. \quad (3.17)$$

Exercise 3.1.9 For the Markov chain Q defined in Eq. (3.16), by Eq. (3.17), we have $\pi_n\lambda = \pi_{n+1}\mu$, for $n = 1, 2, \ldots$. Explain this equality intuitively. (Hint: Consider the total transition rates between states in $\{0, 1, \ldots, n\}$ and states in $\{n+1, n+2, \ldots\}$.)

The *uniformization* technique is used widely to deal with continuous time Markov chain. For the continuous time Markov chain Q, we define a discrete time Markov chain $P = I + Q/(\lambda+\mu)$. Then we have the following observations.

(1) P is a discrete time birth-and-death process.
(2) If limiting probabilities exist for P, then they are also the limiting probabilities for Q, and vice versa, i.e., $\boldsymbol{\pi} Q = 0$ and $\boldsymbol{\pi}\mathbf{e} = 1$ if and only if $\boldsymbol{\pi} P = \boldsymbol{\pi}$ and $\boldsymbol{\pi}\mathbf{e} = 1$.

Thus, finding the limiting probabilities for Q is equivalent to finding limiting probabilities for P. The term "uniformization" comes from the following interpretation of P within the process Q. In fact, Q can be considered as a stochastic process

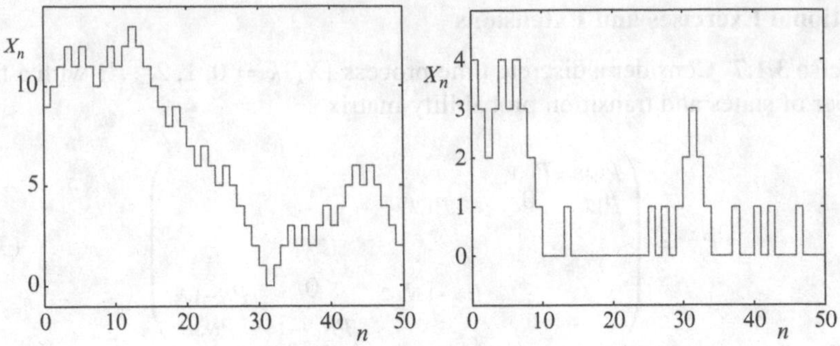

Fig. 3.3 Two sample paths of a birth-death process with $p = 0.4$

with the sojourn time in any state having *a common exponential distribution* with parameter $\lambda+\mu$, and an embedded Markov chain with transition probability matrix P. Thus, the occurrences of transitions follow a Poisson process with parameter $\lambda+\mu$. Since some of the transitions in the Poisson process are not accompanied by a change of state, they are *invisible*. For any state $j > 0$, after an exponential time with parameter $\lambda+\mu$, with probability $\mu/(\lambda+\mu)$, the process enters state $j-1$; and with probability $\lambda/(\lambda+\mu)$, the process enters state $j+1$. For state $j = 0$, after an exponential time with parameter $\lambda+\mu$, with probability $\mu/(\lambda+\mu)$, the process stays in state 0; and with probability $\lambda/(\lambda+\mu)$, the process leaves state 0 and enters state 1. Thus, the total time that the process spends in state 0, once it is in state 0, is the geometric sum of independent exponential times with common parameter $\lambda+\mu$, which is in fact an exponential time with parameter λ (see Proposition 1.1.4).

Exercise 3.1.10 Use the uniformization method and the solution in Eq. (3.9) to prove Eq. (3.17).

Exercise 3.1.11 Write a simulation program to simulate the birth-and-death process defined in Exercise 3.1.4. Calculate the percentage of time the process is in a state and compare the results to the geometric solution in Eq. (3.9). Figure 3.3 plots two sample paths of the Markov chain.

Exercise 3.1.12 For positive numbers $\{\lambda_0, \lambda_1, \ldots, \lambda_K\}$ and $\{\mu_1, \mu_2, \ldots, \mu_K\}$, define a continuous time birth-and-death process $\{X(t), t \geq 0\}$ with infinitesimal generator

$$Q = \begin{matrix} 0 \\ 1 \\ \vdots \\ K \\ K+1 \\ \vdots \end{matrix} \begin{pmatrix} -\lambda_0 & \lambda_0 & & & & \\ \mu_1 & -(\lambda_1+\mu_1) & \lambda_1 & & & \\ & \ddots & \ddots & \ddots & & \\ & & \mu_K & -(\lambda_K+\mu_K) & \lambda_K & \\ & & & \mu_K & -(\lambda_K+\mu_K) & \lambda_K \\ & & & & \ddots & \ddots & \ddots \end{pmatrix}. \tag{3.18}$$

Find condition(s) for the limiting probabilities to exist and be positive. Find the limiting probabilities.

Exercise 3.1.13 For positive numbers $\{\lambda_n,\ n = 0,\ 1,\ 2,\ 3,\ \ldots\}$ and $\{\mu_n,\ n = 1, 2, 3, \ldots\}$, define a continuous time birth-and-death process $\{X(t),\ t \geq 0\}$ with infinitesimal generator

$$Q = \begin{pmatrix} -\lambda_0 & \lambda_0 & & & \\ \mu_1 & -(\lambda_1+\mu_1) & \lambda_1 & & \\ & \mu_2 & -(\lambda_2+\mu_2) & \lambda_2 & \\ & & \ddots & \ddots & \ddots \end{pmatrix}. \tag{3.19}$$

Assume that $\max_n\{\lambda_n,\ \mu_n\}$ is finite. Find conditions for the existence of the limiting probabilities and to be positive. If limiting probabilities exist and are positive, find them. (Hint: Use the uniformization method.) (Note: If $\max_n\{\lambda_n,\ \mu_n\} = \infty$, the problem becomes much more complicated as the process may be *explosive* (see Asmussen (2003)).)

Exercise 3.1.14 A continuous time birth-and-death process $\{X(t),\ t \geq 0\}$ is defined as a continuous time Markov chain with infinitesimal generator Q given by, for $\lambda > 0$ and $\mu > 0$,

$$Q = \begin{pmatrix} -\lambda & \lambda & & & \\ \mu & -(\lambda+\mu) & \lambda & & \\ & 2\mu & -(\lambda+2\mu) & \lambda & \\ & & 3\mu & -(\lambda+3\mu) & \lambda \\ & & & \ddots & \ddots & \ddots \end{pmatrix}. \tag{3.20}$$

Find the limiting probabilities for $\{X(t), t \geq 0\}$.

Exercise 3.1.15 A continuous time birth-and-death process $\{X(t), t \geq 0\}$ is defined as a continuous time Markov chain with infinitesimal generator Q given by, for $\lambda > 0$,

$$Q = \begin{matrix} 0 \\ 1 \\ \vdots \\ m-1 \\ m \end{matrix} \begin{pmatrix} -\lambda & \lambda & & & \\ \lambda & -2\lambda & \lambda & & \\ & \ddots & \ddots & \ddots & \\ & & \lambda & -2\lambda & \lambda \\ & & & \lambda & -\lambda \end{pmatrix}_{(m+1)\times(m+1)}. \tag{3.21}$$

Find the stationary distribution $\pi(m)$ (limiting probabilities) for $\{X(t),\ t \geq 0\}$. Show that $\lim_{m\to\infty}\pi_n(m) = 0$, for $n = 0,\ 1,\ 2, \ldots, m$.

Exercise 3.1.16 Consider the continuous time birth-and-death process $\{X(t), t \geq 0\}$ defined by Eq. (3.16).

(i) Define τ_n as the first passage time to reach state $n-1$, given that $X(0) = n > 0$. Explain intuitively that τ_n has a *PH*-distribution with matrix representation

$$\alpha = (1,\ 0,\ 0,\ \ldots),$$
$$T = \begin{pmatrix} -(\lambda+\mu) & \lambda & & & \\ \mu & -(\lambda+\mu) & \lambda & & \\ & \mu & -(\lambda+\mu) & \lambda & \\ & & \ddots & \ddots & \ddots \end{pmatrix}. \tag{3.22}$$

Note that (1) the *PH*-representation has an infinite number of phases; and (2) the *PH*-representation is independent of n.

(ii) Show that random variables $\{\tau_n, n = 1, 2, \ldots\}$ are independent and identically distributed random variables.

(iii) Show that the first passage time from state n to state 0 is $\tau_n + \tau_{n-1} + \ldots + \tau_1$.

(iv) Let $f^*(s)$ be the LST of τ_n. Show that $f^*(s)$ satisfies the equation: for $s \geq 0$,

$$f^*(s) = \frac{\mu}{\lambda+\mu+s} + \frac{\lambda}{\lambda+\mu+s}(f^*(s))^2. \tag{3.23}$$

(Hint: Decompose the first passage time into two parts: the sojourn time in the current state and future time.)

(v) Show that $E[\tau_n] = 1/(\mu-\lambda)$, if $\mu > \lambda$. What is $E[\tau_n]$, if $\mu \leq \lambda$?

3.2 Quasi-Birth-and-Death Processes: Matrix-Geometric Solution

A discrete time *quasi-birth-and-death* (QBD) process $\{(X_k, J_k), k = 0, 1, 2, \ldots\}$ is a discrete time Markov chain with state space $\{(0, 1), (0, 2), \ldots, (0, m_0)\} \cup \{\{1, 2, \ldots\} \times \{1, 2, \ldots, m\}\}$, where m_0 and m are positive integers, and the variable X_k increases or decreases its value by at most one at each transition. We call X_k the *level variable* and J_k the *phase variable*. The set of states $\{(n, j), j = 1, 2, \ldots, m\}$ is called the *level n*. We consider a *level independent* QBD process with transition probability matrix

$$P = \begin{pmatrix} A_{0,0} & A_{0,1} & & & \\ A_{1,0} & A_{1,1} & A_0 & & \\ & A_2 & A_1 & A_0 & \\ & & \ddots & \ddots & \ddots \\ & & & \ddots & \ddots \end{pmatrix}, \tag{3.24}$$

where $A_{0,0}$ is an $m_0 \times m_0$ matrix, $A_{0,1}$ is an $m_0 \times m$ matrix, $A_{1,0}$ is an $m \times m_0$ matrix, and $\{A_0, A_1, A_2\}$ are nonnegative matrices of order m. Matrices $\{A_0, A_1, A_2, A_{0,0}, A_{0,1}, A_{1,0}, A_{1,1}\}$ satisfy $(A_0 + A_1 + A_2)\mathbf{e} = \mathbf{e}$, $A_{1,0}\mathbf{e} + (A_{1,1} + A_0)\mathbf{e} = \mathbf{e}$, and $A_{0,0}\mathbf{e} + A_{0,1}\mathbf{e} = \mathbf{e}$. Let $A = A_0 + A_1 + A_2$. Then A is a stochastic matrix of order m, which governs the transitions of the phase variable, given that the level variable is 2 or greater. The stochastic matrix A and its related *invariant measures* (a left eigenvector) and *invariant vectors* (a right eigenvector) play an important role in the study of structured Markov chains.

The QBD process $\{(X_k, J_k), k = 0, 1, 2, \ldots\}$ can be represented by a single variable process $\{Z_k, k = 0, 1, 2, \ldots\}$, defined as $Z_k = J_k$, if $X_k = 0$; $Z_k = (X_k - 1)m + m_0 + J_k$, if $X_k \geq 1$. On the other hand, any process $\{Z_k, k = 0, 1, 2, \ldots\}$ with state space $\{0, 1, 2, \ldots\}$ can be represented by $\{(X_k, J_k), k = 0, 1, 2, \ldots\}$ as follows:

$$\begin{aligned} Z_k \leq m_0: &\quad J_k = Z_k,\ X_k = 0; \\ Z_k > m_0: &\quad J_k = (Z_k - m_0) \bmod m,\ X_k = \frac{Z_k - m_0 - J_k}{m} + 1. \end{aligned} \tag{3.25}$$

Note that, for integers a and b, a mod b is the remainder of division a/b. For example, if $a = 15$ and $b = 6$, then a mod $b = 3$. In general, the two-dimensional process $\{(X_k, J_k), k = 0, 1, 2, \ldots\}$ may not have the QBD structure.

Example 3.2.1 The transition probability matrix P given in Eq. (3.26) defines a discrete time Markov chain $\{Z_k, k = 0, 1, 2, \ldots\}$. If the state space is reorganized by: states $\{1, 2\}$ to be level zero, $\{3, 4\}$ to be level 1, $\{5, 6\}$ to be level 2, ..., then the Markov chain becomes a QBD process for which each level has two phases.

$$P = \begin{array}{c} 1 \\ 2 \\ 3 \\ 4 \\ 5 \\ 6 \\ \vdots \end{array} \begin{pmatrix} 0.5 & 0.1 & 0.3 & 0.1 & & & & \\ 0 & 0.4 & 0.4 & 0.2 & & & & \\ 0.3 & 0.2 & 0 & 0.2 & 0.1 & 0.2 & & \\ 0.2 & 0.3 & 0.2 & 0 & 0.1 & 0.2 & & \\ & & 0.3 & 0.2 & 0 & 0.2 & 0.1 & 0.2 \\ & & 0.2 & 0.3 & 0.2 & 0 & 0.1 & 0.2 \\ & & & \ddots & \ddots & \ddots & \ddots & \ddots \end{pmatrix} \begin{array}{l} \to (0,1) \\ \to (0,2) \\ \to (1,1) \\ \to (1,2) \\ \to (2,1) \\ \to (2,2) \\ \quad \vdots \end{array} \begin{array}{l} \left.\begin{array}{l} \\ \\ \end{array}\right\} \text{level 1} \\ \left.\begin{array}{l} \\ \\ \end{array}\right\} \text{level 2} \\ \left.\begin{array}{l} \\ \\ \end{array}\right\} \text{level 3} \\ \quad \vdots \end{array} \tag{3.26}$$

For this example, the relationship between the variables $\{X_k, J_k, Z_k\}$ is $Z_k = 2X_k + J_k$, $k = 0, 1, 2, \ldots$. By blocking, the transition probability matrix P can be written in the form given in Eq. (3.24) with

$$A_{0,0}=\begin{pmatrix}0.5 & 0.1\\ 0 & 0.4\end{pmatrix},\ A_{0,1}=\begin{pmatrix}0.3 & 0.1\\ 0.4 & 0.2\end{pmatrix},$$
$$A_{1,0}=A_2=\begin{pmatrix}0.3 & 0.2\\ 0.2 & 0.3\end{pmatrix},\ A_{1,1}=A_1=\begin{pmatrix}0 & 0.2\\ 0.2 & 0\end{pmatrix},\ A_0=\begin{pmatrix}0.1 & 0.2\\ 0.1 & 0.2\end{pmatrix}. \tag{3.27}$$

We assume that the QBD process $\{(X_k, J_k),\ k = 0, 1, 2, \ldots\}$ defined by Eq. (3.24) is irreducible and positive recurrent (i.e., ergodic). Then its limiting probabilities exist and are positive. Define

$$\begin{aligned}\pi_{n,j} &= \lim_{k\to\infty} P\{X_k = n, J_k = j|(X_0, J_0)\}, \text{for}\\ &\qquad n = 0,\ 1,\ 2, \ldots,\ j = 1,\ 2, \ldots,\ m;\\ \boldsymbol{\pi}_0 &= (\pi_{0,1}, \cdots, \pi_{0,m_0});\\ \boldsymbol{\pi}_n &= (\pi_{n,1}, \cdots, \pi_{n,m}),\quad n = 1,\ 2,\ \ldots;\\ \boldsymbol{\pi} &= (\boldsymbol{\pi}_0,\ \boldsymbol{\pi}_1,\ \boldsymbol{\pi}_2, \ldots).\end{aligned} \tag{3.28}$$

The vector form of the limiting probabilities brings convenience in analysis. To find and analyze $\boldsymbol{\pi}$, we begin with the following *Neuts condition* for ergodicity.

Theorem 3.2.1 *Irreducible QBD process* $\{(X_k, J_k),\ k = 0, 1, 2, \ldots\}$ *is ergodic if and only if* $\boldsymbol{\theta}A_0\mathbf{e} < \boldsymbol{\theta}A_2\mathbf{e}$, *where* $\boldsymbol{\theta}$ *satisfies* $\boldsymbol{\theta}A = \boldsymbol{\theta}$ *and* $\boldsymbol{\theta}\mathbf{e} = 1$.

Proof. Proofs of this theorem can be found in Neuts (1981) and Latouche and Ramaswami (1999). We give a proof to the sufficiency of the conditions to demonstrate the usefulness of the *mean-drift method* in stochastic modeling.

The proof of sufficiency is based on Foster's criteria for the ergodicity of Markov chains (see Chap. 2 in Cohen (1982)). More specifically, the idea is to find the mean-drift of a Lyapunov function defined on the states of the Markov chain. Under the conditions, we show that the mean drift is negative. First, we construct a Lyapunov function. Define

$$A^*(z) = A_0 + zA_1 + z^2A_2,\quad z \geq 0. \tag{3.29}$$

It is clear that $A^*(1) = A$. Denote by $\rho(z)$ the Perron-Frobenius eigenvalue of the nonnegative matrix $A^*(z)$ (i.e., the eigenvalue of $A^*(z)$ with the largest real part). Denote by $\mathbf{u}(z)$ and $\mathbf{v}(z)$ the left and right eigenvectors of $A^*(z)$, i.e., $\mathbf{u}(z)A^*(z) = \rho(z)\mathbf{u}(z)$ and $A^*(z)\mathbf{v}(z) = \rho(z)\mathbf{v}(z)$, normalized by $\mathbf{u}(z)\mathbf{v}(z) = 1$ and $\mathbf{u}(z)\mathbf{e} = 1$. By the Perron-Frobenius theory on nonnegative matrices (Seneta (2006)), both $\mathbf{u}(z)$ and $\mathbf{v}(z)$ are nonnegative. Taking derivatives of both sides of $\mathbf{u}(z)A^*(z) = \rho(z)\mathbf{u}(z)$ and postmultiplying by $\mathbf{v}(z)$ on both sides, yield $\mathbf{u}(z)A^{*(1)}(z)\mathbf{v}(z) + \mathbf{u}^{(1)}(z)A^*(z)\mathbf{v}(z) = \rho^{(1)}(z)\mathbf{u}(z)\mathbf{v}(z) + \rho(z)\mathbf{u}^{(1)}(z)\mathbf{v}(z)$. Note that $A^{*(1)}(z)$, $\mathbf{u}^{(1)}(z)$, and $\rho^{(1)}(z)$ are the derivatives of $A^*(z)$, $\mathbf{u}(z)$, and $\rho(z)$, respectively. Also note that $\mathbf{u}(1) = \boldsymbol{\theta}$ and $\mathbf{v}(1) = \mathbf{e}$.

Letting $z = 1$, we obtain $\mathbf{u}(1)A^{*(1)}(1)\mathbf{v}(1) + \mathbf{u}^{(1)}(1)A^*(1)\mathbf{v}(1) = \rho^{(1)}(1)\mathbf{u}(1)\mathbf{v}(1) + \rho(1)\mathbf{u}^{(1)}(1)\mathbf{v}(1)$. By the normalization conditions, we obtain $\mathbf{u}(1)\mathbf{v}(1) = 1$ and

$\mathbf{u}^{(1)}(1)\mathbf{v}(1) = 0$. Then we obtain $\mathbf{u}(1)A^{*(1)}(1)\mathbf{e} = \rho^{(1)}(1)$, i.e., $\rho^{(1)}(1) = \boldsymbol{\theta}A^{*(1)}(1)\mathbf{e} = 1 + \boldsymbol{\theta}A_2\mathbf{e} - \boldsymbol{\theta}A_0\mathbf{e} > 1$. Thus, there exists z such that $0 < \rho(z) < z < 1$, and $\mathbf{v}(z)$ is positive elementwise. Define a (vector) Lyapunov function for states in level n as

$$\mathbf{f}^*(n) = z^{-n}\mathbf{v}(z), \; n = 0, \; 1, \; 2, \; .... \tag{3.30}$$

It is clear that $\mathbf{f}^*(n)$ goes to positive infinity elementwise, if n goes to infinity. The difference $\mathbf{f}^*(X_{k+1}) - \mathbf{f}^*(X_k)$ is called the drift of the Markov chain at X_k. Then, the mean-drift can be calculated as follows, for $n > 2$,

$$\begin{aligned} &E[\mathbf{f}^*(X_{k+1}) - \mathbf{f}^*(X_k)|X_k = n] \\ &= z^{-n-1}A_0\mathbf{v}(z) + z^{-n}A_1\mathbf{v}(z) + z^{-n+1}A_2\mathbf{v}(z) - z^{-n}\mathbf{v}(z) \\ &= z^{-(n+1)}A^*(z)\mathbf{v}(z) - z^{-n}\mathbf{v}(z) \\ &= z^{-(n+1)}(\rho(z) - z)\mathbf{v}(z) \xrightarrow{n\to\infty} -\infty. \end{aligned} \tag{3.31}$$

It is easy to see that the last expression in Eq. (3.31) is less than or equal to $(\rho(z) - z)\,\mathbf{v}(z) < 0$. For $n \le 2$, the mean drift is apparently finite. By Foster's criteria, the Markov chain is ergodic. This completes the proof of Theorem 3.2.1.

Exercise 3.2.1 Explain the condition $\boldsymbol{\theta}A_0\mathbf{e} < \boldsymbol{\theta}A_2\mathbf{e}$ in Theorem 3.2.1 for ergodicity intuitively. (Hint: Suppose that the Markov chain is in level n. If n is sufficiently large, the distribution of the phase J_k is approximately $\boldsymbol{\theta}$. Consider the drift of the Markov chain towards level zero.)

Under Neuts condition $\boldsymbol{\theta}A_0\mathbf{e} < \boldsymbol{\theta}A_2\mathbf{e}$, the QBD process $\{(X_k, J_k), k = 0, 1, 2, \ldots\}$ is ergodic, and its limiting probabilities exist. Next, we solve linear system $\boldsymbol{\pi}P = \boldsymbol{\pi}$ and $\boldsymbol{\pi}\mathbf{e} = 1$ to find the limiting probabilities. Taking the advantage of the block structure in P, we expand the linear system $\boldsymbol{\pi}P = \boldsymbol{\pi}$ according to the level variable to obtain

$$\begin{aligned} \boldsymbol{\pi}_0 &= \boldsymbol{\pi}_0 A_{0,0} + \boldsymbol{\pi}_1 A_{1,0}; \\ \boldsymbol{\pi}_1 &= \boldsymbol{\pi}_0 A_{0,1} + \boldsymbol{\pi}_1 A_{1,1} + \boldsymbol{\pi}_2 A_2; \\ \boldsymbol{\pi}_n &= \boldsymbol{\pi}_{n-1} A_0 + \boldsymbol{\pi}_n A_1 + \boldsymbol{\pi}_{n+1} A_2, \quad n = 2, \; 3, \; .... \end{aligned} \tag{3.32}$$

To solve Eq. (3.32), we conjecture that the solution has the *matrix-geometric* form $\boldsymbol{\pi}_n = \boldsymbol{\pi}_1 R^{n-1}$ for $n = 1, 2, \ldots$. Although details for getting the final solution are quite involved and technical, the idea is straightforward. Putting the suggested solution into equations in (3.32), we obtain, for $n \ge 2$, $\boldsymbol{\pi}_1 R^{n-1}(R - A_0 - RA_1 - R^2A_2) = 0$. If a nonnegative matrix R can be found for equation $R = A_0 + RA_1 + R^2A_2$, and proper $\boldsymbol{\pi}_0$ and $\boldsymbol{\pi}_1$ can be found, then a solution $\boldsymbol{\pi}$ can be found.

The matrix-geometric solution can be found by using either an algebraic or probabilistic approach, the latter offers more insight into the variables involved. Next, we present the matrix-geometric solution first. Then we give a probabilistic interpretation/proof of that solution.

Theorem 3.2.2 (Neuts (1981)) *If the QBD process* $\{(X_k, J_k), k = 0, 1, 2, \ldots\}$ *is ergodic, its limiting probabilities are given by*

$$\boldsymbol{\pi}_n = \boldsymbol{\pi}_1 R^{n-1}, \quad n = 1, 2, \ldots, \tag{3.33}$$

where the rate matrix R is the minimal nonnegative solution to nonlinear equation

$$R = A_0 + RA_1 + R^2 A_2, \tag{3.34}$$

and the vectors $\boldsymbol{\pi}_0$ *and* $\boldsymbol{\pi}_1$ *are the unique positive solution to the linear system*

$$\begin{aligned} \boldsymbol{\pi}_0 &= \boldsymbol{\pi}_0 A_{0,0} + \boldsymbol{\pi}_1 A_{1,0}; \\ \boldsymbol{\pi}_1 &= \boldsymbol{\pi}_0 A_{0,1} + \boldsymbol{\pi}_1 (A_{1,1} + RA_2); \\ 1 &= \boldsymbol{\pi}_0 \mathbf{e} + \boldsymbol{\pi}_1 (I - R)^{-1}\mathbf{e}. \end{aligned} \quad \text{or} \quad \begin{aligned} (\boldsymbol{\pi}_0, \boldsymbol{\pi}_1) &= (\boldsymbol{\pi}_0, \boldsymbol{\pi}_1) \begin{pmatrix} A_{0,0} & A_{0,1} \\ A_{1,0} & A_{1,1} + RA_2 \end{pmatrix}; \\ 1 &= \boldsymbol{\pi}_0 \mathbf{e} + \boldsymbol{\pi}_1 (I - R)^{-1}\mathbf{e}. \end{aligned} \tag{3.35}$$

The matrix-geometric solution given in Theorem 3.2.2 is a fundamental result in matrix-analytic methods. The solution given by (3.33), (3.34), and (3.35) is a matrix generalization of the geometric solution given in Section 3.1 for the birth-and-death process. In fact, it is easy to verify that the geometric solution is indeed a special case of the matrix-geometric solution.

Exercise 3.2.2 Use Theorem 3.2.2 to find the geometric solution (3.9).

In general, there is no explicit solution to the matrix R. The following simple algorithm can be used for computing R recursively. Let $R[0] = 0$, and

$$R[k+1] = A_0 + R[k]A_1 + (R[k])^2 A_2, \quad k = 0, 1, 2, \ldots. \tag{3.36}$$

Exercise 3.2.3 Assume that R exists and is finite. Show that (i) the sequence $\{R[k], k = 0, 1, 2, \ldots\}$ is non-decreasing; (ii) $R[k] \le R$, for $k = 0, 1, 2, \ldots$; and (iii) $\{R[k], k = 0, 1, 2, \ldots\}$ converges to R.

Exercise 3.2.3 indicates that the algorithm based on (3.36) for computing R is stable.

Exercise 3.2.4 Assume that R exists and is finite. Show that the following sequence $\{R[k], k = 0, 1, 2, \ldots\}$ is non-decreasing and converges to R. Let $R[0] = 0$, and

$$R[k+1] = A_0(I - A_1)^{-1} + (R[k])^2 A_2 (I - A_1)^{-1}, \quad k = 0, 1, 2, \ldots. \tag{3.37}$$

Exercise 3.2.5* Consider the matrix function $A^*(z)$ defined in Eq. (3.29). Recall that $\rho(z)$ is defined as the Perron-Frobenius eigenvalue of $A^*(z)$. It can be shown that there is a unique z^* in (0, 1) such that $\rho(z) = z$ if and only if $\rho^{(1)}(1) > 1$.

(See Exercise 3.4.11 and Chap. 1 in Neuts (1981) for more about the issue.) Show that $sp(R) = z^*$. (Note: $sp(R)$ is the spectrum of R, i.e., the largest modulus of all the eigenvalues of R. Since R is nonnegative, $sp(R)$ is also the Perron-Frobenius eigenvalue of R.)

Exercise 3.2.6 Assume that the QBD process $\{(X_k, J_k), k = 0, 1, 2, \ldots\}$ is ergodic. Show that (i) $sp(R) < 1$; and (ii) the matrix

$$\begin{pmatrix} A_{0,0} & A_{0,1} \\ A_{1,0} & A_{1,1} + RA_2 \end{pmatrix} \tag{3.38}$$

is stochastic.

Based on Theorems 3.2.1 and 3.2.2 and the above discussion, the following computational procedure can be used for computing the limiting probabilities of a QBD process, if they exist and are positive.

Step 1: Input transition blocks $\{A_0, A_1, A_2, A_{0,0}, A_{0,1}, A_{1,0}, A_{1,1}\}$.
Step 2: Compute $\boldsymbol{\theta}$ satisfying $\boldsymbol{\theta}(A_0 + A_1 + A_2) = \boldsymbol{\theta}$ and $\boldsymbol{\theta}\mathbf{e} = 1$. Apply Theorem 3.2.1 to check ergodicity.
Step 3: Use (3.36) to compute R iteratively. Stop the iteration if $\|R(n+1) - R(n)\|_1$ is sufficiently small. (Note: $\|A\|_1 = \max_{1 \le j \le n}\left\{\sum_{i=1}^{n} |a_{i,j}|\right\}$ for matrix $A = (a_{i,j})$.)
Step 4: Solve Eq. (3.35) for $\boldsymbol{\pi}_0$ and $\boldsymbol{\pi}_1$.
Step 5: Use $\{R, \boldsymbol{\pi}_0, \boldsymbol{\pi}_1\}$ to compute limiting probabilities.

Exercise 3.2.7 Use the above computational procedure to determine whether or not the Markov chain in Example 3.2.1 is ergodic. If it is ergodic, find $\{R, \boldsymbol{\pi}_0, \boldsymbol{\pi}_1\}$ numerically.

In the rest of the section, we concentrate on the matrix R and give the matrix-geometric solution (3.33) a probabilistic interpretation. We begin with a probabilistic interpretation to R, which can be used as an alternative definition of R. Define

- ${}_iP^{(k)}_{i,j;i+n,v}$: The *taboo probability* that, given it is initially in state (i, j), the Markov chain reaches $(i + n, v)$ at time k without visiting levels $\{0, 1, \ldots, i\}$ in between, for $k \ge 0$, $i \ge 0$, $n \ge 1$, $v = 1, \ldots, m$, and $j = 1, \ldots, m$.
- $R^{(n)}_{j,v} = \sum_{k=0}^{\infty} {}_iP^{(k)}_{i,j;i+n,v}$: The expected number of visits to state $(i + n, v)$ before the first return to level i, given that the Markov chain starts in state (i, j).

Note that $R^{(n)}_{j,v}$ is actually independent of the level i (>0). Define ${}_iI^{(k)}_{i,j;i+n,v} = 1$ if the Markov chain visits state $(i + n, v)$ at time k without visiting levels $\{0, 1, \ldots, i\}$ in between, given that the Markov chain begins in (i, j); otherwise, ${}_iI^{(k)}_{i,j;i+n,v} = 0$. Then $\sum_{k=0}^{\infty} {}_iI^{(k)}_{i,j;i+n,v}$ is the total number of visits to $(i + n, v)$ before returning to level i, given that the Markov chain begins in (i, j). Then $R^{(n)}_{j,v}$ is the expectation of $\sum_{k=0}^{\infty} {}_iI^{(k)}_{i,j;i+n,v}$.

Proposition 3.2.1 $R^{(n)} = (R^{(1)})^n$, *for* $n = 1, 2, \ldots$.

Proof. We partition the transitions from state (i, j) to $(i + n, v)$ into two parts: from (i, j) to $(i + 1, w)$, in which there is no visit to levels lower than $i + 1$, and from $(i + 1, w)$ to $(i + n, v)$, in which there is no visit to levels lower than $i + 2$. That is, the end of the first part is the last visit to level $i + 1$. Conditioning on the time t and the phase w of the last visit to level $i + 1$, we obtain ${}_iP^{(k)}_{i,j;i+n,v} = \sum_{t=1}^{k}\sum_{w=1}^{m} {}_iP^{(t)}_{i,j;i+1,w}\, {}_{i+1}P^{(k-t)}_{i+1,w;i+n,v}$. Summing up both sides of the equation on k leads to $R^{(n)} = R^{(1)}R^{(n-1)}$. By induction, $R^{(n)} = (R^{(1)})^n$ holds. This completes the proof of Proposition 3.2.1.

Proposition 3.2.2 $R^{(1)} = A_0 + R^{(1)}A_1 + R^{(2)}A_2$.

Proof. It is clear that the last state before visiting state $(i + 1, v)$ can only be in level i, $i + 1$, or $i + 2$. Conditioning on the last state before visiting $(i + 1, v)$, without returning to level i, we obtain

$$
{}_iP^{(k)}_{i,j;i+1,v} = \begin{cases} (A_0)_{j,v}, & \text{if } k = 1; \\ \sum_{u=1}^{m}\left({}_iP^{(k-1)}_{i,j;i+1,u}(A_1)_{u,v} + {}_iP^{(k-1)}_{i,j;i+2,u}(A_2)_{u,v}\right), & \text{if } k \geq 2. \end{cases} \tag{3.39}
$$

Then we have

$$
\begin{aligned}
R^{(1)}_{j,v} &= \sum_{k=1}^{\infty} {}_iP^{(k)}_{i,j;i+1,v} \\
&= (A_0)_{j,v} + \sum_{k=2}^{\infty} {}_iP^{(k)}_{i,j;i+1,v} \\
&= (A_0)_{j,v} + \sum_{k=2}^{\infty}\left(\sum_{u=1}^{m} {}_iP^{(k-1)}_{i,j;i+1,u}(A_1)_{u,v} + \sum_{u=1}^{m} {}_iP^{(k-1)}_{i,j;i+2,u}(A_2)_{u,v}\right) \\
&= (A_0)_{j,v} + \sum_{u=1}^{m}\left(\sum_{k=2}^{\infty} {}_iP^{(k-1)}_{i,j;i+1,u}\right)(A_1)_{u,v} + \sum_{u=1}^{m}\left(\sum_{k=2}^{\infty} {}_iP^{(k-1)}_{i,j;i+2,u}\right)(A_2)_{u,v} \\
&= (A_0)_{j,v} + (R^{(1)}A_1)_{j,v} + (R^{(2)}A_2)_{j,v}.
\end{aligned} \tag{3.40}
$$

This completes the proof of Proposition 3.2.2.

Proposition 3.2.3 $R^{(1)} = R$, *where* R *is the minimal nonnegative solution to Eq.* (3.34).

Proof. Since R is defined as the minimal nonnegative solution to Eq. (3.34), by Propositions 3.2.1 and 3.2.2, we must have $R^{(1)} \geq R$.

To show $R \geq R^{(1)}$, we use the iterative algorithm given in Eq. (3.36). Equation (3.36) indicates that the matrix R includes all the expected visits that the Markov chain reaches state $(i + 1, v)$ after any finite number of transitions, without visiting any state in level i in between, given that the Markov chain is in state (i, j) initially (see Exercise 3.2.8). Therefore, we must have $R \geq R^{(1)}$. This completes the proof of Proposition 3.2.3.

Exercise 3.2.8 Use Eq. (3.36) to show that $\sum_{k=0}^{\infty} {}_iI_{i,j;i+n,v}^{(k)}$ includes all the visits that the Markov chain reaches state $(i+n, v)$ after any finite number of transitions, without visiting any state in level i in between, given that the Markov chain is in state (i, j) initially. (See Neuts (1981) for more details.)

Propositions 3.2.1 to 3.2.3 give the elements in the matrix R a probabilistic interpretation, which is significant both theoretically and practically. Based on the definitions, the (i, j)-th element of R can be interpreted as the expected number of visits to $(n+1, j)$ before the Markov chain returns to levels 0, 1, ..., and n, given that the Markov chain starts in (n, i). Propositions 3.2.1 to 3.2.3 also give Eq. (3.34) an intuitive interpretation. Visits to states in level $n+1$ can be categorized into three mutually exclusive groups: visits immediately from level n (A_0), visits from level $n+1$ (RA_1), and visits from level $n+2$ (R^2A_2). Based on the interpretations of the elements of R, the following results can be proved.

Proposition 3.2.4 *For* $n = 1, 2, \ldots,$ *we have* $\boldsymbol{\pi}_{n+1} = \boldsymbol{\pi}_n R$. *Consequently,* $\boldsymbol{\pi}_n = \boldsymbol{\pi}_1 R^{n-1}$, *for* $n = 1, 2, \ldots$.

Proof. Recall that the limiting probability can be interpreted as the long-run average of the time to be in a state. Based on the interpretation, it is easy to see

$$\pi_{n+1,j} = \lim_{k\to\infty} \frac{\sum_{t=1}^{k} p_{(0,1),(n+1,j)}^{(t)}}{k}. \tag{3.41}$$

Note that the initial state is assumed to be (0, 1). By conditioning on the last time the Markov chain is in level n and the corresponding phase v, we obtain

$$\begin{aligned}
\pi_{n+1,j} &= \lim_{k\to\infty} \frac{1}{k}\left(\sum_{t=1}^{k}\sum_{s=1}^{t-1}\sum_{v=1}^{m} p_{(0,1),(n,v)}^{(s)}\, {}_n p_{(n,v),(n+1,j)}^{(t-s)}\right)\\
&= \lim_{k\to\infty}\sum_{v=1}^{m}\frac{1}{k}\sum_{s=1}^{k} p_{(0,1),(n,v)}^{(s)}\sum_{t=1}^{k-s} {}_n p_{(n,v),(n+1,j)}^{(t)}\\
&= \sum_{v=1}^{m}\left(\lim_{k\to\infty}\frac{1}{k}\sum_{s=1}^{k} p_{(0,1),(n,v)}^{(s)}\right) R_{v,j}\\
&= \sum_{v=1}^{m} \pi_{n,v} R_{v,j}.
\end{aligned} \tag{3.42}$$

Note that to reach a state in level $n+1$ from state (0, 1), the process must visit a state in level n. The results are obtained by writing equation (3.42) in matrix form. This completes the proof of Proposition 3.2.4.

The rate matrix R provides information on the ergodicity of the Markov chain, which is clear from its probabilistic interpretation. While some elements of R can be large numbers, the Perron-Frobenius eigenvalue of R is always in [0, 1].

Proposition 3.2.5 (Neuts (1981)) *The matrix R is the unique nonnegative solution to Eq.* (3.34) *such that* $sp(R) < 1$ *if the Markov chain is ergodic. If the Markov chain is nonergodic, then* $sp(R) = 1$.

See Chap. 1 in Neuts (1981) for a proof of Proposition 3.2.5. Proposition 3.2.5 indicates that, if the QBD process is ergodic, then the matrix $I-R$ is invertible. Proposition 3.2.5 also indicates that Eq. (3.34) always has a nonnegative solution that satisfies $sp(R) \le 1$.

Marginal distributions, moments, and *tail asymptotics* can be obtained from the matrix-geometric solution immediately.

Exercise 3.2.9 For the QBD process $\{(X_k, J_k), k = 0, 1, 2, \ldots\}$, use the matrix-geometric solution to find the distributions of $X_\infty = \lim_{k\to\infty} X_k$ and $J_\infty = \lim_{k\to\infty} J_k$ (assume $m_0 = m$).

Corollary 3.2.1 *The mean of the limiting distribution of* $\{X_k, k = 0, 1, 2, \ldots\}$ *is given by* $\boldsymbol{\pi}_1(I-R)^{-2}\mathbf{e}$.

Proof. By definition,

$$E[X_\infty] = \sum_{n=1}^{\infty} n\boldsymbol{\pi}_n\mathbf{e} = \sum_{n=1}^{\infty} n\boldsymbol{\pi}_1 R^{n-1}\mathbf{e} = \boldsymbol{\pi}_1 \sum_{n=1}^{\infty} nR^{n-1}\mathbf{e} = \boldsymbol{\pi}_1(I-R)^{-2}\mathbf{e}. \quad (3.43)$$

This completes the proof of Corollary 3.2.1.

For nonnegative matrix R, let $\mathbf{v}$ and $\mathbf{u}$ be the left and right Perron-Frobenius eigenvectors of R, respectively, i.e., $\mathbf{v}R = sp(R)\mathbf{v}$ and $R\mathbf{u} = sp(R)\mathbf{u}$, which are normalized by $\mathbf{ve} = \mathbf{vu} = 1$.

Corollary 3.2.2 *If R is irreducible, the limiting probabilities have a geometric decay with decay rate* $sp(R)$ (<1). *That is* $\boldsymbol{\pi}_n = \boldsymbol{\pi}_1 R^{n-1} \approx (\boldsymbol{\pi}_1\mathbf{u})\mathbf{v}\,(sp(R))^{n-1} + o((sp(R))^{n-1})$, *if n is large. (See Neuts* (1986) *for more details.)*

Commentary The matrix-geometric solution (3.33) is one of the most important results in matrix-analytic methods. The matrix-geometric solution is presented in its current form in Neuts (1978, 1980, 1981), and its idea may be traced back to Evans (1967) and Wallace (1969). Computing the rate matrix R is one of the main issues for the matrix-geometric solution. As shown by Exercises 3.2.3 and 3.2.4, R can be computed by using Eq. (3.36) effectively. Another efficient method is Latouche and Ramaswami's logarithmic reduction algorithm for computing R, which is presented and analyzed in detail in Latouche and Ramaswami (1993, 1999) and Ye (2002) (See Exercise 3.3.15). In Gail et al. (1994, 1997), all power bounded solutions to Eq. (3.34) are found.

Additional Exercises and Extensions

With the understanding on R offered in Propositions 3.2.1 to 3.2.5, it is much easier to obtain results on the limiting probabilities of more general Markov chains. Here is an example.

Exercise 3.2.10 Consider a *level dependent* QBD process $\{(X_k, J_k), k = 0, 1, 2, \ldots\}$ with

$$P = \begin{pmatrix} A_{0,0} & A_{0,1} & & & \\ A_{1,0} & A_{1,1} & A_{1,2} & & \\ & A_{2,1} & A_{2,2} & A_{2,3} & \\ & & \ddots & \ddots & \ddots \\ & & & \ddots & \ddots \end{pmatrix}. \tag{3.44}$$

For this QBD process, the transition law of the phase variable J_k depends on the level. Define $R(n)$ as $R_{j,v}(n) = \sum_{k=0}^{\infty} {}_nP^{(k)}_{n-1,j;n,v}$: the expected number of visits to state (n, v) before the first return to the level $n-1$, given that the chain starts in state $(n-1, j)$. Assume that the Markov chain is ergodic. Denote the limiting probabilities of the Markov chain as $\mathbf{p} = (\mathbf{p}_0, \mathbf{p}_1, \mathbf{p}_2, \ldots)$. Prove the following results.

(1) $R(n) = A_{n-1,n} + R(n)A_{n,n} + R(n)R(n+1)A_{n+1,n}$, for $n = 1, 2, \ldots$.
(2) $\mathbf{p}_{n+1} = \mathbf{p}_nR(n+1)$ for $n \geq 0$. Consequently, $\mathbf{p}_n = \mathbf{p}_0R(1)R(2)\ldots R(n)$, for $n = 1, 2, \ldots$.
(3) Vector $\mathbf{p}_0$ satisfies $\mathbf{p}_0(A_{0,0} + R(1)A_{1,0}) = \mathbf{p}_0$ and $\mathbf{p}_0(\mathbf{e} + R(1)\mathbf{e} + R(1)R(2)\mathbf{e} + R(1)R(2)\ldots R(n)\mathbf{e} + \ldots) = 1$.

Exercise 3.2.10 gives a theoretical solution to level dependent QBDs. However, we still need to find matrices $\{R(n), n = 1, 2, \ldots\}$. In general, we have to use approximations or truncations (Bright and Taylor (1995)). For some special cases, explicit solutions can be obtained. For example, in the case the Markov chain becomes level independent at higher levels, is actually level independent after reblocking (to be discussed in detail later). A second case is discussed in the following exercise.

Exercise 3.2.11 Show that all the results obtained in Exercise 3.2.10 holds for QBD type process P defined in Eq. (3.45) with N levels, where N is a finite positive integer. Show that $R(N)$ can be found explicitly. Outline an algorithm for computing the limiting probabilities.

$$P = \begin{pmatrix} A_{0,0} & A_{0,1} & & & \\ A_{1,0} & A_{1,1} & A_{1,2} & & \\ & \ddots & \ddots & \ddots & \\ & & A_{N-1,N-2} & A_{N-1,N-1} & A_{N-1,N} \\ & & & A_{N,N-1} & A_{N,N} \end{pmatrix}. \tag{3.45}$$

We remark that several algorithms have been developed for computing the limiting probabilities for the Markov chain P defined in Eq. (3.45). For example, Naoumov (1996) develops a *modified matrix-geometric solution* and Ye and Li (1994) develops a *folding algorithm* for computing the limiting probabilities of P. We also remark that the *GTH algorithm* (Grassmann et al. (1985)) is a widely used method for computing the limiting probabilities of Markov chains with a finite number of states.

Exercise 3.2.12 Give details on how to solve Eq. (3.35) for vectors $\boldsymbol{\pi}_0$ and $\boldsymbol{\pi}_1$.

Exercise 3.2.13 Check the ergodicity of the following QBD process. Find R and $(\boldsymbol{\pi}_0, \boldsymbol{\pi}_1)$.

$$\begin{aligned} A_{0,0} &= \begin{pmatrix} 0.1 & 0.1 \\ 0.1 & 0.1 \end{pmatrix}, \quad A_{0,1} = \begin{pmatrix} 0.4 & 0.4 \\ 0.4 & 0.4 \end{pmatrix}, \quad A_{1,0} = \begin{pmatrix} 0.2 & 0.3 \\ 0.3 & 0.3 \end{pmatrix}, \\ A_{1,1} = A_1 &= \begin{pmatrix} 0.1 & 0.1 \\ 0 & 0.1 \end{pmatrix}, \quad A_0 = \begin{pmatrix} 0.2 & 0.1 \\ 0.3 & 0 \end{pmatrix}, \quad A_2 = \begin{pmatrix} 0.5 & 0 \\ 0.3 & 0.3 \end{pmatrix}. \end{aligned} \tag{3.46}$$

In addition, find the limiting probability distributions of the level variable $X_\infty = \lim_{k\to\infty} X_k$ and the phase variable $J_\infty = \lim_{k\to\infty} J_k$, respectively.

Exercise 3.2.14 In Eq. (3.24), assume that $A_{0,0} = A_2 + A_1$ and $A_{1,0} = A_2$. Show that $\{J_k, k = 0, 1, 2, \ldots\}$ is a Markov chain. Find the transition probability matrix of $\{J_k, k = 0, 1, 2, \ldots\}$.

Exercise 3.2.15 Consider irreducible continuous time QBD process $\{(X(t), J(t)), t \geq 0\}$ with infinitesimal generator Q having the same structure as P given in Eq. (3.24). Assume that $A_0 + A_1 + A_2$ is irreducible.

(1) Show that the QBD process is ergodic if and only if $\boldsymbol{\theta} A_0 \mathbf{e} < \boldsymbol{\theta} A_2 \mathbf{e}$, where $\boldsymbol{\theta}$ satisfies $\boldsymbol{\theta}(A_0 + A_1 + A_2) = 0$ and $\boldsymbol{\theta}\mathbf{e} = 1$. Explain the condition intuitively.
(2) Show that the stationary distribution (limiting probabilities) is given by

$$\mathbf{p}_n = \mathbf{p}_1 R^{n-1}, \quad n = 1, 2, \ldots. \tag{3.47}$$

where the *rate matrix* R is the minimal nonnegative solution to nonlinear equation

$$0 = A_0 + RA_1 + R^2 A_2. \tag{3.48}$$

Also, $(\mathbf{p}_0, \mathbf{p}_1)$ is the unique positive solution to

$$\begin{aligned} 0 &= \mathbf{p}_0 A_{0,0} + \mathbf{p}_1 A_{1,0}; \\ 0 &= \mathbf{p}_0 A_{0,1} + \mathbf{p}_1 (A_{1,1} + RA_{2,1}); \\ 1 &= \mathbf{p}_0 \mathbf{e} + \mathbf{p}_1 (I - R)^{-1} \mathbf{e}. \end{aligned} \tag{3.49}$$

Show that matrix

$$\begin{pmatrix} A_{0,0} & A_{0,1} \\ A_{1,0} & A_{1,1} + RA_2 \end{pmatrix} \tag{3.50}$$

is an infinitesimal generator.

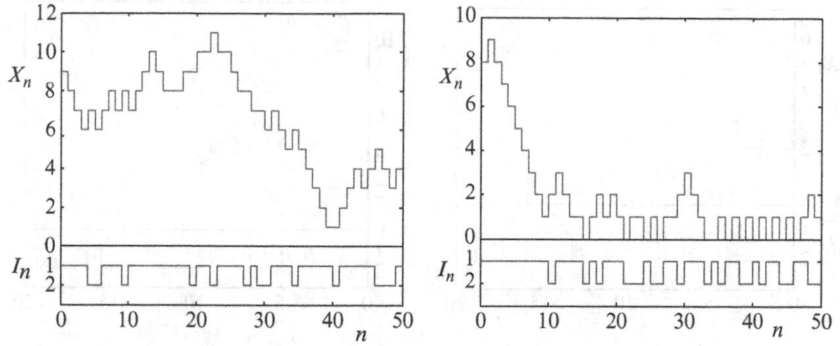

Fig. 3.4 Sample paths of $\{X_k, k = 0, 1, \ldots\}$ and $\{J_k, k = 0, 1, \ldots\}$ defined in Exercise 3.2.13

We remark that, for the continuous time case, the interpretation of R is more complicated. See the last part of Chap. 1 in Neuts (1981) for details.

The uniformization method can be used to find the solutions in Exercise 3.2.15. Let $v = \max\{(-A_{0,0})_{i,i},\ i = 1, 2, \ldots, m_0;\ (-A_{1,1})_{i,i},\ (-A_1)_{i,i},\ i = 1, 2, \ldots, m\}$. Define $P_1 = I + Q/v$. It is easy to verify that P_1 is the transition probability matrix of a discrete time QBD process. It is readily seen that P_1 and Q have the same limiting probabilities (if they exist). Apply Theorems 3.2.1 and 3.2.2, solutions in Exercise 3.2.15 are obtained. With uniformization, the sojourn time in each state is exponentially distributed with parameter v in the Markov chain Q. Then P_1 can be interpreted as the (discrete time) *embedded Markov chain* at the transition epochs. Some of the transitions are within the same state, which are invisible. Then the continuous time Markov chain Q can be defined as a Markov chain with (i) the sojourn time in each state is exponentially distributed with parameter v; and (ii) the embedded Markov chain at transition epochs is P_1.

Here is another way to transform the continuous time Markov chain Q into a discrete time Markov chain. Let $D = \text{diag}(-A_{0,0}{}^{-1}, -A_{1,1}{}^{-1}, -A_1{}^{-1}, -A_1{}^{-1}, \ldots)$. Note that D is nonnegative since $-A_{0,0}$, $-A_{1,1}$, and $-A_1$ are M-matrices (Minc (1988)). Define $P_2 = I + DQ$. Intuitively, P_2 is an embedded Markov chain of Q at the transition epochs where the level variable changes its value (either increasing by one or decreasing by one).

Exercise 3.2.16 Show that P_2 is a discrete time QBD process. Denote by $\mathbf{p}$ the vector for the limiting probabilities of P_2 and by $\boldsymbol{\pi}$ the vector for the limiting probabilities of Q. Show that $\mathbf{p}D = \boldsymbol{\pi}$.

Exercise 3.2.17 Write a simulation program to visualize the QBD process given in Exercise 3.2.13. Plot $\{X_k, k = 0, 1, 2, \ldots\}$ and $\{J_k, k = 0, 1, 2, \ldots\}$. Use the simulation program to estimate the limiting probabilities. (Hint: Determine the next phase and the next level by using matrices $L_0 = (A_{0,0}, A_{0,1})J$, $L_1 = (A_{1,0}, A_{1,1}, A_0)J$, and $L_2 = (A_2, A_1, A_0)J$, where J is a matrix with elements in upper triangular part and on the diagonal being one and the lower triangular part being zero, if the level is 0, 1, or 2 and higher than 2, respectively. See Fig. 3.4 for two sample paths of a QBD process.)

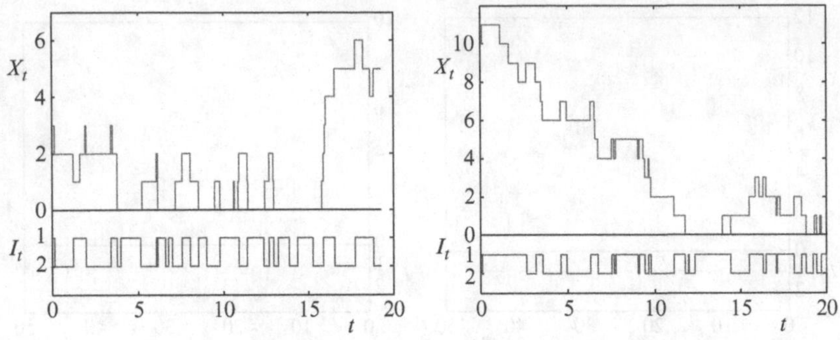

Fig. 3.5 Sample paths of a continuous time QBD process

Exercise 3.2.18 Write a simulation program for the continuous time QBD process defined in Exercise 3.2.15. See Fig. 3.5 for two sample paths.

Reblocking is a useful technique in stochastic modeling. By reblocking the transition probabilities in a transition probability matrix, the structure of the transition probability matrix may be changed (note: transitions themselves are not changed). The new structure might be more suitable for analysis or numerical computation. Here is an example.

Exercise 3.2.19 Consider a Markov chain with the following transition probability matrix:

$$P = \begin{array}{c} 0 \\ 1 \\ 2 \\ 3 \\ 4 \\ \vdots \end{array} \begin{pmatrix} A_{0,0} & A_{0,1} & A_{0,2} & & & & \\ A_{1,0} & A_{1,1} & A_{1,2} & A_{1,3} & & & \\ A_{2,0} & A_{2,1} & A_{2,2} & A_{2,3} & A_{2,4} & & \\ & A_{3,1} & A_{3,2} & A_{3.3} & A_{3,4} & A_{3,5} & \\ & & A_{3,1} & A_{3,2} & A_{3,3} & A_{3,4} & A_{3,5} \\ & & & \ddots & \ddots & \ddots & \ddots \end{pmatrix}. \tag{3.51}$$

Combine levels $\{0, 1\}, \{2, 3\}, \{4, 5\}, \ldots$ to generate new levels $0, 1, 2, \ldots$. After the consolidation of levels, what does the transition probability matrix of the Markov chain look like? Assuming that the Markov chain is ergodic, outline a method for computing its limiting probabilities.

If we combine levels $\{0, 1, 2\}, \{3, 4, 5\}, \{6, 7, 8\}, \ldots$ to generate new levels 0, 1, 2, …, what does the transition probability matrix of the Markov chain look like?

Exercise 3.2.20 (Structure of R) Assume that the last k rows of A_0 are zero. Show that the last k rows of R are zero as well. Prove the result probabilistically without mathematical calculations. Show that a row in A_0 is zero if and only if the corresponding row in R is zero.

Exercise 3.2.21 (Structure of R) Show that (i) if A is reducible, then R is reducible; and (ii) if A is upper/lower triangular, then R is upper/lower triangular.

Exercise 3.2.22 Show that the variance of the stationary distribution of $\{X_k, k = 0, 1, \ldots\}$ is given by

$$Var(X_\infty) = \boldsymbol{\pi}_1\left(2(I-R)^{-3} + (I-R)^{-2}\right)\mathbf{e} - \left(\boldsymbol{\pi}_1(I-R)^{-2}\mathbf{e}\right)^2. \tag{3.52}$$

Exercise 3.2.23 Consider a QBD process with

$$A_0 = \begin{pmatrix} 0.2 & 0.1 \\ 0.3 & 0 \end{pmatrix}, \quad A_1 = \begin{pmatrix} 0.1 & 0.1 \\ 0 & 0.1 \end{pmatrix}, \quad A_2 = \begin{pmatrix} 0.5 & 0 \\ 0.3 & 0.3 \end{pmatrix}. \tag{3.53}$$

Compute the matrix R. Find the Perron-Frobenius eigenvalue $sp(R)$ of R. Is the Markov chain ergodic? (Hint: Use Neuts condition.) (Note: The matrix R provides information on the ergodicity of the QBD process.)

Exercise 3.2.24 Assume $m_0 = m$. Define $\mathbf{p}^*(z) = \sum_{n=0}^{\infty} \boldsymbol{\pi}_n z^n$. Show that $\mathbf{p}^*(z)$ satisfies the following equation: for $z \geq 0$,

$$\begin{aligned} &\mathbf{p}^*(z)\left(I - (zA_0 + A_1 + z^{-1}A_2)\right) \\ &\quad = \boldsymbol{\pi}_0 A_{0,0} + \boldsymbol{\pi}_1 A_{1,0} + z(\boldsymbol{\pi}_0 A_{0,1} + \boldsymbol{\pi}_1 A_{1,1}) \\ &\qquad - z\boldsymbol{\pi}_0 A_0 - (\boldsymbol{\pi}_0 + z\boldsymbol{\pi}_1)A_1 - z^{-1}(\boldsymbol{\pi}_0 + z\boldsymbol{\pi}_1)A_2. \end{aligned} \tag{3.54}$$

Exercise 3.2.25 Consider a QBD process with

$$A_0 = \begin{pmatrix} 0.2 & 0.1 \\ 0.3 & 0 \end{pmatrix}, \quad A_1 = \begin{pmatrix} 0.1 & 0.1 \\ 0 & 0.1 \end{pmatrix}, \quad A_2 = \begin{pmatrix} 0.5 & 0 \\ 0.3 & 0.3 \end{pmatrix}. \tag{3.55}$$

Compute the matrix R and find its maximum element. Find $sp(R)$. (Note: The maximum element of R can be greater than 1, while $sp(R)$ is less than or equal to one.)

3.3 Quasi-Birth-and-Death Processes: Fundamental Period

We continue to consider the QBD process defined by Eq. (3.24). A *fundamental period* of the QBD process is defined as a period that begins at an epoch the Markov chain is in a state at level n and ends at the first epoch the Markov chain is in level $n-1$. Define matrix G with the (i, j)-th component $g_{i,j}$ defined as the probability that, given that the Markov chain starts in (n, i), the Markov chain goes down to the level $n-1$ for the first time by entering state $(n-1, j)$. Denote by τ the first time epoch the Markov chain enters level $n-1$, given that the Markov chain is in level n at time zero. That is, τ is the length of a fundamental period. We decompose the event associated with $g_{i,j}$ into *disjoint events* as follows:

$$\begin{aligned} &\{(X_\tau, J_\tau) = (n-1, j) | (X_0, J_0) = (n, i)\} \\ &= \bigcup_{k=1}^{\infty} \{(X_k, J_k) = (n-1, j), X_t \geq n, \ t = 1, 2, \ldots, k-1 | (X_0, J_0) = (n, i)\}. \end{aligned} \tag{3.56}$$

Then we obtain, for $i, j = 1, \ldots, m$,

$$g_{i,j} = \sum_{k=1}^{\infty} P\{(X_k, J_k) = (n-1, j), X_t \geq n, \ t = 1, 2, \ldots, k-1 | (X_0, J_0) = (n, i)\}. \tag{3.57}$$

By the structure of P defined in Eq. (3.24), it is readily seen that the definition of G is related to transitions of the following (sub) Markov chain before its absorption:

$$\begin{pmatrix} A_1 & A_0 & & & \\ A_2 & A_1 & A_0 & & \\ & A_2 & A_1 & A_0 & \\ & & \ddots & \ddots & \ddots \\ & & & \ddots & \ddots \end{pmatrix}. \tag{3.58}$$

The above matrix is called a sub-stochastic matrix since $(A_1 + A_0)\mathbf{e} \leq \mathbf{e}$ and $(A_1 + A_0)\mathbf{e} \neq \mathbf{e}$. Thus, the definition of G is independent of the level n (≥ 2).

Lemma 3.3.1 *Define matrix $G^{(2)}$ with the (i, j)-th component $g_{i,j}^{(2)}$ defined as the probability that, given that the Markov chain starts in $(n + 2, i)$, the Markov chain goes down to the level n for the first time by entering state (n, j). Then we have $G^{(2)} = G^2$.*

Proof. The proof is based on two simple observations: (1) the first passage from level $n + 2$ to level n can be decomposed into the first passage from level $n + 2$ to level $n + 1$ and the first passage from level $n + 1$ to level n (because of the QBD structure); and (2) probabilistically, the first passage from level $n + 2$ to level $n + 1$ is conditionally independent of and equivalent to the first passage from level $n + 1$ to level n. Conditioning on the phase that the Markov chain first reaches level $n + 1$, we obtain

$$
\begin{aligned}
g_{i,j}^{(2)} &= \sum_{k=1}^{\infty} P\{(X_k, J_k) = (n, j), X_t \geq n+1, t = 1, 2, \ldots, k-1 | (X_0, J_0) = (n+2, i)\} \\
&= \sum_{k=1}^{\infty}\sum_{s=1}^{k-1}\sum_{v=1}^{m} P\{(X_k, J_k) = (n, j), X_t \geq n+1, t = s+1, \ldots, k-1 | (X_s, J_s) = (n+1, v)\} \\
&\quad \times P\{(X_s, J_s) = (n+1, v), X_t \geq n+2, t = 1, 2, \ldots, s-1 | (X_0, J_0) = (n+2, i)\} \\
&= \sum_{v=1}^{m}\sum_{s=1}^{\infty}\sum_{k=s+1}^{\infty} P\{(X_{k-s}, J_{k-s}) = (n, j), X_t \geq n+1, t = 1, \ldots, k-s | (X_0, J_0) = (n+1, v)\} \\
&\quad \times P\{(X_s, J_s) = (n+1, v), X_t \geq n+2, t = 1, 2, \ldots, s-1 | (X_0, J_0) = (n+2, i)\} \\
&= \sum_{v=1}^{m} g_{i,v} g_{v,j}.
\end{aligned} \tag{3.59}
$$

This completes the proof of Lemma 3.3.1.

By conditioning on the next (first) transition and using Lemma 3.3.1, it can be shown that G satisfies the following equation, for $i, j = 1, \ldots, m$,

$$
\begin{aligned}
g_{i,j} &= (A_2)_{i,j} + \sum_{k=1}^{m} (A_1)_{i,k} g_{k,j} + \sum_{k=1}^{m} (A_0)_{i,k} \sum_{u=1}^{m} g_{k,u} g_{u,j}; \\
G &= A_2 + A_1 G + A_0 G^2.
\end{aligned} \tag{3.60}
$$

In the second equality in Eq. (3.60), A_2 represents the one step transition from level n + 1 to level n (for the first time); A_1G represents all the transitions from level n + 1 to level n + 1 in one step, and then from level n + 1 to level n for the first time; and A_0G^2 represents all the transitions from level n + 1 to level n + 2 in one step, and then from level n + 2 to level n for the first time.

Exercise 3.3.1 Give all the details for the proof of Eq. (3.60). (Hint: Write all the conditional probabilities explicitly.) (Note: Although it is important to be able to give a detailed proof of Lemma 3.3.1 and Eq. (3.60), it is also important to understand the proof intuitively and to be able to see such relationships intuitively.)

Furthermore, the following holds for G.

Proposition 3.3.1 *The matrix G is the minimal nonnegative solution to Eq.* (3.60).

Proof. First, we claim that the minimal nonnegative solution to Eq. (3.60) can be obtained as the limit of the sequence $\{G[k], k = 0, 1, 2, \ldots\}$, where $G[0] = 0$, and

$$
G[k+1] = A_2 + A_1 G[k] + A_0 (G[k])^2, \quad k = 1, 2, \ldots. \tag{3.61}
$$

The sequence $\{G[k], k = 0, 1, \ldots\}$ has a finite limit since its row sums are bounded by $\mathbf{e}$, i.e., $G[k]\mathbf{e} \leq \mathbf{e}$ (elementwise), for $k \geq 0$. It can be shown that the limit of $\{G[k],$

$k = 0, 1, \ldots\}$ is the minimal nonnegative solution to Eq. (3.60). It is easy to see $G \geq G[0] = 0$ (elementwise). Since G satisfies Eq. (3.60), if $G \geq G[k]$, then we obtain

$$G[k+1] = A_2 + A_1 G[k] + A_0 (G[k])^2 \leq A_2 + A_1 G + A_0 G^2 = G. \tag{3.62}$$

By induction, we have $G \geq G[k]$ for $k \geq 0$. Then G is greater than or equal to the minimal solution to Eq. (3.60). On the other hand, by Eq. (3.61), it can be shown that the minimal nonnegative solution to (3.60) includes all the paths from level n to level $n-1$ (see Exercise 3.3.2), which implies that the minimal nonnegative solution is greater than or equal to G. Therefore, G is the minimal nonnegative solution to Eq. (3.60). This completes the proof of Proposition 3.3.1.

The matrix $G[k]$ has an explicit probabilistic interpretation.

Exercise 3.3.2 Use the induction method to show that the matrix $G[k]$ contains all the probabilities that the QBD process reaches level $n-1$ from level n for the first time, in at most $2k-1$ transitions, for $k \geq 1$.

Exercise 3.3.2 implies that the maximum level that all the paths contained in $G[k]$ can reach is $n + k-1$. We remark that such probabilistic interpretations provide insight into various algorithms for computing the matrix G and the matrix R (see Latouche and Ramaswami (1999)).

Proposition 3.3.2 (Neuts (1981)) *If the QBD process is recurrent, the matrix G is stochastic, i.e.,* $G\mathbf{e} = \mathbf{e}$. *If the QBD process is transient, then at least one row sum of G is less than one.*

It is clear that all elements of G must be between 0 and 1, which is different from that of the rate matrix R. We remark that (i) the role played by matrix G in the study of the length of the fundamental period will be investigated further in Sect. 3.5; and (ii) the computation of G has been studied extensively with some stable algorithms developed. See Latouche and Ramaswami (1993, 1999) for details.

For the QBD process, the matrix R and matrix G are intimately connected to each other by the following relationship. Define

$u_{i,j}$: the probability that the Markov chain returns to level n by visiting state (n, j) without visiting any state in levels $0, 1, \ldots, n-1$, in between, given that the Markov chain starts in state (n, i).

Let $U = (u_{i,j})$.

Proposition 3.3.3 (Latouche (1987)) *For the QBQ process defined in Eq.* (3.24), *we have (i)* $R = A_0 + RU$, *(ii)* $G = A_2 + UG$, *and (iii)* $U = A_1 + RA_2 = A_1 + A_0 G$.

Proof. Proposition 3.3.3 can be proved using conditional probabilities. For (i) $R = A_0 + RU$, we have

$$
\begin{aligned}
R_{j,v}^{(1)} &= \sum_{k=1}^{\infty} {}_iP_{i,j;i+1,v}^{(k)} = (A_0)_{j,v} + \sum_{k=2}^{\infty} {}_iP_{i,j;i+1,v}^{(k)} = (A_0)_{j,v} + \sum_{k=2}^{\infty}\sum_{s=1}^{k-1}\sum_{u=1}^{m} {}_iP_{i,j;i+1,u}^{(s)} \\
&\quad \times P\{X_k = i+1, J_k = v, X_t \geq i+2, t = s+1, \ldots, k-1 | X_s = i+1, J_s = u\} \\
&= (A_0)_{j,v} + \sum_{u=1}^{m} R_{j,u}^{(1)} u_{u,v}.
\end{aligned}
\tag{3.63}
$$

For (ii) $G = A_2 + UG$, we have

$$
\begin{aligned}
g_{i,j} &= \sum_{k=1}^{\infty} P\{(X_k, J_k) = (n-1, j), X_t \geq n, t = 1, 2, \ldots, k-1 | (X_0, J_0) = (n, i)\} \\
&= (A_2)_{i,j} + \sum_{k=2}^{\infty} P\{(X_k, J_k) = (n-1, j), X_t \geq n, t = 1, 2, \ldots, k-1 | (X_0, J_0) = (n, i)\} \\
&= (A_2)_{i,j} + \sum_{k=2}^{\infty}\sum_{s=1}^{k-1}\sum_{v=1}^{m} P\{(X_s, J_s) = (n, v), X_t \geq n+1, t = 1, 2, \ldots, s-1 | (X_0, J_0) = (n, i)\} \\
&\quad \times P\{(X_k, J_k) = (n-1, j), X_t \geq n, t = s+1, \ldots, k-1 | (X_s, J_s) = (n, v)\} \\
&= (A_2)_{i,j} + \sum_{v=1}^{m} u_{i,v} g_{v,j}.
\end{aligned}
\tag{3.64}
$$

Part (iii) is left as an exercise (see Exercise 3.3.3). This completes the proof of Proposition 3.3.3.

An important consequence of Proposition 3.3.3 is that the computation of R can be done by finding G first. Then the relationship in Proposition 3.3.3 can be used to find U and R. The advantage for doing so is that the computation of G is numerically stable, which is due to the fact that the elements of G are less than or equal to one and its row sums are less than or equal to one. On the other hand, there is no simple bound on the elements of R.

Exercise 3.3.3 Similar to the proofs of parts (i) and (ii) in Proposition 3.3.3, prove part (iii) in Proposition 3.3.3 by conditioning on the first/last transition.

Exercise 3.3.4 By parts (i) and (ii) of Proposition 3.3.3, we obtain

$$
A_1 + RA_2 = A_1 + A_0 G = A_1 + A_0 \left(\sum_{n=0}^{\infty} U^n \right) A_2. \tag{3.65}
$$

Use Eq. (3.65) and probabilistic arguments to prove part (iii) of Proposition 3.3.3. It is easy to see that $RA_2 = A_0 G$. Explain $RA_2 = A_0 G$ probabilistically.

Several algorithms have been developed for computing G. For example, define $\{G[k], k = 0, 1, 2, \ldots\}$ as follows: $G[0] = 0$, and

$$G[k+1] = (I - A_1)^{-1}A_2 + (I - A_1)^{-1}A_0(G[k])^2, \quad k = 1, 2, \ldots. \tag{3.66}$$

Exercise 3.3.5 Show that the sequence $\{G[k], k = 0, 1, 2, \ldots\}$ defined by Eq. (3.66) converges monotonically to G. Give the elements in $G[k]$ a probabilistic interpretation. Implement the computational method.

We remark that (i) according to Exercises 3.3.2 and 3.3.5, algorithms for computing G have an explicit probabilistic interpretation; and (ii) similar probabilistic interpretations are valid for algorithms to compute R.

Exercise 3.3.6 Find limiting probabilities for the following QBD process:

$$\begin{aligned} &A_{0,0} = (0.2), \quad A_{0,1} = (0.4,\ 0.4), \quad A_{1,0} = \begin{pmatrix} 0.5 \\ 0.6 \end{pmatrix}, \\ &A_{1,1} = A_1 = \begin{pmatrix} 0.1 & 0.1 \\ 0 & 0.1 \end{pmatrix}, \quad A_0 = \begin{pmatrix} 0.2 & 0.1 \\ 0.3 & 0 \end{pmatrix}, \quad A_2 = \begin{pmatrix} 0.5 & 0 \\ 0.3 & 0.3 \end{pmatrix}. \end{aligned} \tag{3.67}$$

In addition, find the probability distributions of the level variable X_∞; and compute the matrices G and U.

Commentary The matrix G plays an important role in finding the distribution and moments of the length of a fundamental period. Details are postponed to Sect. 3.5. In addition, the matrix G can be used in analyzing all kinds of events that occur in the fundamental period. See Neuts (1981, 1989), Ramaswami (1982), Latouche and Ramaswami (1999), and Sect. 5.3 for more details.

Matrices R and G are the most important pair of matrices in matrix-analytic methods. Both have an explicit probabilistic interpretation. Although they are clearly different in definition: R is for the expected numbers of visits and G is for first passage probabilities, R and G are also intimately related to each other, as demonstrated by Proposition 3.3.3. See Sect. 3.10.2 for a discussion on the duality between R and G.

Additional Exercises and Extensions

Exercise 3.3.7 Consider irreducible continuous time QBD process $\{(X(t), J(t)), t \geq 0\}$ with infinitesimal generator Q having the same structure as P given in Eq. (3.24). Assume that $A_0 + A_1 + A_2$ is irreducible. Let G be the minimal nonnegative solution to

$$0 = A_2 + A_1 G + A_0 G^2. \tag{3.68}$$

The equation can be rewritten as $G = (-A_1^{-1}A_2) + (-A_1^{-1}A_0)G^2$. Use this equation to interpret the elements in G probabilistically. (Hint: See Exercise 3.2.15.)

Exercise 3.3.8 Define matrix $G^{(k)}$ with (i, j)th component $g_{i,j}{}^{(k)}$ defined as the probability that, given that the Markov chain started in $(n + k, i)$, the Markov chain goes down to the level n for the first time and enters into state (n, j). Show that (i) $G^{(k+1)} = GG^{(k)}$; and (ii) $G^{(k)} = G^k$.

Exercise 3.3.9 Consider a QBD process with N levels. Each level may have a different number of phases. Define the (i, j)-th element of $H(k)$ as the probability that the QBD process visits level $k-1$ for the first time in state $(k-1, j)$, given that the QBD process is initially in state (k, i). Outline an algorithm for computing $H(k)$, $k = 1, \ldots, N$.

Exercise 3.3.10 (Structure of G) Assume that the last k columns of A_2 are zero. Show that the last k columns of G are zero as well. Show the result probabilistically without mathematical calculations. Show that a column in A_2 is zero if and only if the corresponding column in G is zero.

Exercise 3.3.11 (Structure of G) Show that (i) if $A = A_0 + A_1 + A_2$ is reducible, then G is reducible; and (ii) if A is upper/lower triangular, then G is upper/lower triangular.

Exercise 3.3.12 Compute G, U and R for the Markov chain with

$$A_{0,0} = (0.2), \quad A_{0,1} = (0.4,\ 0.4), \quad A_{1,0} = \begin{pmatrix} 0.5 \\ 0.6 \end{pmatrix},$$
$$A_{1,1} = A_1 = \begin{pmatrix} 0.1 & 0.1 \\ 0 & 0.1 \end{pmatrix}, \quad A_0 = \begin{pmatrix} 0.6 & 0.1 \\ 0.5 & 0.2 \end{pmatrix}, \quad A_2 = \begin{pmatrix} 0.1 & 0 \\ 0.1 & 0.1 \end{pmatrix}. \tag{3.69}$$

Is the Markov chain irreducible? Is the Markov chain ergodic? Use the matrix R, matrix G, and Neuts condition to answer the second question, respectively.

Exercise 3.3.13 For Exercise 3.2.23, compute the matrix G. Find $sp(G)$. Explain why $sp(G) < 1$ for the example.

Exercise 3.3.14 Use Proposition 3.3.3 to prove the following *Wiener-Hopf factorization* (also called *RG-factorization*):

$$I - \left(z^{-1}A_2 + A_1 + zA_0\right) = (I - zR)(I - U)\left(I - z^{-1}G\right). \tag{3.70}$$

(Note: The *RG*-factorization holds for more generally structured Markov chains that will be presented in Sect. 3.10.4.)

Exercise 3.3.15* (Latouche and Ramaswami (1993)) (*Logarithmic reduction algorithm*) Define

$$\begin{aligned}
H^{[0]} &= (I - A_1)^{-1} A_0; \\
L^{[0]} &= (I - A_1)^{-1} A_2; \\
H^{[k+1]} &= \left(I - U^{[k]}\right)^{-1} \left(H^{[k]}\right)^2, \quad k = 0, 1, \ldots; \\
L^{[k+1]} &= \left(I - U^{[k]}\right)^{-1} \left(L^{[k]}\right)^2, \quad k = 0, 1, \ldots; \\
U^{[k]} &= H^{[k]} L^{[k]} + L^{[k]} H^{[k]}, \quad k = 0, 1, \ldots.
\end{aligned} \tag{3.71}$$

Show probabilistically that

$$G = \sum_{k=0}^{\infty} \left(\prod_{i=0}^{k-1} H^{[i]} \right) L^{[k]}. \tag{3.72}$$

We remark that the logarithmic reduction algorithm not only provides an efficient way to compute the matrix G, but also leads to an alternative probabilistic interpretation of G. A comprehensive treatment of R and G for QBD processes can be found in Latouche and Ramaswami (1999).

3.4 *GI/M/1* Type Markov Chains: Matrix-Geometric Solution

A discrete time *GI/M/1* type Markov chain $\{(X_k, J_k), k = 0, 1, 2, \ldots\}$ is a discrete time Markov chain with state space $\{(0, 1), (0, 2), \ldots, (0, m_0)\} \cup \{\{1, 2, \ldots\} \times \{1, 2, \ldots, m\}\}$, where m_0 and m are positive integers, and transition probability matrix

$$P = \begin{pmatrix}
B_0 & A_{0,1} & & & \\
B_1 & A_{1,1} & A_0 & & \\
B_2 & A_2 & A_1 & A_0 & \\
B_3 & A_3 & A_2 & A_1 & A_0 \\
\vdots & \vdots & \ddots & \ddots & \ddots & \ddots
\end{pmatrix}. \tag{3.73}$$

It is easy to see that the matrix blocks satisfy the following conditions: $B_0\mathbf{e} + A_{0,1}\mathbf{e} = \mathbf{e}$, $B_1\mathbf{e} + (A_{1,1} + A_0)\mathbf{e} = \mathbf{e}$, and $B_n\mathbf{e} + (A_n + \ldots + A_0)\mathbf{e} = \mathbf{e}$, for $n \geq 2$. Let $A = \sum_{n=0}^{\infty} A_n$. Then A is a stochastic/substochastic matrix.

We remark that, since matrix-analytic methods were originally developed for the study of queueing models, many terminologies in matrix-analytic methods are associated with queueing theory. The term "*GI/M/1* type" comes from the fact that the transition probability matrix of the queue length process embedded at arrival epochs for the *GI/M/1* queue has the structure in matrix P (see Fig. 3.6). The Markov chain is also called *a skip-free to the right* process.

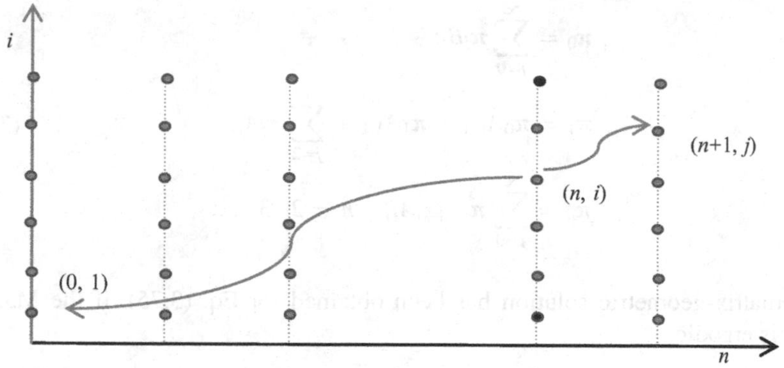

Fig. 3.6 Skip-free to the right transition

Since the analysis of the limiting probabilities of the *GI/M/1* type Markov chains is similar to that of the QBD process, in this section, we skip most of the details on the matrix-geometric solution and only summarize results on the limiting probabilities of the Markov chain.

Theorem 3.4.1 (Neuts (1981)) *Assume that the GI/M/1 type Markov chain* $\{(X_k, J_k),\ k = 0, 1, 2, \ldots\}$ *is irreducible and the matrix A is stochastic. The Markov chain is ergodic if and only if* $1 < \boldsymbol{\theta}\sum_{k=1}^{\infty} kA_k\mathbf{e}$, *where* $\boldsymbol{\theta}$ *satisfies* $\boldsymbol{\theta} = \boldsymbol{\theta}A$ *and* $\boldsymbol{\theta}\mathbf{e} = 1$.

Exercise 3.4.1 Rewrite $1 < \boldsymbol{\theta}\sum_{k=1}^{\infty} kA_k\mathbf{e}$ as $\boldsymbol{\theta}A_0\mathbf{e} < \boldsymbol{\theta}\sum_{k=1}^{\infty}(k-1)A_k\mathbf{e}$. Use the idea of mean-drift to explain the ergodicity condition intuitively.

Exercise 3.4.2 For a *GI/M/1* queueing system, the queue length process at arrival epochs $\{q_n,\ n = 0, 1, 2, \ldots\}$ is a *GI/M/1* type Markov chain with

$$
\begin{aligned}
A_k &= \int_0^{\infty} \frac{\mu^k t^k}{k!} e^{-\mu t} \mathrm{d}F(t), \quad k = 0, 1, 2, \ldots; \\
B_k &= 1 - \sum_{i=0}^{k} A_i, \quad k = 0, 1, 2, \ldots,
\end{aligned} \tag{3.74}
$$

where μ is the service rate of the exponential distribution of the service time, and $F(t)$ is the distribution function of the interarrival time. Note that $A_{0,1} = A_0$ and $A_{1,1} = A_1$ for this case. The Markov chain is called an embedded Markov chain at arrival epochs. Denote by λ^{-1} the mean interarrival time. Use Theorem 3.4.1 to prove that the Markov chain is ergodic if and only if $\lambda/\mu < 1$. (Note: Results for the more general *GI/PH/1* queue can be found in Chap. 4.)

Similar to Sect. 3.2, we use $\boldsymbol{\pi} = (\boldsymbol{\pi}_0, \boldsymbol{\pi}_1, \boldsymbol{\pi}_2, \ldots)$ to denote the limiting probabilities. Expanding the linear system $\boldsymbol{\pi}P = \boldsymbol{\pi}$ and $\boldsymbol{\pi}\mathbf{e} = 1$ for the limiting probabilities yields

$$\boldsymbol{\pi}_0 = \sum_{j=0}^{\infty} \boldsymbol{\pi}_j B_j;$$

$$\boldsymbol{\pi}_1 = \boldsymbol{\pi}_0 A_{0,1} + \boldsymbol{\pi}_1 A_{1,1} + \sum_{j=2}^{\infty} \boldsymbol{\pi}_j A_j; \tag{3.75}$$

$$\boldsymbol{\pi}_n = \sum_{j=0}^{\infty} \boldsymbol{\pi}_{n-1+j} A_j, \quad n = 2,\ 3,\ \ldots.$$

A matrix-geometric solution has been obtained for Eq. (3.75), if the Markov chain is ergodic.

Theorem 3.4.2 (Neuts (1981)) *Assume that the Markov chain* $\{(X_k, J_k),\ k = 0, 1, 2, \ldots\}$ *is ergodic. The limiting probabilities* $\boldsymbol{\pi}$ *are obtained as*

$$\boldsymbol{\pi}_n = \boldsymbol{\pi}_1 R^{n-1}, \quad n = 1, 2, \ldots, \tag{3.76}$$

where the rate matrix R is the minimal nonnegative solution to the nonlinear matrix equation

$$R = \sum_{n=0}^{\infty} R^n A_n, \tag{3.77}$$

and $(\boldsymbol{\pi}_0, \boldsymbol{\pi}_1)$ *is the unique positive solution to*

$$\begin{aligned} \boldsymbol{\pi}_0 &= \boldsymbol{\pi}_0 B_0 + \boldsymbol{\pi}_1 \sum_{n=1}^{\infty} R^{n-1} B_n; \\ \boldsymbol{\pi}_1 &= \boldsymbol{\pi}_0 A_{0,1} + \boldsymbol{\pi}_1 A_{1,1} + \boldsymbol{\pi}_1 \sum_{n=2}^{\infty} R^{n-1} A_n; \\ 1 &= \boldsymbol{\pi}_0 \mathbf{e} + \boldsymbol{\pi}_1 (I - R)^{-1} \mathbf{e}. \end{aligned} \tag{3.78}$$

In matrix form, Eq. (3.78) *can be written as*

$$\begin{aligned} (\boldsymbol{\pi}_0, \boldsymbol{\pi}_1) &= (\boldsymbol{\pi}_0, \boldsymbol{\pi}_1) \begin{pmatrix} B_0 & A_{0,1} \\ \sum_{n=1}^{\infty} R^{n-1} B_n & A_{1,1} + \sum_{n=2}^{\infty} R^{n-1} A_n \end{pmatrix}; \\ 1 &= \boldsymbol{\pi}_0 \mathbf{e} + \boldsymbol{\pi}_1 (I - R)^{-1} \mathbf{e}. \end{aligned} \tag{3.79}$$

The matrix R has the same probabilistic interpretation as that of the matrix R defined in Sect. 3.2. In fact, all the results on R given in Sect. 3.2 are still valid. Marginal distributions and moments of the limiting distribution of $\{(X_k, J_k),\ k = 0, 1, 2, \ldots\}$ can be obtained as well. Details are omitted.

We remark that a fairly comprehensive treatment of R and solutions to Eq. (3.77) can be found in Chap. 1 in Neuts (1981), Gail et al. (1994, 1997), and Bini and Meini (1996). The following fundamental theorem of R is given in Neuts (1981), which is based on Kingman (1961).

Theorem 3.4.3 (Neuts (1981)) *If the Markov chain* $\{(X_k, J_k), k = 0, 1, 2, \ldots\}$ *is ergodic, the rate matrix R is the unique nonnegative solution to Eq.* (3.77) *with* $sp(R) < 1$.

Exercise 3.4.3 Outline an algorithm for computing R and the limiting probabilities for *GI/M/*1 type Markov chains. (Note: Truncation must be used. Alternatively, you can assume that only a finite number of $\{A_0, A_1, A_2, \ldots\}$ and $\{B_0, B_1, B_2, \ldots\}$ are nonzero.)

Exercise 3.4.4 Show that the following *GI/M/*1 type Markov chain is ergodic (i.e., check irreducibility and Neuts condition (Theorem 3.4.1)).

$$\begin{aligned}
&B_0 = \begin{pmatrix} 0.1 & 0.1 \\ 0.1 & 0.1 \end{pmatrix}, \quad A_{0,1} = \begin{pmatrix} 0.4 & 0.4 \\ 0.4 & 0.4 \end{pmatrix}, \\
&B_1 = \begin{pmatrix} 0.2 & 0.3 \\ 0.3 & 0.3 \end{pmatrix}, \quad A_{1,1} = A_1 = \begin{pmatrix} 0.1 & 0.1 \\ 0 & 0.1 \end{pmatrix}, \quad A_0 = \begin{pmatrix} 0.2 & 0.1 \\ 0.3 & 0 \end{pmatrix}, \\
&B_2 = \begin{pmatrix} 0.3 & 0 \\ 0.2 & 0.2 \end{pmatrix}, \quad A_2 = \begin{pmatrix} 0.2 & 0 \\ 0.1 & 0.1 \end{pmatrix}, \\
&B_3 = \begin{pmatrix} 0.1 & 0 \\ 0.1 & 0.1 \end{pmatrix}, \quad A_3 = \begin{pmatrix} 0.2 & 0 \\ 0.1 & 0.1 \end{pmatrix}, \quad A_4 = \begin{pmatrix} 0.1 & 0 \\ 0.1 & 0.1 \end{pmatrix}.
\end{aligned} \tag{3.80}$$

All other transition blocks are zero. Find R and $\{\boldsymbol{\pi}_0, \boldsymbol{\pi}_1\}$ for the Markov chain. Plot the limiting distribution of $\{X_k, k = 0, 1, 2, \ldots\}$.

Exercise 3.4.5 (Exercise 3.4.4 continued) In Exercise 3.4.4, combine the levels $4k + 1, 4k + 2, 4k + 3$, and $4k + 4$ into a single (super) level k, for $k = 0, 1, 2, \ldots$. Then the Markov chain becomes a QBD process. Use the algorithm developed in Sect. 3.2 for computing the limiting probabilities.

Exercise 3.4.6 Give a definition to continuous time *GI/M/*1 type Markov chains. Outline an algorithm for computing the limiting probabilities for such Markov chains. (Hint: Two methods are available: (1) uniformization; and (2) embedded Markov chain at level transition epochs.)

Commentary Chapter 1 in Neuts (1981) gives a comprehensive treatment on the ergodicity and limiting probabilities (the matrix-geometric solution) of the *GI/M/*1 type Markov chains. The rate matrix R plays an important role in the analysis of the *GI/M/*1 type Markov chains. The matrix R is not only related to the limiting probabilities, but also other performance measures of the associated Markov chain (e.g., Ramaswami (1980) and Hsu and He (1991)).

Additional Exercises and Extensions

Exercise 3.4.7 Consider a discrete time Markov chain with the following transition probability matrix:

$$P = \begin{pmatrix} A_{0,0} & A_{0,1} & A_{0,2} & & & & \\ A_{1,0} & A_{1,1} & A_{1,2} & A_{1,3} & & & \\ A_{2,0} & A_{2,1} & A_{2,2} & A_{2,3} & A_{2,4} & & \\ A_{3,0} & A_{-2} & A_{-1} & A_0 & A_1 & A_2 & \\ A_{4,0} & A_{-3} & A_{-2} & A_{-1} & A_0 & A_1 & A_2 \\ \vdots & \vdots & \ddots & \ddots & \ddots & \ddots & \ddots \end{pmatrix} \tag{3.81}$$

Assuming that the Markov chain is ergodic, outline a method for computing its limiting probabilities.

Exercise 3.4.8 Consider a *GI/M/*1 type Markov chain $\{(X_k, J_k), k = 0, 1, 2, \ldots\}$ with

$$\begin{aligned} &B_0 = (0.5), \quad A_{0,1} = \begin{pmatrix} 0.3 & 0.2 \end{pmatrix}, \quad B_1 = B_2 = \begin{pmatrix} 0.5 \\ 0.5 \end{pmatrix}, \\ &A_3 = \begin{pmatrix} 0.2 & 0.3 \\ 0 & 0.5 \end{pmatrix}, \quad A_{1,1} = A_1 = \begin{pmatrix} 0.1 & 0.1 \\ 0 & 0.4 \end{pmatrix}, \quad A_0 = \begin{pmatrix} 0.2 & 0.1 \\ 0 & 0.1 \end{pmatrix}, \end{aligned} \tag{3.82}$$

$B_n = 0$, for $n = 3, 4, \ldots$, $A_2 = 0$, and $A_n = 0$, for $n = 4, 5, \ldots$.

(i) Is the Markov chain irreducible?
(ii) Is the Markov chain ergodic?
(iii) Compute the matrix R. Is the matrix R irreducible?
(iv) Find $(\pi_0, \boldsymbol{\pi}_1)$.
(v) Plot the limiting distribution of $\{X_k, k = 0, 1, 2, \ldots\}$.

Exercise 3.4.9 Assume that a *GI/M/*1 Markov chain is irreducible. If every component of $(A_0 + A_1 + A_2 + \ldots)\mathbf{e}$ is strictly less than 1, is the Markov chain ergodic? Why? Consider the following Markov chain

$$P = \begin{pmatrix} 0 & 1 & & & \\ 0.5 & 0 & 0.5 & & \\ 0.5 & 0 & 0 & 0.5 & \\ \vdots & & \ddots & \ddots & \ddots \end{pmatrix}. \tag{3.83}$$

Find $\boldsymbol{\theta} \sum_{k=1}^{\infty} kA_k\mathbf{e}$. Is the Markov chain ergodic? If so, find its limiting probabilities. (Note: This exercise demonstrates that, if A is not stochastic, the condition in Theorem 3.4.1 may not be necessary for the ergodicity of the Markov chain.)

Exercise 3.4.10 Let $\rho(z) = a_0 + za_1 + z^2a_2 + \ldots$, for $0 \le z \le 1$, where $\{a_n, n = 0, 1, \ldots\}$ are nonnegative, $a_0 > 0$, and $a_0 + a_1 + a_2 + \ldots = 1$. Show that there is a unique z^* in (0, 1) such that $\rho(z) = z$ if and only if $\rho^{(1)}(1) > 1$. Use a graph to demonstrate the result (i.e., draw $\rho(z)$ for $0 \le z \le 1$). Show that there is a unique z^* in $(1, \infty)$ such that $\rho(z) = z$ if and only if $\rho^{(1)}(1) < 1$ and $a_0 + a_1 < 1$. (Note: Recall that $\rho^{(1)}(1)$ is the derivative of $\rho(z)$ at $z = 1$.)

Exercise 3.4.11* (Exercises 3.2.5 and 3.4.10 continued) Let $A^*(z) = A_0 + zA_1 + z^2A_2 + \ldots$, for $0 \le z \le 1$. We assume that $A = A^*(1)$ is irreducible, $A^*(1)\mathbf{e} = \mathbf{e}$, and $sp(A_0) > 0$. Denote by $\rho(z)$ the Perron-Frobenius eigenvalue of $A^*(z)$. Show that (i) $\rho(e^{-s})$ is a concave function of s in $[0, \infty]$; and (ii) there is a unique z^* in (0, 1) such that $\rho(z) = z$ if and only if $\rho^{(1)}(1) > 1$. Show that $sp(R) = z^*$. (See Lemma 1.3.4 in Chap. 1 in Neuts (1981) and see Kingman (1961)) (Note: The results lead to the ergodicity condition for *GI/M/*1 type Markov chains (Theorem 3.4.1). The results are useful for theoretical study of *GI/M/*1 type Markov chains.)

Exercise 3.4.12* (Monotonicity of R) (He (1999)) Assume that $\{A_0, A_1, A_2, \ldots\}$ and $\{C_0, C_1, C_2, \ldots\}$ are two stochastic sequences, i.e., $A_n \ge 0$, $C_n \ge 0$, $n = 0, 1, 2, \ldots$, and $(\sum_{n=1}^{\infty} A_n)\mathbf{e} = \mathbf{e}$ and $(\sum_{n=1}^{\infty} C_n)\mathbf{e} = \mathbf{e}$. Let R_a and R_c be the rate matrices defined by Eq. (3.77) for the two stochastic sequences, respectively. Assume that $A_0 + A_1 + \ldots + A_n \le C_0 + C_1 + \ldots + C_n$ (elementwise) holds for all $n = 0, 1, 2, \ldots$. (This is a generalized *stochastically larger order*.) Show that (i) $sp(R_a) \ge sp(R_c)$; and (ii) $(I - R_a)^{-1}R_a \ge (I - R_c)^{-1}R_c$. (Note: The results provide inequalities and bounds on the rate matrices. On the other hand, conditions for $R_a \ge R_c$ are not yet identified for the general case.)

3.5 *M/G/*1 Type Markov Chains: Fundamental Period

A discrete time *M/G/*1 type Markov chain $\{(X_k, J_k), k = 0, 1, \ldots\}$ is a discrete time Markov chain with state space $\{(0, 1), (0, 2), \ldots, (0, m_0)\} \cup \{\{1, 2, \ldots\} \times \{1, 2, \ldots, m\}\}$, where m_0 and m are positive integers, and transition probability matrix

$$P = \begin{pmatrix} A_{0,0} & A_{0,1} & A_{0,2} & A_{0,3} & \cdots \\ A_{1,0} & A_1 & A_2 & A_3 & \cdots \\ & A_0 & A_1 & A_2 & \ddots \\ & & \ddots & \ddots & \ddots \\ & & & \ddots & \ddots \end{pmatrix}. \tag{3.84}$$

It is easy to see that the matrix blocks satisfy the following conditions: $A_{0,0}\mathbf{e} + (A_{0,1} + A_{0,2} + \ldots)\mathbf{e} = \mathbf{e}$ and $A_{1,0}\mathbf{e} + (A_1 + A_2 + \ldots)\mathbf{e} = \mathbf{e}$. Let $A = \sum_{n=0}^{\infty} A_n$. Then A is a stochastic matrix.

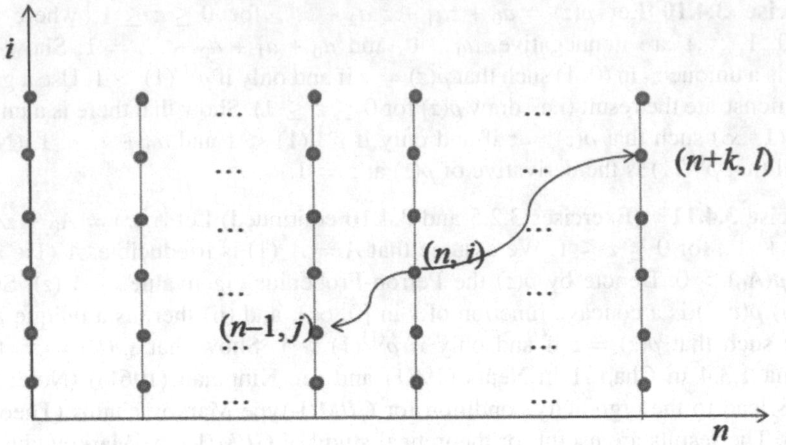

Fig. 3.7 Skip-free to the left transitions

The term "$M/G/1$ type" comes from the fact that the transition probability matrix of the queue length process at departure epochs for the $M/G/1$ queue has the same structure as that in P. It is also called *a skip-free to the left* process (see Fig. 3.7.)

Theorem 3.5.1 (Neuts (1981)) *Assume that the M/G/1 type Markov chain* $\{(X_k, J_k), k = 0, 1, \ldots\}$ *is irreducible and* $\sum_{k=0}^{\infty} kA_{0,k}\mathbf{e}$ *is finite elementwise. Then the Markov chain is ergodic if and only if* $1 > \boldsymbol{\theta}\sum_{k=0}^{\infty} kA_k\mathbf{e}$, *where* $\boldsymbol{\theta}$ *satisfies* $\boldsymbol{\theta} = \boldsymbol{\theta}\sum_{k=0}^{\infty} A_k$ *and* $\boldsymbol{\theta}\mathbf{e} = 1$.

Exercise 3.5.1 Use the idea of mean-drift to explain the ergodicity condition intuitively.

Exercise 3.5.2 For an $M/G/1$ queueing system, the queue length at departure epochs $\{q_n, n = 0, 1, 2, \ldots\}$ is an $M/G/1$ type Markov chain with

$$A_{0,k} = A_k = \int_0^{\infty} \frac{\lambda^k t^k}{k!} e^{-\lambda t} \mathrm{d}F(t), \quad k = 0, 1, 2, \ldots, \tag{3.85}$$

and $A_{1,0} = A_0$, where λ is the arrival rate of the Poisson arrival process and $F(t)$ is the distribution function of the service time. The Markov chain is called an embedded Markov chain at departure epochs. Denote by μ^{-1} the mean service time. Use Neuts condition to prove that the Markov chain is ergodic if and only if $\lambda/\mu < 1$. (Note: Results for the more general $BMAP/G/1$ queue can be found in Chap. 4.)

Denote by $\boldsymbol{\pi} = (\boldsymbol{\pi}_0, \boldsymbol{\pi}_1, \boldsymbol{\pi}_2, \ldots)$ the limiting probabilities of $\{(X_k, J_k), k = 0, 1, \ldots\}$. Expanding the linear system $\boldsymbol{\pi}P = \boldsymbol{\pi}$ and $\boldsymbol{\pi}\mathbf{e} = 1$ for the limiting probabilities yields

$$\begin{aligned}\boldsymbol{\pi}_0 &= \boldsymbol{\pi}_0 A_{0,0} + \boldsymbol{\pi}_1 A_{1,0};\\ \boldsymbol{\pi}_n &= \boldsymbol{\pi}_0 A_{0,n} + \sum_{j=1}^{n+1} \boldsymbol{\pi}_j A_{n+1-j}, \quad n = 1,\ 2,\ \ldots.\end{aligned} \tag{3.86}$$

In general, there is no simple or explicit solution to (3.86). Instead of looking at the limiting probabilities directly, we investigate the first passage probabilities for this Markov chain first. Basically, we generalize the theory developed in Section 3.3 for the fundamental period and the matrix G. We now consider the length of a fundamental period as well as the phase change at the beginning and the end of a fundamental period. Define, for $n \geq 1, k \geq 1, i, j = 1, \ldots, m$, and $0 \leq z \leq 1$,

$$\begin{aligned}g_{i,j}(k) &= P\{(X_k, J_k) = (n-1, j),\ X_l \geq n,\ l = 1, \ldots, k-1 | (X_0, J_0) = (n, i)\},\\ g_{i,j}^*(z) &= \sum_{k=1}^{\infty} z^k g_{i,j}(k).\end{aligned} \tag{3.87}$$

Let $G^*(z) = (g_{i,j}^*(z))$ and $G = (g_{i,j}) = G^*(1)$. The matrix $G^*(z)$ is a generalization of the matrix G by taking the number of transitions during the fundamental period into consideration. (We remark that a similar generalization can be considered for the matrix R for the *GI/M/*1 type Markov chains. See Ramaswami (1980) for details.) Such a generalization leads to information on the length of a fundamental period, in addition to the phase change in a fundamental period.

Before we analyze the above functions and matrices, we offer some observations and explanations on the *M/G/*1 type Markov chain and the matrices defined.

1. The first passage time from level $n + 1$ to level n is probabilistically equivalent for $n \geq 1$.
2. That $g_{i,j} = g_{i,j}^*(1)$ is the probability that, given that the Markov chain starts in (n, i), the Markov chain goes down to the level $n-1$ for the first time by entering state $(n-1, j)$.

Define matrix $G^{(s)}(k)$ with the (i, j)-th component $g_{i,j}^{(s)}(k)$ defined as the probability that, given that the Markov chain starts in $(n + s, i)$, the Markov chain goes down to level n for the first time after exactly k transitions, and enters into state (n, j). Define $G^{*(s)}(z) = \sum_{k=1}^{\infty} z^k G^{(s)}(k)$, for $0 \leq z \leq 1$.

Lemma 3.5.1 *For* $s = 1, 2, \ldots$, $G^{*(s)}(z) = (G^*(z))^s$.

Proof. The proof is similar to the proof of Exercise 3.3.8. First we show that $G^{*(s+1)}(z) = G^*(z)G^{*(s)}(z)$. By definition, we have

$$
\begin{aligned}
&g_{i,j}^{*(s+1)}(z)\\
&=\sum_{k=1}^{\infty} z^k g_{i,j}^{(s+1)}(k)\\
&=\sum_{k=1}^{\infty} z^k P\{X_k=n, J_k=j,\ X_l \ge n+1,\ 1\le l\le k-1|(X_0,J_0)=(n+s+1,i)\}\\
&=\sum_{k=1}^{\infty} z^k \sum_{t=1}^{k-1}\sum_{v=1}^{m} P\{X_t=n+s, J_t=v,\ X_l\ge n+s+1,\ 1\le l\le t-1\\
&\qquad |(X_0,J_0)=(n+s+1,i)\}P\{X_k=n, J_k=j,\ X_l\ge n+1,\ t\le l\le k-1 \qquad (3.88)\\
&\qquad |(X_t,J_t)=(n+s,v)\}\\
&=\sum_{v=1}^{m}\sum_{t=1}^{\infty}\sum_{k=t+1}^{\infty} z^t P\{X_t=n+s, J_t=v,\ X_l\ge n+s+1,\ 1\le l\le t-1|(X_0,J_0)\\
&\qquad =(n+s+1,i)\}z^{k-t}P\{X_k=n, J_k=j,\ X_l\ge n+1,\ t\le l\le k-1|(X_t,J_t)\\
&\qquad =(n+s,v)\}\\
&=\sum_{v=1}^{m} g_{i,v}^*(z) g_{v,j}^{*(s)}(z).
\end{aligned}
$$

By induction, Lemma 3.5.1 is proved.

Using Lemma 3.5.1 and conditioning on the next transition, we obtain

$$
\begin{aligned}
g_{i,j}^*(z) &= \sum_{k=1}^{\infty} z^k g_{i,j}(k)\\
&= z(A_0)_{i,j} + \sum_{k=2}^{\infty} z^k \sum_{n=1}^{\infty}\sum_{v=1}^{m} (A_n)_{i,v} g_{v,j}^{(n)}(k-1) \qquad (3.89)\\
&= z(A_0)_{i,j} + z\sum_{n=1}^{\infty}\left(\sum_{v=1}^{m}(A_n)_{i,v}\sum_{k=2}^{\infty} z^{k-1} g_{v,j}^{(n)}(k-1)\right).
\end{aligned}
$$

Equation (3.89) leads to

$$
\begin{aligned}
G^*(z) &= z\sum_{n=0}^{\infty} A_n (G^*(z))^n;\\
G &= \sum_{n=0}^{\infty} A_n G^n.
\end{aligned} \qquad (3.90)
$$

Proposition 3.5.1 *The matrix* G *is the minimal nonnegative solution to the second equation in* (3.90). *If the M/G/1 type Markov chain is irreducible and recurrent, then* $G\mathbf{e} = \mathbf{e}$ (*i.e.,* $sp(G) = 1$). *Otherwise, at least one row sum of* G *is less than one.*

Now, we study the length of a fundamental period (or the number of transitions in a fundamental period). Assume that the Markov chain is ergodic. We want to find the moments of the length of a fundamental period. Taking derivatives of both sides of the first equation in (3.90) with respect to z yields

$$\frac{dG^*(z)}{dz} = \sum_{n=0}^{\infty} A_n (G^*(z))^n + z \sum_{n=0}^{\infty} A_n \sum_{k=0}^{n-1} (G^*(z))^k \frac{dG^*(z)}{dz} (G^*(z))^{n-k-1}. \tag{3.91}$$

Postmultiplying by **e** on both sides of Eq. (3.91) and setting $z = 1$, we obtain

$$\left.\frac{dG^*(z)}{dz}\right|_{z=1} \mathbf{e} = \mathbf{e} + \left(\sum_{n=0}^{\infty} A_n \sum_{k=0}^{n-1} (G^*(1))^k \right) \left.\frac{dG^*(z)}{dz}\right|_{z=1} \mathbf{e}. \tag{3.92}$$

Note that $G = G^*(1), G\mathbf{e} = \mathbf{e}$, and $(A_0 + A_1 + A_2 + \ldots)\mathbf{e} = \mathbf{e}$. The i-th element of $dG^*(z)/dz|_{z=1}\mathbf{e}$ is $\sum_{k=0}^{\infty} k \left(\sum_{j=1}^{m} g_{i,j}(k) \right)$, the conditional expected number of transitions in a fundamental period.

Theorem 3.5.2 *If the M/G/1 type Markov chain is ergodic and the matrix G is irreducible, then the* conditional expected numbers of transitions *in a fundamental period are given by*

$$\left.\frac{dG^*(z)}{dz}\right|_{z=1} \mathbf{e} = (I - G + \mathbf{eg}) \left(I - A + \mathbf{eg} - \left(\sum_{n=0}^{\infty} nA_n \right) \mathbf{eg} \right)^{-1} \mathbf{e}, \tag{3.93}$$

where **e** *is actually a right eigenvector of G corresponding to eigenvalue one (G***e** $=$ **e**)*, and* **g** *is the invariant measure of G (a left eigenvector) corresponding to eigenvalue one, i.e.,* **g***G* $=$ **g** *and* **ge** $= 1$.

Proof. By Eq. (3.92), we obtain

$$\left(I - \sum_{n=0}^{\infty} A_n \sum_{k=0}^{n-1} G^k \right) \left.\frac{dG^*(z)}{dz}\right|_{z=1} \mathbf{e} = \mathbf{e}. \tag{3.94}$$

Thus, we need to find the inverse of the matrix $I - \sum_{n=0}^{\infty} A_n \sum_{k=0}^{n-1} G^k$. Postmultiplying this matrix by $I - G + \mathbf{eg}$, yields

$$\begin{aligned} &\left(I - \sum_{n=0}^{\infty} A_n \sum_{k=0}^{n-1} G^k \right) (I - G + \mathbf{eg}) \\ &= I - G + \mathbf{eg} - \sum_{n=0}^{\infty} A_n (I - G^n + n\mathbf{eg}) \\ &= I - G + \mathbf{eg} - A + G - \sum_{n=0}^{\infty} nA_n \mathbf{eg} \\ &= I - A + \mathbf{eg} - \left(\sum_{n=0}^{\infty} nA_n \right) \mathbf{eg}. \end{aligned} \tag{3.95}$$

The theorem is proved by showing the following two technical lemmas. The techniques used in their proofs are similar and typical in such cases.

Lemma 3.5.2 *Under the assumptions given in Theorem* 3.5.2, *the matrix* $I-G+\mathbf{eg}$ *is invertible.*

Proof. If the matrix is singular, there exists a nonzero row vector $\mathbf{u}$ such that $\mathbf{u}(I-G+\mathbf{eg})=0$. If $\mathbf{ue}=0$, we must have $\mathbf{u}(I-G)=0$. Since G is nonnegative and irreducible, elements in the vector $\mathbf{u}$ must have the same sign, which contradicts $\mathbf{ue}=0$. If $\mathbf{ue}\neq 0$, without loss of generality, we assume that $\mathbf{ue}=1$. Then we must have $\mathbf{u}(I-G)+\mathbf{g}=0$. Postmultiplying by $\mathbf{e}$ on both sides, we obtain $\mathbf{ge}=0$, which is a contradiction. Therefore, the matrix $I-G+\mathbf{eg}$ must be invertible. This completes the proof of Lemma 3.5.2.

Lemma 3.5.3 *Under the assumptions given in Theorem* 3.5.2, *the matrix* $I-A$ $+\mathbf{eg}-\left(\sum_{n=0}^{\infty}nA_n\right)\mathbf{eg}$ *is invertible.*

Proof. If the matrix is singular, there exists a nonzero row vector $\mathbf{u}$ such that $\mathbf{u}\left(I-A+\mathbf{eg}-\left(\sum_{n=0}^{\infty}nA_n\right)\mathbf{eg}\right)=0$. If $\mathbf{ue}=0$, we must have $\mathbf{u}(I-A)=\mathbf{u}\left(\sum_{n=0}^{\infty}nA_n\right)\mathbf{eg}$. Note that $A\mathbf{e}=\mathbf{e}$. Postmultiplying by $\mathbf{e}$ on both sides yields $\mathbf{u}\left(\sum_{n=0}^{\infty}nA_n\right)\mathbf{e}=0$ and, consequently, $\mathbf{u}(I-A)=0$. The last equation implies $\mathbf{u}$ is either nonnegative or nonpositive (since A can have only one closed set), which contradicts $\mathbf{ue}=0$. If $\mathbf{ue}\neq 0$, without loss of generality, we assume $\mathbf{ue}=1$. Postmultiplying by $\mathbf{e}$ on both sides of $\mathbf{u}(I-A)+\mathbf{g}=\mathbf{u}\left(\sum_{n=0}^{\infty}nA_n\right)\mathbf{eg}$, yields $1=\mathbf{u}\left(\sum_{n=0}^{\infty}nA_n\right)\mathbf{e}$, which leads to $\mathbf{u}(I-A)=0$. Then, by definition, we must have $\mathbf{u}=\boldsymbol{\theta}$. That $\mathbf{u}=\boldsymbol{\theta}$ leads to $1=\boldsymbol{\theta}\left(\sum_{n=0}^{\infty}nA_n\right)\mathbf{e}$, which contradicts the ergodicity condition given in Theorem 3.5.1. Therefore, we have shown that the matrix is invertible. This completes the proof of Lemma 3.5.3.

Finally, Eq. (3.93) is then obtained from Eqs. (3.94) and (3.95). This completes the proof of Theorem 3.5.2.

We remark that higher moments of $G^*(z)$ are of interest in queueing theory and recursive algorithms have been developed for computing them. The results are quite tedious, though. We omit the details.

Exercise 3.5.3 Find the second moments of a fundamental period. (Note: The derivations are long and tedious.)

Theorem 3.5.2 and Exercise 3.5.3 indicate that, for computing moments of fundamental periods, we need to compute G first. By Proposition 3.5.1, G is the minimal nonnegative solution to the second equation in Eq. (3.90), which can be computed recursively as follows: $G[0]=0$, and

$$G[k+1]=\sum_{n=0}^{\infty}A_n(G[k])^n,\quad k=0,\ 1,\ 2,\ \ldots. \tag{3.96}$$

It can be shown that the sequence $\{G[k], k = 0, 1, 2, \ldots\}$ is nondecreasing and converges to G. Therefore, a stable algorithm based on (3.96) can be developed for computing G.

Exercise 3.5.4 For $\{G[k], k = 0, 1, \ldots\}$ generated by Eq. (3.96), show $G[k]\mathbf{e} \le \mathbf{e}$ for all k. Show that $\{G[k], k = 0, 1, \ldots\}$ is nondecreasing and converges to G.

Exercise 3.5.5 Assume that $I - A_1$ is invertible. Define $G[0] = 0$, and

$$G[k+1] = \sum_{n=0:\, n\neq 1}^{\infty} (I - A_1)^{-1} A_n (G[k])^n, \quad k = 0, 1, 2, \ldots. \tag{3.97}$$

Show that the sequence $\{G[k], k = 0, 1, \ldots\}$ is nondecreasing and converges to G. (Hint: Again, you need to show $G[k]\mathbf{e} \le \mathbf{e}$ for all k.)

Similar to that of R, a fairly comprehensive treatment of G and solutions to the second equation in Eq. (3.90) can be found in Gail et al. (1994, 1997).

Exercise 3.5.6 Consider the following $M/G/1$ type Markov chain:

$$A_{0,0} = \begin{pmatrix} 0.1 & 0.1 \\ 0.1 & 0.1 \end{pmatrix}, \quad A_{0,1} = \begin{pmatrix} 0.1 & 0 \\ 0 & 0.1 \end{pmatrix}, \quad A_{0,2} = \begin{pmatrix} 0.1 & 0.1 \\ 0 & 0.1 \end{pmatrix}, \quad A_{0,3} = \begin{pmatrix} 0.2 & 0.3 \\ 0 & 0.6 \end{pmatrix},$$

$$A_{1,0} = A_0 = \begin{pmatrix} 0.5 & 0.2 \\ 0.3 & 0.2 \end{pmatrix}, \quad A_1 = \begin{pmatrix} 0.1 & 0 \\ 0.1 & 0.1 \end{pmatrix}, \quad A_2 = \begin{pmatrix} 0.1 & 0 \\ 0.1 & 0.1 \end{pmatrix}, \quad A_3 = \begin{pmatrix} 0.1 & 0 \\ 0.1 & 0 \end{pmatrix}.$$

All other transition blocks are zero.

(1) Is the Markov chain ergodic?
(2) Find the matrix G.
(3) Find the conditional expected number of transitions in a fundamental period.

Now, we return to the limiting probabilities $\boldsymbol{\pi}$ of the Markov chain. Define

$$\begin{aligned} G_{(1,0)} &= \left(I - \sum_{k=1}^{\infty} A_k G^{k-1} \right)^{-1} A_{1,0}; \\ G_{(0,0)} &= A_{0,0} + \sum_{k=1}^{\infty} A_{0,k} G^{k-1} G_{(1,0)}. \end{aligned} \tag{3.98}$$

The matrix $I - \sum_{k=1}^{\infty} A_k G^{k-1}$ is invertible since $\sum_{k=1}^{\infty} A_k G^{k-1}$ is substochastic and its Perron-Frobenius eigenvalue is less than one. The (i, j)-th element of $G_{(1,0)}$ is the probability that the Markov chain reaches level zero for the first time by visiting state $(0, j)$, given that the Markov chain starts in state $(1, i)$. The (i, j)-th element of $G_{(0,0)}$ is the probability that the Markov chain returns to level zero for the first time by visiting state $(0, j)$, given that the Markov chain starts in state $(0, i)$. If the Markov chain is recurrent, both $G_{(0,0)}$ and $G_{(1,0)}$ are stochastic matrix. It is clear that $G_{(0,0)}$ is the transition probability matrix for transitions between states in level zero, after excising the transitions related to other states. If the original Markov chain is irreducible, $G_{(0,0)}$ is irreducible as well. Denote by $\mathbf{v}$ a left eigenvector of $G_{(0,0)}$ corresponding to eigenvalue one. If we normalize $\mathbf{v}$ by $\mathbf{ve} = 1$, then $\mathbf{v}$ is the stationary distribution of $G_{(0,0)}$. Then we must have $\mathbf{v} = c\boldsymbol{\pi}_0$, where c is a constant (see Kemeny et al. (1976)).

Based on the above idea, a stable recursive algorithm has been developed for computing the limiting probabilities $\boldsymbol{\pi}$.

Theorem 3.5.3 (Ramaswami (1988a)) *Assume that the M/G/1 type Markov chain is ergodic, then its limiting probabilities are given by*

$$\boldsymbol{\pi}_n = \left(\boldsymbol{\pi}_0 \bar{A}_{0,n} + \sum_{k=1}^{n-1} \boldsymbol{\pi}_k \bar{A}_{n+1-k}\right)(I - \bar{A}_1)^{-1}, \quad n = 2, 3, \ldots, \tag{3.99}$$

where

$$\bar{A}_{0,n} = \sum_{k=0}^{\infty} A_{0,n+k} G^k, \quad \bar{A}_n = \sum_{k=0}^{\infty} A_{n+k} G^k, \quad n = 1, 2, \ldots. \tag{3.100}$$

Vector $(\boldsymbol{\pi}_0, \boldsymbol{\pi}_1)$ *satisfies*

$$(\boldsymbol{\pi}_0, \boldsymbol{\pi}_1) = (\boldsymbol{\pi}_0, \boldsymbol{\pi}_1) \begin{pmatrix} A_{0,0} & \sum_{k=0}^{\infty} A_{0,1+k} G^k \\ A_{1,0} & \sum_{k=0}^{\infty} A_{1+k} G^k \end{pmatrix} \tag{3.101}$$

and is normalized so that the total probability is one.

See Ramaswami (1988a) for a proof of Theorem 3.5.3. The proof is based on *censoring* of Markov chains. (Note: Censoring is an important technique to truncate Markov chains with infinite number of states.) Define

$$P_{\le n} = \begin{pmatrix} A_{0,0} & A_{0,1} & A_{0,2} & A_{0,3} & \cdots & A_{0,n-1} & \sum_{k=0}^{\infty} A_{0,n+k} G^k \\ A_{1,0} & A_1 & A_2 & A_3 & \cdots & A_{n-1} & \sum_{k=0}^{\infty} A_{n+k} G^k \\ & A_0 & A_1 & A_2 & \cdots & A_{n-2} & \sum_{k=0}^{\infty} A_{n-1+k} G^k \\ & & \ddots & \ddots & \ddots & \vdots & \vdots \\ & & & & A_0 & A_1 & \sum_{k=0}^{\infty} A_{2+k} G^k \\ & & & & & A_0 & \sum_{k=0}^{\infty} A_{1+k} G^k \end{pmatrix}. \tag{3.102}$$

The Markov chain $P_{\leq n}$ is a censored process of P by eliminating states in level $n+1$ or higher. It can be shown that the limiting probabilities of $P_{\leq n}$ are proportional to the corresponding limiting probabilities of P, which leads to Eq. (3.99). Exercises 3.5.8 and 3.5.16 are designed to help us understand $P_{\leq n}$ and its limiting probabilities.

Exercise 3.5.7 Outline an implementation procedure of Ramaswami's algorithm. Use the procedure to compute limiting probabilities for the *M/G/*1 type Markov chain given in Exercise 3.5.6.

Exercise 3.5.8 By Definition (3.102), we have

$$P_{\leq 1} = \begin{pmatrix} A_{0,0} & \sum_{k=0}^{\infty} A_{0,1+k} G^k \\ A_{1,0} & \sum_{k=0}^{\infty} A_{1+k} G^k \end{pmatrix}. \qquad (3.103)$$

Explain $P_{\leq 1}$ probabilistically. Show that $P_{\leq 1}$ is a stochastic matrix. Let $\mathbf{u}$ be the left eigenvector of $P_{\leq 1}$ corresponding to eigenvalue one. Argue that $\mathbf{u} = c(\boldsymbol{\pi}_0, \boldsymbol{\pi}_1)$ for some constant c.

Commentary Unlike the *GI/M/*1 type Markov chain, there is no matrix-geometric solution to the limiting probabilities of the *M/G/*1 type Markov chain. Nonetheless, there are still plenty of results on the fundamental period and the moments of the stationary distribution (see Neuts (1989)).

Additional Exercises and Extensions

Markov renewal processes The results obtained in this section can be generalized to Markov renewal processes. A continuous time *M/G/*1 type Markov renewal process $\{(X_k, J_k, \tau_k), k = 0, 1, \ldots\}$ is a continuous time process with state space $\{0, 1, 2, \ldots\} \times \{1, 2, \ldots, m\}$, where m is a positive integer, and transition kernel, for $x \geq 0$,

$$P(x) = \begin{pmatrix} A_{0,0}(x) & A_{0,1}(x) & A_{0,2}(x) & A_{0,3}(x) & \cdots \\ A_{1,0}(x) & A_1(x) & A_2(x) & A_3(x) & \cdots \\ & A_0(x) & A_1(x) & A_2(x) & \ddots \\ & & \ddots & \ddots & \ddots \\ & & & \ddots & \ddots \end{pmatrix}. \qquad (3.104)$$

The fundamental periods of this process can be analyzed in detail similarly. We also remark that the *GI/M/*1 type Markov renewal processes can be defined as

$$P(x) = \begin{pmatrix} B_0(x) & A_{0,1}(x) & & & & \\ B_1(x) & A_{1,1}(x) & A_0(x) & & & \\ B_2(x) & A_2(x) & A_1(x) & A_0(x) & & \\ B_3(x) & A_3(x) & A_2(x) & A_1(x) & A_0(x) & \\ \vdots & \vdots & \ddots & \ddots & \ddots & \ddots \end{pmatrix}. \tag{3.105}$$

See Ramaswami (1980a) for a detailed analysis of this process. See Neuts (1981, 1989) for details on the two structured Markov renewal processes.

Exercise 3.5.9 Consider irreducible continuous time *M*/*G*/1 type Markov chain $\{(X(t), J(t)), t \geq 0\}$ with infinitesimal generator Q having the same structure as P given in (3.84). Assume that $A = A_0 + A_1 + A_2 + \ldots$ is irreducible. Use the uniformization method to prove the following results.

(i) Explain intuitively that the *M*/*G*/1 type Markov chain is ergodic if and only if $\boldsymbol{\theta} A_0 \mathbf{e} > \boldsymbol{\theta}(A_2 + 2A_3 + 3A_4 + \ldots)\mathbf{e}$, where $\boldsymbol{\theta}$ satisfies $\boldsymbol{\theta}(A_0 + A_1 + A_2 + \ldots) = 0$ and $\boldsymbol{\theta}\mathbf{e} = 1$.

(ii) Show that the limiting probabilities $\mathbf{p} = (\mathbf{p}_0, \mathbf{p}_1, \mathbf{p}_2, \ldots)$ can be computed as follows

$$\mathbf{p}_n = -\left(\mathbf{p}_0 \bar{A}_{0,n} + \sum_{k=1}^{n-1} \mathbf{p}_k \bar{A}_{n+1-k}\right)\left(\bar{A}_1\right)^{-1}, \quad n = 2, 3, \ldots, \tag{3.106}$$

where

$$\bar{A}_{0,n} = \sum_{k=0}^{\infty} A_{0,n+k} G^k, \quad \bar{A}_n = \sum_{k=0}^{\infty} A_{n+k} G^k, \quad n = 1, 2, \ldots, \tag{3.107}$$

and G is the minimum nonnegative solution to

$$0 = \sum_{n=0}^{\infty} A_n G^n. \tag{3.108}$$

Vector $(\mathbf{p}_0, \mathbf{p}_1)$ satisfies

$$0 = (\mathbf{p}_0, \ \mathbf{p}_1) \begin{pmatrix} A_{0,0} & \sum_{k=0}^{\infty} A_{0,1+k} G^k \\ A_{1,0} & \sum_{k=0}^{\infty} A_{1+k} G^k \end{pmatrix} \tag{3.109}$$

and is normalized so that the total probability is one.

Exercise 3.5.10 (Exercise 3.5.9 continued) Assume that $A = A_0 + A_1 + A_2 + \ldots$ is irreducible. Let $g^*_{i,j}(s)$ be the LST of the length of a fundamental period that reaches level n for the first time in state (n, j), given that the process begins in state $(n + 1, i)$. Let $G^*(s) = \left(g^*_{i,j}(s)\right)$.

(i) Show that $G^*(s)$ satisfies equation: for $s \geq 0$,

$$G^*(s) = \sum_{n=0:\ n\neq 1}^{\infty} (sI - A_1)^{-1} A_n (G^*(s))^n. \tag{3.110}$$

In fact, $G^*(s)$ is the minimal nonnegative solution to Eq. (3.110). Let $G = G^*(0)$. Then G is the minimal nonnegative solution to Eq. (3.108).

(ii) Assume that Neuts condition in part (1) of Exercise 3.5.9 is satisfied. Let $\mathbf{g}$ satisfy $\mathbf{g}G = \mathbf{g}$ and $\mathbf{ge} = 1$. Show that the conditional mean lengths of a fundamental period are given by

$$-\left.\frac{\mathrm{d}G^*(s)}{\mathrm{d}s}\right|_{s=0} \mathbf{e} = (I - G + \mathbf{eg})\left(\left(\sum_{n=0}^{\infty} nA_n\right)\mathbf{eg} - A\right)^{-1} \mathbf{e}. \tag{3.111}$$

Exercise 3.5.11 (Exercise 3.5.6 continued) Using the reblocking method to transform the Markov chain defined in Exercise 3.5.6 into a QBD process, and find its limiting probabilities.

Exercise 3.5.12 (Ergodicity condition) Consider the following Markov chain

$$P = \begin{pmatrix} c_2 & c_2/2^2 & c_2/3^2 & c_2/4^2 & \cdots \\ 0.9 & 0 & 0.1 & & \\ & 0.9 & 0 & 0.1 & \\ \vdots & & \ddots & \ddots & \ddots \end{pmatrix}. \tag{3.112}$$

(i) Find the mean first passage time from state $n + 1$ to state n, for $n \geq 1$. (Hint: Recall Theorem 3.5.2)
(ii) Find the mean first passage time from state $n + 1$ to state 0, for $n \geq 1$.
(iii) Find the mean first passage time from state 0 to state 0.
(iv) Is the Markov chain ergodic? Intuitively, why?
(v) If the transition probabilities from state 0 to other states are propositional to {1, $1/2^3$, $1/3^3$, $1/4^3$, ...}, is the Markov chain ergodic?

Exercise 3.5.13 Consider a vacation *M/M/*1 queue with exponential vacation times (repeating). Let $q(t)$ be the total number of customers in the system at time t. Let $I(t) = 0$, if the server is working; 1, otherwise, at time t. Then $\{(q(t), I(t)), t \geq 0\}$ is a continuous time Markov chain with state space $\{(0, 1), (n, 0), (n, 1), n = 1, 2, \ldots\}$ and infinitesimal generator

$$Q = \begin{pmatrix} A_{0,0} & A_{0,1} & & & \\ A_{1,0} & A_1 & A_0 & & \\ & A_2 & A_1 & A_0 & \\ & & \ddots & \ddots & \ddots \end{pmatrix}, \tag{3.113}$$

where

$$\begin{aligned} &A_{0,0} = -\lambda, \quad A_{0,1} = (0, \quad \lambda), \quad A_{1,0} = \begin{pmatrix} \mu \\ 0 \end{pmatrix}, \\ &A_0 = \begin{pmatrix} \lambda & 0 \\ 0 & \lambda \end{pmatrix}, \quad A_1 = \begin{pmatrix} -\lambda-\mu & 0 \\ \gamma & -\lambda-\gamma \end{pmatrix}, \quad A_2 = \begin{pmatrix} \mu & 0 \\ 0 & 0 \end{pmatrix}. \end{aligned} \tag{3.114}$$

(i) Find the rate matrix R explicitly. Find $sp(R)$ explicitly.
(ii) Find the matrix G explicitly. Find $sp(G)$ explicitly.

Exercise 3.5.14* (Exercises 3.4.10 and 3.4.11 continued) Let $A(z) = A_0 + zA_1 + z^2A_2 + \ldots$, for $0 \le z \le 1$. We assume that $A(1)\mathbf{e} = \mathbf{e}$ and $sp(A_0) > 0$. Denote by $\rho(z)$ the Perron-Frobenius eigenvalue of $A(z)$. Show that (i) $\rho(e^{-s})$ is a concave function in $[0, \infty]$; and (ii) there is no z^* in (0, 1) such that $\rho(z) = z$ if and only if $\rho^{(1)}(1) < 1$.

Define $\mathbf{p}^*(z) = \sum_{n=0}^{\infty} z^n \mathbf{p}_n$ for $0 \le z \le 1$.

Exercise 3.5.15 Show that

$$\begin{aligned} &\mathbf{p}^*(z)\left(zI - \sum_{n=0}^{\infty} z^n A_n\right) \\ &= \mathbf{p}_0\left(z\sum_{n=0}^{\infty} z^n A_{0,n} - \sum_{n=0}^{\infty} z^n A_n\right) + z\mathbf{p}_1(A_{1,0} - A_0). \end{aligned} \tag{3.115}$$

In $P_{\le n}$ defined by Eq. (3.102), the (i, j)-th element of $\sum_{k=0}^{\infty} A_{1+k}G^k$ is the probability that the Markov chain returns to levels $\{0, 1, \ldots, n\}$ for the first time by entering state (n, j), given that the Markov chain is initially in (n, i). Intuitively, the Markov chain transits in one step from level n to level $n + k$ according to A_{1+k}. From level $n + k$, the Markov chain transits to level n for the first time according to G^k. Together, the transition probabilities are given by $\sum_{k=0}^{\infty} A_{1+k}G^k$.

Exercise 3.5.16 (i) Explain $\sum_{k=0}^{\infty} A_{n+k}G^k$ in $P_{\le n}$ probabilistically. (ii) According to Sect. 3.1, the limiting probabilities are the proportion of time the Markov chain stays in a state. For all states in levels $\{0, 1, \ldots, n\}$, explain that the proportions of time in all states of Markov chain $P_{\le n}$ are proportional to that of P.

Exercise 3.5.17 Give a possible explanation to the elements of the matrix $G(z_1, z_2)$ satisfying (Neuts (1991))

$$G(z_1,z_2) = z_1A_0 + A_1G(z_1,z_2) + z_2\sum\nolimits_{n=2}^{\infty} A_n(G(z_1,z_2))^n. \tag{3.116}$$

Exercise 3.5.18 Give a possible explanation to the elements of the matrix $R(z)$ satisfying $R(z) = z\sum_{n=0}^{\infty}(R(z))^n A_n$. (Note: This exercise is for the *GI/M/*1 type Markov chains.)

We remark that Exercises 3.5.17 and 3.5.18 indicate that it is possible to count the occurrences of special events during the fundamental period. More examples can be found in Sect. 5.3, where this idea is used to calculate the total cost incurred during a fundamental period, which can be used to find the mean cost per unit time.

Exercise 3.5.19 Assume that a downward jump (associated with A_0) implies the completion of a service. A service completion generates \$5 profits. Find the probability generating function for the total profits generated in a fundamental period. Derive a method for computing the mean total profit per fundamental period, and mean profit per unit time.

For the QBD process defined in Sect. 3.2, Eq. (3.91) is reduced to

$$\begin{aligned}\frac{\mathrm{d}G^*(z)}{\mathrm{d}z} &= \sum_{n=0}^{2} A_n(G^*(z))^n + zA_1\frac{\mathrm{d}G^*(z)}{\mathrm{d}z} \\ &\quad + zA_2\left(\frac{\mathrm{d}G^*(z)}{\mathrm{d}z}G^*(z) + G^*(z)\frac{\mathrm{d}G^*(z)}{\mathrm{d}z}\right).\end{aligned} \tag{3.117}$$

Let $G^{(1)} = \mathrm{d}G^*(z)/\mathrm{d}z|_{z=1}$. Then Eq. (3.117) is reduced to

$$G^{(1)} = \sum_{n=0}^{2} A_nG^n + A_1G^{(1)} + A_2\left(GG^{(1)} + G^{(1)}G\right). \tag{3.118}$$

Denote by $\phi(A)$ the direct-sum of A, which is obtained by stringing out the components of the rows of A into a single row vector, starting from the first row. It can be shown that $\phi(ABC) = \phi(B)(A' \otimes C)$ for any matrices A, B, and C, as long as the matrix multiplications are valid. By Eq. (3.118), we obtain

$$\phi(G^{(1)}) = \left(I - (A_1)' \otimes I - (A_2G)' \otimes I - (A_2)' \otimes G\right)^{-1}\phi\left(\sum_{n=0}^{2} A_nG^n\right). \tag{3.119}$$

Exercise 3.5.20 Prove Eq. (3.119) by showing that the inverse matrix in Eq. (3.119) exists. Generalize the result in Eq. (3.119) from the QBD processes to the *M/G/*1 type Markov chains.

3.6 QBD Type Markov Chains with a Tree Structure

Sections 3.1, 3.2, 3.3, 3.4, and 3.5 show that, if some special structure exists, the corresponding Markov chain can be analyzed efficiently. They also indirectly imply that the analysis of a Markov chain can be quite challenging if there is nothing special about the Markov chain. Is the transition structure intrinsic or acquired? Apparently, the transition structure is intrinsic for a Markov chain. However, the arrangement of states may make it clear or unclear to identify/utilize the structure of Markov chains.

Example 3.6.1 For the birth-and-death process considered in Sect. 3.1, the states are arranged as {0, 1, 2, 3, 4, ...} (see Eq. (3.3)). If the states are arranged as {0, 2, 1, 4, 3, ...}, the transition probability matrix becomes

$$P = \begin{matrix} 0 \\ 2 \\ 1 \\ 4 \\ \vdots \end{matrix} \overset{\begin{matrix} 0 & 2 & 1 & 4 & \cdots \end{matrix}}{\begin{pmatrix} p_{0,0} & 0 & p_{0,1} & 0 & \\ 0 & 0 & p_{2,1} & 0 & \\ p_{1,0} & p_{1,2} & 0 & 0 & \ddots \\ 0 & 0 & 0 & 0 & \ddots \\ & & & \ddots & \ddots \end{pmatrix}}. \tag{3.120}$$

Compared to Eq. (3.3), the birth-and-death structure is not clear at all in Eq. (3.120).

Example 3.6.1 demonstrates the importance of properly arranging the states of a Markov chain, if some special structure exists. For some cases, a special structure might be identified through the rearrangement of states.

Example 3.6.2 Consider a discrete time Markov chain with state space {0, 1, 2, ...}. The following transitions exist:

- State 0: Only transitions to states 0, 1, and 2 exist;
- State 1: Only transitions to states 0, 1, 3, and 5 exist;
- State 2: Only transitions to states 0, 2, 4, and 6 exist;
- State 3: Only transitions to states 1, 3, 7, and 11 exist;
- State 4: Only transitions to States 2, 4, 8, and 12 exist;
- State 5: Only transitions to States 1, 5, 9, and 13 exist;
- State 6: Only transitions to States 2, 6, 10, and 14 exist;
- ....

Although the transitions are explicitly defined, it is not easy to describe or visualize the transitions intuitively. Next, we rearrange the states as follows. Instead of using a single integer to represent each state, we introduce a string of integers {1, 2} to represent a state (except for state 0). For example, state 1 is represented by string 1, state 2 by 2, state 3 by 11, sate 4 by 21, state 5 by 12, state 6 by 22, state 7 by

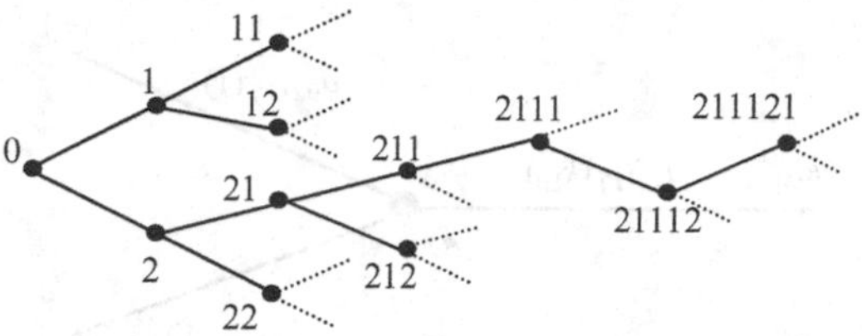

Fig. 3.8 A 2-ary tree

111, state 8 by 211, ... (See Fig. 3.8). In general, state n is represented by string $k_1k_2...k_j$, where k_1, k_2, ..., and k_j are either 1 or 2, and satisfy

$$n = k_1 + 2^{k_2} + \ldots + 2^{(j-2)k_{j-1}} + 2^{(j-1)k_j}. \tag{3.121}$$

It can be shown that the two systems of notation have a one-to-one relationship.

By definition, the Markov chain can go from state $k_1k_2...k_j$ in one step to only four states $\{k_1k_2...k_{j-1}, k_1k_2...k_j, k_1k_2...k_j1, k_1k_2...k_j2\}$. For state 0, the Markov chain can only transit to itself and states 1 and 2 in one step. We call state $k_1k_2...k_{j-1}$ the parent node of $k_1k_2...k_j$, and states $\{k_1k_2...k_j1, k_1k_2...k_j2\}$ the offspring nodes (i.e., children) of $k_1k_2...k_j$. Figure 3.8 demonstrates the transition structure. This type of structure is known as a *tree structure*, in which no loop of transitions exists, in the general literature. The identification of the transition structure of the Markov chain makes it possible to analyze such a Markov chain effectively.

Formally, we define a *K-ary tree* first. The K-ary tree of interest is a tree for which each node has K children, and one parent node, except for node 0. The node 0 is called the *root node*. Strings of integers between 1 and K are used to represent nodes in the tree. For example, the k-th child of the root has a representation k. The l-th child of node k has a representation kl (see Fig. 3.8 for an example with $K = 2$). Node kl is a child of node k, and node k is the parent of node kl. Let $\aleph = \{0\} \cup \{x: x = k_1k_2...k_n, 1 \le k_i \le K, i = 1, \ldots, n, \text{ and } n = 0, 1, 2, \ldots\}$. Any string $x \in \aleph$ is a node in the K-ary tree. The length of string x is defined as the number of integers in the string and is denoted by $|x|$. If $x = 0$, then $|x| = 1$. The following two operations related to strings in $\aleph$ are used.

$$\textit{Addition operation}: \text{ for } x = k_1 \ldots k_n \in \aleph \text{ and } y = h_1 \ldots h_i \in \aleph, \text{ then}$$
$$x + y = k_1 \ldots k_n h_1 \ldots h_i \in \aleph. \text{ (Note : } x + 0 = x.)$$

$$\textit{Subtraction operation}: \text{ for } x = k_1 \ldots k_n \in \aleph,\ y = k_i \ldots k_n \in \aleph,\ i > 0, \text{ then}$$
$$x - y = k_1 \ldots k_{i-1} \in \aleph.$$

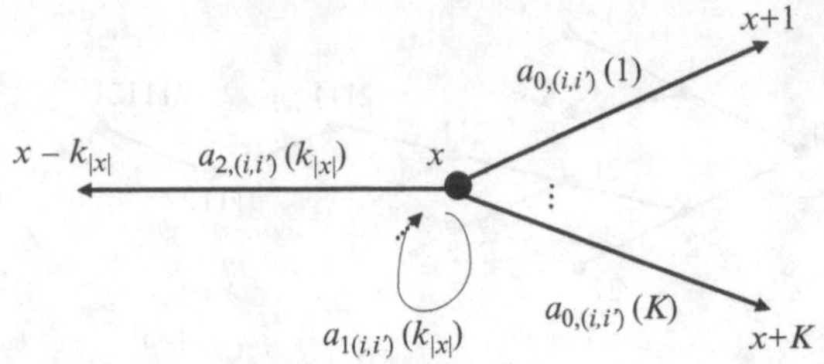

Fig. 3.9 Possible one step transitions when $x \neq 0$

For example, as shown in Fig. 3.8, we have 12 + 1 + 1 = 1211 and 12−2 = 1. The addition $x + y$ of two strings x and y in $\aleph$ is a string obtained by concatenating x and y together. The subtraction operation is valid only for removing the last integer in the string, i.e., to move from a child node to its parent node.

Exercise 3.6.1 Find an one-to-one mapping between nodes in a K-ary tree and the set of nonnegative integers.

Consider a discrete time two-dimensional Markov chain $\{(X_n, J_n), n = 0, 1, 2, \ldots\}$ in which the values of X_n are represented by the nodes of a K-ary tree, and J_n takes integer values between 1 and m. The variable X_n is referred to as the *node* (level) variable and J_n is referred to as the auxiliary (phase) variable of the Markov chain at time n. The Markov chain $\{(X_n, J_n), n = 0, 1, 2, \ldots\}$ takes values in $\aleph\times\{1, 2, \ldots, m\}$. To be called a (homogenous) quasi-birth-and-death Markov chain with a tree structure, $\{(X_n, J_n), n = 0, 1, 2, \ldots\}$ transits at each step to either the current node itself, one of its children, or its parent node. All possible transitions and their corresponding probabilities are given as follows. If $(X_n, J_n) = (x, i)$, the one step transition probabilities are given as: (see Fig. 3.9)

1. $(X_{n+1}, J_{n+1}) = (x + k, i')$ with probability $a_{0,(i,i')}(k)$;
2. $(X_{n+1}, J_{n+1}) = (x, i')$ with probability $a_{1,(i,i')}(k_{|x|})$, if $x \neq 0$;
3. $(X_{n+1}, J_{n+1}) = (x - k_{|x|}, i')$ with probability $a_{2,(i,i')}(k_{|x|})$, if $x \neq 0$; and
4. $(X_{n+1}, J_{n+1}) = (x, i')$ with probability $a_{0,0,(i,i')}$, if $x = 0$.

In matrix form, transition probabilities between nodes are represented by matrix blocks:

1. $A_0(k)$ is an $m\times m$ matrix with elements $a_{0,(i,i')}(k)$;
2. $A_1(k)$ is an $m\times m$ matrix with elements $a_{1,(i,i')}(k)$;
3. $A_2(k)$ is an $m\times m$ matrix with elements $a_{2,(i,i')}(k)$; and
4. $A_{0,0}$ is an $m\times m$ matrix with elements $a_{0,0,(i,i')}$.

Notice that for $A_0(k)$, "k" represents the last integer in the string of the k-th child of the current node (whatever it is); and for $A_1(k)$ and $A_2(k)$, "k" is the last integer of the string of the current node. According to the law of total probability, the matrix blocks satisfy the following equalities

$$\begin{aligned}\left(\sum_{l=1}^{K} A_0(l) + A_1(k) + A_2(k)\right)\mathbf{e} &= \mathbf{e}, \quad k = 1, 2, \ldots, K;\\ \left(\sum_{l=1}^{K} A_0(l) + A_{0,0}\right)\mathbf{e} &= \mathbf{e}.\end{aligned} \tag{3.122}$$

Exercise 3.6.2 (Example 3.6.2 continued) Assume that $m = 1$, $A_{0,0} = 0.7$, $A_0(1) = 0.2$, $A_0(2) = 0.1$, $A_1(1) = 0.5$, $A_1(2) = 0.15$, $A_2(1) = 0.2$, and $A_2(2) = 0.55$. Show that the limiting probabilities of the Markov chain can be obtained as follows. First, find (r_1, r_2) as the minimal nonnegative solution to equations

$$\begin{aligned} r_1 &= A_0(1) + r_1A_1(1) + r_1r_1A_2(1) + r_1r_2A_2(2);\\ r_2 &= A_0(2) + r_2A_1(2) + r_2r_1A_2(1) + r_2r_2A_2(2).\end{aligned} \tag{3.123}$$

Denote by $\pi(x)$ for the limiting probability of state $x = k_1 \ldots k_n \in \aleph$. Show that the limiting probabilities are given by

$$\begin{aligned}\pi(0) &= \frac{1}{1-r_1-r_2};\\ \pi(k_1k_2\ldots k_n) &= \frac{1}{1-r_1-r_2} r_{k_1}r_{k_2}\cdots r_{k_n}, \quad n = 1, 2, \ldots\end{aligned} \tag{3.124}$$

(Hint: Verify the balanced equations for all states.)

Example 3.6.3 The following matrices define a tree structured QBD process with $m = 2$ and $K = 2$:

$$\begin{aligned} A_{0,0} &= \begin{pmatrix} 0.2 & 0.4 \\ 0.2 & 0.3 \end{pmatrix}, A_0(1) = \begin{pmatrix} 0.1 & 0 \\ 0.2 & 0.1 \end{pmatrix}, A_0(2) = \begin{pmatrix} 0 & 0.3 \\ 0.1 & 0.1 \end{pmatrix},\\ A_1(1) &= \begin{pmatrix} 0.1 & 0.1 \\ 0 & 0.2 \end{pmatrix}, A_2(1) = \begin{pmatrix} 0.1 & 0.3 \\ 0.2 & 0.1 \end{pmatrix},\\ A_1(2) &= \begin{pmatrix} 0 & 0.3 \\ 0.2 & 0.1 \end{pmatrix}, A_2(2) = \begin{pmatrix} 0 & 0.3 \\ 0.1 & 0.1 \end{pmatrix}.\end{aligned} \tag{3.125}$$

Having defined the Markov chain of interest, the next step is to analyze the Markov chain and to find its limiting probabilities, if the Markov chain is ergodic. In order to do so, two sets of matrices, $\{R(k), k = 1, \ldots, K\}$ and $\{G(k), k = 1, \ldots, K\}$, are introduced. In essence, the definitions of the two sets of matrices are the same as that of R and G defined in Sections 3.2 and 3.3. We begin with rate matrices $\{R(k), k = 1, 2, \ldots, K\}$.

Matrices $\{R(k), k = 1, 2, \ldots, K\}$ For $x \in \aleph$ and $1 \le k \le K$, define the taboo probability ${}_xP^{(n)}_{(x,i)\,(x+k,j)}$ as the probability that the Markov chain is in state $(x + k, j)$ after n transitions without visiting node x (and its ancestors) in between, given that the Markov chain starts in (x, i). Because of the particular transitional structure, ${}_xP^{(n)}_{(x,i)\,(x+k,j)}$ is independent of x. Define

$$r_{i,j}(k) = \sum_{n=0}^{\infty} {}_xP^{(n)}_{(x,i)\,(x+k,j)}, \quad i,\ j = 1,\ 2, \ldots,\ m, \tag{3.126}$$

and $R(k)$ an $m \times m$ matrix with (i, j)-th element $r_{i,j}(k)$. It can be proved that $\{R(k), k = 1, \ldots, K\}$ are the minimal nonnegative solutions to equations

$$R(k) = A_0(k) + R(k)A_1(k) + \sum_{l=1}^{K} R(k)R(l)A_2(l), \quad k = 1, \ldots, K. \tag{3.127}$$

If the Markov chain is irreducible and positive recurrent, the spectrum (the eigenvalue with the biggest modulus) of matrix $R = R(1) + \ldots + R(K)$ is less than one, i.e., $sp(R) < 1$ (see Yeung and Sengupta (1994)).

Exercise 3.6.3 Define the elements of $R(kl)$ as the expected numbers of visits to node $x + kl$, without visiting node x (and its ancestors) in between, given that the Markov chain starts in node x. (i) Show that $R(kl) = R(k)R(l)$ probabilistically. (ii) Prove Eq. (3.127).

Matrices $\{G(k), k = 1, 2, \ldots, K\}$ For $x \in \aleph$, $x \neq 0$, and $1 \le k \le K$, define the taboo probability $g_{i,j}(k)$ as the probability that the Markov chain $\{(X_n, J_n), n = 0, 1, \ldots\}$ reaches node x for the first time in state (x, j), given that the Markov chain starts in $(x + k, i)$. Let $G(k)$ be an $m \times m$ matrix with elements $g_{i,j}(k)$. It can be proved that $\{G(k), k = 1, \ldots, K\}$ are the minimal nonnegative solutions to the equations

$$G(k) = A_2(k) + A_1(k)G(k) + \sum_{l=1}^{K} A_0(l)G(l)G(k), \quad \text{for } k = 1, \ldots, K. \tag{3.128}$$

If the Markov chain is irreducible and positive recurrent, matrix $G(k)$ is a stochastic matrix, i.e., $G(k)\mathbf{e} = \mathbf{e}$, for $k = 1, \ldots, K$.

Exercise 3.6.4 Define elements of $G(kl)$ as the probabilities that the Markov chain reaches node x from $x + kl$ for the first time. (i) Show that $G(kl) = G(l)G(k)$. (ii) Prove Eq. (3.128).

If $K = 1$, i.e., the classical quasi-birth-and-death Markov chain case, a simple relationship between the matrices R and G has been shown in Latouche (1987) (see Proposition 3.3.3). A similar relationship holds if $K > 1$. Define $U_{i,j}(k)$, for $1 \le k \le K$ and $1 \le i, j \le m$, the probability that the Markov chain will eventually come back to node $x + k$ in state $(x + k, j)$, given that it starts in $(x + k, i)$ and never visits its parent node x in between. Let $U(k)$ be an $m \times m$ matrix with elements $U_{i,j}(k)$. If the Markov chain $\{(X_n, J_n),\ n = 0, 1, 2, \ldots\}$ is irreducible and positive recurrent, the following relationships hold for $\{R(k), G(k), U(k), k = 1, 2, \ldots, K\}$:

$$
\begin{aligned}
R(k) &= A_0(k) + R(k)U(k);\\
G(k) &= A_2(k) + U(k)G(k);\\
U(k) &= A_1(k) + \sum_{l=1}^{K} R(l)A_2(l) = A_1(k) + \sum_{l=1}^{K} A_0(l)G(l).
\end{aligned}
\tag{3.129}
$$

In addition, $R(k)A_2(k) = A_0(k)G(k)$, for $k = 1, 2, \ldots, K$.

Similar to the $K = 1$ case, matrices $\{R(k), k = 1, \ldots, K\}$ are rate matrices and $\{U(k), G(k), k = 1, \ldots, K\}$ are (sub)stochastic matrices. The matrices $\{R(k), G(k), k = 1, \ldots, K\}$ play an important role in the performance analysis for the case with a tree structure.

The limiting probabilities Let, for $x \in \aleph$ and $i = 1, 2, \ldots, m$,

$$
\pi(x, i) = \lim_{n \to \infty} P\{(X_n, J_n) = (x, i) | (X_0, J_0)\},
\tag{3.130}
$$

$\boldsymbol{\pi}(x) = (\pi(x,1), \ldots, \pi(x,m))$. The vectors $\{\boldsymbol{\pi}(x): x \in \aleph\}$ satisfy the following equation, for $x \neq 0$,

$$
\boldsymbol{\pi}(x + k) = \boldsymbol{\pi}(x)A_0(k) + \boldsymbol{\pi}(x + k)A_1(k) + \sum_{l=1}^{K} \boldsymbol{\pi}(x + k + l)A_2(l),
\tag{3.131}
$$

which is useful in understanding the following solution intuitively.

Theorem 3.6.1 (Yeung and Sengupta (1994)) *If the QBD Markov chain* $\{(X_n, J_n), n = 0, 1, 2, \ldots\}$ *is ergodic, its limiting probabilities are given by*

$$
\begin{aligned}
&\boldsymbol{\pi}(x + k) = \boldsymbol{\pi}(x)R(k), \quad x \in \aleph, \quad k = 1, 2, \ldots, K;\\
&\boldsymbol{\pi}(0) = \boldsymbol{\pi}(0)\left(A_{0,0} + \sum_{k=1}^{K} R(k)A_2(k)\right);\\
&\boldsymbol{\pi}(0)(I - R)^{-1}\mathbf{e} = 1.
\end{aligned}
\tag{3.132}
$$

Note that $R = R(1) + \ldots. + R(K)$. Matrices $R(k), k = 1, 2, \ldots, K$ can be calculated using the following simple algorithm. Let $R(k)[0] - 0$, for $k = 1, 2, \ldots, K$, and

$$
R(k)[n + 1] = A_0(k) + R(k)[n]A_1(k) + \sum_{l=1}^{K} R(k)[n]R(l)[n]A_2(l).
\tag{3.133}
$$

It can be shown that $R(k)[n], n = 0, 1, 2, \ldots$ is a monotone sequence that converges to $R(k)$ from below, for $k = 1, 2, \ldots, K$. This algorithm is simple and easy to implement. Matrices $G(k), k = 1, \ldots, K$ can be computed in a similar way.

In summary, the limiting probabilities of the Markov chain $(X_n, J_n), n = 0, 1, 2, \ldots$ can be found using the following procedure.

Step 1: Data input: m, K, $A_{0,0}$, and $A_0(k), A_1(k), A_2(k), k = 1, \ldots, K$.
Step 2: Compute matrices $R(k), G(k), k = 1, \ldots, K$.
Step 3: Compute vectors $\boldsymbol{\pi}(0)$.
Step 4: Compute string distribution $\{\boldsymbol{\pi}(x), x \in \aleph\}$.

Commentary Markov chains with a tree structure and an auxiliary phase is introduced in Yeung and Sengupta (1994) and Takine et al. (1995). The QBD type Markov chains with a tree structure are special cases of the *GI/M/*1 and *M/G/*1 types (Yeung and Alfa (1999) and He (2000b)). Conditions for the existence of the limiting probabilities, i.e., ergodicity of the Markov chain, can be found in Sect. 3.8 (also see He (2003a, b)). Markov chains with a tree structure have found applications in the study of queueing models (see Chap. 4) and communications networks (e.g., van Houdt and Blondia (2001, 2004)).

Additional Exercises and Extensions

Exercise 3.6.5 (Example 3.6.3 continued) Consider the tree structured QBD process defined in Example 3.6.3. Compute $\{R(1), R(2), G(1), G(2), \boldsymbol{\pi}_0\}$. What are the Perron-Frobenius eigenvalues of $R = R(1) + R(2)$, $G(1)$, and $G(2)$?

Exercise 3.6.6 Consider a tree structured QBD process with $m = 2$, $K = 2$, and

$$
\begin{aligned}
A_{0,0} &= \begin{pmatrix} 0.2 & 0.2 \\ 0.2 & 0 \end{pmatrix}, A_0(1) = \begin{pmatrix} 0.1 & 0.2 \\ 0.2 & 0.4 \end{pmatrix}, A_0(2) = \begin{pmatrix} 0 & 0.3 \\ 0.1 & 0.1 \end{pmatrix}, \\
A_1(1) &= \begin{pmatrix} 0.1 & 0.1 \\ 0 & 0 \end{pmatrix}, A_2(1) = \begin{pmatrix} 0.1 & 0.1 \\ 0.2 & 0 \end{pmatrix}, \\
A_1(2) &= \begin{pmatrix} 0 & 0.3 \\ 0.1 & 0.1 \end{pmatrix}, A_2(2) = \begin{pmatrix} 0.1 & 0 \\ 0 & 0 \end{pmatrix}.
\end{aligned}
\tag{3.134}
$$

Compute $\{R(1), R(2), G(1), G(2)\}$. Find $R = R(1) + R(2)$. What are the Perron-Frobenius eigenvalues of R, $G(1)$, and $G(2)$? Comment on the relationship between the Perron-Frobenius eigenvalues and the ergodicity of the Markov chain.

Similar Markov chains with more complicated boundary conditions can be analyzed in the same way (See He (2000b)).

Exercise 3.6.7 Consider a 2-ary tree with a *soil node*, denoted as -1, connected only to the root node 0 (see Fig. 3.10). The transition blocks between nodes 0 and -1 are $\{B_{-1,-1}, B_{-1,0}, B_{0,-1}\}$. Show that the limiting probabilities can be obtained by solving Eqs. (3.127) and (3.135).

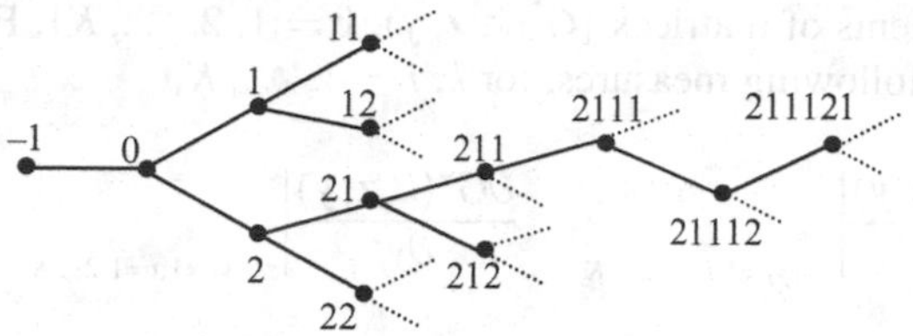

Fig. 3.10 A 2-ary tree with a soil node

$$\pi(x+k) = \pi(x)R(k), \quad x \in \aleph, \quad k = 1, \ldots, K;$$
$$(\pi(-1), \pi(0)) = (\pi(-1), \pi(0)) \begin{pmatrix} B_{-1,-1} & B_{-1,0} \\ B_{0,-1} & A_{0,0} + \sum_{k=1}^{K} R(k)A_2(k) \end{pmatrix}; \quad (3.135)$$
$$\pi(-1)\mathbf{e} + \pi(0)(I-R)^{-1}\mathbf{e} = 1.$$

Exercise 3.6.8 Explain why the matrix-geometric solution still exists for tree structured QBD processes for which the auxiliary variable has a different number of states for boundary nodes -1 and 0.

Exercise 3.6.9 Consider the tree structured QBD process $\{(X_n, J_n), n = 0, 1, 2, \ldots\}$. Define $Y_n = |X_n|$, i.e., the cardinal number of X_n, for $n = 0, 1, 2, \ldots$. We define a new process $\{(Y_n, J_n), n = 0, 1, 2, \ldots\}$. Show that the limiting probabilities of the new process is given by the matrix geometric distribution $\{\pi(0)R^t, t = 0, 1, 2, \ldots\}$. (Note that the new process may not be a Markov chain. Thus, its limiting probabilities can only be obtained from that of the QBD process $\{(X_n, J_n), n = 0, 1, 2, \ldots\}$.)

Exercise 3.6.10 Define $G^*(k, \mathbf{z})$, where $\mathbf{z} = (z_1, z_2, \ldots, z_K)$, satisfying, for $k = 1, \ldots, K$,

$$G^*(k,\mathbf{z}) = z_k A_2(k) + A_1(k)G^*(k,\mathbf{z}) + \sum_{l=1}^{K} A_0(l)G^*(l,\mathbf{z})G^*(k,\mathbf{z}). \quad (3.136)$$

Interpret the elements of matrices $\{G^*(k, \mathbf{z}), k = 1, 2, \ldots, K\}$.

Exercise 3.6.11* Define $G^*(k, \mathbf{z}, \mathbf{y})$, where $\mathbf{z} = (z_1, z_2, \ldots, z_K)$ and $\mathbf{y} = (y_1, y_2, \ldots, y_K)$, satisfying, for $k = 1, \ldots, K$,

$$G^*(k,\mathbf{z},\mathbf{y}) = z_k A_2(k) + A_1(k)G^*(k,\mathbf{z},\mathbf{y}) + \sum_{l=1}^{K} y_l A_0(l)G^*(l,\mathbf{z},\mathbf{y})G^*(k,\mathbf{z},\mathbf{y}). \quad (3.137)$$

Interpret the elements of matrices $\{G^*(k, \mathbf{z}, \mathbf{y}), k = 1, 2, \ldots, K\}$. Find expressions and interpret the following measures, for $k, j = 1, \ldots, K$,

$$\left.\frac{\partial G^*(k,\mathbf{z},\mathbf{y})}{\partial z_j}\right|_{z_i=y_i=1,i=1,2,..,K}, \quad \left.\frac{\partial G^*(k,\mathbf{z},\mathbf{y})}{\partial y_j}\right|_{z_i=y_i=1,i=1,2,..,K}. \tag{3.138}$$

3.7 *GI/M/1* Type Markov Chains with a Tree Structure

Recall that the K-ary tree is defined as $\aleph = \{0\} \cup \{x: x = k_1k_2\ldots k_n, 1 \le k_i \le K, i = 1, \ldots, n, n = 1, 2, \ldots\}$, where K is a positive integer. Node $xk = x + k$ is called a type k node, for $x \in \aleph$.

We consider a Markov chain $\{(X_n, J_n), n = 0, 1, 2, \ldots\}$, where X_n takes values in $\aleph$ and J_n takes integer values from 1 to m, and m is a positive integer. The random variable J_n is an auxiliary variable (also called the phase variable). The transition probabilities of the Markov chain are given as, for x and y in $\aleph$, $1 \le k \le K$, and $i, j = 1, \ldots, m$,

$$\begin{aligned} P\{X_{n+1} = x+k,\ J_{n+1} = j|X_n = x+y,\ J_n = i\} &= a_{i,j}(y,k); \\ P\{X_{n+1} = 0,\ J_{n+1} = j|X_n = y,\ J_n = i\} &= b_{i,j}(y). \end{aligned} \tag{3.139}$$

All other transition probabilities are assumed to be zero. Figure 3.11 shows the possible one step transitions for the level variable $X_n = x = 21$ in a 2-ary tree: $\{0, 1, 2, 21, 22, 211, 212\}$.

Let $A(y, k)$ be an $m\times m$ matrix with (i, j)-th element $a_{i,j}(y, k)$. Let $B(y)$ be an $m\times m$ matrix with (i, j)-th element $b_{i,j}(y)$. If $x = k_1\ldots k_n$, denote by $f(x, i) = k_{n-i+1}\ldots k_n$, for $1 \le i \le n$, and $f(x, 0) = 0$. By the law of total probability, we have

$$\left(B(x) + \sum_{i=0}^{|x|}\sum_{k=1}^{K} A(f(x,i),k)\right)\mathbf{e} = \mathbf{e}, \quad \text{for } x \in \aleph. \tag{3.140}$$

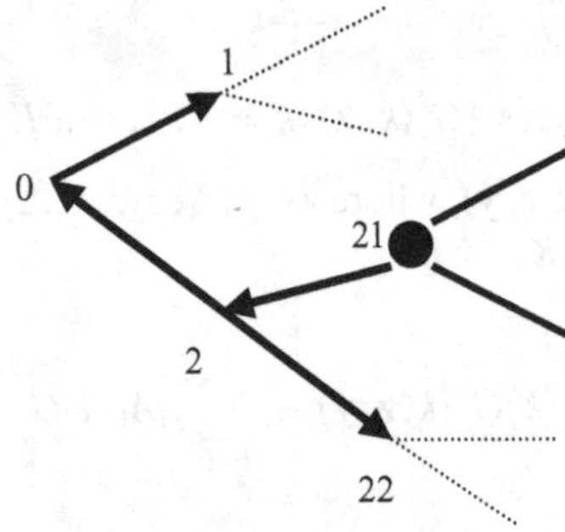

Fig. 3.11 Nodes reachable in one transition from node $x = 21$

From the definition, it is clear that in one transition, the Markov chain can move from the current node to one of its children, its ancestors (parent, parent of parent, ...), or any node that is an immediate child of an ancestor of the current node. The transition probabilities depend on the type of the targeted node. Thus, there are K sets of (matrix) transition probabilities, $\{A(x, k), x \in \aleph\}$, $k = 1, 2, \ldots, K$, plus $\{B(x), x \in \aleph\}$ for transitions to the root node. Also note that the transitions from any node (including the root node) to its immediate children are governed by $\{A(0, k), k = 1, 2, \ldots, K\}$.

We assume that at least one of the matrices $\{A(0, k), k = 1, \ldots, K\}$ is nonzero. Then $\{(X_n, J_n), n = 0, 1, 2, \ldots\}$ is called a *GI/M/*1 *type Markov chain with a tree structure*. The Markov chain $\{(X_n, J_n), n = 0, 1, 2, \ldots\}$ has the classical *GI/M/*1 type Markov chains ($K = 1$) and the QBD processes with a tree structure as its special cases.

Example 3.7.1 The QBD type Markov chain with a tree structure is a special *GI/M/*1 type Markov chain with a tree structure, by letting $A(0, k) = A_0(k)$, $A(k, k) = A_1(k)$, $A(jk, j) = A_2(k)$, $B(0)$ satisfying $B(0)\mathbf{e} + (A_0(1) + \ldots + A_0(K))\mathbf{e} = \mathbf{e}$, and all other matrices zero.

Exercise 3.7.1 Show that a *GI/M/*1 type Markov chain with a tree structure and $K = 1$ is a *GI/M/*1 type Markov chain.

In Sect. 3.6, the notation used for transition probabilities is similar to that for the QBD process in Section 3.2. The notation in this section is different from that for the *GI/M/*1 type Markov chain in Sect. 3.4. The main reason is that the transitions have to be defined for nodes, instead of the levels.

The limiting probabilities of the Markov chain are found to be of the matrix-geometric type.

Theorem 3.7.1 (Yeung and Sengupta (1994)) *Assume that the GI/M/1 type Markov chain with a tree structure is ergodic. Denote by* $\boldsymbol{\pi} = \{\boldsymbol{\pi}(x), x \in \aleph\}$ *the limiting probabilities. Then*

$$\boldsymbol{\pi}(x) = \boldsymbol{\pi}(0)R(x), \quad x \neq 0,\ x \in \aleph, \tag{3.141}$$

where $\boldsymbol{\pi}(0)$ *satisfies*

$$\begin{aligned} \boldsymbol{\pi}(0) &= \boldsymbol{\pi}(0)\left(B(0) + \sum_{x\in\aleph} R(x)B(x)\right); \\ \boldsymbol{\pi}(0)&\left(I - \sum_{k=1}^{K} R(k)\right)^{-1} \mathbf{e} = 1, \end{aligned} \tag{3.142}$$

and

$$R(x) = R(k_1)R(k_2)\cdots R(k_{|x|-1})R(k_{|x|}), \quad x = k_1k_2\cdots k_{|x|-1}k_{|x|} \in \aleph, \tag{3.143}$$

$R(0) = I$, and $\{R(j), j = 1, \ldots, K\}$ are the minimal nonnegative solution to matrix equations

$$R(k) = \sum_{x \in \aleph} R(x)A(x,k), \quad \text{for } k = 1, \ldots, K. \tag{3.144}$$

The (i, j)-th element of $R(k)$ can be interpreted probabilistically as the expected number of visits to $(x + k, j)$ before visiting node x again, given that the Markov chain starts in state (x, i).

Exercise 3.7.2 Let $R(k)[0] = 0$, for $k = 1, \ldots, K$, and $R(0)[n] = I$, for $n = 1, 2, \ldots$. Define

$$R(k)[n+1] = \sum_{x \in \aleph} R(x)[n]A(x,k), \quad \text{for } k = 1, \ldots, K. \tag{3.145}$$

Show that $\{R(k)[n], k = 1, \ldots, K, n = 0, 1, 2, \ldots\}$ are nondecreasing in n, and converge to the minimal nonnegative solution to Eq. (3.144).

Exercise 3.7.3 Let $R = R(1) + R(2) + \ldots + R(K)$. If the Markov chain is ergodic, show

$$\sum_{x \in \aleph:\ |x| = n} \boldsymbol{\pi}(x) = \boldsymbol{\pi}(0)R^n, \quad n = 1, 2, \ldots. \tag{3.146}$$

In addition, show that $sp(R(1) + R(2) + \ldots + R(K)) < 1$.

Commentary The *GI/M/*1 type Markov chain with a tree structure is introduced in Yeung and Sengupta (1994). In He (2003a, b), necessary and sufficient conditions are found for the Markov chain to be ergodic.

Additional Exercises and Extensions

Exercise 3.7.4 Consider a *GI/M/*1 type Markov chain with a tree structure with $K = 2$. The Markov chain can transit in one step from one node to its children, parent, grandparent, and the children of parent or grandparent. That is, one level up and two levels back. To introduce the Markov chain explicitly, we define the following transition blocks for $x \in \aleph$.

Node $y = x + 11$:

$$A(0,1) = A(y, y+1) = \begin{pmatrix} 0.1 & 0 \\ 0.1 & 0 \end{pmatrix}, \quad A(0,2) = A(y, y+2) = \begin{pmatrix} 0.1 & 0.1 \\ 0 & 0.1 \end{pmatrix},$$

$$A(1,1) = A(y, x+1+1) = \begin{pmatrix} 0.1 & 0.1 \\ 0.2 & 0.1 \end{pmatrix}, \quad A(1,2) = A(y, x+1+2) = \begin{pmatrix} 0.2 & 0 \\ 0.1 & 0 \end{pmatrix},$$

$$A(11,1) = A(y, x+1) = \begin{pmatrix} 0.2 & 0 \\ 0.1 & 0 \end{pmatrix}, \quad A(11,2) = A(y, x+2) = \begin{pmatrix} 0 & 0.1 \\ 0 & 0.3 \end{pmatrix};$$

It is clear that elements of $\{A(0, 1), A(0,2)\}$ are the transition probabilities to a child of node y; elements of $\{A(1, 1), A(1,2)\}$ are the transition probabilities to a sibling of node y; elements of $A(11, 1)$ are the transition probabilities to the parent of node y; and elements of $A(11,2)$ are the transition probabilities to a sibling of the parent of node y.

Node $y = x + 12$:

$$A(0,1) = A(y, y+1), \quad A(0,2) = A(y, y+2),$$
$$A(2,1) = A(y, x+1+1) = \begin{pmatrix} 0.2 & 0 \\ 0 & 0.2 \end{pmatrix}, \quad A(2,2) = A(y, x+1+2) = \begin{pmatrix} 0 & 0 \\ 0.1 & 0.1 \end{pmatrix},$$
$$A(12,1) = A(y, x+1) = \begin{pmatrix} 0.1 & 0.2 \\ 0 & 0.1 \end{pmatrix}, \quad A(12,2) = A(y, x+2) = \begin{pmatrix} 0.2 & 0 \\ 0.1 & 0.2 \end{pmatrix};$$

Node $y = x + 21$:

$$A(0,1) = A(y, y+1), \quad A(0,2) = A(y, y+2),$$
$$A(1,1) = A(y, x+2+1), \quad A(1,2) = A(y, x+2+2);$$
$$A(21,1) = A(y, x+1) = \begin{pmatrix} 0 & 0.1 \\ 0.1 & 0.1 \end{pmatrix}, \quad A(21,2) = A(y, x+2) = \begin{pmatrix} 0.2 & 0 \\ 0.2 & 0 \end{pmatrix};$$

Node $y = x + 22$:

$$A(0,1) = A(y, y+1), \quad A(0,2) = A(y, y+2),$$
$$A(2,1) = A(y, x+2+1), \quad A(2,2) = A(y, x+2+2),$$
$$A(22,1) = A(y, x+1) = \begin{pmatrix} 0.3 & 0 \\ 0.1 & 0 \end{pmatrix}, \quad A(22,2) = A(y, x+2) = \begin{pmatrix} 0.2 & 0 \\ 0.2 & 0.1 \end{pmatrix};$$

and

$$B(0) = \begin{pmatrix} 0.1 & 0.6 \\ 0.3 & 0.5 \end{pmatrix}, \quad B(1) = \begin{pmatrix} 0.2 & 0.1 \\ 0.3 & 0.1 \end{pmatrix}, \quad B(2) = \begin{pmatrix} 0.4 & 0.1 \\ 0.2 & 0.2 \end{pmatrix}.$$

All other transition blocks are zero.

(i) Find all the nonzero transition matrices from node $x + 111$ to other nodes.
(ii) Compute $R(1)$ and $R(2)$. Find $sp(R(1) + R(2))$.
(iii) Find π_0.
(iv) Find $\pi(1121)$ and $\pi(2212)$.

Exercise 3.7.5 Explain why the matrix-geometric solution still exists for the tree structured *GI/M/*1 type Markov chain for which the auxiliary variable has a different number of states for boundary node 0. Also explain why that the matrix-geometric solution still exists if the boundary transitions are more complicated.

Exercise 3.7.6 Consider a *GI/M/*1 type Markov chain with a tree structure for which the Markov chain only transits from a node to its child, or its ancestors (not its ancestors' children). Show that all transition matrices are zero except for $\{A(k + x, k), x \in \aleph, k = 1, \ldots, K\}$ and $\{A(0, k), k = 1, \ldots, K\}$. For $K = 2$, consider the following example.

Node $y = x + 111$:

$$A(0,1) = A(y,\ y+1) = \begin{pmatrix} 0.1 & 0.1 \\ 0.1 & 0 \end{pmatrix}, \quad A(0,2) = A(y,\ y+2) = \begin{pmatrix} 0.1 & 0.1 \\ 0 & 0.1 \end{pmatrix},$$
$$A(1,1) = A(y,\ x+111) = \begin{pmatrix} 0.1 & 0.1 \\ 0.2 & 0.1 \end{pmatrix}, \quad A(11,1) = A(y,\ x+11) = \begin{pmatrix} 0.2 & 0 \\ 0.1 & 0 \end{pmatrix},$$
$$A(111,\ 1) = A(y, x+1) = \begin{pmatrix} 0.1 & 0.1 \\ 0.1 & 0.3 \end{pmatrix};$$

Node $y = x + 112$:

$$A(0,1) = A(y, y+1), \quad A(0,2) = A(y, y+2),$$
$$A(2,2) = A(y, x+112) = \begin{pmatrix} 0 & 0 \\ 0.2 & 0.1 \end{pmatrix}, \quad A(12,1) = A(y, x+11) = \begin{pmatrix} 0.1 & 0.2 \\ 0 & 0.1 \end{pmatrix};$$
$$A(112,1) = A(y, x+1) = \begin{pmatrix} 0.1 & 0.2 \\ 0.3 & 0.1 \end{pmatrix};$$

Node $y = x + 121$:

$$A(0,1) = A(y, y+1), \quad A(0,2) = A(y, y+2), \quad A(1,1) = A(y, x+121),$$
$$A(21,2) = A(y, x+12) = \begin{pmatrix} 0.2 & 0 \\ 0.2 & 0 \end{pmatrix}, \quad A(121,1) = A(y, x+1) = \begin{pmatrix} 0.2 & 0 \\ 0.3 & 0 \end{pmatrix};$$

Node $y = x + 122$:

$$A(0,1) = A(y, y+1), \quad A(0,2) = A(y, y+2), \quad A(2,2) = A(y, x+122),$$
$$A(22,2) = A(y, x+12) = \begin{pmatrix} 0.1 & 0.1 \\ 0.2 & 0 \end{pmatrix}, \quad A(122,1) = A(y, x+1) = \begin{pmatrix} 0.2 & 0 \\ 0.3 & 0 \end{pmatrix};$$

Node $y = x + 211$:

$$A(0,1) = A(y, y+1), \quad A(0,2) = A(y, y+2), \quad A(1,1) = A(y, x+211),$$
$$A(11,1) = A(y, x+21), \quad A(211,2) = A(y, x+2) = \begin{pmatrix} 0 & 0.2 \\ 0.2 & 0.2 \end{pmatrix}.$$

Node $y = x + 212$:

$$A(0,1) = A(y, y+1), \quad A(0,2) = A(y, y+2), \quad A(2,2) = A(y, x+212),$$
$$A(12,1) = A(y, x+21), \quad A(212,2) = A(y, x+2) = \begin{pmatrix} 0.3 & 0.1 \\ 0.1 & 0.3 \end{pmatrix};$$

Node $y = x + 221$:

$$A(0,1) = A(y, y+1), \quad A(0,2) = A(y, y+2), \quad A(1,1) = A(y, x+221),$$
$$A(21,2) = A(y, x+22), \quad A(221,2) = A(y, x+2) = \begin{pmatrix} 0.1 & 0.1 \\ 0.2 & 0.1 \end{pmatrix};$$

Node $y = x + 222$:

$$A(0,1) = A(y, y+1), \quad A(0,2) = A(y, y+2), \quad A(2,2) = A(y, x+222),$$
$$A(22,2) = A(y, x+22), \quad A(222,2) = A(x+222, x+2) = \begin{pmatrix} 0.1 & 0.1 \\ 0 & 0.3 \end{pmatrix}.$$

In addition, we have

$$B(0) = \begin{pmatrix} 0.1 & 0.6 \\ 0.3 & 0.5 \end{pmatrix}, \quad B(1) = \begin{pmatrix} 0.2 & 0.1 \\ 0.3 & 0.1 \end{pmatrix}, \quad B(2) = \begin{pmatrix} 0.4 & 0.1 \\ 0.2 & 0.2 \end{pmatrix},$$
$$B(11) = \begin{pmatrix} 0.1 & 0.1 \\ 0.3 & 0.1 \end{pmatrix}, \quad B(12) = \begin{pmatrix} 0.3 & 0.1 \\ 0.3 & 0.1 \end{pmatrix}, \quad B(21) = \begin{pmatrix} 0.1 & 0.1 \\ 0.1 & 0.2 \end{pmatrix},$$
$$B(22) = \begin{pmatrix} 0.1 & 0.1 \\ 0.1 & 0.2 \end{pmatrix}.$$

(i) Find all the nonzero transition matrices from node $x + 1121$ to other nodes.
(ii) Compute $R(1)$ and $R(2)$. Find $sp(R(1) + R(2))$.
(iii) Find $\boldsymbol{\pi}_0$.
(iv) Find $\boldsymbol{\pi}(1121)$ and $\boldsymbol{\pi}(2212)$.

To study the ergodicity of tree structured Markov chains, a *mapping* is introduced. The fixed points of the mapping are used to construct conditions for ergodicity. The *theory of fixed points* is utilized. The following exercises show some properties of the mapping and its fixed points. For simplicity, we assume $m = 1$. Define

$$\begin{aligned} \mathrm{R}_{+}^{K} &= \{\mathbf{u} = (u_1, \ldots, u_K): \quad u_k \geq 0,\ k = 1, \ldots, K\}; \\ \mathrm{R}_{+,\Sigma=1}^{K} &= \left\{\mathbf{u} = (u_1, \ldots, u_K): \quad \sum_{k=1}^{K} u_k = 1,\ u_k \geq 0,\ k = 1, \ldots, K\right\}; \\ \mathrm{R}_{+,\Sigma\leq 1}^{K} &= \left\{\mathbf{u} = (u_1, \ldots, u_K): \quad \sum_{k=1}^{K} u_k \leq 1,\ u_k \geq 0,\ k = 1, \ldots, K\right\}. \end{aligned} \tag{3.147}$$

Define mapping $\mathcal{A}$: $\mathrm{R}_+^K \to \mathrm{R}_+^K$: for $\mathbf{u} \in R_+^K$, $\mathcal{A}(\mathbf{u}) = (a^*{}_1(\mathbf{u}), a^*{}_2(\mathbf{u}), \ldots, a^*{}_K(\mathbf{u}))$, where

$$a_k^*(\mathbf{u}) = \sum_{x \in \aleph} A(x,k) u_1^{N(x,1)} u_2^{N(x,2)} \cdots u_K^{N(x,K)}, \tag{3.148}$$

and $N(x, j)$ is the number of appearances of integer j in the string x. A *fixed point* of the mapping $\mathcal{A}$ is defined as an element $\mathbf{u}$ such that $\mathcal{A}(\mathbf{u}) = \mathbf{u}$.

Exercise 3.7.7* (He (2003a)) Assume that $\lim_{n\to\infty} \max_{x\in\aleph:\,|x|=n}\{B(x)\} = 0$ and $m = 1$. (i) Show that the set $\mathrm{R}_{+,\Sigma=1}^K$ is invariant under $\mathcal{A}$ (i.e., for any $\mathbf{u} \in \mathrm{R}_{+,\Sigma=1}^K$, $\mathcal{A}(\mathbf{u}) \in \mathrm{R}_{+,\Sigma=1}^K$), and there exists at least one fixed point of $\mathcal{A}$ in $\mathrm{R}_{+,\Sigma=1}^K$. (ii) Show that the set $\mathrm{R}_{+,\Sigma\le 1}^K$ is invariant under $\mathcal{A}$. (iii) There exists a minimal fixed point $\mathbf{u}_{\min}$ in $\mathrm{R}_{+,\Sigma\le 1}^K$ such that, if $\mathbf{u} = (u_1, \ldots, u_K)$ is a fixed point of $\mathcal{A}$ in $\mathrm{R}_{+,\Sigma\le 1}^K$, then $u_k \ge (\mathbf{u}_{\min})_k$, $k = 1, 2, \ldots, K$.

Define a $K\times K$ matrix $\mathcal{A}^{(1)}(\mathbf{v})$ whose (k, j)-th element is $\partial a_k^*(\mathbf{u})/\partial u_j\big|_{\mathbf{u}=\mathbf{v}}$ for any $\mathbf{v}$ in $\mathrm{R}_{+,\Sigma=1}^K$.

Exercise 3.7.8* (He (2003a)) Assume that the Markov chain is irreducible, $\lim_{n\to\infty} \max_{x\in\aleph:\,|x|=n}\{B(x)\} = 0$, and $m = 1$. Denote by $\mathbf{u}^*$ a fixed point of $\mathcal{A}$ in $\mathrm{R}_{+,\Sigma=1}^K$. Assume that $\mathcal{A}^{(1)}(\mathbf{u}^*)$ is irreducible. Show that the Markov chain is (i) positive recurrent if $sp(\mathcal{A}^{(1)}(\mathbf{u}^*)) > 1$; (ii) null recurrent if $sp(\mathcal{A}^{(1)}(\mathbf{u}^*)) = 1$; and (iii) transient if $sp(\mathcal{A}^{(1)}(\mathbf{u}^*)) < 1$. Explain the conditions intuitively.

3.8 *M/G/1* Type Markov Chains with a Tree Structure

A discrete time Markov chain $\{(X_n, J_n), n = 0, 1, 2, \ldots\}$ takes values in $\aleph\times\{1, 2, \ldots, m\}$. If the process $\{(X_n, J_n), n = 0, 1, 2, \ldots\}$ transits at each step only to its parent node or a descendent of its parent node, then it is called a homogenous Markov chain of *M/G/1* type with a tree structure. All possible transitions and their corresponding probabilities are given as follows. If $(X_n, J_n) = (x + k, i)$ for $x\in\aleph$, $1 \le k \le K$ and $1 \le i, i' \le m$, then

(1) $(X_{n+1}, J_{n+1}) = (x + y, i')$ with probability $a_{(i,i')}(k, y)$ for $y\in\aleph$, i.e.,

$$a_{(i,i')}(k,y) = P\{X_{n+1} = x + y, J_{n+1} = i' | X_n = x + k, J_n = i\}. \tag{3.149}$$

Note that transition probabilities depend only on the last integer k in the string representing the current node $x + k$. If $(X_n, J_n) = (0, i)$ for $1\le i, i'\le m$, then

(2) $(X_{n+1}, J_{n+1}) = (y, i')$ with probability $b_{(i,i')}(y)$ for $y\in\aleph$, i.e.,

$$b_{(i,i')}(y) = P\{X_{n+1} = y, J_{n+1} = i' | X_n = 0, J_n = i\}. \tag{3.150}$$

In matrix form, the transition probabilities are represented as:

(1)′ $A(k, y)$ is an $m \times m$ matrix with elements $a_{(i,i')}(k, y)$, $k = 1, \ldots, K$, for $y \in \aleph$;
(2)′ $B(y)$ is an $m \times m$ matrix with elements $b_{(i,i')}(y)$ for $y \in \aleph$.

By definition, we must have

$$\sum_{x \in \aleph} A(k,x)\mathbf{e} = \mathbf{e}, \quad k = 1, 2, \ldots, K;$$
$$\sum_{x \in \aleph} B(x)\mathbf{e} = \mathbf{e}. \tag{3.151}$$

Exercise 3.8.1 Show that the QBD process defined in Section 3.6 is a special case of the *M/G/*1 type Markov chain defined in this section.

Let $\mathbf{G} = \{G(1), \ldots, G(K)\}$, where $G(1), G(2), \ldots, G(K)$ are $m \times m$ stochastic or sub-stochastic matrices, i.e., $G(k) \geq 0$ and $G(k)\mathbf{e} \leq \mathbf{e}$, $k = 1, \ldots, K$, and satisfy the following equations, for $k = 1, \ldots, K$,

$$G(k) = A(k,0) + \sum_{x \in \aleph,\ x = k_1 k_2 \cdots k_{|x|} \neq 0} A(k,x)G(k_{|x|})G(k_{|x|-1}) \cdots G(k_1). \tag{3.152}$$

Let $\mathbf{G}^* = \{G^*(1), \ldots, G^*(K)\}$ be the minimal nonnegative solution to Eq. (3.152). According to Takine et al. (1995), the (i, j)-th element of the matrix $G^*(k)$ is the probability that the Markov chain will eventually reach the node x in the state (x, j), given that the Markov chain is in the state $(x + k, i)$ initially. It has been proved in Takine, et al. (1995) that the set $\mathbf{G}^*$ is unique and all matrices in the set $\mathbf{G}^*$ are stochastic if the Markov chain is recurrent. Define a sequence $\{G(x)[n], x \in \aleph\}$ as follows. Let $G(0)[n] = I$, for $n = 0, 1, \ldots, G(x)[0] = 0$ for $x \in \aleph$ and $x \neq 0$, and

$$G(k)[n+1] = \sum_{y = h_1 h_2 \cdots h_{|y|} \in \aleph} A(k,y)G(h_{|y|})[n]G(h_{|y|-1})[n] \cdots G(h_1)[n]. \tag{3.153}$$

It is easy to show that $\{G(k)[n], n = 0, 1, 2, \ldots\}$ is a uniformly bounded and nondecreasing sequence of matrices, for $k = 1, 2, \ldots, K$.

Let $\mathfrak{R}$ be a set of elements $\mathbf{G}$ for which $\{G(1), \ldots, G(K)\}$ are stochastic matrices and satisfy Eq. (3.152). For any set $\mathbf{G} = \{G(1), \ldots, G(K)\} \in \mathfrak{R}$, define the following $m \times m$ matrices, for $k, j = 1, \ldots, K$,

$$
\begin{aligned}
&N(0,j,\mathbf{G}) = 0,; \\
&N(k,j,\mathbf{G}) = \delta(k,j)I; \\
&N(x,j,\mathbf{G}) = \delta(k_{|x|},j)I + \sum_{n=1}^{|x|-1} G(k_{|x|})\cdots G(k_{n+1})\delta(k_n,j), \qquad (3.154)\\
&\qquad\qquad \text{for } x = k_1\cdots k_{|x|},\ |x| \geq 2; \\
&p(k,j,\mathbf{G}) = \sum_{x\in\aleph} A(k,x)N(x,j,\mathbf{G}),
\end{aligned}
$$

where $\delta(k, j) = 1$, if $k = j$; 0, otherwise. Define $(mK)\times(mK)$ matrix $P(\mathbf{G})$ by

$$
P(\mathbf{G}) = \begin{pmatrix} p(1,1,\mathbf{G}) & \cdots & p(1,K,\mathbf{G}) \\ \vdots & \vdots & \vdots \\ p(K,1,\mathbf{G}) & \cdots & p(K,K,\mathbf{G}) \end{pmatrix}. \qquad (3.155)
$$

Intuitively, the matrix $P(\mathbf{G})$ represents the mean distance the Markov chain moves away from the root node in one transition, given that the set $\mathbf{G}$ is used to represent the change of state of the auxiliary variable. For any set $\mathbf{G}$ in $\Re$, we now prove that the Perron-Frobenius eigenvalue of the matrix $P(\mathbf{G})$, $sp(P(\mathbf{G}))$, provides information for a complete classification of the Markov chain of interest.

Theorem 3.8.1 (He (2003a)) *For any set* $\mathbf{G}\in\Re$, *if the matrix* $P(G)$ *is irreducible, then the irreducible M/G/1 type Markov chain with a tree structure* $\{(X_n, J_n), n = 0, 1, 2, \ldots\}$ *is*

1. *Positive recurrent if and only if* $sp(P(\mathbf{G})) < 1$;
2. *Null recurrent if and only if* $sp(P(\mathbf{G})) = 1$; and
3. *Transient if and only if* $sp(P(\mathbf{G})) > 1$.

Example 3.8.1 Consider the case with $m = 1$. For this special case, $\mathbf{G} = \{1, 1, \ldots, 1\}$. Then the matrix $P(\mathbf{G})$ can be constructed directly from transition probabilities $\{A(k, x), k = 1, 2, \ldots, K, x \in\aleph\}$. By Theorem 3.8.1, the ergodicity of the Markov chain is obtained by computing the Perron-Frobenius eigenvalue of $P(\mathbf{G})$.

Commentary The discrete time *M/G/*1 type Markov chain with a tree structure is first introduced in Takine et al. (1995). Equations for the limiting probabilities are established in Takine et al. (1995) as well. The conditions for ergodicity are obtained in He (2000a, 2003a).

Additional Exercises and Extensions

Exercise 3.8.2 Consider the following *M/G/*1 type Markov chain with a tree structure: $K = 2$,

$$B(0) = \begin{pmatrix} 0.1 & 0 \\ 0.1 & 0.1 \end{pmatrix}, \quad B(1) = \begin{pmatrix} 0 & 0.5 \\ 0.2 & 0 \end{pmatrix}, \quad B(2) = \begin{pmatrix} 0.2 & 0.2 \\ 0.1 & 0.5 \end{pmatrix},$$

$$A(1,0) = \begin{pmatrix} 0.1 & 0.1 \\ 0.3 & 0.3 \end{pmatrix}, \quad A(1,1) = \begin{pmatrix} 0 & 0.1 \\ 0 & 0 \end{pmatrix}, \quad A(1,2) = \begin{pmatrix} 0.1 & 0 \\ 0.1 & 0 \end{pmatrix},$$

$$A(1,11) = \begin{pmatrix} 0.1 & 0.1 \\ 0 & 0 \end{pmatrix}, \quad A(1,12) = \begin{pmatrix} 0 & 0 \\ 0 & 0.1 \end{pmatrix}, \quad A(1,21) = \begin{pmatrix} 0.1 & 0 \\ 0.1 & 0 \end{pmatrix},$$

$$A(1,22) = \begin{pmatrix} 0.1 & 0 \\ 0 & 0 \end{pmatrix}, \quad A(1,111) = \begin{pmatrix} 0 & 0.1 \\ 0 & 0 \end{pmatrix}, \quad A(1,212) = \begin{pmatrix} 0.1 & 0 \\ 0.1 & 0 \end{pmatrix},$$

$$A(2,0) = \begin{pmatrix} 0.5 & 0 \\ 0.5 & 0.1 \end{pmatrix}, \quad A(2,11) = \begin{pmatrix} 0 & 0.2 \\ 0 & 0 \end{pmatrix}, \quad A(2,21) = \begin{pmatrix} 0 & 0 \\ 0.1 & 0 \end{pmatrix},$$

$$A(2,12) = \begin{pmatrix} 0.1 & 0 \\ 0.1 & 0.1 \end{pmatrix}, \quad A(2,22) = \begin{pmatrix} 0 & 0.1 \\ 0 & 0 \end{pmatrix}, \quad A(2,222) = \begin{pmatrix} 0.1 & 0 \\ 0.1 & 0 \end{pmatrix},$$

and all other transition blocks are zero.

(i) Find the all the transition matrices from node $x = 11221$ to other nodes.
(ii) Compute a set of stochastic matrices $G(1)$ and $G(2)$ satisfying Eq. (3.152).
(iii) Compute $sp(P(\mathbf{G}))$. Is the Markov chain ergodic?

Exercise 3.8.3 Consider the following *M/G/*1 type Markov chain with a tree structure: $K = 2$,

$$B(0) = \begin{pmatrix} 0.1 & 0 \\ 0.1 & 0.1 \end{pmatrix}, \quad B(1) = \begin{pmatrix} 0 & 0.5 \\ 0.2 & 0 \end{pmatrix}, \quad B(2) = \begin{pmatrix} 0.2 & 0.2 \\ 0.1 & 0.5 \end{pmatrix},$$

$$A(1,0) = \begin{pmatrix} 0 & 0.1 \\ 0.1 & 0.1 \end{pmatrix},, \quad A(1,11) = \begin{pmatrix} 0.1 & 0.1 \\ 0 & 0 \end{pmatrix}, \quad A(1,21) = \begin{pmatrix} 0.1 & 0 \\ 0.1 & 0 \end{pmatrix},$$

$$A(1,22) = \begin{pmatrix} 0.1 & 0 \\ 0 & 0 \end{pmatrix}, \quad A(1,111) = \begin{pmatrix} 0 & 0.1 \\ 0.1 & 0 \end{pmatrix}, \quad A(1,212) = \begin{pmatrix} 0.1 & 0.3 \\ 0.1 & 0.5 \end{pmatrix},$$

$$A(2,0) = \begin{pmatrix} 0.1 & 0 \\ 0.1 & 0.1 \end{pmatrix}, \quad A(2,11) = \begin{pmatrix} 0 & 0.2 \\ 0 & 0 \end{pmatrix}, \quad A(2,21) = \begin{pmatrix} 0 & 0 \\ 0.1 & 0 \end{pmatrix},$$

$$A(2,12) = \begin{pmatrix} 0.1 & 0 \\ 0.1 & 0.1 \end{pmatrix}, \quad A(2,22) = \begin{pmatrix} 0.2 & 0.1 \\ 0.2 & 0 \end{pmatrix}, \quad A(2,222) = \begin{pmatrix} 0.1 & 0.2 \\ 0.1 & 0.2 \end{pmatrix},$$

and all other transition blocks are zero.

(i) Compute a set of stochastic matrices $G(1)$ and $G(2)$ satisfying Eq. (3.152).
(ii) Compute $sp(P(\mathbf{G}))$. Is the Markov chain ergodic?

Exercise 3.8.4 Let $\mathbf{z} = (z_1, \ldots, z_K)$. Define $\mathbf{G}(\mathbf{z}) = (G(1, \mathbf{z}), G(2, \mathbf{z}), \ldots, G(k, \mathbf{z}))$ as the minimal nonnegative solution to

$$G(k, \mathbf{z}) = z_k A(k, 0) + \sum_{x \in \aleph,\ x = k_1 k_2 \cdots k_{|x|} \neq 0} A(k, x) G(k_{|x|}, \mathbf{z}) G(k_{|x|-1}, \mathbf{z}) \cdots G(k_1, \mathbf{z}). \tag{3.156}$$

Explain matrices $\{G(1, \mathbf{z}), G(2, \mathbf{z}), \ldots, G(K, \mathbf{z})\}$ intuitively.

The limiting probabilities of an $M/G/1$ type Markov chain with a tree structure are not easy to obtain. Nevertheless, there are some results similar to that in Sect. 3.5. Define $G(y) = G(h_{|y|})G(h_{|y|-1})\ldots G(h_1)$, for $y = h_1\ldots h_{|y|-1}\, h_{|y|} \in \aleph$. Define

$$\begin{aligned}\tilde{A}^{k,x} &= A(k,x) + \sum_{y\in\aleph, y\neq 0} A(k,x+y)G(y), \quad x \in \aleph;\\ \tilde{B}^{x} &= B(x) + \sum_{y\in\aleph, y\neq 0} B(x+y)G(y), \quad x \in \aleph.\end{aligned} \tag{3.157}$$

Exercise 3.8.5* (Takine et al. (1995)) Show that the limiting probabilities satisfy the following equations

$$\mathbf{x}_{x+k} = \mathbf{x}_0\tilde{B}^{x+k} + \sum_{\sigma=1}^{|x|}\sum_{i=1}^{K} \mathbf{x}_{x^+(\sigma)+i}\tilde{A}^{i,x^-(\sigma)+k} + \sum_{i=1}^{K} \mathbf{x}_{x+i}\tilde{A}^{i,k}, \quad x \in \aleph, \tag{3.158}$$

where $\mathbf{x}_0$ satisfies $\mathbf{x}_0\tilde{B}^{0} = \mathbf{x}_0$ and is so normalized that the total probability is one.

Exercise 3.8.6* (He (2003a)) Assume that $m = 1$. Consider the mapping $\mathcal{A}$ defined in Eq. (3.148) with $A(x, k)$ being replaced with $A(k, x)$. Show that $\mathbf{e}'$ is a fixed point of $\mathcal{A}$, i.e., $\mathcal{A}(\mathbf{e}') = \mathbf{e}'$. Show that

$$p(k,j,\mathbf{e}') = \left.\frac{\partial a_k^*(\mathbf{u})}{\partial u_j}\right|_{\mathbf{u}=\mathbf{e}'}. \tag{3.159}$$

Assume that $P(\mathbf{e}')$ is irreducible. Show that, if $sp(P(\mathbf{e}')) \leq 1$, $\mathbf{e}'$ is the minimal nonnegative fixed point of $\mathcal{A}$ in R_+^K defined in Eq. (3.147).

Exercise 3.8.7 Consider a QBD process with a tree structure defined in Section 3.6. Assume that $K = 2$, $m = 1$, $A_0(1) = 0.1$, $A_0(2) = 0.3$, $A_1(1) = 0.2$, $A_2(1) = 0.4$, $A_1(2) = 0.0$, $A_2(2) = 0.6$, and $A_{0,0} = 0.6$. Find the matrix $P(\mathbf{e}')$ and $sp(P(\mathbf{e}'))$. Is the Markov chain ergodic?

3.9 QBD Type Markov Chains with Infinitely Many Background Phases

We consider the QBD process $\{(X_k, J_k), k = 0, 1, 2, \ldots\}$ with transition probability matrix defined in Eq. (3.24). In this section, we assume $m = \infty$ and $m_0 \leq \infty$. We focus on the limiting probabilities $\boldsymbol{\pi} = (\boldsymbol{\pi}_0, \boldsymbol{\pi}_1, \ldots,)$, especially the tail asymptotics of the limiting probabilities (i.e., limits related to the vector sequence $\{\boldsymbol{\pi}_0, \boldsymbol{\pi}_1, \ldots, \boldsymbol{\pi}_n, \ldots\}$). The analysis of such Markov chains requires advanced mathematical tools. We only present a brief introduction to the issues of interest in this section.

We first summarize some results parallel to the case with $m < \infty$. Details are referred to Sects. 3.2 and 3.3.

1. The rate matrix R and matrix G can be defined similarly. The same probabilistic interpretations hold for R and G.
2. The matrix-geometric solution exists if the Markov chain is ergodic (Theorem 3.2.2 in Sect. 3.2) (see Miller (1981)).

The matrices R and G and the matrix-geometric solution can be found explicitly only for some special cases if $m = \infty$. For the general case, there are a number of new issues to be addressed. We illustrate the issues with a few simple examples.

Example 3.9.1 Consider a *tandem queue* with two stages: $M/M/1 \to /M/1$ (see definition of tandem queues in, e.g., Chao et al. (1999)). Customers arrive from outside to queue 1 and join the queue there. Upon the completion of its service at queue 1, a customer proceeds to queue 2 and joins the queue there. Customers arrive according to a Poisson process with arrival rate λ. Service times are all exponentially distributed with service rate μ_1 at queue 1 and service rate μ_2 at queue 2. It is well-known that the queueing model is a simple *Jackson network* (Jackson (1963)).

Let $q_1(t)$ be the queue length in queue 1 (including the one in service, if there is one), and $q_2(t)$ be the queue length in queue 2. It can be shown that process $\{(q_1(t), q_2(t)), t \geq 0\}$ is a continuous time Markov chain with infinitesimal generator Q given by

$$Q = \begin{pmatrix} -(\lambda+\mu_2)I + \mu_2(J' + \mathbf{e}_1\mathbf{e}'_1) & \lambda I & & \\ \mu_1 J & A_1 & \lambda I & \\ & \mu_1 J & A_1 & \lambda I \\ & & \ddots & \ddots & \ddots \end{pmatrix}, \quad (3.160)$$

where $A_1 = -(\lambda+\mu_1+\mu_2)I + \mu_2(J' + \mathbf{e}_1\mathbf{e}'_1)$, $\mathbf{e}_1 = (1, 0, 0, \ldots)'$,

$$I = \begin{pmatrix} 1 & & & \\ & 1 & & \\ & & \ddots & \\ & & & \ddots \end{pmatrix}, \quad J = \begin{pmatrix} 0 & 1 & & \\ & 0 & 1 & \\ & & \ddots & \ddots \\ & & & \ddots \end{pmatrix}. \quad (3.161)$$

Let $\rho_1 = \lambda/\mu_1$ and $\rho_2 = \lambda/\mu_2$. Assume that $\rho_1 < 1$ and $\rho_2 < 1$. Then the Markov chain $\{(q_1(t), q_2(t)), t \geq 0\}$ is ergodic. The limiting probabilities of $\{(q_1(t), q_2(t)), t \geq 0\}$ has a *product-form solution* given as follows, for $n_1, n_2 = 0, 1, 2, \ldots$,

$$\begin{aligned} \pi(n_1, n_2) &= \lim_{t\to\infty} P\{q_1(t) = n_1, q_1(t) = n_2\} \\ &= (1-\rho_1)\rho_1^{n_1}(1-\rho_2)\rho_2^{n_2}. \end{aligned} \quad (3.162)$$

Define $P = I + Q/(\lambda+\mu_1+\mu_2)$. Then P is a discrete time QBD process with infinitely many background phases. It is easy to see that P and Q have the same limiting probabilities. By Eq. (3.162), it is clear that

$$\boldsymbol{\pi}_0 = (1-\rho_1)(1-\rho_2)\left(1, \rho_2, \rho_2^2, \rho_2^3, \ldots\right) \tag{3.163}$$

and $\boldsymbol{\pi}_n = \rho_1{}^n \boldsymbol{\pi}_0$, $n = 0, 1, 2, \ldots$. Thus, for the Markov chain P, an explicit solution exists for the limiting probabilities satisfying $\boldsymbol{\pi}P = \boldsymbol{\pi}$ and $\boldsymbol{\pi}\mathbf{e} = 1$.

Exercise 3.9.1 By routine calculations, verify that $\lambda\boldsymbol{\pi}_0 + \boldsymbol{\pi}_1 A_1 + \mu_1 \boldsymbol{\pi}_2 J = 0$.

Let R be the minimal nonnegative solution to $\lambda I + RA_1 + \mu_1 R^2 J = 0$.

Exercise 3.9.2 Assume that $I - R$ is invertible, which is true under the condition $\rho_1 < 1$ and $\rho_2 < 1$. Show that $R\mathbf{e} = \rho_1 \mathbf{e}$.

The above results indicate

$$\boldsymbol{\pi}_n = \boldsymbol{\pi}_0 R^n = \boldsymbol{\pi}_0 \rho_1^n, \quad n = 1, 2, \ldots. \tag{3.164}$$

The expression in Eq. (3.164) can be written in a more general form: there exists $\alpha > 1$ such that

$$\lim_{n\to\infty} \alpha^n \boldsymbol{\pi}_n = \mathbf{c}, \tag{3.165}$$

where $\mathbf{c}$ is a positive vector. If Eq. (3.165) holds, we say that the limiting probabilities have a *geometric decay* along the level direction and the *decay rate* is $1/\alpha$.

For QBD processes with a finite number of background phases, it is well-known that Eq. (3.165) holds if the Markov chain is ergodic and R is irreducible. The decay rate is the Perron-Frobenius eigenvalue of R, i.e., $\alpha = 1/sp(R)$.

However, for a QBD process with infinitely many background phases, $\boldsymbol{\pi}_n = \boldsymbol{\pi}_0 R^n$ may hold but Eq. (3.165) may fail. The issue of tail asymptotics becomes far more complicated. There are several fundamental issues that differentiate the cases with finite and infinite numbers of background phases.

(i) Since $m = \infty$, computations cannot be done for general QBD processes, except for special cases such as Example 3.9.1.
(ii) Truncation methods can be applied, but there is no guarantee that the solutions obtained by truncations approximate the original solution (e.g., Bean and Latouche (2010) and Latouche et al. (2011).)
(iii) The spectrum of R can be continuous (e.g., Kroese et al. (2004).)
(iv) Boundary transitions have little effect on the tail asymptotics for the case with finitely many background phases, but may affect the tail asymptotics for the case with infinitely many background phases (e.g., Kroese et al. (2004) or Theorem 3.9.1.)

The following example demonstrates the effect of boundary conditions on the tail asymptotics.

Example 3.9.2 Consider the tandem queue introduced in Example 3.9.1. However, for this example, we assume that the arrival rate to the second server is state dependent, if queue 1 is empty (i.e., $q_1(t) = 0$). That is, in the tandem queue, if the first stage is empty, then an external arrival process to the second stage is turned on and its arrival rate depends on the queue length in the second stage. This external arrival process is turned off if the first server starts to serve customers. More specifically, $A_0 = A_{0,1} = \lambda I$, $A_1 = -(\lambda+\mu_1+\mu_2)I + \mu_2(J' + \mathbf{e}_1\mathbf{e}'_1)$, $A_2 = \mu_1 J$, and

$$A_{0,0} = -\lambda I + \begin{pmatrix} -\lambda_0 & \lambda_0 & & & \\ \mu_2 & -(\lambda_1+\mu_2) & \lambda_1 & & \\ & \mu_2 & -(\lambda_2+\mu_2) & \lambda_2 & \\ & & \ddots & \ddots & \ddots \end{pmatrix}, \tag{3.166}$$

where $\{\lambda_0, \lambda_1, \lambda_2, \ldots\}$ are the arrival rates to server 2, conditioning on the number of customers in the second queue. We assume that the Markov chain is ergodic (or the queueing network is stable). The following results can be obtained (Kroese et al. (2004)). Recall that the rate matrix R, in Example 3.9.1, is the minimal nonnegative solution to equation $\lambda I + R(-(\lambda + \mu_1 + \mu_2)I + \mu_2(J' + \mathbf{e}_1\mathbf{e}'_1)) + \mu_1 R^2 J = 0$. Then the rate matrix R is independent of the arrival rates $\{\lambda_0, \lambda_1, \lambda_2, \ldots\}$.

1. The linear system $\mathbf{w}R = z\mathbf{w}$ has solutions $\mathbf{w} \in l^1$ for all $z \in (0, \min\{1, \mu_1/\mu_2\})$. (Note: $\mathbf{w} \in l^1$ means $|w_1| + |w_2| + |w_3| + \ldots < \infty$.) This indicates that the spectrum of R contains a continuous part.
2. Define $\tau(z) = -\lambda - \mu_1 - \mu_2(1 - z) + 2\sqrt{\lambda\mu_1/z}$. There is a unique value η in (0, 1) with $\tau(\eta) = 0$.
3. When $\mu_1 \le \mu_2$, the linear system $\mathbf{w}R = z\mathbf{w}$ has positive solutions $\mathbf{w} \in l^1$ for all $z \in (\eta, \mu_1/\mu_2)$; when $\mu_1 > \mu_2$, the linear system $\mathbf{w}R = z\mathbf{w}$ has positive solutions $\mathbf{w} \in l^1$ for all $z \in (\rho_2, 1)$.
4. Assume $\mu_1 > \mu_2$. For $z \in (\rho_2, 1)$, define

$$\begin{aligned} \lambda_0 &= \mu_2 z, \\ \lambda_i &= \lambda_{i-1}\frac{w_{i-1}}{w_i} + \mu_2 z - \mu_1, \quad i = 1,\ 2,\ \ldots. \end{aligned} \tag{3.167}$$

The sequence $\{\lambda_0, \lambda_1, \lambda_2, \ldots\}$ is strictly positive. Further, $\mathbf{w}(A_{0,0} + RA_2) = 0$. Based on the above results, we obtain the following theorem.

Theorem 3.9.1 (Kroese et al. (2004)) *In Example* 3.9.2, *assume that* $\rho_1 < 1$, $\rho_2 < 1$, *and* $\mu_1 > \mu_2$. *For* $z \in (\rho_2, 1)$, *define a Markov chain P using* $\{\lambda_0, \lambda_1, \lambda_2, \ldots\}$ *in* $A_{0,0}$. *Then the limiting probabilities are given by* $\boldsymbol{\pi}_n = c\mathbf{w}R^n = z^n c\mathbf{w}$, $n = 0, 1, 2, \ldots$, *for some normalizing constant c*.

Theorem 3.9.1 indicates that the decay rate of the limiting probabilities may change if the transitions within level zero change. It also shows that the decay rate is not uniquely determined by R, which is different from the case for which Markov chains have a finite number of background phases.

Due to the complexity of the issues involved, we only present two sufficient conditions for Eq. (3.165) for QBD Markov chains with infinitely many background states. We begin with some new concepts. A matrix A is said to be α-positive for positive number α if αA has nonnegative nonzero right and left invariant vectors $\mathbf{u}$ and $\mathbf{v}$, respectively, such that $\mathbf{uv} < \infty$. Define, for $z > 0$,

$$A^*(z) = z^{-1}A_2 + A_1 + zA_0. \tag{3.168}$$

We impose the following conditions on the Markov chains.

(i) $A = A_2 + A_1 + A_0$ is irreducible and aperiodic.
(ii) $\{A_n, n = 0, 1, 2\}$ is 1-*arithmetic* in the sense that for every pair (i, j), the greatest common divisor of $\{n\text{: } (A_n)_{(i,j)} > 0, n = 1, 2, \ldots\}$ is one.
(iii) There exist $\alpha > 1$, a positive row vector $\mathbf{x}$, and a positive column vector $\mathbf{y}$ such that $\mathbf{x}A^*(\alpha) = \mathbf{x}$ and $A^*(\alpha)\mathbf{y} = \mathbf{y}$.
(iv) $\mathbf{xy} < \infty$ for $\mathbf{x}$ and $\mathbf{y}$ in (iii).
(v) $\boldsymbol{\pi}_0 A_{0,0}\mathbf{y} < \infty$ for $\mathbf{y}$ in (iii).

Theorem 3.9.2 (Li et al. (2007)) *Under conditions* (i) – (v), *we have*

$$\lim_{n\to\infty} \alpha^n \boldsymbol{\pi}_n = \frac{\alpha\boldsymbol{\pi}_1\mathbf{r}}{\mathbf{xr}}\mathbf{x} < \infty, \tag{3.169}$$

where $\mathbf{r} = (I - A_1 - RA_2 - \alpha^{-1}A_2)\mathbf{y}$.

We have a look at the limit in Eq. (3.169) intuitively. Define

$$\tilde{G}_+^{(\alpha)} = (\text{diag}(\mathbf{x}))^{-1}(\alpha R')\text{diag}(\mathbf{x}). \tag{3.170}$$

Exercise 3.9.3 Show that $\tilde{G}_+^{(\alpha)}$ is a stochastic matrix. Suppose that the stochastic matrix $\tilde{G}_+^{(\alpha)}$ has a single irreducible class. Then the limiting probabilities are given by $(\mathbf{xy})^{-1}\mathbf{y}'\text{diag}(\mathbf{x})$, and

$$\lim_{n\to\infty} (\text{diag}(\mathbf{x}))^{-1}(\alpha R')^n\text{diag}(\mathbf{x}) = \lim_{n\to\infty} \left(\tilde{G}_+^{(\alpha)}\right)^n = \frac{1}{\mathbf{xy}}\mathbf{e}\mathbf{y}'\text{diag}(\mathbf{x}). \tag{3.171}$$

Then

$$\lim_{n\to\infty} \alpha^n \boldsymbol{\pi}_n = \alpha \lim_{n\to\infty} \left(\text{diag}(\mathbf{x})\tilde{G}_+^{(\alpha)}\text{diag}^{-1}(\mathbf{x})\boldsymbol{\pi}'_1\right)^n = \frac{\alpha\boldsymbol{\pi}_1\mathbf{y}}{\mathbf{xy}}\mathbf{x} \tag{3.172}$$

Verifying conditions for Theorem 3.9.2 for Markov chains is usually not easy. The following theorem provides a little more details on the boundary conditions.

Theorem 3.9.3 (Li et al. (2007)) *Assume that there exist* $\alpha > 1$, *a positive row vector* $\mathbf{x}$ *such that* $\mathbf{x}A^*(\alpha) = \mathbf{x}$. *Also assume*

$$\lim_{k\to\infty} \frac{\pi_{1,k}}{x_k} = c. \tag{3.173}$$

Then, for any nonnegative column vector $\mathbf{h}$ *satisfying* $\mathbf{xh} < \infty$,

$$\lim_{n\to\infty} \alpha^n \boldsymbol{\pi}_n \mathbf{h} = \alpha c \mathbf{xh}. \tag{3.174}$$

In Theorem 3.9.3, if c is finite, then $\boldsymbol{\pi}_n\mathbf{h}$ decays geometrically with rate $1/\alpha$. In particular, choose $\mathbf{h} = (0, \ldots, 0, 1, 0, \ldots)'$. Then Theorem 3.9.3 indicates that $\{\pi_{n,j}, n = 0, 1, 2, \ldots\}$ decays geometrically for any fixed j.

A useful tool in the study of tail asymptotics using matrix-analytic methods is given in the following exercise.

Exercise 3.9.4* (Ramaswami and Taylor (1996)) Assume that $\mathbf{w} \in l^1$ and $|z| < 1$. The vector $\mathbf{w}$ and scalar z satisfy $\mathbf{w}R = z\mathbf{w}$ if and only if $\mathbf{w} = \mathbf{w}A^*(z)$.

Commentary Corollary 3.2.2 gives tail asymptotics results if there are a finite number of background phases. For that case, the solution is fairly complete. For structured Markov chains with an infinite number of background phases, the issue is far more complicated. This issue is further discussed in Sect. 3.10.6.

Additional Exercises and Extensions

Example 3.9.3 (Priority queues) Consider a priority $M/M/1$ queue with two types of customers and a preemptive repeat service discipline. The arrivals of the customers are Poisson processes with rates of high and low priority customers being λ_1 and λ_2, respectively. The service times are exponential with rates of high and low priority customers being μ_1 and μ_2, respectively. Let $q_1(t)$ be the total number of high priority customers in the system (including the one in service, if there is one), and $q_2(t)$ be the total number of low priority customers. The process $\{(q_1(t), q_2(t)), t \geq 0\}$ is a continuous time Markov chain with infinitesimal generator Q given by

$$Q = \begin{pmatrix} -(\lambda_1+\lambda_2+\mu_2)I + \lambda_2 J + \mu_2(J' + \mathbf{e}_1\mathbf{e}'_1) & \lambda_1 I & & \\ \mu_1 I & A_1 & \lambda_1 I & \\ & \mu_1 I & A_1 & \lambda_1 I \\ & & \ddots & \ddots & \ddots \end{pmatrix}. \tag{3.175}$$

where $A_1 = -(\lambda + \mu_1 + \lambda_2)I + \lambda_2 J$. It is well-known that the Markov chain is ergodic if and only if $\rho_1 + \rho_2 < 1$. Define $P = I + Q/(\lambda_1+\lambda_2+\mu_1+\mu_2)$, which defines a discrete time QBD process with an infinite number of background phases.

Exercise 3.9.5 (Example 3.9.3 continued) For P, explain intuitively that the rate matrix R is an upper triangular matrix. Further, show

$$R = \begin{pmatrix} r_0 & r_1 & r_2 & \cdots \\ & r_0 & r_1 & \ddots \\ & & \ddots & \ddots \\ & & & \ddots \end{pmatrix}, \tag{3.176}$$

where $\{r_0, r_1, r_2, \ldots\}$ are positive and satisfy the following equations

$$\begin{aligned} &\lambda_1 - (\lambda_1 + \mu_1 + \lambda_2) r_0 + \mu_1 r_0^2 = 0; \\ &-(\lambda_1 + \mu_1 + \lambda_2) r_n + \mu_1 \sum_{i=0}^{n} r_i r_{n-i} = 0, \quad n = 1, 2, 3, \ldots. \end{aligned} \tag{3.177}$$

Exercise 3.9.6 (Exercise 3.9.5 continued) Let $r = r_0 + r_1 + r_2 + \ldots$. Show that

$$\begin{aligned} r_0 &= \frac{\lambda_1 + \mu_1 + \lambda_2 - \sqrt{(\lambda_1 + \mu_1 + \lambda_2)^2 - 4\lambda_1 \mu_1}}{2\mu_1}; \\ r &= \frac{\lambda_1 + \mu_1 + \lambda_2 + \sqrt{(\lambda_1 + \mu_1 + \lambda_2)^2 - 4\lambda_1 \mu_1}}{2\mu_1}. \end{aligned} \tag{3.178}$$

(Note: R is the minimal nonnegative solution whose elements satisfy Eq. (3.177).)

3.10 Additional Topics

The study of structured Markov chains has been so extensive that only a small number of basic results are collected in this chapter. As demonstrated in this chapter, the study in this area typically involves (i) introduction of a special structure; (ii) defining matrices of R or G type; (iii) basic theory; (iv) probabilistic interpretation; and (v) computational methods. In this section, we first have some discussion on further issues related to R and G. Then we briefly discuss studies related to more general Markov chains.

3.10.1 Computation of the Rate Matrix R and the Matrix G

Both R and G play an important role in the theory of matrix-analytic methods and in the computation of performance measures. A number of efficient methods have been developed for computing the two matrices.

1. The monotonic iterative methods given in Sect. 3.3 are among the early methods used for computing R and G. See Ramaswami (1988b) for more details on such methods.
2. The logarithmic reduction method developed in Latouche and Ramaswami (1993) is particularly efficient for the QBD case. It is worth to mention that the probabilistic interpretation of the logarithmic reduction method is particularly intriguing.
3. The cyclic reduction method is summarized in Bini et al. (2005).

4. In addition, for special stochastic models such as the $MAP/G/1$ queue and $GI/PH/1$ queue, special methods are developed for computing R and G, especially for queueing models (Lucantoni and Ramaswami (1985) and Sengupta (1989)).

3.10.2 Duality Between R and G

Consider a sequence of nonnegative matrices $\{A_0, A_1, A_2, \ldots\}$. Assume that $A = A_0 + A_1 + A_2 + \ldots$ is stochastic. According to Sects. 3.4 and 3.5, the sequence can be used to construct $M/G/1$ type and $GI/M/1$ type Markov chains. The relationship between the matrices R and G corresponding to the Markov chains is called the duality between R and G. A specific dual relationship between R and G is shown as follows.

Assume that A is an irreducible stochastic matrix. Then there is positive $\boldsymbol{\theta}$ such that $\boldsymbol{\theta}A = \boldsymbol{\theta}$ and $\boldsymbol{\theta}\mathbf{e} = 1$.

Exercise 3.10.1 Let $C_n = (\text{diag}(\boldsymbol{\theta}))^{-1}A_n{}'\text{diag}(\boldsymbol{\theta})$, $n = 0, 1, 2, \ldots$. Show that the sum of $\{C_0, C_1, C_2, \ldots\}$ is a stochastic matrix.

Suppose that R is the minimal nonnegative solution to Eq. (3.77). Then it is easy to show

$$\begin{aligned}&\text{diag}^{-1}(\boldsymbol{\theta})R'\text{diag}(\boldsymbol{\theta})\\ &\quad=\sum_{n=0}^{\infty}\text{diag}^{-1}(\boldsymbol{\theta})(A_n)'\text{diag}(\boldsymbol{\theta})\left(\text{diag}^{-1}(\boldsymbol{\theta})R'\text{diag}(\boldsymbol{\theta})\right)^n.\end{aligned} \tag{3.179}$$

That is, if R is the rate matrix corresponding to $\{A_0, A_1, A_2, \ldots\}$, then $(\text{diag}(\boldsymbol{\theta}))^{-1}R'\text{diag}(\boldsymbol{\theta})$ is the G matrix of the sequence $\{C_0, C_1, C_2, \ldots\}$, which is called the dual of R.

According to the duality theory, the computation of R and G is reduced to the computation of one of them. Usually, G is computed first since the matrix G is stochastic or substochastic, and it is easier to do error control in computation. There are different types of duality between R and G. Readers are referred to Asmussen and Ramaswami (1990) and Ramaswami (1990) for the duality theory on R and G.

3.10.3 Level Dependent QBD (LDQBD) Processes

Level dependent QBD processes have wide applications. However, how to compute the limiting probabilities for such QBD processes is a challenging problem. A key issue is to determine the truncation point. In Bright and Taylor (1995), an algorithm is proposed for computing the limiting probabilities.

3.10.4 The *GI/G*/1 Type Markov Chains and *RG*-Factorization (Wiener-Hopf Factorization)

A *GI/G*/1 type Markov chain has the following structure

$$P = \begin{pmatrix} B_0 & B_1 & B_2 & B_3 & B_4 & \cdots \\ B_{-1} & A_0 & A_1 & A_2 & A_3 & \cdots \\ B_{-2} & A_{-1} & A_0 & A_1 & A_2 & \cdots \\ B_{-3} & A_{-2} & A_{-1} & A_0 & A_1 & \cdots \\ \vdots & \vdots & \ddots & \ddots & \ddots & \ddots \end{pmatrix}. \tag{3.180}$$

Define $L_n = \{(n, 1), (n, 2), \ldots, (n, m)\}$, $L_{\le n} = \bigcup_{l=0}^{n} L_n$, and $L_{\ge n} = \bigcup_{l=n}^{\infty} L_n$, for $n = 0, 1, 2, \ldots$. We define R-measures and G-measures as follows: $R_{l,n}$, for $l < n$, and $G_{l,n}$, for $l > n$. That $R_{l,n}$ is an $m \times m$ matrix whose (i, j)-th element is the expected number visits to state (n, j) before hitting any state in $L_{\le(n-1)}$, given that the process starts in state (l, i). That $G_{l,n}$ is an $m \times m$ matrix whose (i, j)th element is the probability of hitting state (n, j) when the process enters $L_{\le(n-1)}$ for the first time, given that the process starts in state (l, i). By the repeating structure of the Markov chain, we can write $R_{n-l} = R_{l,n}$ and $G_{n-l} = G_{n,l}$. Thus, there are only two sequences of matrices $\{R_n, n = 1, 2, \ldots\}$ and $\{G_n, n = 1, 2, \ldots\}$ involved in the study of the *GI/G*/1 type Markov chains. Let Φ_0 be an $m \times m$ matrix whose (i, j)-th element is the probability of hitting state $(0, j)$ when the process enters level zero for the first time, given that the process starts in state $(0, i)$. The following Wiener-Hopf factorization is a fundamental and useful result

$$I - \sum_{n=-\infty}^{\infty} z^n A_n = \left(I - \sum_{n=1}^{\infty} z^n R_n\right)(I - \Phi_0)\left(I - \sum_{n=1}^{\infty} z^{-n} G_n\right). \tag{3.181}$$

For the limiting probabilities $\mathbf{p} = (\mathbf{p}_0, \mathbf{p}_1, \mathbf{p}_2, \ldots)$, if they exist, define $\mathbf{p}^*(z) = \mathbf{p}_0 + z\mathbf{p}_1 + z^2\mathbf{p}_2 + \ldots$, for $0 \le z \le 1$. It can be shown

$$\mathbf{p}^*(z)\left(I - \sum_{n=1}^{\infty} z^n R_n\right) = \mathbf{p}_0 \sum_{n=1}^{\infty} z^n R_{0,n}. \tag{3.182}$$

which is also useful in the study of the *GI/G*/1 type Markov chains. More details on *GI/G*/1 type Markov chains can be found in Zhao et al. (1998, 2003). The *RG*-factorization has been generalized to *GI/G*/1 type Markov chains with infinitely many background phases. Such relationships can be used in tail asymptotics of the limiting probabilities (Miyazawa and Zhao (2004)). A comprehensive treatment of *RG*-factorization in matrix-analytic methods can be found in Li (2010).

3.10.5 Matrix-Exponential Solution: *GI/M/1* Type and *M/G/1* Type Markov Chains

In Sengupta (1989), the continuous version of the Markov chain studied in Sect. 3.4 (*GI/M/*1 type) is introduced and investigated. For the Markov process considered in Sengupta (1989), the time is continuous and the state space of the level variable is continuous as well.

Define a bivariate continuous time Markov process $\{(X_t, N_t), t \geq 0\}$ with the following properties.

1. The stochastic process $\{X_t, t \geq 0\}$ is skip-free to the right and takes values in $[0, \infty]$. It increases at a linear rate of 1, provided there are no downward jumps.
2. Changes in state of $\{(X_t, N_t), t \geq 0\}$ can also take place in one of the following two ways.
   (i) If $(X_t, N_t) = (x, i)$, it can change its state to somewhere between $(x-u, j)$ and $(x-u + \mathrm{d}u, j)$ at a rate of $\mathrm{d}A_{i,j}(u)$, where $0 \leq u < x$ and $1 \leq i, j \leq m$. Let $A(u)$ be an $m \times m$ matrix whose (i, j)-th element is $A_{i,j}(u)$. Let $A = A(\infty)$ with $A_{i,j} = A_{i,j}(\infty)$. We assume that A is an irreducible matrix. To avoid trivialities, we assume that $A_{i,i}(0) = 0$ for $1 \leq i \leq m$.
   (ii) If $(X_t, N_t) = (x, i)$, it can change its state to $(0, j)$ at a rate of $B_{i,j}(x)$, where $1 \leq i, j \leq m$ and $x > 0$. Let $B(x)$ be an $m \times m$ matrix whose (i, j)-th element is $B_{i,j}(x)$. We assume that $B(\infty) = 0$.
3. The matrices $A(x)$ and $B(x)$ satisfy the condition

$$\sum_{j=1}^{m} A_{i,j}(x) + \sum_{j=1}^{m} B_{i,j}(x) = -D_{0,i}, \tag{3.183}$$

for all $x > 0$ and $1 \leq i \leq m$. Let $D_0 = \mathrm{diag}(D_{0,1}, \ldots, D_{0,m})$.

Let $\pi_1(x)$ be the steady state density (if it exists) of $\{(X_t, N_t), t \geq 0\}$, i.e., for $i = 1, \ldots, m$,

$$\pi_i(x)\mathrm{d}x = \lim_{t \to \infty} \frac{1}{t} \int_0^t P\{x < X_u \leq x + \mathrm{d}x, N_u = i\}\mathrm{d}u, \tag{3.184}$$

with $\boldsymbol{\pi}(x) = (\pi_1(x), \pi_2(x), \ldots, \pi_m(x))$. If the Markov process is irreducible and positive recurrent, the following results have been shown.

Theorem 3.10.1 (Sengupta (1989)) *For the continuous time Markov chain* $\{(X_t, N_t), t \geq 0\}$, *we have*

$$\begin{aligned} \boldsymbol{\pi}(x) &= \boldsymbol{\pi}(0)\exp(Tx), \quad x \geq 0; \\ T &= D_0 + \int_0^\infty \exp(Tu)\mathrm{d}A(u), \end{aligned} \tag{3.185}$$

which is called a matrix-exponential solution.

In Takine (1996), a continuous-time bivariate Markov process with the skip-free to the left property ($M/G/1$ type) is introduced. A necessary and sufficient condition for the Markov chain to be positive recurrent has been obtained.

3.10.6 Structured Markov Chains with Infinitely Many Background Phases and Tail Asymptotics

For many applications (e.g., Examples 3.9.1–3.9.3), the number of phases in a level is infinite. This type of Markov chains is called structured Markov chains with infinitely many *(but countable)* background states. Such Markov chains are further categorized into *QBD, GI/M/1, M/G/1,* and *GI/G/1* types.

For the QBD and $GI/M/1$ type Markov chains with infinitely many background phases, the matrix-geometric solution exists for the limiting probabilities (Miller (1981)). The rate matrix R is infinite in size. Ramaswami and Taylor (1996) offer some studies on R. For QBD and $M/G/1$ type Markov chains with infinite blocks, the matrix G can be defined. In the probabilistic sense, the theory developed for structured Markov chains with finite background phases can be generalized to the case with infinite background phases. However, the two cases are fundamentally different in the computational sense. In essence, the matrix-geometric solution with an infinite matrix R is not computable. To make things even worse, truncation may or may not work for computing limiting probabilities. Thus, tail asymptotics becomes an important topic in this area.

The matrix-geometric solution leads to a geometric tail with decay rate $sp(R)$ (see Corollary 3.2.2) for the case with finitely many background phases. For the case with infinitely many background phases, the issue is far more complicated. The tail can be in different forms such as (i) *geometric/nongeometric*; and (ii) *light-tailed/heavy-tailed*. The tail asymptotics can be *exact* or *non-exact geometric*. The tail asymptotics not only depends on the spectrum of R, but also depends on the transitions associated with boundary states. Recent results can be found in Takahashi et al. (2001), Kroese et al. (2004), Miyazawa and Zhao (2004), Tang and Zhao (2008), He et al. (2009), Miyazawa (2009), etc. Interesting issues and results can be summarized in the following topics.

1. Boundary effects on tail asympotics (Kroese et al. (2004)).
2. Double QBD (Miyazawa (2009)), in which the tail asymptotics are found for all cases.
3. QBD with irreducible/reducible $A = A_0 + A_1 + A_2$ (Li et al. (2007) and Ozawa (2012)).
4. Conditions for the exact geometric decay for $GI/G/1$ type Markov chains with infinitely many background phases (Miyazawa and Zhao (2004)).

There are other approaches to the tail asymptotics of Markov chains with infinitely many background phases. For instance, a probabilistic approach is taken by Foley and McDonald (2001).

Nielsen and Ramaswami (1997) consider QBD processes with a continuous phase variable.

3.10.7 Fluid Processes and Markov Modulated Fluid Processes

For many applications, including queueing and risk models, fluid processes are used as input processes. Structures can be introduced into fluid processes. Thus, a natural generalization of *MAP*s is Markov modulated fluid process. Research in this area has flourished in the past decade. Readers are referred to Ramaswami (1999) and Ahn and Ramaswami (2003) for details.

3.10.8 Time Dependent Structured Markov Chains

When the parameters of a Markov chain are time dependent, we call it a time dependent Markov chain. In general, it is difficult to analyze such Markov chains. For some special cases (e.g., periodic), such Markov chains have been investigated (Masuyama and Takine (2005) and Margolius (2007)).

References

Ahn S, Ramaswami V (2003) Fluid flow models and queues: a connection by stochastic coupling. Stoch Models 19:325–348

Asmussen S (2003) Applied probability and queues, 2nd edn. Springer, New York

Asmussen S, Ramaswami V (1990) Probabilistic interpretation of some duality results of the matrix paradigms in queueing theory. Stoch Models 6:715–733

Bean N, Latouche G (2010) Approximations to quasi-birth-and-death processes with infinite blocks. Adv Appl Prob 42:1102–1125

Bini DA, Meini B (1996) On the solution of a nonlinear matrix equation arising in queueing problems. SIAM J Matrix Anal Appl 17:906–926

Bini DA, Latouche G, Meini B (2005) Numerical methods for structured Markov chains. Oxford University Press, Oxford

Bright LW, Taylor PG (1995) Calculating the equilibrium distribution in level dependent quasi-birth-and-death processes. Stoch Models 11:497–525

Chao XL, Miyazawa M, Pinedo M (1999) Queueing networks: customers, signals, and product form solutions. Wiley, New York

Cohen JW (1982) The single server queues. North-Holland, Amsterdam

Evans RV (1967) Geometric distribution in some two-dimensional queueing systems. Oper Res 15:830–846

Foley R, McDonald D (2001) Join the shortest queue: stability and exact asymptotics. Ann Appl Prob 11:569–607

Gail HR, Hantler SL, Taylor BA (1994) Solutions of the basic matrix equation for *M*/*G*/1 and *G*/*M*/1 type Markov chains. Stoch Models 10:1–43

Gail HR, Hantler SL, Taylor BA (1997) Non-skip-free *M/G/*1 and *G/M/*1 type Markov chains. Adv Appl Prob 29:733–758

Grassmann WK (2000) Computational probability. Kluwer, Boston

Grassmann WK, Taksar MI, Heyman DP (1985) Regenerative analysis and steady state distributions for Markov chains. Oper Res 33:1107–1116

He QM (1999) Partial orders and the matrices *R* in matrix analytic methods. SIAM J Matrix Anal Appl 20:871–885

He QM (2000a) Classification of Markov chains of *M/G/*1 type with a tree structure and its applications to queueing models. Oper Res Lett 26:67–80

He QM (2000b) Quasi-birth-and-death Markov processes with a tree structure and the *MMAP*[*K*]/*PH*[*K*]/1 queue. Eur J Oper Res 120:641–656

He QM (2003a) A fixed point approach to the classification of Markov chains with a tree structure. Stoch Models 19:76–114

He QM (2003b) The classification of Markov chains of matrix *GI/M/*1 type with a tree structure and its queueing applications. J Appl Prob 40:1087–1102

He Q-M, Li H, Zhao Y (2009) Light-tailed behaviour in QBD processes with countably many phases. Stoch Models 25:50–75

Hsu GH, He QM (1991) The distribution of the first passage time for the Markov processes of *GI/M/*1 type. Stoch Models 7:397–417

Jackson JR (1963) Jobshop-like queueing systems. Manag Sci 10:131–142

Karlin S, Taylor HM (1975) A first course in stochastic processes. Academic, San Diego

Karlin S, Taylor HM (1981) A second course in stochastic processes. Academic, San Diego

Kemeny JG, Snell JL, Knapp AW (1976) Denumerable Markov chains. Springer, New York

Kingman JFC (1961) A convexity property of positive matrices. Q J Math 12:283–284

Kroese DP, Scheinhardt WRW, Taylor PG (2004) Spectral properties of the tandem Jackson network, seen as a quasi-birth-and-death process. Ann Appl Prob 14:2057–2089

Latouche G (1987) A note on two matrices occurring in the solution of quasi-birth-and-death processes. Stoch Models 3:251–257

Latouche G, Ramaswami V (1993) A logarithmic reduction algorithm for quasi-birth-and-death process. J Appl Prob 30:650–674

Latouche G, Ramaswami V (1999) Introduction to matrix analytic methods in stochastic modeling. ASA & SIAM, Philadelphia

Latouche G, Nguyen GT, Taylor PG (2011) Queues with boundary assistance: the effects of truncation. Queueing Syst 69:175–197

Li QL (2010) Constructive computation in stochastic models with applications: the RG-factorizations. Tsinghua University Press/Springer-Verlag, Beijing/Berlin Heidelberg

Li H, Miyazawa M, Zhao YQ (2007) Geometric decay in a QBD process with countable background states with applications to a join-the-shortest-queue model. Stoch Models 23:413–438

Lucantoni DM, Ramaswami V (1985) Efficient algorithms for solving the non-linear matrix equations arising in phase type queues. Stoch Models 1:29–51

Margolius BH (2007) Transient and periodic solution to the time-inhomogeneous quasi-birth death process. Queueing Syst 56:183–194

Masuyama H, Takine T (2005) Algorithmic computation of the time-dependent solution of structured Markov chains and its application to queues. Stoch Models 21:885–912

Meyn SP, Tweedie RL (2009) Markov chains and stochastic stability, 2nd edn. Cambridge University Press, London

Miller DG (1981) Computation of steady-state probabilities for *M/M/*1 priority queues. Oper Res 29:945–958

Minc H (1988) Non-negative matrices. Wiley, New York

Miyazawa M (2009) Tail decay rates in double QBD processes and related reflected random walks. Math Oper Res 34:547–575

Miyazawa M, Zhao YQ (2004) The stationary tail asymptotics in the *GI/G/1* type queue with countably many background states. Adv Appl Prob 36:1231–1251

Naoumov V (1996) Matrix-multiplicative approach to quasi-birth-and-death processes analysis. In: Chakravarthy SR, Alfa AS (eds) Matrix-analytic methods in stochastic models. Marcel Dekker, New York

Neuts MF (1978) Markov chains with applications in queueing theory, which have a matrix-geometric invariant probability vector. Adv Appl Prob 10:185–212

Neuts MF (1980) The probabilistic significance of the rate matrix in matrix-geometric invariant vectors. J Appl Prob 17:291–296

Neuts MF (1981) Matrix-geometric solutions in stochastic models – an algorithmic approach. The Johns Hopkins University Press, Baltimore

Neuts MF (1986) The caudal characteristic curve of queues. Adv Appl Prob 18:221–254

Neuts MF (1989) Structured stochastic matrices of *M/G/1* type and their applications. Marcel Dekker, New York

Neuts MF (1991) The joint distribution of arrivals and departures in quasi-birth-and-death processes. In: Stewart WJ (ed) Numerical solution of Markov chains. Marcel Dekker, New York, pp 147–159

Nielsen BF, Ramaswami V (1997) A computational framework for a quasi birth and death process with a continuous phase variable. In: Teletrafic contributions for the information age. Elsevier, New York, pp 477–486

Ozawa T (2012) Asymptotics for the stationary distribution in a discrete-time two-dimensional quasi-birth-and-death process. Queueing Syst. doi:10.1007/s11134 -012-9323-9

Ramaswami V (1980) The *N/G/1* queue and its detailed analysis. Adv Appl Prob 12:222–261

Ramaswami V (1982) The busy period of queues which have a matrix-geometric steady state probability vector. Opsearch 19:265–281

Ramaswami V (1988a) Stable recursion for the steady state vector in Markov chains of *M/G/1* type. Stoch Models 4:183–188

Ramaswami V (1988b) Nonlinear matrix equations in applied probability – solution techniques and open problems. SIAM Rev 30:256–263

Ramaswami V (1990) A duality theorem for the matrix paradigms in queueing theory. Stoch Models 5:151–161

Ramaswami V (1999) Matrix analytic methods for stochastic fluid flows. In: Smith D, Hey P (eds) Teletraffic engineering in a competitive world (proceedings of the 16th international teletraffic congress), Elsevier Science B.V., Edinburgh, pp 1019–1030

Ramaswami V, Taylor PG (1996) Some properties of the rate operators in level dependent quasi-birth-and-death processes with a countable number of phases. Stoch Models 12:143–164

Resnick SI (1992) Adventure in stochastic processes. Birkhäuser, Cambridge

Revuz D (1975) Markov chains. North-Holland, Amsterdam

Ross SM (2010) Introduction to probability models, 10th edn. Academic, New York

Seneta E (2006) Non-negative matrices and Markov chains, 2nd edn. Springer, New York

Sengupta B (1989) Markov processes whose steady state distribution is matrix-exponential with an application to the *GI/PH/1* queue. Adv Appl Prob 21:159–180

Takahashi Y, Fujimoto K, Makimoto N (2001) Geometric decay of the steady-state probabilities in a quasi-birth-and-death process with a countable number of phases. Stoch Models 17:1–24

Takine T (1996) A continuous version of matrix-analytic methods with the skip-free to the left property. Stoch Models 12:673–682

Takine T, Sengupta B, Yeung RW (1995) A generalization of the matrix *M/G/1* paradigm for Markov chains with a tree structure. Stoch Models 11:411–421

Tang J, Zhao YQ (2008) Stationary tail asymptotics of a tandem queue with feedback. Ann Oper Res 160:173–189

van Houdt B, Blondia C (2001) Stability and performance of stack algorithms for random access communication modeled as a tree structured QBD Markov chain. Stoch Models 17:247–270 van Houdt B, Blondia C (2004) Robustness of Q-ary collision resolution algorithms in random access systems. Perform Eval 57:357–377

Wallace V (1969) The solution of quasi birth and death processes arising from multiple access computer systems. Ph.D. thesis, Systems Engineering Laboratory, University of Michigan, Ann Arbor

Ye Q (2002) On Latouche-Ramaswami's logarithmic reduction algorithm for quasi-birth-and-death processes. Stoch Models 18:449–467

Ye JD, Li SQ (1994) Folding algorithm: a computational method for finite QBD processes with level-dependent transitions. IEEE Trans Commun 42:625–639

Yeung RW, Alfa AS (1999) The quasi-birth-death type Markov chain with a tree structure. Stoch Models 15:639–659

Yeung RW, Sengupta B (1994) Matrix product-form solutions for Markov chains with a tree structure. Adv Appl Prob 26:965–987

Zhao YQ, Li W, Braun WJ (1998) Infinite block-structured transition matrices and their properties. Adv Appl Prob 30:365–384

Zhao YQ, Li W, Braun WJ (2003) Censoring, factorizations, and spectral analysis for transition matrices with block – repeating entries. Methodol Comput Appl Prob 5:35–58

Chapter 4
Applications in Queueing Theory

Abstract This chapter focuses on the applications of the matrix-analytic methods developed in Chaps. 1, 2, and 3 in queueing theory. The emphasis is on both the introduction of analytically and numerically tractable stochastic models and the analysis of such models. The first part of this chapter deals with a number of simple and classical queueing models. The second part analyzes a few queueing models with multiple types of customers. Algorithms for computing performance measures are developed.

Queueing phenomena can be found everywhere (e.g., in airport check-in systems, supermarket check-out counters, traffic intersections, restaurants, bank branches, telecommunications systems, manufacturing systems) Queueing models have received considerable attention from both researchers and practitioners in the past one-hundred years, resulting in a large collection of papers and books (e.g., Erlang (1917), Kendal (1953), Kleinrock (1975), Neuts (1981, 1989b), Cohen (1982), Chaudhry and Templeton (1983), Hsu (1988), Buzacott and Shanthikumar (1993), Prabhu (1998), Chen and Yao (2001), Whitt (2002), Asmussen (2003), Tian and Zhang (2006)).

A typical queueing system consists of a queue (or queues) and a server (or servers) (see Fig. 4.1). Customers arrive in the system from outside and join the queue in a certain way (e.g., joining the queue at its head or its tail). The server picks up customers and serves them according to certain service discipline (e.g., first-come-first-served (FCFS) and first-come-last-served (FCLS)). Customers leave the system immediately after their service is completed. Figure 4.1 illustrates the flow of customers in a queueing system. To define a queueing system, typically, we need to specify the following:

1. The arrival process of customers;
2. The queueing discipline: how customers join the queue(s) and how they are picked up by the server(s); and
3. The service process: the number of servers and the service times.

Q.-M. He, *Fundamentals of Matrix-Analytic Methods*,
DOI 10.1007/978-1-4614-7330-5_4, © Springer Science+Business Media New York 2014

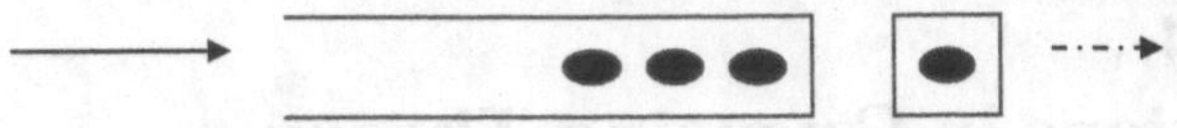

Fig. 4.1 The structure of a simple queueing model

In this chapter, we adopt Kendal's notation A/B/C to denote queueing systems, where A represents the type of the arrival process, B represents the type of the service time, and C represents the number of servers. By default, we assume that customers are served on a FCFS basis, unless otherwise stated. If the notation A/B/C/D is used, then D is for the service discipline or the waiting space (i.e., the capacity of queues).

For queueing systems, *queue lengths*, *waiting times*, and *busy periods* are of primary interest to applications. The most important performance measures are the traffic intensity, mean queue length, distribution of the queue length, mean waiting time, distribution of the waiting time, mean busy period, etc. In this chapter, we shall investigate a number of classical queueing models and develop algorithms for computing their performance measures. (Note: To be consistent with the convention in queueing theory, in this chapter, we use "mean" for the mathematical expectation of random variables.)

The following two queueing models help us understand the structure of queueing systems.

Consider a $D/D/1$ queue defined as: (i) customers arrives at discrete time epochs $\{nd, n \geq 0\}$, one at a time; The interarrival time is constant d; (ii) the service time of a customer is constant s; The server serves one customer at a time; (iii) customers are served on a FCFS basis; and (iv) there are a single queue and a single server.

Exercise 4.1 Assume that the $D/D/1$ queue is empty at time 0. Find (i) the queue length at any time t; and (ii) the waiting times of customers. (Hint: Consider two cases: $s \leq d$ and $s > d$.)

For this special case, explicit solutions can be obtained for all performance measures since both interarrival times and service times are constants. However, for real-world queueing systems, there is randomness in the arrival process, the service process, or both. Thus, it is more interesting to consider queueing systems with randomness. As a consequence, queueing systems become difficult to analyze.

Consider a $GI/G/1$ queue defined as follows: (i) customers arrive according to a renewal process (i.e., the interarrival times are independent and identically distributed random variables); (ii) the service times are independent and identically distributed random variables, and independent of the arrival process; (iii) customers are served on a FCFS basis; and (iv) there are a single queue and a single server.

Let $q(t)$ be the number of customers in the system at time t (the *queue length*). Let $V(t)$ be the *virtual waiting time* (total workload in the system) at time t. In Fig. 4.2, a sample path of the queue length $q(t)$ and a sample path of the virtual waiting time $V(t)$ are drawn. In Fig. 4.2a, $q(t)$ increases by one when a customer arrives and decreases by one when a service completes. In Fig. 4.2b, $V(t)$ jumps up by the amount of the workload (service time) of a customer at its arrival epoch and decreases linearly at rate one. Note that, in Fig. 4.2, plots (a) and (b) are for different samples.

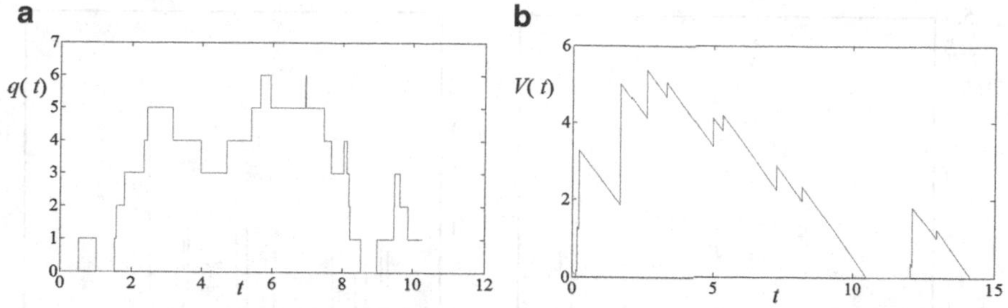

Fig. 4.2 Sample paths of the queue length and virtual waiting time for the *GI/G/1* queue

Exercise 4.2 Answer the following questions.

(i) Mark the waiting time of the third customer in Fig. 4.2a.
(ii) Identify busy periods in Fig. 4.2a.
(iii) Identify idle periods in Fig. 4.2b.

In this chapter, the matrix-analytic methods developed in Chaps. 1, 2, and 3 are utilized in analyzing a serial of queueing models. By taking advantage of the versatility of *PH*-distributions and *MAP*s, a number of queueing models that are both analytically and numerically tractable are introduced. By taking advantage of the algorithmic approach, efficient computational methods are developed for all the queueing models considered in this chapter, which is a highlight of matrix-analytic methods in stochastic modeling.

The first part of this chapter mainly focuses on the modeling of a number of simple and typical queueing models, i.e., the introduction of Markov chains for the queueing systems. Corresponding performance measures can be obtained by applying the methods developed in Chap. 3. In the second part, some technical details are presented for a few queueing models with multiple types of customers. Readers should be familiar with basic concepts in queueing theory such as *arrival process, arrival epoch, service time, service completion epoch, service discipline, queue length, waiting time, virtual waiting time, sojourn time, workload, idle period, busy period, busy cycle, embedded Markov chain, system stability, FCFS, LCFS, RS*, etc. We recommend Kleinrock (1975), Cohen (1982), and Asmussen (2003) for basic knowledge about queueing theory.

4.1 The *M/M/1* Queue

The *M/M/1* queue is defined as a single server queueing system with a Poisson arrival process and exponential service times. Figure 4.1 shows the flow of customers in the queueing system. We assume that the arrival process and the

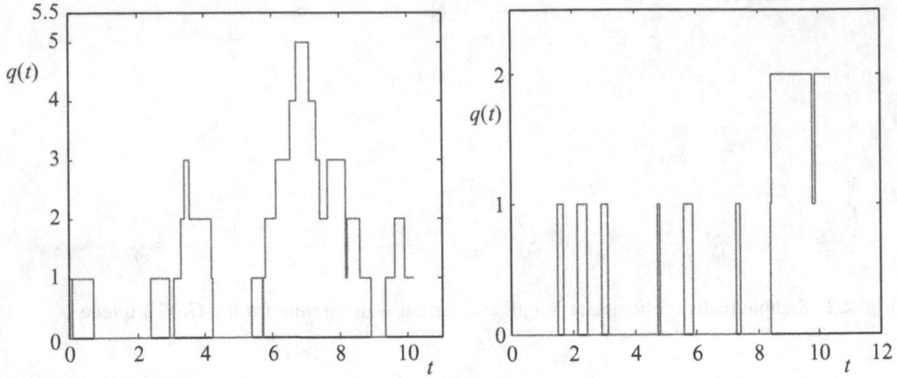

Fig. 4.3 Two sample paths of the queue length for an *M*/*M*/1 queue with $\lambda = 1$ and $\mu = 1.5$

service process are independent. The service times are independent of each other as well. Customers join a single queue and are served on a first-come-first-served (FCFS) basis. The waiting space is infinite in size. In summary, the *M*/*M*/1 queue has the following characteristics and parameters.

1. Poisson arrival process with parameter λ.
2. A single queue.
3. A single server.
4. Exponential service times with common parameter μ.
5. FCFS service discipline.
6. Infinite waiting space.

Let $q(t)$ be the queue length (i.e., the total number of customers in the queueing system) at time t. Two sample paths of the queue length $q(t)$ are depicted in Fig. 4.3. As is indicated, the queueing behavior can be dramatically different for a queue at different time periods. Figure 4.3 not only demonstrates the randomness of the queueing process, but also indicates the challenge in analyzing such a stochastic model.

The rest of this section focuses on the queue length and waiting times of the queueing model. Explicit results have been obtained for the queue length and waiting times at *any* specific time (i.e., transient state) and at an *arbitrary* time (i.e., steady state or $t \to \infty$). We focus on the steady state behavior of the system.

The stochastic process $\{q(t), t \geq 0\}$ is a continuous time Markov chain with state space $\{0, 1, 2, \ldots\}$ for the following reasons:

(i) Since the arrival process is Poisson, the time until the next arrival at time t is always exponentially distributed with parameter λ, independent of the queue length and service time;

(ii) At time t, if the server is idle, there is no service; otherwise, the server is busy and the remaining service time is exponentially distributed with parameter μ.

Thus, the current queue length is the only information needed to predict when the next arrival occurs (in fact, they are independent) and when the next service completion occurs. Since the queue length depends solely on the arrival and service processes, the current queue length is the only information needed to predict future queue length probabilistically. Therefore, $\{q(t), t \geq 0\}$ is a continuous time Markov chain.

Next, we find the infinitesimal generator for $\{q(t), t \geq 0\}$. If $q(t) = 0$, the next event is an arrival after an exponential time with parameter λ. Thus, the total transition rate in this state is λ and the transition is from state 0 to 1 with probability one. If $q(t) = q > 0$, the next event is either an arrival after an exponential time with parameter λ (denoted as X) or a service completion after an exponential time with parameter μ (denoted as Y). Thus, the time until the next event of $\{q(t), t \geq 0\}$ is $\min\{X, Y\}$, which is exponentially distributed with parameter $\lambda + \mu$. Then the total transition rate in this state is $\lambda + \mu$ and the transition is from state q to $q + 1$ with probability $\lambda/(\lambda + \mu)$ (or rate λ) (i.e., $X < Y$ or the arrival occurs first) or from state q to $q - 1$ with probability $\mu/(\lambda + \mu)$ (or rate μ) (i.e., $X > Y$ or the service is completed first). In summary, we obtain the following infinitesimal generator for $\{q(t), t \geq 0\}$

$$Q = \begin{array}{c} 0 \\ 1 \\ 2 \\ \vdots \\ \vdots \end{array} \begin{pmatrix} -\lambda & \lambda & & & \\ \mu & -(\lambda+\mu) & \lambda & & \\ & \mu & -(\lambda+\mu) & \lambda & \\ & & \ddots & \ddots & \ddots \\ & & & \ddots & \ddots \end{pmatrix}. \tag{4.1}$$

Alternatively, we can define the continuous time Markov chain $\{q(t), t \geq 0\}$ by specifying its sojourn time in each state and its embedded Markov chain at transitions.

(i) State space $\{0, 1, 2, \ldots\}$.
(ii) Sojourn time in each state is exponentially distributed with parameter λ for state 0, and $\lambda + \mu$ for all other states.
(iii) At the end of the stay in each state, the next transition is determined by embedded Markov chain $\{q_n, n = 0, 1, 2, \ldots\}$, where q_n is the queue length right after the n-th transition (an arrival or a departure), with transition probability matrix

$$P = \begin{array}{c} 0 \\ 1 \\ 2 \\ \vdots \\ \vdots \end{array} \begin{pmatrix} 0 & 1 & & & \\ \frac{\mu}{\lambda+\mu} & 0 & \frac{\lambda}{\lambda+\mu} & & \\ & \frac{\mu}{\lambda+\mu} & 0 & \frac{\lambda}{\lambda+\mu} & \\ & & \ddots & \ddots & \ddots \\ & & & \ddots & \ddots \end{pmatrix}. \tag{4.2}$$

We remark that the two definitions for the continuous time Markov chain $\{q(t), t \geq 0\}$ are equivalent. In fact, any continuous time Markov chain can be defined by the two methods, which are both useful in stochastic modeling. However, the definition in Eq. (4.1) is more commonly used.

Apparently, $\{q(t), t \geq 0\}$ is an irreducible birth-and-death process. The Markov chain is irreducible since every state communicates with state 0. We remark that, to show that a Markov chain is irreducible, we usually choose a special state (e.g., state 0) and show that any other state communicates with it. Define, if the limit exists,

$$\pi_n = \lim_{t\to\infty} P\{q(t) = n|q(0)\}, \quad n = 0, 1, 2, ... \tag{4.3}$$

Let $\boldsymbol{\pi} = (\pi_0, \pi_1, \ldots, \pi_n, \ldots)$. Then we have $\boldsymbol{\pi}Q = 0$ and $\boldsymbol{\pi}\mathbf{e} = 1$. Solving the linear equations yields, if $\lambda/\mu < 1$, (see Exercise 3.1.8)

$$\pi_n = \left(1 - \frac{\lambda}{\mu}\right)\left(\frac{\lambda}{\mu}\right)^n, \quad n = 0, 1, 2, ... \tag{4.4}$$

We present some basic results for the *M*/*M*/1 queue in the following proposition.

Proposition 4.1.1 *Consider an M/M/1 queue.*

1. *Define Traffic intensity $\rho = \lambda/\mu$. The queueing system* is stable *(i.e., the Markov chain $\{q(t), t \geq 0\}$ is ergodic) if and only if $\rho < 1$.*
2. *In steady state, the probability that the server is busy is $\rho = \lambda/\mu$, and the probability that the server is idle is $1 - \rho$.*
3. *In steady state, the stationary distribution of the queue length q is given by Eq.* (4.4).
4. *In steady state, the distribution of the number of waiting customers q_w, which is called the* actual queue length, *is given by*

$$P\{q_w = n\} = \begin{cases} 1-\rho^2, & n = 0; \\ (1-\rho)\rho^{n+1}, & n = 1, 2, ... \end{cases} \tag{4.5}$$

5. *In steady state, the mean queue length is $E[q] = \lambda/(\mu - \lambda) = \rho/(1-\rho)$. The mean actual queue length is $E[q_w] = \lambda^2/(\mu(\mu-\lambda)) = \rho^2/(1-\rho)$. In addition, we have $E[q] = E[q_w] + \rho$.*
6. *In steady state, the* actual waiting *time W_a (i.e., the time spent in queue) has the following distribution:*

$$P\{W_a \leq t\} = 1 - \rho + \rho(1 - \exp(-(\mu - \lambda)t)), \; t \geq 0. \tag{4.6}$$

Then we have $E[W_a] = \rho/(\mu(1-\rho))$. *The mean of the* waiting time *of a customer (i.e., the total time in the system or the* sojourn time) W *is given by* $E[W] = 1/(\mu-\lambda) = 1/(\mu(1-\rho))$. *Thus, we have* $E[W] = E[W_a] + 1/\mu$.

7. *The* Little's law: $E[q_w] = \lambda E[W_a]$ *and* $E[q] = \lambda E[W]$. *(Note: Little's law applies to more general queueing models such as the GI/G/1 queue. The Little's law is useful for finding performance measures and checking the accuracy of numerical solutions.)*
8. *PASTA (Poisson arrival see time average): The stationary distribution of the queue length q is the distribution of the queue length at an arbitrary time. For the M/M/1 queue, it is also the queue length distribution at an (arbitrary) arrival epoch. Furthermore, it is the queue length distribution at an (arbitrary) departure epoch. (This property holds for queueing systems with a Poisson arrival process.)*
9. *A* busy period *is defined as a time period during which the server is busy all the time. Apparently, a busy period begins when a customer arrives to an empty system and ends when the system becomes empty again. Denote by* τ_B *the length of a busy period. The LST of* τ_B *is*

$$G^*(s) \equiv E[e^{-s\tau_B}] = \frac{\lambda + \mu + s - \sqrt{(\lambda+\mu+s)^2 - 4\lambda\mu}}{2\lambda}, \quad s \geq 0. \tag{4.7}$$

The mean length of a busy period is $E[\tau_B] = 1/(\mu - \lambda)$.

Proof. Part 1 is clear that the stationary distribution of the queue length exists if and only if $\rho < 1$. Part 2 is true since $\pi_0 = 1 - \rho$. Part 4 and part 5 are obtained from part 3. Part 6 can be proved by using part 8 and the following calculations. Conditioning on the queue length at an arbitrary time, we have

$$\begin{aligned} P\{W_a < t\} &= \sum_{n=0}^{\infty} P\{W_a < t | q = n\}\pi_n \\ &= \sum_{n=0}^{\infty} P\{S_n < t\}(1-\rho)\rho^n \\ &= 1 - \rho \exp(-(\mu-\lambda)t) \\ &= 1 - \rho + \rho(1 - \exp(-(\mu-\lambda)t)), \end{aligned} \tag{4.8}$$

where S_n is an Erlang random variable with parameter (n, μ). The third equality in Eq. (4.8) can be proved by finding the LST (see the proof of Proposition 1.1.4) or by using density functions of the random variables involved (scc Eq. (1.5)). Part 7 follows part 5 and part 6. Part 8 can be shown by the memoryless property of the Poisson process or by introducing an embedded Markov chain at arrival epochs (see Exercise 4.1.6). The proof of part 9 is similar to that of the fundamental period for QBD processes in Chap. 3. A busy period can be decomposed into two independent periods: the sojourn time in state 1 (i.e., $q(t) = 1$) and subsequent time. The LST of the sojourn time in state 1 is $(\lambda + \mu)/(\lambda + \mu + s)$. The LST of the subsequent time depends on the next transition. By conditioning on the next transition, it is easy to establish

$$\begin{aligned} G^*(s) &= \frac{\lambda+\mu}{\lambda+\mu+s}\left(\frac{\mu}{\lambda+\mu}+\frac{\lambda}{\lambda+\mu}(G^*(s))^2\right) \\ &= \frac{\mu}{\lambda+\mu+s}+\frac{\lambda}{\lambda+\mu+s}(G^*(s))^2. \end{aligned} \tag{4.9}$$

The rest of the proof is routine. This completes the proof of Proposition 4.1.1.

Exercise 4.1.1 From the service capacity point of view, explain intuitively why the queueing system is stable if $\lambda < \mu$, and unstable if $\lambda > \mu$.

Exercise 4.1.2 The following questions refer to Proposition 4.1.1.

(i) Plot the queue length distribution for $\lambda = 1$ and $\mu = 1.5$.
(ii) Plot the mean queue length as a function of ρ for $0 < \rho < 1$.
(iii) For $\mu = 1.5$, plot the mean waiting time as a function of ρ.
(iv) Complete the proof of Eq. (4.6) using an LST approach.
(v) Complete the proof of Eq. (4.6) using the density function of the Erlang distribution.
(vi) Explain Little's law intuitively.
(vii) Find the mean length of a busy period and the mean number of customers served in a busy period.

Commentary The *M*/*M*/1 queue is a classical queueing model that has been thoroughly investigated. Basic results on the *M*/*M*/1 queue can be found in most of the introductory books on queueing theory. See Kleinrock (1975), Cohen (1982), Gross and Harris (1985), and Ross (2010) for more details.

For performance analysis of queueing models, a number of measures and their corresponding notation are used. For convenience, the commonly used notations are summarized in the following Table 4.1. A summary of the queueing models investigated in this chapter is given in Table 4.2.

Additional Exercises and Extensions

Exercise 4.1.3 For an *M*/*M*/1 queue with $\rho < 1$, show that $E[z^q] = 1 - \rho + (1 - \rho)\rho z(1 - z\rho)^{-1}$, for $0 \le z \le 1$, and $E[e^{-sW_a}] = 1 - \rho + \rho(\mu - \lambda)(s + \mu - \lambda)^{-1}$, for $s \ge 0$.

Exercise 4.1.4 (Exercise 4.1.3 continued) For the *M*/*M*/1 queue, show that, in steady state, variances of the queue length and the waiting time are given by

Table 4.1 Commonly used notations for performance measures

Notation	Definition
$q(t)$	The queue length at time t, including the one(s) in service, if there is any
q	The queue length in steady state (i.e., $q = \lim_{t\to\infty} q(t)$)
q_{w}	The number of waiting customers in steady state
W_n	The waiting time of the n-th customer
W	The waiting (sojourn) time in steady state
W_{a}	The actual waiting time of a customer in steady state
$V(t)$	The workload (virtual waiting time) at time t
V	The workload in steady state
τ_{B}	The length of a busy period

Table 4.2 A summary of queueing models investigated in the chapter

Type	Specific queueing model
GI/*G*/1	*M*/*M*/1 (Sect. 4.1) *PH*/*M*/1 (Sect. 4.2)
	M/*PH*/1 (Sect. 4.3) *GI*/*PH*/1 (Sect. 4.7)
MAP/*G*/1	*MAP*/*PH*/1 (Sect. 4.4.1)
	MAP/*G*/1 (Sect. 4.8)
Bulk queue	*BMAP*/*PH*/1 (Sect. 4.4.2) *MAP*/$PH^{[2]}$/1 (Sect. 4.4.3)
Vacation	*M*/*M*/1 vacation (Sect. 4.4.5)
Network	*M*/*M*/1 with transfers (Sect. 4.5)
Multi-types of customers	LCFS: *MMAP*[*K*]/*PH*[*K*]/1 (Sect. 4.6)
	FCFS: *MMAP*[*K*]/*G*[*K*]/1 (Sect. 4.9)

$$
\begin{aligned}
Var(q) &= \frac{\rho}{(1-\rho)^2}; \qquad & Var(q_{\mathrm{w}}) &= \frac{\rho^2(1+\rho-\rho^2)}{(1-\rho)^2};\\
Var[W] &= \frac{1}{\mu^2(1-\rho)^2}; \qquad & Var[W_{\mathrm{a}}] &= \frac{(2-\rho)\rho}{\mu^2(1-\rho)^2}.
\end{aligned}
\tag{4.10}
$$

Exercise 4.1.5 Consider an *M*/*M*/*s* queue, in which there are s identical servers. Define $\rho = \lambda/(s\mu)$ and assume $\rho < 1$.

(i) Find the infinitesimal generator of the continuous time Markov chain $\{q(t), t \geq 0\}$ for the queue length.

(ii) Show that the stationary distribution of the queue length is given by

$$
\pi_n = \begin{cases}
\dfrac{(s\rho)^n/n!}{\sum\limits_{i=0}^{s-1}\frac{(s\rho)^i}{i!}+\frac{(s\rho)^s}{s!(1-\rho)}}, & n = 0, 1, \ldots, s-1;\\[3ex]
\dfrac{(s\rho)^s\rho^{n-s}/s!}{\sum\limits_{i=0}^{s-1}\frac{(s\rho)^i}{i!}+\frac{(s\rho)^s}{s!(1-\rho)}}, & n = s,\ s+1, \cdots s+2,\ \ldots.
\end{cases}
\tag{4.11}
$$

(iii) Plot the distribution of queue length for the $M/M/3$ queue with $\lambda = 2.5$ and $\mu = 1$.
(iv) Explain intuitively why the $M/M/s$ queue is stable if $\lambda/(s\mu) < 1$.

The *embedded Markov chain* technique was introduced by Kendall in 1953 to study the $M/G/1$ queue. This method and the *supplementary variable method* (see Cohen (1982)) are the most widely used classical methods in queueing theory. Today, the two methods are still being used, especially in combination with other advanced methods, in the study of queueing and other stochastic models. We shall use the embedded Markov chains in the study of $GI/M/1$ and $M/G/1$ types of queueing models. As an example, we introduce embedded Markov chains at arrival epochs and departure epochs for the $M/M/1$ queue.

In the $M/M/1$ queue, define $q_{a,n}$ as the queue length *right before the arrival* of the n-th customer, i.e., the queue length seen by the n-th customer. It is clear that $q_{a,n}$ can increase its value by at most one at each transition, but it can decrease to zero at one transition. Since the service times are exponentially distributed, $\{q_{a,n}, n = 0, 1, 2, \ldots\}$ is a $GI/M/1$ type Markov chain with transition probability matrix

$$P = \begin{pmatrix} b_0 & a_0 & & & & \\ b_1 & a_1 & a_0 & & & \\ b_2 & a_2 & a_1 & a_0 & & \\ b_3 & a_3 & a_2 & a_1 & a_0 & \\ \vdots & \vdots & \ddots & \ddots & \ddots & \ddots \end{pmatrix}, \tag{4.12}$$

where

$$\begin{aligned} a_k &= \int_0^\infty \frac{e^{-\mu t}(\mu t)^k}{k!} \lambda e^{-\lambda t} \mathrm{d}t, \quad k = 0, 1, 2, \ldots; \\ b_k &= \int_0^\infty \left(\sum_{n=k+1}^\infty \frac{e^{-\mu t}(\mu t)^n}{n!} \right) \lambda e^{-\lambda t} \mathrm{d}t, \quad k = 0, 1, 2, \ldots. \end{aligned} \tag{4.13}$$

Exercise 4.1.6 Explain intuitively why $\{q_{a,n}, n = 0, 1, 2, \ldots\}$ is a Markov chain. Find the limiting probabilities of the embedded Markov chain at arrival epochs defined in Eq. (4.12). Prove PASTA for the $M/M/1$ queue. (Hint: Use $\mathbf{p}P = \mathbf{p}$ and $\mathbf{pe} = 1$. Alternatively, you can use results in Sect. 3.4.)

Exercise 4.1.7 Define $q_{d,n}$ as the queue length *right after the departure* of the n-th customer, i.e., the number of customers left behind by the n-th departing customer. Show that $\{q_{d,n}, n = 0, 1, 2, \ldots\}$ is an $M/G/1$ type Markov chain with transition probability matrix

$$P = \begin{pmatrix} a_0 & a_1 & a_2 & a_3 & a_4 & \cdots \\ a_0 & a_1 & a_2 & a_3 & a_4 & \cdots \\ & a_0 & a_1 & a_2 & a_3 & \cdots \\ & & a_0 & a_1 & a_2 & \cdots \\ & & & \ddots & \ddots & \ddots \end{pmatrix}, \tag{4.14}$$

where $a_k = \int_0^{\infty} e^{-\lambda t}(\lambda t)^k \mu e^{-\mu t}\mathrm{d}t/k!$, $k = 0, 1, 2, \ldots$. We call $\{q_{\mathrm{d},n}, n = 0, 1, 2, \ldots\}$ an embedded Markov chain at departure epochs. Find the limiting probabilities of the embedded Markov chain.

Exercise 4.1.8 Let N be the number of customers served in a busy period. Define $G^*(z) = E[z^N]$. Show that, for $0 \le z \le 1$,

$$G^*(z) = z\frac{\mu}{\lambda+\mu} + \frac{\lambda}{\lambda+\mu}(G^*(z))^2. \tag{4.15}$$

Find the mean number of customers served in the busy period. Find the variance of the number of customers served in the busy period.

4.2 The *PH/M/1* Queue

The flow of customers in the *PH/M/1* queue is the same as that in the *M/M/1* queue (see Fig. 4.1). Customers arrive according to a *PH*-renewal process (see Example 2.2.3 in Chap. 2). The interarrival times have a common *PH*-distribution with *PH*-representation $(\boldsymbol{\alpha}, T)$ of order m_{a} (see Sect. 1.2 in Chap. 1). Assume that $\boldsymbol{\alpha}\mathbf{e} = 1$ so that the interarrival times are always positive. Let $\lambda = -1/(\boldsymbol{\alpha}T^{-1}\mathbf{e})$, the arrival rate (i.e., the mean number of arrivals per unit time). The service times are exponentially distributed with parameter μ. The service times and the arrival process are independent. Define

$q(t)$: the queue length at time t, i.e., the total number of customers in the system.
$I_{\mathrm{a}}(t)$: the phase of the underlying Markov chain of the arrival process at time t.

The process $\{q(t), t \ge 0\}$ may not be a Markov chain since $q(t)$ alone does not provide enough information for predicting the arrival of the next customer. Thus, we need to add variable $I_{\mathrm{a}}(t)$ to construct a Markov chain. Note that the underlying process $\{I_{\mathrm{a}}(t): t \ge 0\}$ of the arrival process is a continuous time Markov chain with infinitesimal generator $T + \mathbf{T}^0\boldsymbol{\alpha}$, where $\mathbf{T}^0 = -T\mathbf{e}$ (see Lemma 1.3.1 and Example 2.2.3).

Theorem 4.2.1 *The stochastic process* $\{(q(t), I_{\mathrm{a}}(t)), t \ge 0\}$ *is a continuous time Markov chain with state space* $\{0, 1, 2, \ldots\} \times \{1, 2, \ldots, m_{\mathrm{a}}\}$ *and infinitesimal generator*

$$Q = \begin{pmatrix} T & \mathbf{T}^0\boldsymbol{\alpha} & & & \\ \mu I & -\mu I + T & \mathbf{T}^0\boldsymbol{\alpha} & & \\ & \mu I & -\mu I + T & \mathbf{T}^0\boldsymbol{\alpha} & \\ & & \ddots & \ddots & \ddots \\ & & & \ddots & \ddots \end{pmatrix}. \tag{4.16}$$

Proof. Since service times are exponential, future services can be predicted based on $q(t)$ at the current time t. Based on $I_{\mathrm{a}}(t)$, future arrivals can be predicted. Thus, if

$q(t)$ and $I_a(t)$ are known at time t, the future arrival and service processes can be determined probabilistically. Consequently, $\{(q(t), I_a(t)), t \geq 0\}$ is a continuous time Markov chain.

To construct Q, recall the probabilistic interpretation of PH-representations (Sect. 1.2) and the probabilistic interpretation of MAPs (Sect. 2.2). Suppose $q(t) = 1$. There are three possibilities for the next transition: (i) the arrival of a customer and a change in the phase of the underlying Markov chain; (ii) no arrival and a change in the phase; and (iii) a service completion. If a change of phase without absorption in the underlying Markov chain $\{I_a(t): t \geq 0\}$ occurs first, $q(t)$ remains the same. The change of phase follows the transition rates given in T. If an arrival occurs first, i.e., the underlying process is absorbed, then $q(t)$ becomes two. The rates for such events are given in $\mathbf{T}^0$ and a new phase is selected according to $\boldsymbol{\alpha}$. Then the rates for such transitions are given by $\mathbf{T}^0\boldsymbol{\alpha}$. If the service is completed first, then $q(t)$ becomes zero. The rate for such an event is μ, regardless of the phase. The rates for such events are given by μI for any given $q(t) > 0$. The total rate of transition in each underlying state is μ plus the absolute value of the corresponding diagonal element in T. For example, if $I_a(t) = 2$, the time until the next event in the process is $\min\{X, Z_2\}$, where X is the service time that is exponentially distributed with parameter μ, and Z_2 is exponentially distributed with parameter $-t_{2,2}$. Note that $T = (t_{i,j})$. Thus, the total transition rate for state $(q(t) = 1, I_a(t) = 2)$ is $\mu - t_{2,2}$. In summary, the transition rates for $q(t)$ increasing to 2 are given by $\mathbf{T}^0\boldsymbol{\alpha}$; for $q(t)$ decreasing to 0 are given by μI; and for $q(t)$ remaining to be one are given by $-\mu I + T$.

The transition rates for $q(t) > 1$ can be interpreted similarly. For $q(t) = 0$, there is no service and the next event is an arrival. The transition rates for $q(t)$ increasing to 1 are given by $\mathbf{T}^0\boldsymbol{\alpha}$; and for $q(t)$ remaining to be zero are given by T. Thus, Eq. (4.16) is obtained. This completes the proof of Theorem 4.2.1.

We remark that the above arguments used to obtain the infinitesimal generator can be applied to almost all other Markovian models in this book. In fact, the arguments have been used in Sect. 1.3 and Chaps. 2 and 3. In the rest of this chapter and Chap. 5, the arguments are used without presenting the details explicitly.

The process $\{(q(t), I_a(t)), t \geq 0\}$ is a quasi-birth-and-death (QBD) process. The theory developed in Chap. 3 can be applied.

Theorem 4.2.2 *Assume that the matrix $T + \mathbf{T}^0\boldsymbol{\alpha}$ is irreducible. The Markov chain $\{(q(t), I_a(t)), t \geq 0\}$ is ergodic if and only if $\rho = \lambda/\mu < 1$ (i.e., the queueing system is stable if and only if $\rho < 1$).*

Proof. Since $\{I_a(t), t \geq 0\}$ is irreducible, any state communicates with state (0, 1). Then $\{(q(t), I_a(t)), t \geq 0\}$ is irreducible. Next, we only need to check Neuts condition for ergodicity (Theorem 3.2.1). Let $\boldsymbol{\theta}_a$ satisfy $\boldsymbol{\theta}_a(T + \mathbf{T}^0\boldsymbol{\alpha}) = 0$ and $\boldsymbol{\theta}_a\mathbf{e} = 1$. It can be shown (see Lemma 1.3.1) that $\boldsymbol{\theta}_a\mathbf{T}^0 = \lambda$. Then Neuts condition is equivalent to

$$\boldsymbol{\theta}_a(\mu I)\mathbf{e} > \boldsymbol{\theta}_a(\mathbf{T}^0\boldsymbol{\alpha})\mathbf{e} \Leftrightarrow \mu > \boldsymbol{\theta}_a\mathbf{T}^0 \Leftrightarrow \mu > \lambda. \tag{4.17}$$

This completes the proof of Theorem 4.2.2.

We now assume that $\lambda/\mu < 1$ so that the queueing system is stable. That implies that the stationary distributions of the queue length, waiting times, and other performance measures exist. Define

$$\begin{aligned}\pi_{n,j} &= \lim_{t\to\infty} P\{q(t)=n,\ I_a(t)=j|q(0),I_a(0)\},\\ &\qquad\qquad n=0,\ 1,\ 2,\ldots,\ j=1,\ 2,\ldots,\ m_a;\\ \boldsymbol{\pi}_n &= (\pi_{n,1},\cdots,\pi_{n,m_a}),\qquad n=0,\ 1,\ 2,\ \ldots;\\ \boldsymbol{\pi} &= (\boldsymbol{\pi}_0,\ \boldsymbol{\pi}_1,\ \boldsymbol{\pi}_2,\ \ldots).\end{aligned} \tag{4.18}$$

Note that $\boldsymbol{\pi}$ gives us the queue length distribution at an arbitrary time epoch. Based on the general theory of continuous time Markov chains, $\boldsymbol{\pi}$ satisfies $\boldsymbol{\pi}Q = 0$ and $\boldsymbol{\pi}\mathbf{e} = 1$. The linear system $\boldsymbol{\pi}Q = 0$ and $\boldsymbol{\pi}\mathbf{e} = 1$ can be expanded as follows.

$$\begin{aligned}&\boldsymbol{\pi}_0 T + \boldsymbol{\pi}_1\mu = 0;\\ &\boldsymbol{\pi}_n \mathbf{T}^0\boldsymbol{\alpha} + \boldsymbol{\pi}_{n+1}(T-\mu I) + \boldsymbol{\pi}_{n+2}\mu = 0,\quad n=0,\ 1,\ 2,\ \ldots.\end{aligned} \tag{4.19}$$

It can be verified by Theorem 3.2.2 or Exercise 3.2.15 that Eq. (4.19) has the following matrix-geometric solution

$$\boldsymbol{\pi}_n = \boldsymbol{\pi}_0 R^n,\quad n=0,\ 1,\ 2,\ldots, \tag{4.20}$$

where the matrix R is the minimal nonnegative solution to the nonlinear equation

$$\mathbf{T}^0\boldsymbol{\alpha} + R(T-\mu I) + R^2\mu = 0. \tag{4.21}$$

Also, $\boldsymbol{\pi}_0$ is the unique positive solution to

$$\begin{aligned}&\boldsymbol{\pi}_0(T+\mu R) = 0;\\ &\boldsymbol{\pi}_0(I-R)^{-1}\mathbf{e} = 1.\end{aligned} \tag{4.22}$$

With the matrix-geometric solution (4.20) in hand, it is possible to obtain other performance measures.

Proposition 4.2.1 *Assume that* $\lambda/\mu < 1$.

1. *In steady state, the distribution of the queue length is given by* $\{\boldsymbol{\pi}_0 R^n\mathbf{e},\ n=1,\ 2,\ldots\}$. *The mean queue length is* $\boldsymbol{\pi}_0 R(I-R)^{-2}\mathbf{e}$.
2. *We have* $\boldsymbol{\pi}_0(I-R)^{-1} = \boldsymbol{\theta}_a$. *(This is useful for checking computation accuracy.)*

Exercise 4.2.1 Show Proposition 4.2.1. Explain part 2 intuitively.

Example 4.2.1 We have $\boldsymbol{\pi}_0\mathbf{e} = 1-\rho$, which can be proved as follows. Postmultiplying by $\mathbf{e}$ on both sides of the equalities in Eq. (4.19) yields

$$\boldsymbol{\pi}_n \mathbf{T}^0 = \boldsymbol{\pi}_{n+1} \mathbf{e}\mu, \quad n = 0, 1, 2, \ldots \tag{4.23}$$

Postmultiplying by $T^{-1}\mathbf{e}$ on both sides of the equations in Eq. (4.19), using equalities in Eq. (4.23), we obtain

$$\begin{aligned} \boldsymbol{\pi}_0 \mathbf{e} &= (-\boldsymbol{\pi}_1 T^{-1}\mathbf{e})\mu; \\ \boldsymbol{\pi}_n \mathbf{e}(1 + \mu\boldsymbol{\alpha} T^{-1}\mathbf{e}) - \mu\boldsymbol{\pi}_n T^{-1}\mathbf{e} + \mu\boldsymbol{\pi}_{n+1} T^{-1}\mathbf{e} &= 0, \quad n = 1, 2, \ldots \end{aligned} \tag{4.24}$$

Then

$$\begin{aligned} (1 + \mu\boldsymbol{\alpha} T^{-1}\mathbf{e}) \sum_{n=1}^{\infty} \boldsymbol{\pi}_n \mathbf{e} - \mu\boldsymbol{\pi}_1 T^{-1}\mathbf{e} = 0 \quad &\Rightarrow \\ (1 + \mu\boldsymbol{\alpha} T^{-1}\mathbf{e})(1 - \boldsymbol{\pi}_0 \mathbf{e}) + \boldsymbol{\pi}_0 \mathbf{e} = 0 \quad &\Rightarrow \\ \boldsymbol{\pi}_0 \mathbf{e} = 1 + \frac{1}{\mu\boldsymbol{\alpha} T^{-1}\mathbf{e}} = 1 - \rho. & \end{aligned} \tag{4.25}$$

Consider the queueing process in [0, t]. The average number of arrivals is approximately λt and the average time needed by the server to serve all the arrived customers is $\lambda t/\mu$. Approximately, the server idle time in [0, t] can be $t - \lambda t/\mu$. Then the percentage of time the server is idle is $(t - \lambda t/\mu)/t = 1 - \rho$, which is consistent with the long-run average interpretation of $\boldsymbol{\pi}_0 \mathbf{e}$.

Exercise 4.2.2 Consider a *PH*/*M*/1 queue with

$$\left(\boldsymbol{\alpha} = (0.5,\ 0.5), \quad T = \begin{pmatrix} -0.6 & 0 \\ 0.4 & -0.5 \end{pmatrix} \right) \tag{4.26}$$

and $\mu = 1$. (1) Compute λ and ρ. (2) Compute R, $\boldsymbol{\pi}_0$, and $\boldsymbol{\pi}_1$. (3) Compute the distribution of the queue length. (4) Compute the mean queue length. (5) Find the probability that the system is empty.

Commentary The *PH*/*M*/1 queue can approximate the more general *GI*/*M*/1 queue. The study of the busy period for the *PH*/*M*/1 queue can be done as a special case of the *MAP*/*G*/1 queue in Sect. 4.8.

Using the *PH*/*M*/1 queue as an example, we describe the main steps for introducing Markov chains for stochastic systems.

1. Introduce system variables (e.g., $q(t)$);
2. Introduce auxiliary variables for a Markov chain (e.g., $I_a(t)$);
3. Identify the state space (e.g., $\{(q, i), q = 0, 1, 2, \ldots, i = 1, 2, \ldots, m_a\}$);
4. Check the Markovian property for individual states (e.g., state (2, 3)):

   4.1 Identify activities associated with a state (e.g., arrival and service);
   4.2 Find the times of associated activities (e.g., exponential times);
   4.3 Find the times for individual events (e.g., exponential times);

5. Find the infinitesimal generator Q: for each state,
   5.1 Find the transition rates for individual activities associated with the state;
   5.2 Identify the next state associated with each transition rate;
   5.3 Calculate the total transition rate.

Additional Exercises and Extensions

Define vector generating function $\mathbf{p}^*(z) = \sum_{n=0}^{\infty} z^n \boldsymbol{\pi}_n$, for $0 \le z \le 1$.

Exercise 4.2.3 Show that $\mathbf{p}^*(z) = \boldsymbol{\pi}_0(I - zR)^{-1}$. Use the generating function $\mathbf{p}^*(z)$ to find the mean and variance of the queue length.

Exercise 4.2.4 The LST of the virtual waiting time (i.e., the total workload in the system at arbitrary time t) in steady state, denoted by V, is given as

$$E[e^{-sV}] = \boldsymbol{\pi}_0\mathbf{e} + \boldsymbol{\pi}_0\mu((s+\mu)I - R\mu)^{-1}\mathbf{e}. \tag{4.27}$$

The mean waiting time is given by $E[V] = \boldsymbol{\pi}_0(I-R)^{-2}R\mathbf{e}/\mu$. Find $Var(V)$. Explain the relationship between $E[V]$ and the mean queue length given in Proposition 4.2.1 intuitively. (Hint: (1) The distribution of the queue length at an arbitrary time is given in Proposition 4.2.1; and (2) The service time has the memoryless property.)

Define $\Delta = \mu^{-1}\text{diag}((I-R)^{-1}\mathbf{e})$, $T_{\text{v}} = \mu\Delta^{-1}(R-I)\Delta$, and $\boldsymbol{\alpha}_{\text{v}} = \boldsymbol{\pi}_1\mu\Delta$. Verify that $(\boldsymbol{\alpha}_{\text{v}}, T_{\text{v}})$ is a *PH*-representation. (Note: This result shows that the virtual waiting time V has a *PH*-distribution. Similar result holds for queues with a QBD process for its queue length. See Ozawa (2006) for details. Thus, the waiting time and sojourn time in the queues investigated in Sects. 4.2, 4.3, and 4.4 have *PH*-distributions. Their corresponding *PH*-representations can be found explicitly.)

Exercise 4.2.5 Consider the *PH/M/s* queue. (1) Write down the infinitesimal generator of the process $\{(q(t), I_{\text{a}}(t)), t \ge 0\}$. (2) Find the stationary distribution of $\{(q(t), I_{\text{a}}(t)), t \ge 0\}$. (3) Find the stationary distribution of the queue length.

Exercise 4.2.6 For the *PH/M/s* queue, if $T + \mathbf{T}^0\boldsymbol{\alpha}$ is reducible, is the Markov chain $\{(q(t), I_{\text{a}}(t)), t \ge 0\}$ irreducible?

Let $q_{\text{a},n}$ be the queue length seen by the n-th arrival. It is easy to see that $\{q_{\text{a},n}, n = 0, 1, 2, \ldots\}$ is a Markov chain since the service time has an exponential distribution.

Exercise 4.2.7 Use the renewal theory (C'inlar (1969)) to show that the queue length at an arbitrary arrival epoch is given by

$$\lim_{n\to\infty} P\{q_{a,n} = k\} = \frac{\boldsymbol{\pi}_k \mathbf{T}^0}{\sum_{t=0}^{\infty} \boldsymbol{\pi}_t \mathbf{T}^0} = \frac{\boldsymbol{\pi}_0 R^k \mathbf{T}^0}{\boldsymbol{\pi}_0 (I-R)^{-1} \mathbf{T}^0}, \quad k = 0, 1, 2, \ldots. \tag{4.28}$$

Show that the actual waiting time of an arbitrary customer has LST, for $s \geq 0$,

$$\lim_{n\to\infty} E[e^{-sW_{a,n}}] = \frac{\boldsymbol{\pi}_0 \mathbf{T}^0 + \boldsymbol{\pi}_1 \mu((s+\mu)I - \mu R)^{-1} \mathbf{T}^0}{\boldsymbol{\pi}_0 (I-R)^{-1} \mathbf{T}^0}. \tag{4.29}$$

Similar to Exercise 4.2.4, show that the actual waiting time is *PH*-distributed and find a *PH*-representation for it. Show that the mean actual waiting time is given by

$$E[W_a] = \boldsymbol{\pi}_1 (I-R)^{-2} \mathbf{T}^0 / (\boldsymbol{\pi}_0 (I-R)^{-1} \mathbf{T}^0 \mu). \tag{4.30}$$

We briefly discuss the difference between $V(t)$ and $W_{a,n}$. Let $\{a_n, n = 1, 2, \ldots\}$ be the arriving time epochs of customers. Then we have $W_{a,n} = V(a_n)$. That is, the actual waiting times are the workloads in the system at specific (i.e., arriving) time epochs. In general, the limiting distributions of the virtual waiting times and the actual waiting times can be different, as shown by the *PH/M/1* queue, but can be the same, as shown by the *M/M/1* queue.

Exercise 4.2.8 Show that the time until the queue length becomes five has a *PH*-distribution, given that the queueing system is empty at time zero. Find a *PH*-representation of this time. (Hint: You need to specify the distribution of the phase at time zero.)

Define $G^*(s)$ as the (matrix) LST of the length of a fundamental period of Q defined in Eq. (4.16), i.e., the (i, j)-th element of $G^*(s)$ is the LST of the length of a busy period for which the phase at the end of the busy period is j, given that the underlying Markov chain is initially in phase i, for $i, j = 1, 2, \ldots, m_a$.

Exercise 4.2.9 For the *PH/M/1* queue,

(i) Show $G^*(s) = (sI - T - \mu I)^{-1} \left(\mu I + \mathbf{T}^0 \boldsymbol{\alpha} (G^*(s))^2\right)$, $s \geq 0$.
(ii) What is the distribution of the phase at the beginning of a busy period?
(iii) Find the mean length of a busy period.
(iv) Show that $G_c^*(s) = G^*(s)(sI - T)^{-1}\mathbf{T}^0$ is the LST of the length of a busy cycle (i.e., busy period + idle period).
(v) Find the mean length of a busy cycle.
(vi) $G^*(s, z)$ is the transform of the joint distribution of the number of customers served in a busy period and the length of the busy period. Derive an equation for $G^*(s, z)$. (Note: For more about the use of transform of joint distributions in queueing theory, see Prabhu (1998).)

Exercise 4.2.10 Assume that a busy period begins with five customers. Find the joint transform of the length of the busy period and the total number of customers served during the period. (Hint: Since there are five customers in the system at time $t = 0$, the server has to work for five consecutive busy periods, each with one customer initially, to become idle.)

Exercise 4.2.11* Assume that initially the total workload of a busy period is $V(0) = 10$. Find the joint transform of the length of the busy period and the total number of customers served in the busy period.

4.3 The $M/PH/1$ Queue

The flow of customers in the $M/PH/1$ queue is the same as that in the $M/M/1$ queue (see Fig. 4.1). The arrival process is a Poisson process with parameter λ. The service times have a common PH-distribution function with PH-representation $(\boldsymbol{\beta}, S)$ of order m_s. We assume that $\boldsymbol{\beta}\mathbf{e} = 1$. Let $\mu = -1/(\boldsymbol{\beta}S^{-1}\mathbf{e})$, the service rate. Define

$q(t)$: the queue length at time t, i.e., the total number of customers in the system.
$I_s(t)$: the phase of the service process at time t, if the server is busy at time t; 0, otherwise.

Exercise 4.3.1 Explain intuitively why $\{q(t), t \geq 0\}$ may not be a continuous time Markov chain.

Consider the stochastic process $\{(q(t), I_s(t)), t \geq 0\}$. Since the arrival process is a Poisson process, the time until the next arrival is exponentially distributed with parameter λ at any time t. Furthermore, since the service phase $I_s(t)$ is recorded, the residual service time (time until the next service completion) is known probabilistically. Then it is easy to see that $\{(q(t), I_s(t)), t \geq 0\}$ is a continuous time Markov chain with state space $\{0\}\cup\{\{1, 2, \ldots\} \times \{1, 2, \ldots, m_s\}\}$ and infinitesimal generator

$$Q = \begin{pmatrix} -\lambda & \lambda\boldsymbol{\beta} & & & \\ \mathbf{S}^0 & -\lambda I + S & \lambda I & & \\ & \mathbf{S}^0\boldsymbol{\beta} & -\lambda I + S & \lambda I & \\ & & \ddots & \ddots & \ddots \\ & & & \ddots & \ddots \end{pmatrix}, \tag{4.31}$$

where $\mathbf{S}^0 = -S\mathbf{e}$. It is clear that $\{(q(t), I_s(t)), t \geq 0\}$ is a QBD process. The transition rates in Q can be interpreted in a way similar to that in Eq. (4.16). Note that level zero has only one state since the server is idle when the system is empty.

Theorem 4.3.1 *Assume that the matrix $S + \mathbf{S}^0\boldsymbol{\beta}$ is irreducible. The Markov chain $\{(q(t), I_s(t)), t \geq 0\}$ is ergodic if and only if $\rho = \lambda/\mu < 1$ (i.e., the queueing system is stable if and only if $\rho < 1$.)*

Proof. By the assumption on $S + \mathbf{S}^0\boldsymbol{\beta}$, the Markov chain is irreducible. Then we only need to check Neuts condition for ergodicity. Let $\boldsymbol{\theta}_s$ satisfy $\boldsymbol{\theta}_s(S + \mathbf{S}^0\boldsymbol{\beta}) = 0$ and $\boldsymbol{\theta}_s\mathbf{e} = 1$. It can be shown that $\boldsymbol{\theta}_s\mathbf{S}^0 = \mu$ (see Lemma 1.3.1). Then Neuts condition (Theorem 3.2.1) is equivalent to

$$\boldsymbol{\theta}_s(\lambda I)\mathbf{e} < \boldsymbol{\theta}_s(\mathbf{S}^0\boldsymbol{\beta})\mathbf{e} \Leftrightarrow \lambda < \boldsymbol{\theta}_s\mathbf{S}^0 \Leftrightarrow \mu > \lambda. \tag{4.32}$$

This completes the proof of Theorem 4.3.1.

Now, we assume $\lambda/\mu < 1$ for a stable system. Define

$$\begin{aligned} \pi_{n,j} &= \lim_{t\to\infty} P\{q(t) = n,\ I_s(t) = j | q(0), I_s(0)\}, \\ &\qquad\qquad n = 0,\ 1,\ 2, \ldots,\ j = 1,\ 2, \ldots,\ m_s; \\ \boldsymbol{\pi}_n &= (\pi_{n,1}, \cdots, \pi_{n,m_s}), \quad n = 1,\ 2,\ \ldots; \\ \boldsymbol{\pi} &= (\pi_0,\ \boldsymbol{\pi}_1,\ \boldsymbol{\pi}_2,\ \ldots). \end{aligned} \tag{4.33}$$

Assume that the limiting probabilities exist and are positive. According to the general theory of continuous time Markov chains, $\boldsymbol{\pi}$ satisfies $\boldsymbol{\pi}Q = 0$ and $\boldsymbol{\pi}\mathbf{e} = 1$, which can be expanded as follows

$$\begin{aligned} &\pi_0(-\lambda) + \boldsymbol{\pi}_1\mathbf{S}^0 = 0; \\ &\pi_0\lambda\boldsymbol{\beta} + \boldsymbol{\pi}_1(S - \lambda I) + \boldsymbol{\pi}_2\mathbf{S}^0\boldsymbol{\beta} = 0; \\ &\boldsymbol{\pi}_n\lambda + \boldsymbol{\pi}_{n+1}(S - \lambda I) + \boldsymbol{\pi}_{n+2}\mathbf{S}^0\boldsymbol{\beta} = 0, \quad n = 1,\ 2,\ \ldots. \end{aligned} \tag{4.34}$$

From Eq. (4.34), it is easy to obtain

$$\begin{aligned} \pi_0\lambda &= \boldsymbol{\pi}_1\mathbf{S}^0; \\ \lambda\boldsymbol{\pi}_n\mathbf{e} &= \boldsymbol{\pi}_{n+1}\mathbf{S}^0,\ n = 1,\ 2,\ \ldots. \end{aligned} \tag{4.35}$$

Exercise 4.3.2 Prove equalities in Eq. (4.35) and explain them intuitively.

Postmultiplying by $\boldsymbol{\beta}$ on both sides of Eq. (4.35), we obtain $\lambda\boldsymbol{\pi}_n\mathbf{e}\boldsymbol{\beta} = \boldsymbol{\pi}_{n+1}\mathbf{S}^0\boldsymbol{\beta}$, for $n = 1, 2, \ldots$. Together with Eq. (4.34), we obtain

$$\begin{aligned} \pi_0\lambda\boldsymbol{\beta} &= \boldsymbol{\pi}_1(\lambda I - \lambda\mathbf{e}\boldsymbol{\beta} - S); \\ \lambda\boldsymbol{\pi}_n &= \boldsymbol{\pi}_{n+1}(\lambda I - \lambda\mathbf{e}\boldsymbol{\beta} - S),\ n = 1,\ 2,\ \ldots. \end{aligned} \tag{4.36}$$

Based on Eq. (4.36), the following explicit matrix-geometric solution can be obtained.

Theorem 4.3.2 *Assume that $\lambda/\mu < 1$ and $S + S^0\boldsymbol{\beta}$ is irreducible. Then we have*

$$\boldsymbol{\pi}_n = \pi_0\boldsymbol{\beta}\lambda^n(\lambda I - \lambda\mathbf{e}\boldsymbol{\beta} - S)^{-n},\ n = 1,\ 2,\ \ldots, \tag{4.37}$$

where $\pi_0 = 1 - \lambda/\mu$. The rate matrix R in the matrix-geometric solution for the QBD process Q is found explicitly as $R = \lambda(\lambda I - \lambda\mathbf{e}\boldsymbol{\beta} - S)^{-1}$.

Proof. By Eq. (4.36), the solution in Eq. (4.37) can be obtained if the matrix $\lambda I - \lambda \mathbf{e}\boldsymbol{\beta} - S$ is invertible. If the matrix is singular, there exists nonzero vector $\mathbf{u}$ such that $\mathbf{u}(\lambda I - \lambda \mathbf{e}\boldsymbol{\beta} - S) = 0$. If $\mathbf{ue} = 0$, then $\mathbf{u}(\lambda I - S) = 0$, which is impossible since all eigenvalues of S have a negative real part. If $\mathbf{ue} \neq 0$, without loss of generality, we assume $\mathbf{ue} = 1$. Then $\mathbf{u}(\lambda I - S) = \lambda\boldsymbol{\beta}$. By Exercise 1.3.25, the *PH*-distribution $(\boldsymbol{\beta}, S - \lambda I)$ is stochastically smaller than the *PH*-distribution $(\boldsymbol{\beta}, S)$. The equality leads to $\mathbf{ue} = -\lambda\boldsymbol{\beta}(S - \lambda I)^{-1}\mathbf{e} \leq \lambda\boldsymbol{\beta}(-S)^{-1}\mathbf{e} < 1$, which contradicts $\mathbf{ue} = 1$. Therefore, the matrix $\lambda I - \lambda \mathbf{e}\boldsymbol{\beta} - S$ is invertible. Then Eq. (4.37) is obtained.

Next, to show $R = \lambda(\lambda I - \lambda \mathbf{e}\boldsymbol{\beta} - S)^{-1}$, we need to prove that the inverse matrix of $\lambda I - \lambda \mathbf{e}\boldsymbol{\beta} - S$ is nonnegative. Note that $\lambda I - S$ is an *M*-matrix and its inverse matrix is nonnegative. By definition, we have

$$\begin{aligned}
&(\lambda I - \lambda \mathbf{e}\boldsymbol{\beta} - S)^{-1} \\
&\quad = (\lambda I - S)^{-1}\left(I - \lambda \mathbf{e}\boldsymbol{\beta}(\lambda I - S)^{-1}\right)^{-1} \\
&\quad = (\lambda I - S)^{-1}\sum_{n=0}^{\infty}\left(\lambda \mathbf{e}\boldsymbol{\beta}(\lambda I - S)^{-1}\right)^{n} \\
&\quad = (\lambda I - S)^{-1}\sum_{n=0}^{\infty}\lambda^{n}\mathbf{e}\left(\boldsymbol{\beta}(\lambda I - S)^{-1}\mathbf{e}\right)^{n-1}\boldsymbol{\beta}(\lambda I - S)^{-1} \\
&\quad \geq 0.
\end{aligned} \tag{4.38}$$

It is easy to verify that $\lambda(\lambda I - \lambda \mathbf{e}\boldsymbol{\beta} - S)^{-1}$ satisfies matrix equation

$$\lambda I + X(S - \lambda I) + X^2\mathbf{S}^0\boldsymbol{\beta} = 0. \tag{4.39}$$

By the definition of R, we must have $R \leq \lambda(\lambda I - \lambda \mathbf{e}\boldsymbol{\beta} - S)^{-1}$. Since we have $\boldsymbol{\pi}_n = \boldsymbol{\pi}_1 R^{n-1}$, for $n = 1, 2, \ldots$, and $\boldsymbol{\pi}_1$ is positive, we must have $R = \lambda(\lambda I - \lambda \mathbf{e}\boldsymbol{\beta} - S)^{-1}$.

Lastly, using $\pi_0 + \boldsymbol{\pi}_1(I - R)^{-1}\mathbf{e} = 1$, we show that $\pi_0 = 1 - \lambda/\mu = 1 - \rho$ as follows. Since the total probability is one, we must have

$$\begin{aligned}
1 &= \pi_0 + \pi_0\lambda\boldsymbol{\beta}(\lambda I - \lambda \mathbf{e}\boldsymbol{\beta} - S)^{-1}\left(I - \lambda(\lambda I - \lambda \mathbf{e}\boldsymbol{\beta} - S)^{-1}\right)^{-1}\mathbf{e} \\
&= \pi_0 + \pi_0\lambda\boldsymbol{\beta}(-\lambda \mathbf{e}\boldsymbol{\beta} - S)^{-1}\mathbf{e} \\
&= \pi_0 - \pi_0\lambda\boldsymbol{\beta}S^{-1}\mathbf{e}\sum_{n=0}^{\infty}(-\lambda\boldsymbol{\beta}S^{-1}\mathbf{e})^{n} \\
&= \pi_0 + \pi_0\sum_{n=1}^{\infty}(-\lambda\boldsymbol{\beta}S^{-1}\mathbf{e})^{n} \\
&= \pi_0 + \pi_0\sum_{n=1}^{\infty}(\lambda/\mu)^{n} \\
&= \pi_0(1 - \lambda/\mu)^{-1}.
\end{aligned} \tag{4.40}$$

This completes the proof of Theorem 4.3.2.

The results in Theorem 4.3.2 are explicit (in terms of the original system parameters). They can be rewritten as

$$\begin{aligned} \pi_0 &= 1 - \lambda/\mu = 1 - \rho; \\ \boldsymbol{\pi}_n &= (1-\rho)\boldsymbol{\beta} R^n, \quad n = 1,\ 2,\ \dots. \end{aligned} \tag{4.41}$$

Exercise 4.3.3 Show that the mean queue length is given as $(1-\rho)\boldsymbol{\beta} R(I-R)^{-2}\mathbf{e}$.

Exercise 4.3.4 Assume that $\lambda = 2$,

$$\boldsymbol{\beta} = (0.2, 0.8),\ S = \begin{pmatrix} -2 & 1 \\ 1 & -10 \end{pmatrix}. \tag{4.42}$$

Find π_0, R, $\boldsymbol{\pi}_{10}$, and the mean queue length. Note that $\pi_0 + \boldsymbol{\pi}_1\mathbf{e} + \boldsymbol{\pi}_2\mathbf{e} + \boldsymbol{\pi}_3\mathbf{e} + \dots = 1$, which can be used to verify the correctness of the results.

Commentary We note that both the *PH*/*M*/1 queue and the *M*/*PH*/1 queue have been studied in Neuts (1981). They are the queueing models for which results can be obtained for many performance measures. For *M*/*PH*/1, π_0 and R are obtained explicitly. For *PH*/*M*/1, there are no explicit formulas for π_0 and R, and iterative methods have been developed for computing them.

Additional Exercises and Extensions

Exercise 4.3.5 If $\lambda/\mu < 1$, show that $I - R$ is invertible.

Let $q_{d,n}$ be the queue length left behind by the n-th departed customer. It is easy to see that $\{q_{d,n},\ n = 0,\ 1,\ 2,\ \dots\}$ is a Markov chain since the arrival process is Poisson.

Exercise 4.3.6 Use the renewal theory to show that the queue length at an arbitrary departure epoch is given by, for $k = 0,\ 1,\ 2,\ \dots$,

$$\lim_{n\to\infty} P\{q_{d,n} = k\} = \frac{\boldsymbol{\pi}_{k+1}\mathbf{S}^0}{\sum_{t=1}^{\infty} \boldsymbol{\pi}_t \mathbf{S}^0} = \frac{\boldsymbol{\beta} R^{k+1}\mathbf{S}^0}{\boldsymbol{\beta} R(I-R)^{-1}\mathbf{S}^0}. \tag{4.43}$$

Since the Poisson process possesses the memoryless property, an actual arrival can be considered as an arrival at an arbitrary time. Thus, the PASTA property applies to the *M*/*PH*/1 queue.

Exercise 4.3.7 Show that the actual waiting time of an arbitrary customer has LST, for $s \geq 0$,

$$\lim_{n\to\infty} E[e^{-sW_{a,n}}] = 1 - \rho + (1-\rho)\boldsymbol{\beta}\sum_{k=1}^{\infty} R^k (sI-S)^{-1}\mathbf{S}^0 \left(\boldsymbol{\beta}(sI-S)^{-1}\mathbf{S}^0\right)^{k-1}. \tag{4.44}$$

Show that, in steady state, the mean actual waiting time is

$$E[W_a] = (1-\rho)\boldsymbol{\beta}R(I-R)^{-1}\left(-S^{-1} + \frac{1}{\mu}R(I-R)^{-1}\right)\mathbf{e}. \qquad (4.45)$$

Exercise 4.3.8 Find the variance of the queue length in steady state.

Exercise 4.3.9 Define $G^*(s)$ as the (matrix) LST of the busy period. Show that

$$G^*(s) = \boldsymbol{\beta}(sI - S + \lambda I)^{-1}\left(\mathbf{S}^0 + \lambda(G_1^*(s))^2\mathbf{e}\right), \qquad (4.46)$$

where $G_1^*(s) = (sI - S + \lambda I)^{-1}\left(\mathbf{S}^0\boldsymbol{\beta} + \lambda(G_1^*(s))^2\right)$, $s \geq 0$.

(i) Find the mean busy period.
(ii) Find the distribution of each idle period.
(iii) Find the LST of a busy cycle.
(iv) Find the mean busy cycle.
(v) Derive an equation for $G^*(s, z)$, where $G^*(s, z)$ is the transform of the joint distribution of the number of customers served in a busy period and the length of a busy period.

Exercise 4.3.10 Assume that a busy period begins with five customers. Find the joint transform of the length of the busy period and the total number of customers served during the period.

Exercise 4.3.11 Assume that the workload at the beginning of a busy period is $V(0) = 10$. Find the joint transform of the busy period and the total number of customers served during the period.

Exercise 4.3.12 Consider an *M/PH/*2 queue in which there are two identical servers. (1) Introduce a continuous time Markov chain to represent the queueing system. (2) Write down the infinitesimal generator for the Markov chain. (3) Outline a method for computing the stationary distribution of the queue length.

For many queueing models such as the *GI/G/*1 queue, it is well-known that, in steady state and under fairly general conditions, the queue lengths at arrival epochs and departure epochs have the same distribution (e.g., Chaudhry and Templeton (1983)).

Exercise 4.3.13 Argue that, for the *M/PH/*1 queue, in steady state, the distributions of queue lengths at arrival epochs, an arbitrary time, and departure epochs are the same. Show that the distribution given in Eq. (4.43) and $\{(1-\rho)\boldsymbol{\beta}R^n\mathbf{e}, n = 0, 1, 2, \ldots\}$ are consistent mathematically. (Hint: Use the proof of Eq. (4.40) to simplify the denominator in Eq. (4.43). Also note $(\lambda I - \lambda\mathbf{e}\boldsymbol{\beta} - S)\mathbf{e} = \mathbf{S}^0$.)

Exercise 4.3.14* (Exercise 4.3.7 continued) Show that, for $s > 0$,

$$\lim_{n\to\infty} E[e^{-sW_{a,n}}] = 1 - \rho + (1-\rho)\phi(I)\left(sI - I\otimes S - R'\otimes(\mathbf{S}^0\boldsymbol{\beta})\right)^{-1}((\boldsymbol{\beta}R)'\otimes\mathbf{S}^0), \quad (4.47)$$

where $\phi(I)$ is the direct-sum of the matrix I, i.e., row vector $\phi(I)$ is obtained by stringing out the rows of I starting from the first row. See Eq. (3.119) in Sect. 3.5. Equation (4.47) shows that the actual waiting time has a matrix-exponential distribution. Further, show that the actual waiting time is *PH*-distributed. (Hint: See Ozawa (2006) and Exercise 4.2.4.)

4.4 The *MAP*/*PH*/1 Queue and Its Variants

In this section, we extend the ideas used in Sects. 4.2 and 4.3 to the construction of structured Markov chains for more general queueing models. We demonstrate further the power of the use of *PH*-distributions and *MAPs* in stochastic modeling. We shall mainly focus on the stochastic modeling part. For all the examples, the theory developed in Chap. 3 can be applied to compute performance measures. In all the queues, the flow of customers is the same as that in the *M*/*M*/1 queue (see Fig. 4.1).

4.4.1 *The MAP/PH/1 Queue*

The arrival process is a Markovian arrival process (*MAP*) with matrix representation (D_0, D_1) of order m_a. Let $\boldsymbol{\theta}_a$ satisfy $\boldsymbol{\theta}_a(D_0 + D_1) = 0$ and $\boldsymbol{\theta}_a\mathbf{e} = 1$. The arrival rate is given by $\lambda = \boldsymbol{\theta}_a D_1\mathbf{e}$. The service times have a common *PH*-distribution with *PH*-representation $(\boldsymbol{\beta}, S)$ of order m_s. We assume $\boldsymbol{\beta}\mathbf{e} = 1$. Let $\mu = -1/(\boldsymbol{\beta}S^{-1}\mathbf{e})$, the service rate. Define

$q(t)$: the queue length at time t, i.e., the total number of customers in the system.
$I_a(t)$: the phase of the *MAP* at time t.
$I_s(t)$: the phase of the service process at time t, if the server is busy at time t; 0, otherwise.

Consider the process $\{(q(t), I_a(t), I_s(t)), t \geq 0\}$. Since $I_a(t)$ keeps track of the phase of the underlying Markov chain for the arrival process and $I_s(t)$ keeps track of the phase of the service process, if the server is working, it is easy to see that $\{(q(t), I_a(t), I_s(t)), t \geq 0\}$ is a continuous time Markov chain with state space $\{\{0\}\times\{1, 2, \ldots, m_a\}\}$ $\cup\{\{1, 2, \ldots\}\times\{1, 2, \ldots, m_a\}\times\{1, 2, \ldots, m_s\}\}$ and infinitesimal generator

$$Q = \begin{pmatrix} D_0 & D_1 \otimes \boldsymbol{\beta} & & & \\ I \otimes \mathbf{S}^0 & D_0 \otimes I + I \otimes S & D_1 \otimes I & & \\ & I \otimes \mathbf{S}^0\boldsymbol{\beta} & D_0 \otimes I + I \otimes S & D_1 \otimes I & \\ & & \ddots & \ddots & \ddots \\ & & & \ddots & \ddots \end{pmatrix}. \tag{4.48}$$

The transition of the process $\{(I_a(t), I_s(t)), t \geq 0\}$ is governed by $(D_0 + D_1)\otimes I + I\otimes(\mathbf{S}^0\boldsymbol{\beta} + S)$ (if $q(t) > 1$), which can be decomposed into arrival (matrix) rate

$D_1 \otimes I$, service completion rate $I \otimes (\mathbf{S}^0 \boldsymbol{\beta})$, and "only phase change" rate $D_0 \otimes I + I \otimes S$. Note that the queue length increases or decreases by at most one if the state of the Markov chain changes. Thus, $\{(q(t), I_a(t), I_s(t)), t \geq 0\}$ is a QBD process. The theory developed in Chap. 3 can be applied to study the stationary distribution of the queue length and the busy period of the system directly. We omit the details.

Exercise 4.4.1 Assume that $D_0 + D_1$ and $S + \mathbf{S}^0\boldsymbol{\beta}$ are irreducible. Show that the QBD process $\{(q(t), I_a(t), I_s(t)), t \geq 0\}$ is ergodic (i.e., the queueing system is stable) if and only if $\rho = \lambda/\mu < 1$. What happens to the QBD process if $D_0 + D_1$ is reducible? What happens to the QBD process if $S + \mathbf{S}^0\boldsymbol{\beta}$ is reducible?

Exercise 4.4.2 Outline a method for computing the matrix R, G, the stationary distribution of the queue length, the mean queue length, and the probability that the queue is empty for the *MAP/PH/*1 queue with

$$\begin{aligned} D_0 &= \begin{pmatrix} -2 & 1 \\ 0 & -1.5 \end{pmatrix}, \quad D_1 = \begin{pmatrix} 1 & 0 \\ 1 & 0.5 \end{pmatrix}, \\ \boldsymbol{\beta} &= (0.8,\ 0.2), \quad S = \begin{pmatrix} -5 & 0 \\ 0.5 & -1 \end{pmatrix}. \end{aligned} \tag{4.49}$$

Note: Check system stability first.

Exercise 4.4.3 In the definition, we assume $\boldsymbol{\beta}\mathbf{e} = 1$ for our queueing system. Consider an *MAP/PH/*1 queue with $0 < \boldsymbol{\beta}\mathbf{e} < 1$. (i) Explain intuitively what might be the consequence of $\boldsymbol{\beta}\mathbf{e} < 1$. (ii) Construct a Markov chain for the queueing process.

Exercise 4.4.4 For the *MAP/PH/*1 queue, denote by $\boldsymbol{\pi} = (\boldsymbol{\pi}_0, \boldsymbol{\pi}_1, \ldots)$ the stationary distribution of $\{(q(t), I_a(t), I_s(t)), t \geq 0\}$, where $\boldsymbol{\pi}_0 = (\pi_{0,1}, \ldots, \pi_{0,m_a})$, $\boldsymbol{\pi}_n = (\boldsymbol{\pi}_{n,1}, \boldsymbol{\pi}_{n,2}, \ldots, \boldsymbol{\pi}_{n,m_a}), \ldots$, for $n = 1, 2, \ldots$, and $\boldsymbol{\pi}_{n,j} = (\pi_{n,j,1}, \pi_{n,j,2}, \ldots, \pi_{n,j,m_s})$, for $n = 1, 2, \ldots$, and $j = 1, 2, \ldots, m_a$. Show that

$$\boldsymbol{\theta}_a = \boldsymbol{\pi}_0 + \sum_{n=1}^{\infty} (\boldsymbol{\pi}_{n,1}\mathbf{e},\ \boldsymbol{\pi}_{n,2}\mathbf{e}, \ldots,\ \boldsymbol{\pi}_{n,m_a}\mathbf{e}). \tag{4.50}$$

Explain Eq. (4.50) intuitively. Let $\mathbf{u} = \sum_{n=1}^{\infty} (\boldsymbol{\pi}_{n,1} + \boldsymbol{\pi}_{n,2} + \ldots + \boldsymbol{\pi}_{n,m_a})$. Is $\mathbf{u}$ the stationary distribution of $S + \mathbf{S}^0\boldsymbol{\beta}$? Why or why not?

4.4.2 The *BMAP/PH/*1 Queue

The arrival process is a batch Markovian arrival process (*BMAP*) with matrix representation $(D_0, D_1, D_2, \ldots, D_N)$ of order m_a, where N is a positive integer. Let $\boldsymbol{\theta}_a$ satisfy $\boldsymbol{\theta}_a(D_0 + D_1 + \ldots + D_N) = 0$ and $\boldsymbol{\theta}_a\mathbf{e} = 1$. The arrival rate is defined as $\lambda = \boldsymbol{\theta}_a(D_1 + 2D_2 + \ldots + ND_N)\mathbf{e}$. The service times have a common

PH-distribution with PH-representation ($\boldsymbol{\beta}$, S) of order m_s. We assume $\boldsymbol{\beta}\mathbf{e} = 1$. Let $\mu = -1/(\boldsymbol{\beta}S^{-1}\mathbf{e})$, the service rate.

Define $q(t)$, $I_a(t)$, and $I_s(t)$ the same as for the $MAP/PH/1$ queue. It is easy to see that $\{(q(t), I_a(t), I_s(t)), t \geq 0\}$ is a continuous time Markov chain with state space $\{\{0\}\times\{1, 2, \ldots, m_a\}\}\cup\{\{1, 2, \ldots\}\times\{1, 2, \ldots, m_a\}\times\{1, 2, \ldots, m_s\}\}$ and infinitesimal generator

$$Q = \begin{pmatrix} D_0 & D_1 \otimes \boldsymbol{\beta} & \cdots & D_N \otimes \boldsymbol{\beta} & & \\ I \otimes \mathbf{S}^0 & D_0 \otimes I + I \otimes S & D_1 \otimes I & \cdots & D_N \otimes I & \\ & I \otimes \mathbf{S}^0\boldsymbol{\beta} & D_0 \otimes I + I \otimes S & D_1 \otimes I & \cdots & D_N \otimes I \\ & & \ddots & \ddots & \ddots & \ddots \end{pmatrix}. \tag{4.51}$$

Note that the queue length may increase by at most N or decreases by at most one when the state of the Markov chain changes. Apparently, $\{(q(t), I_a(t), I_s(t)), t \geq 0\}$ is an $M/G/1$ type Markov chain. The theory developed in Chap. 3 can be applied to study the stationary distribution of the queue length and the busy period directly. We omit the details.

Exercise 4.4.5 For the $BMAP/PH/1$ queue, assume that $D_0 + D_1 + D_2 + \ldots + D_N$ and $S + \mathbf{S}^0\boldsymbol{\beta}$ are irreducible. Show that the $M/G/1$ type Markov chain $\{(q(t), I_a(t), I_s(t)), t \geq 0\}$ is ergodic (i.e., the queueing system is stable) if and only if $\rho = \lambda/\mu < 1$.

Exercise 4.4.6 Based on Theorem 3.5.3, outline a method for computing the stationary distribution of the queue length, the mean queue length, and the probability that the queue is empty for the $BMAP/PH/1$ queue. (Hint: You need to define G_{10} for the transition from level 1 to level 0 since the number of phases in level 0 is less than that of other levels.) Consider the following example:

$$D_0 = \begin{pmatrix} -2 & 1 \\ 0 & -1.5 \end{pmatrix}, \quad D_1 = \begin{pmatrix} 0.8 & 0 \\ 1 & 0.4 \end{pmatrix}, \quad D_2 = \begin{pmatrix} 0.2 & 0 \\ 0 & 0.1 \end{pmatrix},$$

$$\boldsymbol{\beta} = (0.8,\ 0.2), \quad S = \begin{pmatrix} -5 & 0 \\ 0.5 & -1 \end{pmatrix}. \tag{4.52}$$

If N is finite, the $M/G/1$ type Markov chain can be transformed into a QBD process by *re-blocking*, which is a quite useful technique in stochastic modeling.

Exercise 4.4.7 Reblock the matrix Q in Eq. (4.51) and outline a method to find the stationary distribution of the queue length for the $BMAP/PH/1$ queue. Solve the numerical example given in Exercise 4.4.6 by re-blocking. (Hint: There are different ways for re-blocking. They are different in how they define a new level 0. Choose the one you feel most comfortable to work with. For instance, you can group levels $3k + 1$, $3k + 2$, and $3k + 3$ together for $k = 0, 1, 2, \ldots$, if $N = 3$.)

Exercise 4.4.8 For the *BMAP/PH/*1 queue, denote by $\boldsymbol{\pi} = (\boldsymbol{\pi}_0, \boldsymbol{\pi}_1, \ldots)$ the stationary distribution of $\{(q(t), I_a(t), I_s(t)), t \geq 0\}$, where $\boldsymbol{\pi}_0 = (\pi_{0,1}, \ldots, \pi_{0,m_a})$, $\boldsymbol{\pi}_1 = (\boldsymbol{\pi}_{1,1}, \ldots, \boldsymbol{\pi}_{1,m_a})$, $\boldsymbol{\pi}_2 = (\boldsymbol{\pi}_{2,1}, \ldots, \boldsymbol{\pi}_{2,m_a})$, …. Show that

$$\boldsymbol{\theta}_a = \boldsymbol{\pi}_0 + \sum_{n=1}^{\infty} (\boldsymbol{\pi}_{n,1}\mathbf{e}, \boldsymbol{\pi}_{n,2}\mathbf{e}, \ldots, \boldsymbol{\pi}_{n,m_a}\mathbf{e}). \tag{4.53}$$

Explain Eq. (4.53) intuitively. Let $\mathbf{u} = \sum_{n=1}^{\infty} (\boldsymbol{\pi}_{n,1} + \boldsymbol{\pi}_{n,2} + \ldots + \boldsymbol{\pi}_{n,m_a})$. Is $\mathbf{u}$ the stationary distribution of $S + \mathbf{S}^0\boldsymbol{\beta}$? Why or why not?

4.4.3 The *MAP/PH*$^{[2]}$*/1 Queue*

In the *MAP/PH*$^{[2]}$/1 queue, two customers are served simultaneously. Specifically, the server begins a service if there is at least one customer in the system. If there are two or more customers in the system, then two of them enter the server and receive service. If there is one customer in the system, the server serves the customer. However, if a new customer arrives during the service of a single customer, then the new customer also enters the server to receive service upon arrival, and the two customers in service complete their service together.

The arrival process is a Markovian arrival process (*MAP*) with irreducible matrix representation (D_0, D_1) of order m_a. Let $\boldsymbol{\theta}_a$ satisfy $\boldsymbol{\theta}_a(D_0 + D_1) = 0$ and $\boldsymbol{\theta}_a\mathbf{e} = 1$. The arrival rate is defined as $\lambda = \boldsymbol{\theta}_a D_1\mathbf{e}$. The service times have a common *PH*-distribution function with *PH*-representation $(\boldsymbol{\beta}, S)$ of order m_s. We assume $\boldsymbol{\beta}\mathbf{e} = 1$. Let $\mu = -2/(\boldsymbol{\beta}S^{-1}\mathbf{e})$, the maximum service rate, since each service may complete two customers.

Define $q(t)$, $I_a(t)$, and $I_s(t)$ the same as that for the *MAP/PH/*1 queue. It is easy to see that $\{(q(t), I_a(t), I_s(t)), t \geq 0\}$ is a continuous time Markov chain with state space $\{\{0\}\times\{1, 2, \ldots, m_a\}\}\cup\{\{1, 2, \ldots\}\times\{1, 2, \ldots, m_a\}\times\{1, 2, \ldots, m_s\}\}$ and infinitesimal generator

$$Q = \begin{pmatrix} D_0 & D_1 \otimes \boldsymbol{\beta} & & & \\ I \otimes \mathbf{S}^0 & D_0 \otimes I + I \otimes S & D_1 \otimes I & & \\ I \otimes \mathbf{S}^0 & 0 & D_0 \otimes I + I \otimes S & D_1 \otimes I & \\ & I \otimes \mathbf{S}^0\boldsymbol{\beta} & 0 & D_0 \otimes I + I \otimes S & D_1 \otimes I \\ & & \ddots & \ddots & \ddots & \ddots \end{pmatrix}. \tag{4.54}$$

Note that the queue length increases by at most one or decreases by at most two when the state of the Markov chain changes. It is apparent that $\{(q(t), I_a(t), I_s(t)), t \geq 0\}$ is a *GI/M/*1 type Markov chain. The theory developed in Chap. 3 can be applied to study the stationary distribution of the queue length and the busy period directly. We omit the details.

Exercise 4.4.9 Assume that $D_0 + D_1$ and $S + \mathbf{S}^0\boldsymbol{\beta}$ are irreducible for the *MAP/PH*$^{[2]}$/1 queue. Show that the *GI/M/*1 type Markov chain $\{(q(t), I_a(t), I_s(t)), t \geq 0\}$ is ergodic (i.e., the queueing system is stable) if and only if $\rho = \lambda/\mu < 1$.

Exercise 4.4.10 Outline a method for computing the matrix R, the stationary distribution of the queue length, and the probability that the queue is empty for the *MAP/PH*$^{[2]}$/1 queue. Consider an example with parameters given by Eq. (4.49), but the service batch size is two.

Exercise 4.4.10 can also be solved by using the reblocking technique to transform the *GI/M/*1 type Markov chain into a QBD process.

Exercise 4.4.11 Consider a *BMAP/PH*$^{[2]}$/1 queue with batch arrivals and batch services. System parameters are given in Eq. (4.52). Use the idea of reblocking (into a QBD process) to develop a method for finding the stationary distribution of the queue length.

Exercise 4.4.12 Consider an $M^{[2]}/M^{[2]}/1$ queue, where customers arrive, in batches of size 2, according to a Poisson process with parameter $\lambda = 1$; customers are served in batches of size 1 or 2 (i.e., if there are two or more than two customers in the system at a service completion epoch, two customers leaves the system.). Service times are exponentially distributed with parameter $\mu = 1.5$. Is the queueing system stable? Explain your answer intuitively. If your answer is yes, find the stationary distribution of the queue length.

Exercise 4.4.13 (The *MAP/PH/1/N* queue with limited waiting space) In an *MAP/PH/*1 queue, if an arriving customer finds that the server is busy and there are N customers waiting, the customer leaves the system immediately without service. Introduce a Markov chain for this queueing model. Outline an algorithm for computing the stationary distribution of its queue length.

Exercise 4.4.14 For the *MAP/PH*$^{[2]}$/1 queue, denote by $\boldsymbol{\pi} = (\boldsymbol{\pi}_0, \boldsymbol{\pi}_1, \ldots)$ the stationary distribution of $\{(q(t), I_a(t), I_s(t)), t \geq 0\}$, where $\boldsymbol{\pi}_0 = (\pi_{0,1}, \ldots, \pi_{0,m_a})$, $\boldsymbol{\pi}_1 = (\boldsymbol{\pi}_{1,1}, \ldots, \boldsymbol{\pi}_{1,m_a})$, $\boldsymbol{\pi}_2 = (\boldsymbol{\pi}_{2,1}, \boldsymbol{\pi}_{2,2}, \ldots, \boldsymbol{\pi}_{2,m_a})$, ... Show that

$$\boldsymbol{\theta}_a = \boldsymbol{\pi}_0 + \sum_{n=1}^{\infty} (\boldsymbol{\pi}_{n,1}\mathbf{e},\ \boldsymbol{\pi}_{n,2}\mathbf{e}, \ldots,\ \boldsymbol{\pi}_{n,m_a}\mathbf{e}). \tag{4.55}$$

Explain Eq. (4.55) intuitively. Let $\mathbf{u} = \sum_{n=1}^{\infty} (\boldsymbol{\pi}_{n,1} + \boldsymbol{\pi}_{n,2} + \ldots + \boldsymbol{\pi}_{n,m_a})$. Is $\mathbf{u}$ the stationary distribution of $S + \mathbf{S}^0\boldsymbol{\beta}$? Why or why not?

4.4.4 The *MAP/MAP/*1 Queue

The flow of customers in the *MAP/MAP/*1 queue is the same as that in the *M/M/*1 queue. The arrival process is a Markovian arrival process (*MAP*) with matrix representation (D_0, D_1) of order m_a. The service process is also a Markovian arrival process, called a *Markovian service process*, with matrix representation (C_0, C_1) of order m_s, where C_0 includes the rates of transitions without a service completion and C_1 includes the rates of transitions with a service completion. When the queueing system starts a new busy period, the service phase may have to be initialized. There can be many ways to initialize the service phase at the beginning of a busy period. Here are three possibilities:

Case 1 The service phase is initialized according to stochastic vector $\boldsymbol{\beta}$ at the beginning of a busy period.

Case 2 The service phase at the beginning of a busy period is the same as the phase at the end of the proceeding busy period (i.e., the service phase is frozen during a system idle period.)

Case 3 The service process continues during the system idle period, although there is no customer served during this period of time. Thus, the service phase at the beginning of a busy period is the phase of the service process at the end of the last idle period.

Define $q(t)$, $I_a(t)$, and $I_s(t)$ the same as that for the *MAP/PH/*1 queue. It is easy to see that $\{(q(t), I_a(t), I_s(t)), t \geq 0\}$ is a continuous time Markov chain with state space $\{\{0\}\times\{1, 2, \ldots, m_a\}\times\{1, 2, \ldots, m_0\}\}\cup\{\{1, 2, \ldots\}\times\{1, 2, \ldots, m_a\}\times\{1, 2, \ldots, m_s\}\}$, where m_0 depends on the service process during an idle period. The infinitesimal generator of the Markov chain is

$$Q=\begin{pmatrix} Q_{0,0} & Q_{0,1} & & & \\ Q_{1,0} & D_0\otimes I+I\otimes C_0 & D_1\otimes I & & \\ & I\otimes C_1 & D_0\otimes I+I\otimes C_0 & D_1\otimes I & \\ & & I\otimes C_1 & D_0\otimes I+I\otimes C_0 & D_1\otimes I \\ & & & \ddots & \ddots & \ddots \end{pmatrix}, \tag{4.56}$$

where $Q_{0,0}$, $Q_{0,1}$, and $Q_{1,0}$ are specified as follows.

For Case 1, $Q_{0,0} = D_0$, $Q_{0,1} = D_1\otimes\boldsymbol{\beta}$, and $Q_{1,0} = I \otimes(C_1\mathbf{e})$.
For Case 2, $Q_{0,0} = D_0\otimes I$, $Q_{0,1} = D_1\otimes I$, and $Q_{1,0} = I \otimes C_1$.
For Case 3, $Q_{0,0} = D_0\otimes I + I\otimes (C_0 + C_1)$, $Q_{0,1} = D_1\otimes I$, and $Q_{1,0} = I \otimes C_1$.

Exercise 4.4.15 Implement an algorithm for computing the mean queue length for the *MAP/MAP/*1 queue.

Exercise 4.4.16 For the *MAP/MAP/*1 queue, denote by $\boldsymbol{\pi} = (\boldsymbol{\pi}_0, \boldsymbol{\pi}_1, \ldots)$ the stationary distribution of $\{(q(t), I_a(t), I_s(t)), t \geq 0\}$, where $\boldsymbol{\pi}_0 = (\pi_1, \ldots, \boldsymbol{\pi}_{m_a})$, $\boldsymbol{\pi}_1 = (\boldsymbol{\pi}_{1,1}, \ldots, \boldsymbol{\pi}_{1,m_a})$, $\boldsymbol{\pi}_2 = (\boldsymbol{\pi}_{2,1}, \boldsymbol{\pi}_{2,2}, \ldots, \boldsymbol{\pi}_{2,m_a})$, ... Show that

$$\boldsymbol{\theta}_a = \boldsymbol{\pi}_0 + \sum_{n=1}^{\infty} (\boldsymbol{\pi}_{n,1}\mathbf{e},\ \boldsymbol{\pi}_{n,2}\mathbf{e}, \ldots,\ \boldsymbol{\pi}_{n,m_a}\mathbf{e}). \tag{4.57}$$

Explain Eq. (4.57) intuitively. Let $\mathbf{u} = \sum_{n=1}^{\infty}(\boldsymbol{\pi}_{n,1} + \boldsymbol{\pi}_{n,2} + \ldots + \boldsymbol{\pi}_{n,m_a})$. Is $\mathbf{u}$ the stationary distribution of $C_0 + C_1$? Why or why not?

Although the matrix-geometric solution can be computed efficiently for small and moderate block size m, the method may be less efficient if m is large (e.g., greater than 10,000), especially if the boundary condition is complex. For such cases, classical methods such as Gaussian elimination method and Gauss-Seidel method can be used for computing the stationary distribution of the queue length.

Exercise 4.4.17 (Gauss-Seidel method) Construct a case with $m_a = 100$ and $m_s = 100$. Compute the limiting probabilities by using the matrix-geometric solution and by using Gauss-Seidel method as follows: (i) truncate the Markov chain at level N to obtain matrix Q_N; (ii) replace the first column of Q_N with $\mathbf{e}$, and denote the matrix as $Q_{N,1}$; (iii) decompose $Q_{N,1}$ into $Q_{N,1} = D - L - U$, where D is a diagonal matrix, L is a lower triangular matrix with zeros on the diagonal, and U is an upper triangular matrix with zeros on the diagonal; (iv) $\mathbf{b}$ is a column vector with the first element being one and the rest is zero; and (v) let $\mathbf{x}_0 = 0$, and, for $k = 1, 2, \ldots$,

$$\mathbf{x}_{k+1} = D^{-1}L\mathbf{x}_{k+1} + D^{-1}U\mathbf{x}_k + D^{-1}\mathbf{b}. \tag{4.58}$$

It can be shown that $\{\mathbf{x}_0, \mathbf{x}_1, \ldots\}$ converges to $\mathbf{x}^*$, which is an approximate of the stationary distribution $\boldsymbol{\pi}$. Implement the algorithm for computing $\boldsymbol{\pi}$ approximately. Comment on the efficiency of the Gauss-Seidel approach and the matrix-geometric solution approach numerically.

4.4.5 The *M/M/*1 Vacation Queue

Consider an *M/M/*1 queue with multiple vacations. If the system is empty, the server goes on vacation. When the server returns from a vacation and there is no customer in the system, the server goes on vacation again; otherwise, the server begins to serve customers until the system becomes empty. Assume that the vacation times have Erlang distributions with common parameters (m, v). We also assume that the vacation times, service times, and the arrival process are mutually independent.

Let $I(t)$ be the phase of the Erlang vacation time at time t, if the server is on vacation. Assume that $I(t) = 0$ if the server is busy at time t. The state space of $\{(q(t), I(t)), t \geq 0\}$ is $\{(0, 1), \ldots, (0, m)\} \cup \{(q, 0), (q, 1), \ldots, (q, m), q \geq 1\}$. Then it is easy to see that $\{(q(t), I(t)), t \geq 0\}$ is a continuous time Markov chain with infinitesimal generator

$$P = \begin{pmatrix} A_{0,0} & A_{0,1} & & & \\ A_{1,0} & A_1 & A_0 & & \\ & A_2 & A_1 & A_0 & \\ & & \ddots & \ddots & \ddots \\ & & & \ddots & \ddots \end{pmatrix}, \tag{4.59}$$

where

$$A_{0,0} = \begin{matrix} (0,1) \\ \vdots \\ (0,m-1) \\ (0,m) \end{matrix} \begin{pmatrix} -(\lambda+\nu) & \nu & & \\ & \ddots & \ddots & \\ & & -(\lambda+\nu) & \nu \\ \nu & & & -(\lambda+\nu) \end{pmatrix}; \quad A_{0,1} = (0, \lambda I); \quad (4.60)$$

$$A_{1,0} = \begin{matrix} (1,0) \\ (1,1) \\ \vdots \\ (1,m-1) \\ (1,m) \end{matrix} \begin{pmatrix} \mu & & & \\ 0 & 0 & & \\ & \ddots & \ddots & \\ & & 0 & 0 \\ & & & 0 \end{pmatrix}; \quad A_2 = \begin{matrix} (q,0) \\ (q,1) \\ \vdots \\ (q,m-1) \\ (q,m) \end{matrix} \begin{pmatrix} \mu & & & \\ & 0 & & \\ & & \ddots & \\ & & & \ddots \\ & & & & 0 \end{pmatrix};$$

$$A_0 = \lambda I; \quad A_1 = \begin{matrix} (q,0) \\ (q,1) \\ \vdots \\ (q,m-1) \\ (q,m) \end{matrix} \begin{pmatrix} -(\lambda+\mu) & 0 & & & \\ & -(\lambda+\nu) & \nu & & \\ & & \ddots & \ddots & \\ & & & -(\lambda+\nu) & \nu \\ & \nu & & & -(\lambda+\nu) \end{pmatrix},$$

for $q \geq 1$.

If $\lambda < \mu$, the Markov chain is ergodic (i.e., the queueing system is stable). The matrix-geometric solution can be found for the stationary distribution of the queue length: $\boldsymbol{\pi}_n = \boldsymbol{\pi}_1 R^{n-1}$, $n = 1, 2, \ldots$. For this special case, we study the tail asymptotics of the matrix-geometric distribution (see Corollary 3.2.2).

Suppose that ξ is the Perron-Frobenius eigenvalue of R. Define $A^*(z) = A_0 + zA_1 + z^2A_2$.

Exercise 4.4.18 Show that zero is an eigenvalue of matrix $A^*(\xi)$.

It is easy to find

$$A^*(z) = \begin{pmatrix} \lambda-(\lambda+\mu)z+z^2\mu & 0 & & & \\ & \lambda-(\lambda+\nu)z & \nu z & & \\ & & \ddots & \ddots & \\ & & & \lambda-(\lambda+\nu)z & \nu z \\ & \nu z & & & \lambda-(\lambda+\nu)z \end{pmatrix}. \quad (4.61)$$

Exercise 4.4.19 Show $\det(A^*(z)) = (\lambda - (\lambda+\mu)z + z^2\mu)(\lambda - (\lambda+\nu)z)^m$, where det(.) is determinant of matrix. The root of $\det(A^*(z))$ with the largest modulus is

$$\omega = \begin{cases} \lambda/\mu, & \text{if } \mu \le \lambda + \nu; \\ \lambda/(\lambda+\nu), & \text{if } \mu > \lambda + \nu. \end{cases} \tag{4.62}$$

By Exercise 4.4.19 and Ramaswami and Taylor (1996), $\xi = \omega$ holds. We also know that ξ is an eigenvalue of R of multiplicity one, if $\mu < \lambda + \nu$; of multiplicity $m + 1$, if $\mu = \lambda + \nu$; and of multiplicity m, if $\mu > \lambda + \nu$.

Exercise 4.4.20* For the stationary distribution, show that

$$\boldsymbol{\pi}_n = \boldsymbol{\pi}_1 R^{n-1} = \begin{cases} \xi^n \mathbf{u} + o(\xi^n), & \text{if } \mu < \lambda + \nu; \\ n^{m+1}\xi^n \mathbf{u} + o(n^{m+1}\xi^n), & \text{if } \mu = \lambda + \nu; \\ n^m \xi^n \mathbf{u} + o(n^m \xi^n), & \text{if } \mu > \lambda + \nu, \end{cases} \tag{4.63}$$

where **u** is a positive constant vector.

Commentary A special case of the *MAP*/*PH*/1 queue is the *PH*/*PH*/1 queue, which is a queue with a *PH*-arrival process and a *PH*-service time (see Takahashi (1981)). It is clear that many variants of the queueing models presented in this section can be analyzed in a similar way. In fact, almost all queueing models with *MAPs* for arrivals and *PHs* for service times can be analyzed similarly. Based on Ozawa (2006), the distributions of the waiting and sojourn times in all the queues considered in this section, i.e., queues associated with a QBD process, are of phase-type, and their *PH*-representations can be found easily from the stationary distribution of the queue length.

Additional Exercises and Extensions

Exercise 4.4.21 Consider an $M/PH^{[X]}/1$ queue, where X has a general discrete distribution on positive integers, and X is the maximum number of customers to be served simultaneously in each service. Introduce a continuous time Markov chain for this queueing model. Find the condition under which the queueing model is stable. (Note: The number of customers served in a service is determined at the completion of the service by X and the queue length.)

Exercise 4.4.22 Customers arrive at a single server queue according to a Poisson process with parameter λ. The service times have a common exponential distribution with parameter μ. However, an arrival that finds n customers already in the system will only join the system with probability $1/(n + 1)$. That is, with probability $n/(n + 1)$ the newly arrived customer will not join the system. Let $X(t)$ be the number of customers in the system (including the one in service) at time t.

1. Argue that the arrival rate of customers (actually) joining the queue is $\lambda/(n + 1)$, if there are n customers in the system. Show that $\{X(t), t \geq 0\}$ is a continuous time Markov chain and find its infinitesimal generator. Is the continuous time Markov chain $\{X(t), t \geq 0\}$ *time reversible*? (Note: For definition of time reversibility of continuous time Markov chains, see Chap. 6 in Ross (2010).)
2. Find the limiting probability $p_j = \lim_{t\to\infty} P\{X(t) = j|X(0)\}$, for $j = 0, 1, 2, \ldots$. Find the probability that an arbitrary customer will join the queue. (Note: The final results must be simple and explicit.)
3. Assume that customers arrive according to the Markovian arrival process with matrix representation (D_0, D_1). The service times have a common *PH*-distribution with *PH*-representation $(\boldsymbol{\alpha}, T)$. Introduce a continuous time Markov chain for $X(t)$ and find its infinitesimal generator.

Exercise 4.4.23 Consider the *MAP/PH/1/N* queue with a finite waiting space. There can be at most N customers waiting in the system. That is, if an arriving customer finds N waiting customers, that customer is lost. Customers leaving the system can be categorized into two types: lost and served.

(1) Introduce Markov chain $\{(q(t), I_a(t), I_s(t)), t \geq 0\}$ for this queueing system and find its infinitesimal generator.
(2) Define an *MMAP* with two types of customers, lost or served customers, for the departure process of the queue.
(3) Define an *MMAP* with $N + 2$ types of customers such that a type j customer leaves the queueing system with j customers in the system, $j = 0, 1, 2, \ldots, N + 1$. (Note: If a customer is lost, then the customer leaves the queueing system with $N + 1$ customers.)
(4) Define an *MMAP* with $N + 2$ types of customers such that a type j customer finds j customers in the system at its arrival epoch, $j = 0, 1, 2, \ldots, N + 1$.
(5) Define an *MMAP* with $2N + 4$ types of customers, which is the superposition of the *MMAP*s defined in (3) and (4).

Exercise 4.4.24 Consider the *MAP/PH/1* queue. Show that the time until the queue length becomes five has a *PH*-distribution, given that the queueing system is empty at time zero and the distribution of the phase of the underlying Markov chain of the arrival process is $\boldsymbol{\alpha}$ at time zero. Present a *PH*-representation of this time.

Exercise 4.4.25 Consider the *MAP/PH/1* queue. Show that the number of customers who arrive during the time period until the queue length becomes five has a discrete *PH*-distribution, given that the queueing system is empty at time zero and the distribution of the phase of the underlying Markov chain of the arrival process is $\boldsymbol{\alpha}$ at time zero. Present a matrix-representation of the *PH*-distribution.

Exercise 4.4.26 Consider the *MAP/PH/1* queue. Find the distributions of the waiting time and the sojourn time of an arbitrary customer. (Hint: Ozawa (2006).)

4.5 Two *M*/*M*/1 Queues with Customer Transfers

This section considers a system consisting of two *M*/*M*/1 queues with transfers of customers between them. The two queues are denoted as queue 1 and queue 2. Customers of queue i arrive according to a Poisson process with parameters λ_i, $i = 1, 2$. The service times of queue i have a common exponential distribution with parameter μ_i, $i = 1, 2$. We assume that the service times and the arrival processes are mutually independent. The two queues operate independently, except for the transfers of customers between them. Whenever the lengths of the two queues differ by L, K customers are transferred from the longer to the shorter queue, where L and K are positive integers with $K < L$ (See Fig. 4.4).

For the queueing system of interest, the technique used in the last few sections does not generate a QBD, *M*/*G*/1, or *GI*/*M*/1 type Markov chain directly. In this section, we demonstrate how to transform a non-QBD process into a QBD process.

Let $q_i(t)$ be the length of queue i at time t, $i = 1, 2$. Since the arrival processes are Poisson and service times are exponential, $\{(q_1(t), q_2(t)), t \geq 0\}$ is a continuous time Markov chain with state space $\{(n_1, n_2): n_1 \geq 0, n_2 \geq 0, |n_1 - n_2| < L\}$. The Markov chain is irreducible since every state communicates with state (0, 0).

Exercise 4.5.1 Show that the transition rates for the Markov chain $\{(q_1(t), q_2(t)), t \geq 0\}$ are given as follows

$$q_{(n_1,n_2),(y_1,y_2)} = \begin{cases} \lambda_1, & \begin{cases} \text{if } y_1 = n_1+1 < n_2+L,\ y_2 = n_2; \\ \text{if } y_1 = n_1+1-K,\ y_2 = n_2+K,\ n_1 = n_2+L-1; \end{cases} \\ \lambda_2, & \begin{cases} \text{if } y_2 = n_2+1 < n_1+L,\ y_1 = n_1; \\ \text{if } y_2 = n_2+1-K,\ y_1 = n_1+K,\ n_2 = n_1+L-1; \end{cases} \\ \mu_1, & \begin{cases} \text{if } y_1 = n_1-1 > n_2-L,\ y_2 = n_2,\ n_1 \geq 1; \\ \text{if } y_1 = n_1-1+K,\ y_2 = n_2-K,\ n_1 = n_2-L+1, n_1 \geq 1; \end{cases} \\ \mu_2, & \begin{cases} \text{if } y_2 = n_2-1 > n_1-L,\ y_1 = n_1,\ n_2 \geq 1; \\ \text{if } y_2 = n_2-1+K,\ y_1 = n_1-K,\ n_2 = n_1-L+1,\ n_2 \geq 1; \end{cases} \\ 0, & \text{Otherwise.} \end{cases} \tag{4.64}$$

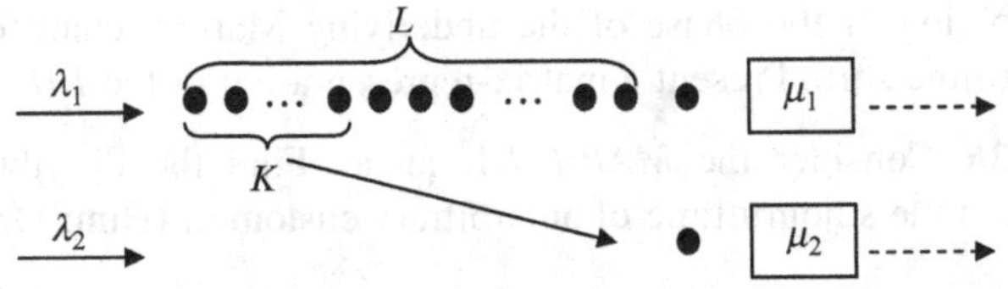

Fig. 4.4 Two *M*/*M*/1 queues with customer transfers

The queueing system is stable if and only if the Markov chain $\{(q_1(t), q_2(t)), t \geq 0\}$ is positive recurrent. Define

$$\rho = \frac{\lambda_1 + \lambda_2}{\mu_1 + \mu_2}. \tag{4.65}$$

Theorem 4.5.1 (He and Neuts (2002)) *The queueing system is stable if and only if* $\rho < 1$.

Exercise 4.5.2 Explain intuitively why the queueing system is stable if and only if $\rho < 1$ for ρ defined in Eq. (4.65). (Note: A proof of Theorem 4.5.1 can be found in He and Neuts (2002). The proof is based on Foster's criterion and a special Lyaponov function.)

The process $\{(q_1(t), q_2(t)), t \geq 0\}$ is a two dimensional Markov chain. Suppose that we use $q_1(t)$ as the level variable. The level variable may decrease or increase by K, depending on the value of $q_2(t)$. Thus, the process is not of QBD type. In addition, for each value of $q_1(t)$, $q_2(t)$ takes only a finite number of values, but the values are changing. In order to analyze $\{(q_1(t), q_2(t)), t \geq 0\}$ effectively, we transform the process into a QBD process. Let

$$q(t) = q_1(t) + q_2(t), \quad J(t) = q_1(t) - q_2(t), \quad t \geq 0. \tag{4.66}$$

Then $q(t)$ is the total number of customers in the system and $J(t)$ represents the difference between the two queue lengths. Clearly, $\{(q_1(t), q_2(t)), t \geq 0\}$ and $\{(q(t), J(t)), t \geq 0\}$ determine each other uniquely. In fact, we have

$$q_1(t) = \frac{q(t) + J(t)}{2}; \quad q_2(t) = \frac{q(t) - J(t)}{2}, \quad t \geq 0. \tag{4.67}$$

It is readily seen that $\{(q(t), J(t)), t \geq 0\}$ is a continuous time Markov chain. Since the state of the queueing system changes only at customer arrival or service completion epochs, $q(t)$ changes its value by one at its transition epochs. The variable $J(t)$ takes finite values $\{L-1, L-2, \ldots, 1, 0, -1, \ldots, -(L-2), -(L-1)\}$. Therefore, $\{(q(t), J(t)), t \geq 0\}$ is a QBD process.

The state space of the Markov chain $\{(q(t), J(t)), t \geq 0\}$ can be divided into levels according to the value of $q(t)$. The states in each level are explicitly given as follows.

(i) For $0 \leq n \leq L - 1$, level n has $n + 1$ states: (n, n), $(n, n - 2)$, …, $(n, -(n - 2))$, and $(n, -n)$.

(ii) For $n \geq L$, there are two cases:

(ii.1) If $n - L$ is odd, level n has L states: $(n, L - 1)$, $(n, L - 3)$, …, $(n, -(L - 3))$, and $(n, -(L - 1))$.

(ii.2) If $n - L$ is even, level n has $L - 1$ states: $(n, L - 2)$, $(n, L - 4)$, …, $(n, -(L - 4))$, and $(n, -(L - 2))$.

Example 4.5.1 Assume $L = 3$. The state space of $\{(q(t), J(t)), t \geq 0\}$ is

$n = 0$: (0, 0);
$n = 1$: (1, 1), (1, −1);
$n = 2$: (2, 2), (2, 0), (2, −2);
$n = 3$: (3, 1), (3, −1);
$n = 4$: (4, 2), (4, 0), (4, −2);
$n = 3 + 2s$: $(n, 1), (n, -1), s = 1, 2, \ldots$;
$n = 3 + 2s + 1$: $(n, 2), (n, 0), (n, -2), s = 1, 2, \ldots$.

Based on the above arrangement of states, the infinitesimal generator of the Markov chain $\{(q(t), J(t)), t \geq 0\}$ is given as:

$$Q_1 = \begin{pmatrix} A_{0,1} & A_{0,0} & & & & & \\ A_{1,2} & A_{1,1} & A_{1,0} & & & & \\ & \ddots & \ddots & \ddots & & & \\ & & A_{L-1,2} & A_{L-1,1} & A_{L-1,0} & & \\ & & & A_{L,2} & A_{L,1} & A_{L,0} & \\ & & & & A_{L+1,2} & A_{L+1,1} & A_{L+1,0} \\ & & & & \ddots & \ddots & \ddots \end{pmatrix}, \quad (4.68)$$

where $A_{L+2n,i} = A_{L,i}$ and $A_{L+2n+1,i} = A_{L+1,i}$, for $i = 0, 1, 2$ and $n = 0, 1, 2, \ldots$.

Exercise 4.5.3 Define all the transition block matrices in Q_1 explicitly.

The construction shows that the QBD process $\{(q(t), J(t)), t \geq 0\}$ is level dependent. The analysis of level dependent QBD processes is, in general, difficult. We need to further reorganize the state space of $\{(q(t), J(t)), t \geq 0\}$ to generate a level independent QBD process.

According to (ii.1) and (ii.2), for $n \geq 0$, the level $L + 2n$ has $L - 1$ states and the level $L + 2n + 1$ has L states. If we regroup the states in the levels $L + 2n$ and $L + 2n + 1$, we obtain a new set with $2L - 1$ states for all $n \geq 0$. We call the new set the level $L + n$, whose states are arranged as: $(L + 2n, L - 2)$, $(L + 2n, L - 4)$, …, $(L + 2n, -(L - 4))$, $(L + 2n, -(L - 2))$, $(L + 2n + 1, L - 1)$, $(L + 2n + 1, L - 3)$, …, $(L + 2n + 1, -(L - 3))$, and $(L + 2n + 1, -(L - 1))$.

Example 4.5.2 (Example 4.5.1 continued) Assume $L = 3$. Combining the levels $L + 2s$ and $L + 2s + 1$ in the state space of $\{(q(t), J(t)), t \geq 0\}$ yields

$n = 0$: (0, 0);
$n = 1$: (1, 1), (0, −1);
$n = 2$: (2, 2), (2, 0), (2, −2);
$n = 3$: (3, 1), (3, −1), (3, 2), (3, 0), (3, −2);
$n \geq 4$: $(n, 1), (n, -1), (n, 2), (n, 0), (n, -2)$;

As it is shown in Example 4.5.2, after regrouping the states, the resulting QBD process is level independent. Corresponding to the new partition of the state space, we introduce new Markov chain $\{(X(t), J(t)), t \geq 0\}$ as

$$X(t) = \begin{cases} q(t), & \text{if } q(t) \le L-1; \\ L + \left\lfloor \dfrac{q(t)-L}{2} \right\rfloor, & \text{if } q(t) \ge L, \end{cases} \tag{4.69}$$

and $J(t)$ given in Eq. (4.66), where "$\lfloor x \rfloor$" is the largest integer that is equal to or smaller than x. Clearly, the continuous time Markov chain $\{(X(t), J(t)), t \ge 0\}$ is irreducible and level independent. The infinitesimal generator Q_2 of $\{(X(t), J(t)), t \ge 0\}$ is given by

$$Q_2 = \begin{pmatrix} A_{0,1} & A_{0,0} & & & & & & \\ A_{1,2} & A_{1,1} & A_{1,0} & & & & & \\ & \ddots & \ddots & \ddots & & & & \\ & & A_{L-1,2} & A_{L-1,1} & A^*_{L-1,0} & & & \\ & & & A^*_{L,2} & A_1 & A_0 & & \\ & & & & A_2 & A_1 & A_0 & \\ & & & & & A_2 & A_1 & A_0 \\ & & & & & & \ddots & \ddots & \ddots \end{pmatrix}. \tag{4.70}$$

Again, the blocks in Q_2 can be given explicitly. We omit the details.

Exercise 4.5.4 Assume $L = 3$ and $K = 2$. Find all the blocks in the infinitesimal generator Q_2.

Since $\{(X(t), J(t)), t \ge 0\}$ is a level independent QBD process, its stationary distribution (if it exists) has a matrix geometric solution. Let R be the minimal nonnegative (matrix) solution to the equation

$$A_0 + RA_1 + R^2A_2 = 0. \tag{4.71}$$

Define matrices $\{R_L, R_{L-1}, \ldots, R_1\}$ as

$$\begin{aligned} R_L &= -A^*_{L-1,0}(A_1 + RA_2)^{-1}; \\ R_{L-1} &= -A_{L-2,0}(A_{L-1,1} + R_L A^*_{L,2})^{-1}; \\ R_n &= -A_{n-1,0}(A_{n,1} + R_{n+1}A_{n+1,2})^{-1}, \quad n = 1, 2, \ldots, L-2. \end{aligned} \tag{4.72}$$

Denote by $\boldsymbol{\pi}_n$ the stationary probability vector of the Markov chain $\{(X(t), J(t)), t \ge 0\}$ corresponding to level n. Note that the elements of the vector $\boldsymbol{\pi}_n$ are indexed according to the actual value of $J(t)$ (e.g., $\boldsymbol{\pi}_n = (\pi_{n,L-2}, \pi_{n,L-4}, \ldots, \pi_{n,-(L-4)}, \pi_{n,-(L-2)}, \pi_{n,L-1}, \pi_{n,L-3}, \ldots, \pi_{n,-(L-3)}, \pi_{n,-(L-1)})$, for $n \ge L$). The stationary probability vectors are given as

$$\begin{aligned} \pi_0 &= \left(1 + R_1\mathbf{e} + R_1R_2\mathbf{e} + \cdots + R_1R_2\cdots R_{L-1}\mathbf{e} + R_1R_2\cdots R_L(I-R)^{-1}\mathbf{e}\right)^{-1}; \\ \boldsymbol{\pi}_n &= \pi_0 R_1 R_2 \cdots R_n, \qquad n = 1, \ldots, L; \\ \boldsymbol{\pi}_n &= \pi_0 R_1 R_2 \cdots R_L R^{n-L}, \qquad n = L+1, L+2, \ldots. \end{aligned} \tag{4.73}$$

Based on the stationary distribution, a number of performance measures for a stable queueing system can be derived.

Exercise 4.5.5 Consider the queueing model with $L = 3$, $K = 2$, $\lambda_1 = 1$, $\lambda_2 = 2$, $\mu_1 = 1.5$, and $\mu_2 = 2.5$. Find the joint distribution of the queue lengths in the following two steps.

(i) Use Eq. (4.73) to find the stationary distribution of $\{(X(t), J(t)), t \geq 0\}$.
(ii) Find the joint distribution of $\{(q_1(t), q_2(t)), t \geq 0\}$.

Commentary More details on the queueing model presented in this section can be found in He and Neuts (2002). Queueing systems with customer transfers have been studied extensively. Both explicit and approximation solutions have been found for various models (e.g., Zhao and Grassmann (1990, 1995), Hassin and Haviv (1999), and references therein).

Additional Exercises and Extensions

Exercise 4.5.6 Show that $sp(R) = \rho^2$. (Note: This result holds for any L and K satisfying $K < L$. That indicates that the tail asymptotics property of $\{(X(t), J(t)), t \geq 0\}$ is independent of L and K. Intuitively, the result should hold since $X(t)$ is the total number of customers in the system, which is not affected by the specific values of L and K.)

Exercise 4.5.7 Let $q(t) = \min\{q_1(t), q_2(t)\}$, $J(t) = q_1(t) - q_2(t)$, $t \geq 0$. Show that $\{(q(t), J(t)), t \geq 0\}$ is a continuous time $M/G/1$ type Markov chain. (Note: This is another approach to transforming the original process into a structured Markov chain.)

4.6 The *MMAP*[*K*]/*PH*[*K*]/1/LCFS Non-Preemption Queue

Consider a single server queueing system with a Markovian arrival process with marked arrivals (*MMAP*[*K*]) and phase-type service times. Customers are distinguished into K types. The service times of different types of customers may have different distribution functions. All types of customers join a single queue and are served on a last-come-first-served (LCFS) non-preemptive basis.

In Sect. 2.5, a Markovian arrival process with marked arrivals is defined by a set of $m_a \times m_a$ matrices $\{D_k, k = 0, 1, \ldots, K\}$, where m_a is a positive integer. The matrices $\{D_k, k = 1, \ldots, K\}$ are nonnegative and are the arrival rates of type k customers. The matrix D_0 has negative diagonal elements, nonnegative off-diagonal elements, and is invertible. Let

$$D = D_0 + \sum_{k=1}^{K} D_k. \tag{4.74}$$

Then matrix D is the infinitesimal generator of the underlying Markov chain. Let $\boldsymbol{\theta}_a$ be the stationary distribution of D. The stationary arrival rate of type k customers is given by $\lambda_k = \boldsymbol{\theta}_a D_k \mathbf{e}$, $k = 1, 2, \ldots, K$.

The service times of type k customers have a common *PH*-distribution with a matrix representation $(\boldsymbol{\alpha}_k, T_k)$ of order m_k. Let $\mathbf{T}^0{}_k = -T_k\mathbf{e}$. The mean service time is given by $1/\mu_k = -\boldsymbol{\alpha}_k T^{-1}{}_k\mathbf{e}$. Then μ_k is the average service rate of type k customers.

The traffic intensity of the system is defined as $\rho = \lambda_1/\mu_1 + \cdots + \lambda_K/\mu_K$. We assume $\rho < 1$ to ensure system stability. Intuitively, $\rho < 1$ implies that the system has enough capacity to serve all customers. Consider the interval $(0, t)$. On average, $\lambda_1 t$ type $1, \cdots$, and $\lambda_K t$ type K customers arrive in $(0, t)$. On average, $\mu_1 t_1$ type $1, \cdots$, and $\mu_K t_K$ type K customers are served in $(0, t)$ if t_1 units of time, $\cdots$, and t_K units of time are used to serve type 1 customers, $\cdots$, and type K customers in $(0, t)$, respectively. We must have $t_1 + \cdots + t_K < t$ since there are idle periods when the system is stable. If the system is positive recurrent, there must be a set $\{t, t_1, \cdots, t_K\}$ such that $\lambda_1 t \le \mu_1 t_1, \cdots$, and $\lambda_K t \le \mu_K t_K$. This leads to $\rho \le (t_1 + t_2 + \cdots + t_K)/t < 1$.

The queueing system is represented by the following four variables:

$q(t)$: the string of customers in queue (excluding the one in service, if there is any) at time t, $q(t) \in \aleph \cup \{-1\}$;
$I_a(t)$: the state of the underlying Markov chain D at time t, $1 \le I_a(t) \le m_a$;
$I_1(t)$: the type of the customer in service (if any) at time t, $1 \le I_1(t) \le K$;
$I_s(t)$: the phase of the *PH*-distribution of the current service (if any) at time t, $1 \le I_s(t) \le m_{I_1(t)}$.

Note that $\aleph = \{0\} \cup \{x: x = k_1 k_2 \cdots k_n, 1 \le k_i \le K, i = 1, \ldots, n, n = 1, 2, \ldots\}$. Any string $x \in \aleph$ is a node in the K-ary tree (see definition in Sect. 3.6). String addition and subtraction are defined in Sect. 3.6.

If there is no customer in the system at time t, denote by $q(t) = -1$. If there is one customer in the system at time t, $q(t) = 0$. If there are customers waiting at time t, $q(t)$ is a string in $\aleph$. For example (for $K = 2$), $q(t) = 122$ implies that there are three customers waiting in the system at time t: the customer who arrived first is of type 1; the customer who arrived second is of type 2; and the customer who arrived last is of type 2. If a new customer of type k arrives, $q(t)$ becomes $122k$ since the service discipline is LCFS. The current service is not affected by the arrival of the type k customer. If the current service is completed before the next arrival, the type k customer enters service and $q(t)$ returns to 122.

It is easy to see that $(q(t), I_a(t), I_1(t), I_s(t))$ is a continuous time Markov chain with state space: $\{-1\} \times \{1, 2, \cdots, m_a\}$ plus $\aleph \times \{1, 2, \cdots, m_a\} \times \cup^K{}_{k=1}(\{k\} \times \{1, 2, \cdots, m_k\})$. This is a QBD Markov chain with a tree structure if $(I_a(t), I_1(t), I_s(t))$ is defined as the auxiliary random variable with $m_a \bar{m}$ states, where $\bar{m} = m_1 + \cdots + m_K$, except for $q(t) = -1$, where the auxiliary variable takes values $\{1, 2, \cdots, m_a\}$. Furthermore, the infinitesimal generator of the QBD Markov chain is defined by the following transition blocks.

For $x = k_1 \cdots k_{n-1}k \in \aleph$ and $1 \le k \le K$, we have

$$A_0(k) = D_k \otimes I_{\bar{m}\times\bar{m}}; \quad \text{(a type } k \text{ customer arrives)}$$

$$A_1(k) = D_0 \otimes I_{\bar{m}\times\bar{m}} + I_{m_a\times m_a} \otimes \begin{pmatrix} T_1 & & \\ & \ddots & \\ & & T_K \end{pmatrix}; \quad \text{(no service completed and no arrival)} \tag{4.75}$$

$$A_2(k) = I_{m_a\times m_a} \otimes \left(\begin{pmatrix} \mathbf{T}_1^0 \boldsymbol{\alpha}_k \\ \vdots \\ \mathbf{T}_K^0 \boldsymbol{\alpha}_k \end{pmatrix} (0, \cdots, 0, I_{m_k\times m_k}, 0, \cdots, 0) \right).$$

(a service is completed and the next (last in queue) is of type k)

For $x = 0$,

$$\begin{aligned} A_0(k) &= D_k \otimes I_{\bar{m}\times\bar{m}}; \quad \text{(a type } k \text{ customer arrives)} \\ A_{0,0} &= A_1(1); \quad \text{(no service completed and no arrival)} \\ B_{0,-1} &= I_{m_a\times m_a} \otimes \begin{pmatrix} \mathbf{T}_1^0 \\ \vdots \\ \mathbf{T}_K^0 \end{pmatrix}. \quad \text{(a service is completed)} \end{aligned} \tag{4.76}$$

For $x = -1$,

$$\begin{aligned} B_{-1,0} &= (D_1 \otimes \boldsymbol{\alpha}_1 \cdots D_K \otimes \boldsymbol{\alpha}_K); \quad \text{(a customer arrives)} \\ B_{-1,-1} &= D_0. \quad \text{(no arrival and no service)} \end{aligned} \tag{4.77}$$

The QBD Markov chain that describes the queueing system of interest is defined explicitly. The stationary distribution of this QBD process is presented next. Let, $x \in \aleph$,

$$\begin{aligned} \pi(x,i,k,j) &= \lim_{t\to\infty} P\{(q(t), I_a(t), I_1(t), I_s(t)) = (x,i,k,j)\}; \\ \pi(-1,i) &= \lim_{t\to\infty} P\{(q(t), I_a(t)) = (-1,i)\}, \end{aligned} \tag{4.78}$$

and

$$\begin{aligned} \boldsymbol{\pi}(x,i,k) &= (\pi(x,i,k,1), \cdots, \pi(x,i,k,m_k)); \\ \boldsymbol{\pi}(x,i) &= (\boldsymbol{\pi}(x,i,1), \cdots, \boldsymbol{\pi}(x,i,K)); \end{aligned} \tag{4.79}$$

$$\boldsymbol{\pi}(x) = (\boldsymbol{\pi}(x,\ 1), \cdots, \boldsymbol{\pi}(x, m_a)),\ x \neq -1;$$
$$\boldsymbol{\pi}(-1) = (\boldsymbol{\pi}(-1, 1), \cdots, \boldsymbol{\pi}(-1, m_a)).$$

Some basic results on the limiting probabilities are collected in the following theorem.

Theorem 4.6.1 (He (2000)) *For the queueing system of interest, assume that* $\rho < 1, D$ *is irreducible, and* $T_k + \mathbf{T}_k^0 \boldsymbol{\alpha}_k$ *is irreducible, for* $k = 1, 2, \ldots, K$. *The Markov chain* $\{(q(t),\ I_a(t),\ I_1(t),\ I_s(t)),\ t \geq 0\}$ *is irreducible. In steady state, for* $k = 1, 2, \ldots, K$,

(a) *The rate of starting to serve a type k customer is given by*

$$\sum_{j=1}^{m_a} \sum_{i=1}^{m_a} \pi(-1, j)(D_k)_{j,i} + \sum_{x \in \aleph:\ x \neq -1} \sum_{l=1}^{K} \sum_{i=1}^{m_a} \boldsymbol{\pi}(x + k, i, l)\mathbf{T}_l^0 = \lambda_k; \tag{4.80}$$

(b) *The probability that a type k customer is in service is*

$$\sum_{x \in \aleph:\ x \neq -1} \sum_{i=1}^{m_a} \boldsymbol{\pi}(x, i, k)\mathbf{e} = \lambda_k / \mu_k; \tag{4.81}$$

(c) *The probability that the queueing system is busy is*

$$\rho = \sum_{x \in \aleph:\ x \neq -1} \boldsymbol{\pi}(x)\mathbf{e} = \sum_{k=1}^{K} \lambda_k / \mu_k; \tag{4.82}$$

(d) *The probability that the queueing system is empty is* $\boldsymbol{\pi}(-1)\mathbf{e} = 1 - \rho$.

Furthermore, the Markov chain of interest is positive recurrent, i.e., the queueing system is stable, if and only if $\rho < 1$.

Exercise 4.6.1* Give all necessary details for a proof of Theorem 4.6.1.

When the Markov chain is positive recurrent, using formulas presented in Sect. 3.6, the following theorem can be obtained.

Theorem 4.6.2 *If the Markov chain* $\{(q(t), I_a(t), I_1(t), I_s(t)),\ t \geq 0\}$ *is positive recurrent and irreducible, its stationary distribution is given by*

$$\boldsymbol{\pi}(x + k) = \boldsymbol{\pi}(x)R(k),\ \ x \in \aleph,\ \ k = 1, \ldots, K;$$

$$(\boldsymbol{\pi}(-1),\ \boldsymbol{\pi}(0)) \begin{pmatrix} D_0 & (D_1 \otimes \boldsymbol{\alpha}_1, \cdots, D_K \otimes \boldsymbol{\alpha}_K) \\ I \otimes \begin{pmatrix} \mathbf{T}_1^0 \\ \vdots \\ \mathbf{T}_K^0 \end{pmatrix} & A_{0,0} + \sum_{k=1}^{K} R(k)A_2(k) \end{pmatrix} = 0; \tag{4.83}$$

$$\boldsymbol{\pi}(-1)\mathbf{e} + \boldsymbol{\pi}(0)(I - R)^{-1}\mathbf{e} = 1,$$

where $R = R(1) + \cdots + R(K)$, *and* $\{R(k), k = 1, \ldots, K\}$ *are the minimal nonnegative solutions to*

$$0 = D_k \otimes I + R(k)\left(D_0 \otimes I + I \otimes \begin{pmatrix} T_1 & & \\ & \ddots & \\ & & T_K \end{pmatrix}\right)$$

$$+ R(k)\sum_{l=1}^{K} R(l) I \otimes \left(\begin{pmatrix} \mathbf{T}_1^0 \boldsymbol{\alpha}_k \\ \vdots \\ \mathbf{T}_K^0 \boldsymbol{\alpha}_k \end{pmatrix} (0, \cdots, 0, I, 0, \cdots, 0)\right). \tag{4.84}$$

The computation of $\{R(k), k = 1, \ldots, K\}$ *can be done using the algorithm given in Sect.* 3.6.

Let $L = |q(t)|$, *i.e., the number of customers waiting at arbitrary time t, then*

$$P\{L = n\} = \sum_{x:\ |x|=n} \boldsymbol{\pi}(x)\mathbf{e} = \boldsymbol{\pi}(0)R^n\mathbf{e}, \quad n = 0,\ 1,\ 2,\ \ldots. \tag{4.85}$$

Clearly, Eq. (4.85) shows the exponential decay of the queue length in the queueing system of interest, regardless of the types of customers. By Eq. (4.85), it is easy to obtain $E[L] = \boldsymbol{\pi}(0)R(I - R)^{-2}\boldsymbol{e}$. Define $\overline{L}$ as the total number of customers in the system (customers in queue and in service). The mean number of customers in the system can be computed by using the formula

$$E[\overline{L}] = \boldsymbol{\pi}(0)R(I - R)^{-2}\mathbf{e} + \boldsymbol{\pi}(0)(I - R)^{-1}\mathbf{e} = \boldsymbol{\pi}(0)(I - R)^{-2}\mathbf{e}. \tag{4.86}$$

Commentary Busy periods of the queue or fundamental periods of the Markov chain can be analyzed using the theory developed in Sect. 3.6. Using results on the fundamental periods, the waiting times of different types of customers in such a queueing model can be found (see He (2000)).

Additional Exercises and Extensions

Exercise 4.6.2 Consider the *MMAP*[2]/*PH*[2]/1/LCFS queue with

$$D_0 = \begin{pmatrix} -2 & 1 \\ 0 & -5 \end{pmatrix}, \quad D_1 = \begin{pmatrix} 1 & 0 \\ 0 & 2 \end{pmatrix}, \quad D_2 = \begin{pmatrix} 0 & 0 \\ 2 & 1 \end{pmatrix}; \tag{4.87}$$

$$\boldsymbol{\alpha}_1 = (0.2, 0.8), \quad T_1 = \begin{pmatrix} -5 & 1 \\ 5 & -7 \end{pmatrix};$$
$$\boldsymbol{\alpha}_2 = (0.5, 0.5), \quad T_2 = \begin{pmatrix} -15 & 2 \\ 4 & -10 \end{pmatrix}.$$

(i) Find ρ.
(ii) Find matrices $R(1)$ and $R(2)$.
(iii) Find the stationary distribution of the queue string $q(t)$. Find the stationary distribution of the queue length and the mean queue length.

Exercise 4.6.3* Explain that the waiting time of a type k customer is the first passage time from node $x + k$ to node x. (i) Similar to Sect. 3.6, find the LST of the waiting time of a type k customer. (ii) Find the joint transform of the waiting time of a type k customer and the number of customers served during its waiting period.

Exercise 4.6.4* Use the approach developed in this section to analyze the *MMAP*[*K*]/*PH*[*K*]/1/LCFS preemption-repeat Queue. (Note: Preemption-repeat means that (i) a newly arrived customer pushes the customer in service, if there is any, back to the queue, and takes over the server to receive service; and (ii) when a customer reenters the server, it is treated like a new customer (i.e., the service time has the same probability distribution each time a customer receives service). For the preemption-repeat case, $q(t)$ can include the customer currently in service. Thus, there is no need to include the soil node $x = -1$ in the state space.)

Exercise 4.6.5* Consider the *MMAP*[*K*]/*PH*[*K*]/1/LCFS preemption-resume queue. Explain the difficulty in using the approach developed in this section to analyze the queueing model. (Note: For the preemption-repeat case, when a customer reenters the server, it resumes its service at the interrupted point. For *PH*-service time, one needs to keep track of service phases of all customers waiting in the queue, so that their service can be resumed properly when they reenter the server.)

4.7 The *GI*/*PH*/1 Queue

The flow of customers in the *GI*/*PH*/1 queue is the same as that of the *M*/*M*/1 queue (see Fig. 4.1). Customers arrive according to a *renewal process* with interarrival times having common distribution function $F(t)$ satisfying $F(0+) = 0$. The mean interarrival time is $1/\lambda$. Service times have a common *PH*-distribution function with *PH*-representation $(\boldsymbol{\beta}, S)$ of order m_s. We assume that $\boldsymbol{\beta}\mathbf{e} = 1$. Let $\mu = -1/(\boldsymbol{\beta}S^{-1}\mathbf{e})$, which is the service rate.

Since the arrival process may not be Markovian, it is not possible to introduce a simple continuous time Markov chain to describe the queueing process. Following the embedded Markov chain approach introduced by Kendall (1953), one can construct a Markov renewal process at the arrival epochs to describe the queue length process. Let

- q_n be the queue length just before the arrival of the n-th customer, i.e., the queue length seen by the n-th arriving customer;
- τ_n be the interarrival time between the $(n - 1)$-st and n-th customers;
- J_n be the phase of the service process right after the arrival of the n-th customer.

Then $\{(q_n, J_n, \tau_n), n = 0, 1, 2, \ldots\}$ is a Markov renewal process. The state space of the process is $\{(q, j, x), q = 0, 1, 2, \ldots, j = 1, \ldots, m_s, x \geq 0\}$. The transition probability matrix of the Markov renewal process is given by, for $x \geq 0$,

$$P(x) = \begin{pmatrix} B_0(x) & A_0(x) & & & \\ B_1(x) & A_1(x) & A_0(x) & & \\ B_2(x) & A_2(x) & A_1(x) & A_0(x) & \\ B_3(x) & A_3(x) & A_2(x) & A_1(x) & A_0(x) \\ \vdots & \vdots & \ddots & \ddots & \ddots & \ddots \end{pmatrix}, \tag{4.88}$$

where

$$\begin{aligned} A_k(x) &= \int_0^x P(k, t)\mathrm{d}F(t), \quad k = 0,\ 1,\ 2,\ \ldots; \\ B_k(x) &= \int_0^x \sum\nolimits_{n=k+1}^{\infty} P(n, t)\mathrm{d}F(t)\,\mathbf{e}\boldsymbol{\beta}, \quad k = 0,\ 1,\ 2,\ \ldots; \\ \frac{\mathrm{d}P(0,t)}{\mathrm{d}t} &= P(0,t)S; \\ \frac{\mathrm{d}P(n,t)}{\mathrm{d}t} &= P(n,t)S + P(n-1,t)\mathbf{S}^0\boldsymbol{\beta}, \ n = 1,\ 2,\ \ldots. \end{aligned} \tag{4.89}$$

Note that $P(n, t)$ is for the service process, which is the (matrix) probability that n customers are served in $[0, t]$, i.e., the (i, j)-th element of $P(n, t)$ is the probability that n customers are served in $[0, t]$, the phase of the underlying Markov chain of the service process is j at time t, given that the underlying Markov chain is in phase i at time 0. We have $P(0, 0) = I$ and $P(n, 0) = 0$, for $n = 1, 2, \ldots$. The matrices $\{A_k(x), k = 0, 1, 2, \ldots\}$ are for the conditional distributions of the number of customers served between two consecutive arrivals.

We note that there is a service phase even if $q_n = 0$, since the phase variable is for the phase after the arrival. If $q_n = 0$, a new service begins immediately and the phase is initialized based on $\boldsymbol{\beta}$. Next, we show how to use this Markov renewal process to obtain system performance measures.

Exercise 4.7.1 Is it possible to introduce an embedded Markov chain at the departure epochs for the *GI*/*PH*/1 queue?

The stationary queue length at arrivals We consider the embedded Markov chain (at the arrival epochs) $\{(q_n, J_n), n = 0, 1, 2, \ldots\}$ to find the stationary queue length distribution at arrivals. That Markov chain has transition probability matrix $P = P(\infty)$, which is a *GI*/*M*/1 type Markov chain. Let $B_n = B_n(\infty)$ and $A_n = A_n(\infty)$, $n = 0, 1, 2, \ldots$. According to Sect. 3.4, the limiting probabilities of the Markov chain have a matrix-geometric solution. For the specific model, the matrix-geometric solution can be made more explicit.

First, we examine ergodicity condition for the Markov chain, which is also the stability condition of the queueing system. It can be shown that:

$$\begin{aligned} A(x) &= \sum_{k=0}^{\infty} A_k(x) = \int_0^x \sum_{k=0}^{\infty} P(k,t)\mathrm{d}F(t) \\ &= \int_0^x \exp\left((S + \mathbf{S}^0\boldsymbol{\beta})t\right)\mathrm{d}F(t); \\ A = A(\infty) &= \int_0^\infty \exp\left((S + \mathbf{S}^0\boldsymbol{\beta})t\right)\mathrm{d}F(t); \\ \boldsymbol{\theta}_s A = \boldsymbol{\theta}_s, \quad & \boldsymbol{\theta}_s\mathbf{e} = 1. \quad (\boldsymbol{\theta}_s(S + \mathbf{S}^0\boldsymbol{\beta}) = 0.) \end{aligned} \tag{4.90}$$

Then it is easy to verify that

$$\begin{aligned} A^*(z,x) &\equiv \sum_{k=0}^{\infty} z^k A_k(x) = \int_0^x \sum_{k=0}^{\infty} z^k P(k,t)\mathrm{d}F(t) \\ &= \int_0^x \exp\left((S + z\mathbf{S}^0\boldsymbol{\beta})t\right)\mathrm{d}F(t); \\ \boldsymbol{\gamma}^* &\equiv \sum_{k=0}^{\infty} kA_k(\infty)\mathbf{e} = \mu\mathbf{e}/\lambda + (I - A)S^{-1}\mathbf{e}/\lambda. \end{aligned} \tag{4.91}$$

The generating function $A^*(z, x)$ is for the number of customers served between two consecutive arrivals. Then $\boldsymbol{\gamma}^*$ is for the conditional mean numbers of customers served between two consecutive arrivals.

Exercise 4.7.2 Show the equalities in Eq. (4.91). In addition, show that $\boldsymbol{\theta}_s \sum_{k=0}^{\infty} kA_k\mathbf{e} > 1$ is equivalent to $\lambda/\mu < 1$.

Theorem 4.7.1 (Neuts (1981)) *Assume* $\lambda/\mu < 1$. *The limiting probabilities of the embedded Markov chain* $\{(q_n, J_n), n = 0, 1, 2, \ldots\}$ *is given by*

$$\boldsymbol{\pi}_k = (\boldsymbol{\beta}(I - R)^{-1}\mathbf{e})\boldsymbol{\beta}R^k, \quad k = 0, 1, 2, \ldots, \tag{4.92}$$

where R is the minimal nonnegative solution to $R = \sum_{k=0}^{\infty} R^k A_k$. *The computation of the matrix R can be done by using an iterative algorithm developed in Lucantoni and Ramaswami* (1985) *or by using Eq.* (4.93), *which does not require computing the matrices* $\{A_0, A_1, \ldots, A_n, \ldots\}$ *explicitly.*

The matrix-geometric solution (4.92) can be used to find the waiting time distribution as well. The results are tedious and yet all details can be worked out. Sengupta (1989) has shown that the matrix R satisfies an equation with an exponential function. That result generalizes the matrix-geometric solution to the *matrix-exponential solution* (see Sect. 3.10.5). The matrix-exponential solution gives simple and elegant expressions for the distributions of the virtual and actual waiting times.

Theorem 4.7.2 (Sengupta (1989)) *The matrix R is the minimal nonnegative solution to the following exponential equation:*

$$R = \int_0^\infty \exp\left((S + R\mathbf{S}^0\boldsymbol{\beta})x\right)\mathrm{d}F(x). \tag{4.93}$$

Equation (4.93) leads to an algorithm for computing R without computing matrices $\{A_0, A_1, \ldots, A_n, \ldots\}$ explicitly. For some special cases, the computation of R can be simplified further.

Example 4.7.1 Assume that the interarrival times have a common discrete distribution given by $P\{X = x_i\} = p_i$, $i = 1, 2, \ldots, N$, where $x_i > 0$, $p_i > 0$, $i = 1, 2, \ldots, N$, and $p_1 + p_2 + \ldots + p_N = 1$. Define $R[0] = 0$, and

$$R[k+1] = \sum_{i=1}^{N} \exp\left((S + R[k]\mathbf{S}^0\boldsymbol{\beta})x_i\right)p_i, \quad k = 0, 1, 2, \ldots. \tag{4.94}$$

Then it can be shown that $\{R[k], k = 0, 1, 2, \ldots\}$ converges to R monotonically.

Example 4.7.2 Assume that the interarrival times have a common PH-distribution with PH-representation $(\boldsymbol{\alpha}, T)$. Then Eq. (4.93) becomes

$$\begin{aligned} R &= \int_0^\infty \exp\left((S + R\mathbf{S}^0\boldsymbol{\beta})x\right)\boldsymbol{\alpha}\exp(Tx)\mathbf{T}^0\mathrm{d}x \\ &= (I \otimes \boldsymbol{\alpha})\int_0^\infty \exp\left(((S + R\mathbf{S}^0\boldsymbol{\beta}) \otimes I + I \otimes T)x\right)(I \otimes \mathbf{T}^0)\mathrm{d}x \\ &= -(I \otimes \boldsymbol{\alpha})\left((S + R\mathbf{S}^0\boldsymbol{\beta}) \otimes I + I \otimes T\right)^{-1}(I \otimes \mathbf{T}^0). \end{aligned} \tag{4.95}$$

Then an iterative algorithm can be developed for computing R by using Eq. (4.95).

Exercise 4.7.3 Based on Eq. (4.95), outline an iterative method for computing R.

Let $W(x)$ be the steady state sojourn time distribution and $W_q(x)$ be the steady state waiting time distribution of a customer.

Theorem 4.7.3 (Sengupta (1989)) *If the queue is stable, we have*

$$\begin{aligned} W(x) &= 1 - \boldsymbol{\gamma}\exp(Tx)\mathbf{e}, \quad x \geq 0; \\ W_q(x) &= 1 - \boldsymbol{\gamma}R\exp(Tx)\mathbf{e}, \quad x \geq 0, \end{aligned} \tag{4.96}$$

where $\boldsymbol{\gamma} = -\mu^{-1}\boldsymbol{\theta}_s T(I-R)^{-1}$, *and* T *satisfies*

$$T = S + \int_0^\infty \exp(Tt)\mathrm{d}F(t)\mathbf{S}^0\boldsymbol{\beta} = S + R\mathbf{S}^0\boldsymbol{\beta}. \tag{4.97}$$

We remark that Theorem 4.7.3 shows that the sojourn time and the actual waiting time in a *GI/PH/*1 queue have a matrix-exponential distribution. Theorem 4.7.3 does not show that they are *PH*-distributions since T may not be a *PH*-generator. In Sengupta (1990a), it is shown that $W_q(x)$ has a *PH*-distribution and a *PH*-representation is given. Then $W(x)$ must be a *PH*-distribution since the sojourn time is the sum of the actual waiting time and the service time, which are independent of each other. A *PH*-representation can be found for $W(x)$ as well. Further generalization can be found in Asmussen and O'Cinnedei (1998).

Commentary The *GI/PH/*1 queue is a quite general queue. Neuts (1978, 1981) and Sengupta (1989) give fairly complete answers to both queue length and waiting time distributions. More general models are considered in Neuts (1981) and Sengupta (1990b), and some discussion on those generalizations can be found in Sect. 4.10.1.

Additional Exercises and Extensions

Exercise 4.7.4 Develop a simple algorithm for computing the matrix-geometric solution for the queue length and the matrix-exponential solution for the waiting times, for the *PH/PH/*1 queue. Use Ozawa (2006) to find the *PH*-representations for the waiting times. Comment on the matrix representations obtained by using the two different approaches.

Exercise 4.7.5 Based on Theorem 4.7.1, find the mean queue length at arrivals. Based on Theorem 4.7.3, find the mean sojourn time and the mean actual waiting time for a *GI/PH/*1 queue.

Exercise 4.7.6 Consider an *GI/PH/*1 queue with interarrival time $X = 0.3$, w.p. 0.4; 1.5, w.p. 0.4; and 3.5, w.p. 0.2. The service time has a *PH*-distribution with *PH*-representation

$$\boldsymbol{\alpha} = (0.2,\ 0.8), \quad T = \begin{pmatrix} -2 & 1 \\ 0 & -5 \end{pmatrix}. \tag{4.98}$$

Find the matrix R. Find the mean queue length and the mean waiting time.

Exercise 4.7.7* (Neuts (1981)) For the *GI*/*PH*/1 queue, find the distribution of the queue length at an arbitrary time. Find the distribution of the queue length right after an arbitrary departure.

4.8 The *MAP*/*G*/1 Queue

The flow of customers in the *MAP*/*G*/1 queue is the same as that in the *M*/*M*/1 queue (see Fig. 4.1). Customers arrive according to Markovian arrival process (D_0, D_1) of order m_a, and service times are independent and identically distributed random variables with common distribution function $F(t)$. Denote by λ the arrival rate (see Sect. 4.4) and μ the service rate (i.e., the reciprocal of the mean service time). The service times and the arrival process are independent of each other. We focus on the busy period and analyze it in three steps.

(i) First, we introduce an embedded Markov chain for the queue length right after (customer) departures. Based on results obtained in Sect. 3.5, the Markov chain can be used to find the mean number of customers served in a busy period. The Markov chain can also be used to find the stationary distribution of the queue length at departures.
(ii) Second, we introduce a Markov renewal process for the queue length right after departures. The Markov renewal process can be used to learn more about the queueing model. For instance, it can be used to analyze the lengths of busy periods.
(iii) Using the Markov renewal process, we go from matrix-geometric solutions to matrix-exponential solutions. We show that the matrix G is the minimal nonnegative solution to a matrix-exponential equation. With the matrix-exponential formalism, a number of formulas for performance measures can be simplified.

In addition, we collect results on the queue lengths and waiting times in exercises.

4.8.1 *Analysis of the Embedded Markov Chain at Departure Epochs*

To analyze the queueing system, we first introduce an embedded Markov chain for the queue length at departure epochs. Let

- q_n be the queue length right after the departure of the n-th customer;
- J_n be the phase of the arrival process right after the n-th departure epoch.

Since J_n keeps track of the phase of the arrival process, then $\{(q_n, J_n), n = 0, 1, 2, \ldots\}$ is an *M*/*G*/1 type Markov chain. Note that the term "*M*/*G*/1 type" comes from this queueing model. The state space of the process is $\{(q, j), q = 0, 1, 2, \ldots,$

$j = 1, \ldots, m_a\}$. Between two consecutive departures, there can be any number of arrivals. Since the phase of the arrival process is recorded at the epoch right after a departure, the distribution of the number of arrivals between two consecutive departures can be found. We need to consider two cases: (i) the queueing system is nonempty right after the departure; and (ii) the queueing system is empty right after the departure. For case (i), the number of customers in the system at the next departure epoch is the current number plus the number of customers arrived during the next service less by one. For case (ii), we have to wait until a new customer arrives. Then a service begins. The number of customers at the next departure epoch is the number of customers who arrive during the service. Define, for $i, j = 1, 2, \ldots, m_a$,

$$
\begin{aligned}
(A_k)_{i,j} &= P\{q_{n+1} = q_n - 1 + k, J_{n+1} = j | q_n > 0, J_n = i\}; \\
(B_k)_{i,j} &= P\{q_{n+1} = k, J_{n+1} = j | q_n = 0, J_n = i\}.
\end{aligned} \tag{4.99}
$$

Conditioning on the length of the service time, we have

$$
\begin{aligned}
A_k &= \int_0^\infty P(k,t)\mathrm{d}F(t), \quad k = 0,\ 1,\ 2,\ \ldots; \\
B_k &= \int_0^\infty \exp(D_0 t)\mathrm{d}t D_1 \int_0^\infty P(k,t)\mathrm{d}F(t) \\
&= \left(-D_0^{-1}D_1\right)\int_0^\infty P(k,t)\mathrm{d}F(t) = \left(-D_0^{-1}D_1\right)A_k, \quad k = 0,\ 1,\ 2,\ \ldots.
\end{aligned} \tag{4.100}
$$

The function $P(k, t)$ is defined in Sect. 2.3 for *MAP*s, which is the probability that exactly k customers arrived in $[0, t]$. Similar to Eq. (4.89), the functions $\{P(n, t), n = 0, 1, 2, \ldots\}$ satisfy the following differential equations: (Note: This equation is the same as Eq. (2.48).)

$$
\begin{aligned}
\frac{\mathrm{d}P(0,t)}{\mathrm{d}t} &= P(0,t)D_0; \\
\frac{\mathrm{d}P(n,t)}{\mathrm{d}t} &= P(n,t)D_0 + P(n-1,t)D_1, \quad n = 1,\ 2,\ \ldots.
\end{aligned} \tag{4.101}
$$

Exercise 4.8.1 Explain $B_k = (-D_0{}^{-1}D_1)A_k$ intuitively.

The transition probability matrix of the Markov chain $\{(q_n, J_n), n = 0, 1, 2, \ldots\}$ is given by

$$
P = \begin{pmatrix}
B_0 & B_1 & B_2 & B_3 & B_4 & \cdots \\
A_0 & A_1 & A_2 & A_3 & A_4 & \cdots \\
 & A_0 & A_1 & A_2 & A_3 & \cdots \\
 & & A_0 & A_1 & A_2 & \cdots \\
 & & & \ddots & \ddots & \ddots
\end{pmatrix}. \tag{4.102}
$$

Exercise 4.8.2 Is it possible to introduce an embedded discrete time Markov chain with a countable state space at the arrival epochs for the *MAP*/*G*/1 queue?

Based on the results given in Sect. 3.5, the number of customers served in a busy period can be analyzed and Theorem 3.5.3 can be used to compute the distribution of the queue length at departure epochs. Details are omitted since the results can be obtained by using refined methods.

Exercise 4.8.3 Define matrix $G^*(z)$ as the minimal nonnegative solution to equation, for $0 \leq z \leq 1$,

$$G^*(z) = z\sum_{n=0}^{\infty} A_n (G^*(z))^n. \tag{4.103}$$

Explain the relationship between the elements of $G^*(z)$ and the busy period of the queue. Let $G = G^*(1)$. Explain G probabilistically.

The matrix G is the minimal nonnegative solution to Eq. (4.103) with $z = 1$. See Sect. 3.5 for more about G.

Exercise 4.8.4 Assume that the Markov chain $\{(q_n, J_n), n = 0, 1, 2, \ldots\}$ is irreducible. Show that the Markov chain is ergodic if and only if $\rho = \lambda/\mu < 1$.

Exercise 4.8.5 Denote by $\mathbf{p} = (\mathbf{p}_0, \mathbf{p}_1, \ldots)$ the stationary distribution of the Markov chain $\{(q_n, J_n), n = 0, 1, 2, \ldots\}$. Is it possible that $\boldsymbol{\theta}_a = \mathbf{p}_0 + \mathbf{p}_1 + \ldots$? How can the vector $\mathbf{p}_0 + \mathbf{p}_1 + \ldots$ be interpreted? (Hint: See Exercise 2.2.12.)

4.8.2 Embedded Markov Renewal Process at Departure Epochs

Next, we add the time factor into the embedded Markov chain and introduce a Markov renewal process. Let

- τ_n be the time between the $(n-1)$-st and n-th departures.

Then $\{(q_n, J_n, \tau_n), n = 0, 1, 2, \ldots\}$ is a Markov renewal process. The state space of the process is $\{(q, j, x), q = 0, 1, 2, \ldots, j = 1, 2, \ldots, m_a, x \geq 0\}$. The transition probability matrix of the Markov renewal process is given by (or the transition kernel for this case)

$$P(x) = \begin{pmatrix} B_0(x) & B_1(x) & B_2(x) & B_3(x) & B_4(x) & \cdots \\ A_0(x) & A_1(x) & A_2(x) & A_3(x) & A_4(x) & \cdots \\ & A_0(x) & A_1(x) & A_2(x) & A_3(x) & \cdots \\ & & A_0(x) & A_1(x) & A_2(x) & \cdots \\ & & & \ddots & \ddots & \ddots \end{pmatrix}, \tag{4.104}$$

where

$$
\begin{aligned}
A_k(x) &= \int_0^x P(k,t)\mathrm{d}F(t), \quad k=0,\ 1,\ \ldots; \\
B_k(x) &= \int_0^x \exp(D_0 t)D_1\left(\int_0^{x-t} P(k,v)\mathrm{d}F(v)\right)\mathrm{d}t, \quad k=0,\ 1,\ \ldots.
\end{aligned}
\tag{4.105}
$$

We define ξ as the first passage time from level $n+1$ to n (Recall the fundamental period defined in Sect. 3.5). Define, for $i, j = 1, 2, \ldots, m_a$, and $k = 0, 1, 2, \ldots$,

$$
\begin{aligned}
g_{i,j}(t,k) &= P\{q_k = n-1, J_k = j,\ \xi < t, q_l \geq n,\ 1 \leq l \leq k-1 | (q_0, J_0) = (n,i)\}, \\
g^*_{i,j}(s,z) &= \int_0^\infty e^{-st}\mathrm{d}_t\left(\sum_{k=0}^{\infty} z^k g_{i,j}(t,k)\right).
\end{aligned}
\tag{4.106}
$$

Let $G^*(s, z) = (g^*_{i,j}(s, z))$, which is the joint transform of the number of customers served in a busy period (a fundamental period) and the length of the busy period. It is easy to see $G^*(s, 0) = 0$, $G^*(0, z) = G^*(z)$, and $G^*(0, 1) = G$. If the queueing system is stable or the Markov chain is ergodic (which is guaranteed by $\lambda/\mu < 1$), then G is stochastic. It can be shown that $G^*(s, z)$ satisfies the following equation and is the minimal nonnegative solution to it.

Theorem 4.8.1 (Lucantoni (1991)) *For* $0 \leq z \leq 1$ *and* $s \geq 0$, $G^*(s, z)$ *is the minimal nonnegative solution to equation*

$$
\begin{aligned}
G^*(s,z) &= z\sum_{n=0}^{\infty} A_n^*(s)(G^*(s,z))^n; \\
A_n^*(s) &= \int_0^\infty e^{-sx}\mathrm{d}A_n(x), \quad n=0,\ 1,\ 2,\ \ldots.
\end{aligned}
\tag{4.107}
$$

Exercise 4.8.6 Explain Eq. (4.107) for $G^*(s, z)$ intuitively.

Conditioning on the initial phase of the Markov chain at the beginning of a busy period, we can find the mean number of customers served in a busy period. Using Eq. (4.107), by Theorem 3.5.2 and routine calculations, we obtain

$$
\begin{aligned}
\mathbf{u}_z &\equiv \frac{\mathrm{d}G^*(s,z)}{\mathrm{d}z}\mathbf{e}\bigg|_{z=1-,s=0+} = (I - G + \mathbf{eg})\left(I - A + \mathbf{eg} - \left(\sum_{n=0}^{\infty} nA_n\right)\mathbf{cg}\right)^{-1}\mathbf{e}; \\
\mathbf{u}_s &\equiv -\frac{\mathrm{d}G^*(s,z)}{\mathrm{d}s}\mathbf{e}\bigg|_{z=1-,s=0+} = \mathbf{u}_z/\mu,
\end{aligned}
\tag{4.108}
$$

which are the conditional mean number of customers served in a busy period and the conditional mean length of a busy period, respectively. Note that the vector **g** satisfies $\mathbf{g}G = \mathbf{g}$, $\mathbf{ge} = 1$, and **g** is nonnegative.

Exercise 4.8.7 Explain the relationship between $\mathbf{u}_z$ and $\mathbf{u}_s$ intuitively.

The *busy cycle* (a busy period plus the idle period immediately after it) can be analyzed. Define, for $k = 0, 1, 2, \ldots, i, j = 1, \ldots, m_a$,

$$g_{i,j}^{(0)}(t,k) = P\{q_k = 0, J_k = j, \xi < t, q_l \neq 0, 1 \leq l \leq k-1 | (q_0, J_0) = (0,i)\}. \quad (4.109)$$

Denote by $G^{*(0)}(s,z)$ the joint transform of $\left(g_{i,j}^{(0)}(t,k)\right)$. Then it can be shown

$$G^{*(0)}(s,z) = zB_0^*(s) + z\sum_{n=1}^{\infty} B_n^*(s)(G^*(s,z))^n, \quad (4.110)$$

where $B_n^*(s) = \int_0^\infty e^{-sx} dB_n(x)$, for $n = 0, 1, 2, \ldots$. Intuitively, Eq. (4.110) can be explained as follows. The busy cycle is decomposed into two parts: the time until the first departure and the time from the first departure to the first time the queue becomes empty. The time until the first departure is determined by $\{B^*_n(s), n = 0, 1, 2, \ldots\}$. If there are n customers in the queue at the first departure epoch, the system will return to the empty status after n consecutive busy periods (since each customer in the system at the first departure epoch corresponds to a busy period), whose distribution is determined by $(G^*(s, z))^n$. By conditioning on the number of customers at the first departure epoch, Eq. (4.110) can be established.

The mean number of customers served in a busy cycle and the mean length of a busy cycle can be obtained.

$$\begin{aligned} \mathbf{u}_z^{(0)} &\equiv \left.\frac{dG^{*(0)}(s,z)}{dz}\mathbf{e}\right|_{z=1-,s=0+} = (-D_0^{-1}D_1)\mathbf{u}_z; \\ \mathbf{u}_s^{(0)} &\equiv -\left.\frac{dG^{*(0)}(s,z)}{ds}\mathbf{e}\right|_{z=1-,s=0+} = (-D_0^{-1})\mathbf{e} + (-D_0^{-1}D_1)\mathbf{u}_s. \end{aligned} \quad (4.111)$$

Exercise 4.8.8 Show Eq. (4.111). Explain the results in Eq. (4.111) intuitively.

4.8.3 Matrix-Exponential Solutions

The matrix $G^*(s, z)$ satisfies Eq. (4.107), which can be called a matrix-geometric equation. Next, we show that $G^*(s, z)$ also satisfies a matrix-exponential equation (see Neuts (1989a)). To establish the equation, we consider an actual busy period and condition on the service time of the first customer in the busy period. Let

$\phi_{i,j}(t,\ k,\ x)$ be the probability that the initial workload of a busy period is x, the busy period ends before t, there are k arrives in this period, and the phase of the arrival process is j at the end of this period, given that the phase is i at the beginning of the busy period;

$\phi^*_{i,j}(s,z,x) = \int\limits_0^{\infty} e^{-st}\mathrm{d}_t\left(\sum\limits_{k=0}^{\infty} z^k \phi_{i,j}(t,k,x)\right)$, $i,\ j = 1,\ 2,\ \ldots,\ m_a$, which is the joint transform of $\phi_{i,j}(t,\ k,\ x)$; and

$\Phi^*(s,z,x) = (\phi^*_{i,j}(s,\ z,\ x))_{m\times m}$.

Lemma 4.8.1 *The following relationships hold for $s \geq 0$, $0 \leq z \leq 1$, $x, y \geq 0$:*

$$(1)\ \Phi^*(s,z,x+y) = \Phi^*(s,z,x)\Phi^*(s,z,y);$$

$$(2)\ G^*(s,z) = z\int\limits_0^{\infty} \Phi^*(s,z,x)\mathrm{d}F(x); \tag{4.112}$$

$$(3)\ \Phi^*(s,z,x) = \exp((-sI + D_0 + D_1G^*(s,z))x).$$

Proof. Note that the workload in the system is decreasing continuously at a constant rate of one, if the workload is positive. If the workload at time zero is $x + y$, there must be a time that the workload becomes y for the first time. This period of time is probabilistically equivalent to a period of time in which the workload goes from x to 0. Then the workload goes from y to 0. Thus, the busy period with initial workload $x + y$ can be decomposed in two periods: the workload goes from $x + y$ to y, and the workload goes from y to zero. By keeping track of the phase at the point the workload becomes y for the first time, the random variables in the two periods (i.e., lengths and numbers of customers served) become conditionally independent, which leads to

$$\begin{aligned}
&\phi^*_{i,j}(s,z,x+y)\\
&\quad= \int\limits_0^{\infty} e^{-st}\mathrm{d}_t\left(\sum_{k=0}^{\infty} z^k \int_0^t \sum_{u=0}^{k}\sum_{l=1}^{m} (\phi_{i,l}(\mathrm{d}v,u,x)\phi_{l,j}(t-v,k-u,y))\right)\\
&\quad= \sum_{l=1}^{m}\left(\int\limits_0^{\infty} e^{-sv}\mathrm{d}_v\left(\sum_{u=0}^{\infty} z^u\phi_{i,l}(v,u,x)\right)\right)\left(\int\limits_0^{\infty} e^{-st}\mathrm{d}_t\left(\sum_{k=0}^{\infty} z^k\phi_{l,j}(t,k,x)\right)\right)\\
&\quad= \sum_{l=1}^{m} \phi^*_{i,l}(s,z,x)\phi^*_{l,j}(s,z,y).
\end{aligned} \tag{4.113}$$

This proves result (1).

Result (2) is obtained by conditioning on the service time of the first customer in a busy period.

For (3), by conditioning on the arriving time of the first customer after time zero, we obtain

$$\Phi(t,0,x) = \delta(t \geq x)\exp(D_0 t);$$

$$\Phi(t,k,x) = \int_0^t \exp(D_0 v)D_1 \int_0^{t-x-v} \Phi(t-v,k-1,x-v+w)\mathrm{d}F(w)\mathrm{d}v, \quad k = 1,\ 2,\ \ldots;$$

$$\Phi^*(s,z,x) = \exp(-(sI - D_0)x)\left\{ I + \int_0^x \exp((sI - D_0)v)D_1 G^*(s,z)\Phi^*(s,z,v)\mathrm{d}v \right\}. \tag{4.114}$$

Taking the derivative of $\Phi^*(s, z, x)$ with respect to x, we obtain

$$\begin{aligned}\frac{\mathrm{d}\Phi^*(s,z,x)}{\mathrm{d}x} &= -(sI - D_0)\Phi^*(s,z,x) + D_1 G^*(s,z)\Phi^*(s,z,v) \\ &= (-(sI - D_0) + D_1 G^*(s,z))\Phi^*(s,z,x).\end{aligned} \tag{4.115}$$

Solving the above differential equation with initial condition $\Phi^*(s, z, 0) = I$, result (3) is obtained. This completes the proof of Lemma 4.8.1.

Combining (2) and (3) in Lemma 4.8.1 yields the following matrix exponential form for matrices $G^*(s, z)$ and G.

Theorem 4.8.2 (Neuts (1989a)) *For the MAP/G/1 queue, we have, for* $s \geq 0$, $0 \leq z \leq 1$,

$$\begin{aligned} G^*(s,z) &= z\int_0^\infty \exp((-sI + D_0 + D_1 G^*(s,z))x)\mathrm{d}F(x); \\ G &= \int_0^\infty \exp((D_0 + D_1 G)x)\mathrm{d}F(x). \end{aligned} \tag{4.116}$$

One of the consequences of the above equalities is that G and $D_0 + D_1 G$ commute. In addition, the second equality in Eq. (4.116) leads to a new iterative algorithm for computing the matrix G without computing matrices $\{A_0, A_1, \ldots\}$, which is required previously in Eq. (4.107).

Exercise 4.8.9 Use Eq. (4.116) to show that matrices G and $D_0 + D_1 G$ are *commutable*.

Exercise 4.8.10 For the *MAP/PH/1* queue defined in Sect. 4.4.1, use the second equation in (4.116) to develop an algorithm for computing G. Hint: For this case, we have the following calculations.

$$
\begin{aligned}
G &= \int_0^\infty \exp((D_0 + D_1G)x)\boldsymbol{\beta}\exp(Sx)\mathbf{S}^0\mathrm{d}x \\
&= (I \otimes \boldsymbol{\beta}) \int_0^\infty \exp((D_0 + D_1G)x) \otimes \exp(Sx)\mathrm{d}x(I \otimes \mathbf{S}^0) \\
&= (I \otimes \boldsymbol{\beta}) \int_0^\infty \exp(((D_0 + D_1G) \otimes I + I \otimes S)x)\mathrm{d}x(I \otimes \mathbf{S}^0) \\
&= -(I \otimes \boldsymbol{\beta})((D_0 + D_1G) \otimes I + I \otimes S)^{-1}(I \otimes \mathbf{S}^0).
\end{aligned}
\tag{4.117}
$$

See the proof of Proposition 1.3.1 for a proof for the third line in Eq. (4.117).

In general, the matrix G can be computed by the following iterative method: $G[0] = 0$,

$$
G[k+1] = \int_0^\infty \exp((D_0 + D_1G[k])x)\mathrm{d}F(x), \quad k = 0,\ 1,\ 2,\ \ldots \tag{4.118}
$$

Then $\{G[k],\ k = 0,\ 1,\ 2,\ \ldots\}$ is a nondecreasing sequence and converges to G. Another consequence of Eq. (4.116) is that $\mathbf{u}_z$ defined in Eq. (4.108) has the following simpler expression.

Theorem 4.8.3 *If $\rho < 1$ and D is irreducible, then*

$$
\mathbf{u}_z = (\mathbf{eg}(I - D_1/\mu) - D - D_1G + GD_1)^{-1}\mathbf{e}. \tag{4.119}
$$

Proof. Taking derivatives of both sides of Eq. (4.116) with respect to z, letting $s = 0$ and $z = 1$, and postmultiplying by $\mathbf{e}$ on both sides, yields

$$
\mathbf{u}_z^* = \mathbf{e} + \int_0^\infty \sum_{n=1}^\infty \frac{x^n}{n!}(D_0 + D_1G)^{n-1}\mathrm{d}F(x)D_1\mathbf{u}_z^* \tag{4.120}
$$

Recall that vector $\mathbf{g}$ satisfies $\mathbf{g}G = \mathbf{g}$ and $\mathbf{ge} = 1$. Also note that $(D_0 + D_1G)\mathbf{e} = 0$. By Eq. (4.116), $\mathbf{g}$ also satisfies $\mathbf{g}(D_0 + D_1G) = 0$. In fact, it is easier to see that, if any vector $\mathbf{y}$ satisfies $\mathbf{y}(D_0 + D_1G) = 0$ and $\mathbf{ye} = 1$, by Eq. (4.116), $\mathbf{y}$ must satisfy $\mathbf{y}G = \mathbf{y}$. By the uniqueness of $\mathbf{g}$, we must have $\mathbf{g} = \mathbf{y}$. For the integrand in Eq. (4.120), we have

$$
\begin{aligned}
&(\mathbf{eg} - (D_0 + D_1G)) \sum_{n=1}^\infty \frac{x^n}{n!}(D_0 + D_1G)^{n-1} \\
&= x\mathbf{eg} + I - \exp((D_0 + D_1G)x).
\end{aligned}
\tag{4.121}
$$

It can be shown that the matrix $\mathbf{eg} - (D_0 + D_1G)$ is invertible (see the proof of Theorem 2.3.2). Using Eq. (4.121), Eq. (4.120) becomes

$$\left(I - (\mathbf{eg} - (D_0 + D_1G))^{-1}\left((I - G)D_1 + \frac{\mathbf{eg}}{\mu}D_1\right)\right)\mathbf{u}_z^* = \mathbf{e}, \tag{4.122}$$

which leads to, using $(\mathbf{eg} - (D_0 + D_1G))\mathbf{e} = \mathbf{e}$,

$$(\mathbf{eg}(I - D_1/\mu) - D_0 - D_1 - D_1G + GD_1)\mathbf{u}_z^* = \mathbf{e}. \tag{4.123}$$

Then Eq. (4.119) is proved if the matrix on the left hand side of Eq. (4.123), denoted as M, is invertible.

If M is not invertible, there exists a nonzero vector $\mathbf{v}$ such that $\mathbf{v}M = 0$. We consider two cases: $\mathbf{ve} = 0$ and $\mathbf{ve} \neq 0$. If $\mathbf{ve} = 0$, $\mathbf{v}M = 0$ becomes $\mathbf{v}(GD_1 - D_1G - D) = 0$. Postmultiplying by G on both sides and using $G(D_0 + D_1G) = (D_0 + D_1G)G$, we obtain $\mathbf{v}(GD_0 - D_1G) = 0$. Combining the last two equations yields $\mathbf{v}(G - I)D = 0$. If $\mathbf{v}(G - I) = 0$, we must have $\mathbf{v} = c\mathbf{g}$ for some constant c. Since $\mathbf{ge} = 1$, we must have $c = 0$ and, consequently, $\mathbf{v} = 0$, which contradicts the assumption that $\mathbf{v}$ is nonzero. If $\mathbf{ve} \neq 0$, without loss of generality, we assume $\mathbf{ve} = 1$. Then $\mathbf{v}M = 0$ becomes

$$\mathbf{g}(I - D_1/\mu) = \mathbf{v}(D + D_1G - GD_1). \tag{4.124}$$

Postmultiplying by G on both sides of Eq. (4.124) yields $\mathbf{g}(I - D_1G/\mu) = \mathbf{v}(GD_0 + D_1G)$. Combining the last two equations, we obtain $(\mathbf{g}/\mu - \mathbf{v}(I - G))D = 0$. If $\mathbf{g}/\mu - \mathbf{v}(I - G) = 0$, we obtain $(\mathbf{g}/\mu - \mathbf{v}(I - G))\mathbf{e} = 0$, which leads to $1/\mu = 0$, which is impossible. If $(\mathbf{g}/\mu - \mathbf{v}(I - G)) \neq 0$, then we must have $\mathbf{g}/\mu - \mathbf{v}(I - G) = c\boldsymbol{\theta}$, for some constant c. Postmultiplying by $\mathbf{e}$ on both sides, yields $c = 1/\mu$ and $(\mathbf{g} - \boldsymbol{\theta})/\mu = -\mathbf{v}(I - G)$. Postmultiplying by $\mathbf{e}$ on both sides of Eq. (4.124), yields $1 - \mathbf{g}D_1\mathbf{e}/\mu = \mathbf{v}(I - G)D_1\mathbf{e} = (\boldsymbol{\theta} - \mathbf{g})D_1\mathbf{e}/\mu = \rho - \mathbf{g}D_1\mathbf{e}/\mu$, which contradicts $\rho < 1$.

In summary, $\mathbf{v}$ does not exist and M is invertible. Therefore, Eq. (4.119) is obtained. This completes the proof of Theorem 4.8.3.

Compared to Eq. (4.108), the expression in Eq. (4.119) does not involve the matrix A and the sum of $\{nA_n, n = 0, 1, 2, \ldots\}$, which is good for computation. Equations (4.116) and (4.119) show that performance measures can be computed by using system parameters directly, without computing intermediate matrices such as A. This simplification can be quite useful in application.

Commentary The *MAP*/*G*/1 queue is a generalization of the classical *M*/*G*/1 queue (Cohen (1982)). Classical results include (i) queue length distributions at arrival, departure, and arbitrary epochs; (ii) distributions of the actual and virtual waiting times (LST) (e.g., Pollaczek-Khinchine formula for the waiting time Neuts (1986a, b)); and (iii) busy period analysis. We focused on the busy period analysis for the *MAP*/*G*/1 queue. For the *BMAP*/*G*/1 queue, extensive studies on the queue lengths and waiting times have been carried out (e.g., Ramaswami (1980), Lucantoni et al. (1990), Sengupta (1990a), Lucantoni (1991)). The more general *SM/PH/*1 and *MAP/SM/*1 queues have been analyzed as well (e.g., Sengupta (1990b), Lucantoni and Neuts (1994)).

Additional Exercises and Extensions

Exercise 4.8.11 Consider the busy period and busy cycle for the *MAP/PH/*1 queues. Compute the matrix G and the mean number of customers served in a busy period, the mean length of a busy period, the mean number of customers served in a busy cycle, and the mean length of a busy cycle for Exercises 4.2.2, 4.3.4, and 4.4.2, respectively. (Hint: Use Eqs. (4.108), (4.111), and (4.119).)

Exercise 4.8.12 Consider the *BMAP/G/*1 queue. The arrival process has matrix representation $(D_0, D_1, D_2, \ldots, D_N)$. Explain the following generalization of Eq. (4.116) intuitively:

$$\begin{aligned} G^*(s,z) &= z\int_0^\infty \exp\left(\left(-sI + D_0 + \sum_{n=1}^N D_n (G^*(s,z))^n\right)x\right)\mathrm{d}F(x); \\ G &= \int_0^\infty \exp\left(D_0 + \sum_{n=1}^N D_n G^n\right)\mathrm{d}F(x). \end{aligned} \tag{4.125}$$

By Eq. (4.125), for *BMAP/G/*1, Eq. (4.119) becomes

$$\mathbf{u}_z = \left(\mathbf{eg}(I - \tilde{D}_1/\mu) - (I - G)\tilde{D}_1 - \sum_{n=0}^N D_n G^n\right)^{-1}\mathbf{e}, \tag{4.126}$$

where

$$\begin{aligned} \tilde{D}_1 &= \sum_{n=1}^N D_n \sum_{j=0}^{n-1} G^j \\ &= \left(D - \sum_{n=0}^N D_n G^n + \left(\sum_{n=0}^N nD_n\right)\mathbf{eg}\right)(I - G + \mathbf{eg})^{-1}, \end{aligned} \tag{4.127}$$

Similarly, Eqs. (4.108) and 4.111) become

$$\begin{aligned} \mathbf{u}_s &= \mathbf{u}_z/\mu; \\ \mathbf{u}_z^{(0)} &\equiv \frac{\mathrm{d}G^{*(0)}(s,z)}{\mathrm{d}z}\mathbf{e}\bigg|_{z=1-,s=0+} = (-D_0^{-1}\tilde{D}_1)\mathbf{u}_z; \\ \mathbf{u}_s^{(0)} &\equiv -\frac{\mathrm{d}G^{*(0)}(s,z)}{\mathrm{d}s}\mathbf{e}\bigg|_{z=1-,s=0+} = (-D_0^{-1})\mathbf{e} + (-D_0^{-1}\tilde{D}_1)\mathbf{u}_s. \end{aligned} \tag{4.128}$$

Exercise 4.8.13 Outline a method for computing the matrix G, the mean number of customers served in a busy period, the mean length of a busy period, the mean number of customers served in a busy cycle, and the mean length of a busy cycle for the following *BMAP*/*PH*/1 queue.

$$D_0 = \begin{pmatrix} -2 & 1 \\ 0 & -1.5 \end{pmatrix}, \quad D_1 = \begin{pmatrix} 0.8 & 0 \\ 1 & 0.4 \end{pmatrix}, \quad D_2 = \begin{pmatrix} 0.2 & 0 \\ 0 & 0.1 \end{pmatrix},$$
$$\boldsymbol{\beta} = (0.8,\ 0.2), \quad S = \begin{pmatrix} -5 & 0 \\ 0.5 & -1 \end{pmatrix}. \tag{4.129}$$

For the *MAP*/*G*/1 queue, define matrix Q satisfying

$$Q = D_0 + D_1 \int_0^{\infty} \exp(Qt)\mathrm{d}F(t). \tag{4.130}$$

It has been shown in Takine and Hasegawa (1994) that Q is an infinitesimal generator of the underlying Markov chain after excising busy periods. Denote by $\mathbf{y}_0 = (y_{0,1}, \ldots, y_{0,m_a})$, the joint probabilities of the empty queueing system and phase of the underlying Markov chain at an arbitrary time.

Exercise 4.8.14* (Lucantoni (1991)) Show that $\mathbf{y}_0 Q = 0$ and $\mathbf{y}_0 \mathbf{e} = 1 - \rho$.

Let $v^*_i(s)$ be the LST of the workload in the queueing system at an arbitrary time if the phase of the underlying Markov chain is i. Let $\mathbf{v}^*(s) = (v^*_1(s), v^*_2(s), \ldots, v^*_{m_a}(s))$.

Exercise 4.8.15* (Lucantoni (1991)) Show that, for $s \geq 0$,

$$\mathbf{v}^*(s) = s\mathbf{y}_0(sI + D_0 + D_1 f^*(s))^{-1}. \tag{4.131}$$

Consequently, the LST of the actual waiting time is given by

$$\mathbf{w}^*(s) = \frac{s\mathbf{y}_0}{\lambda}(sI + D_0 + D_1 f^*(s))^{-1} D_1. \tag{4.132}$$

For the *M*/*G*/1 queue, show that (4.132) can be reduced to

$$w^*(s) = \frac{s(1-\rho)}{s - \lambda + \lambda f^*(s)}, \tag{4.133}$$

which is the well-known Pollaczek-Khinchine formula. We remark that Eqs. (4.132) and (4.133) can also be derived by utilizing the *level crossing method* (He and Jewkes (1996) and Brill (2008)).

Exercise 4.8.16* (Takine (2001a)) Let $\mathbf{x}_k$ be a vector of size m_a whose j-th element is the probability that the queue length is k and the arrival process is in phase j right after the departure of an arbitrary customer. Let $\mathbf{y}_k$ be a vector of order m_a whose

j-th element is the probability that the queue length is k and the arrival process is in phase j at an arbitrary time. Denote by $\mathbf{x}^*(z)$ the probability generating function of $\{\mathbf{x}_0, \mathbf{x}_1, \ldots\}$. Denote by $\mathbf{y}^*(z)$ the probability generating function of $\{\mathbf{y}_0, \mathbf{y}_1, \ldots\}$. Denote by $A^*(z)$ the probability generating function of $\{A_0, A_1, \ldots\}$. Assume $0 \leq z \leq 1$.

1. Show that $A^*(z) = \int_0^\infty \exp((D_0 + zD_1)x)\mathrm{d}F(x)$.
2. Show that $\mathbf{y}_0 = \lambda \mathbf{x}_0(-D_0)^{-1}$ (Lucantoni 1991).
3. Show that $\mathbf{x}^*(z)(zI - A^*(z)) = \mathbf{x}_0(-D_0)^{-1}(D_0 + zD_1)A^*(z)$.
4. Show that $\mathbf{y}^*(z)(zI - A^*(z)) = (z - 1)\mathbf{y}_0 A^*(z)$.
5. Show that $\mathbf{y}^*(z)(D_0 + zD_1) = \lambda(z - 1)\mathbf{x}^*(z)$.
6. For the $M/G/1$ queue, show that $x^*(z)(z - A^*(z)) = (1 - \rho)(z - 1)A^*(z)$.

4.9 The *MMAP*[*K*]/*G*[*K*]/1 Queue

The flow of customers in the *MMAP*[*K*]/*G*[*K*]/1 queue is the same as that in the $M/M/1$ queue (see Fig. 4.1). The system has a single server that serves K types of customers, where K is a positive integer. The K types of customers arrive according to Markovian arrival process with marked arrivals $(D_0, D_1, \ldots, D_K)$ of order m_a (see Sect. 2.5). All customers join a single queue and are served on a first-come-first-served basis. Service times of type k customers have common distribution function $F_k(t)$, for $k = 1, 2, \ldots, K$. The arrival process and the service times are independent.

Denote by $\{I_a(t), t \geq 0\}$ the underlying continuous time Markov chain of the arrival process. The infinitesimal generator of $\{I_a(t), t \geq 0\}$ is given by $D = D_0 + D_1 + \ldots + D_K$, which is assumed to be irreducible. Let $\boldsymbol{\theta}_a$ satisfy $\boldsymbol{\theta}_a D = 0$ and $\boldsymbol{\theta}_a \mathbf{e} = 1$. The arrival rate of type k customers is given by $\lambda_k = \boldsymbol{\theta}_a D_k \mathbf{e}$, $k = 1, 2, \ldots, K$. The service rate of type k customer is denoted by μ_k, which is the reciprocal of the mean service time of type k customers, $k = 1, 2, \ldots, K$. Define $\rho_k = \lambda_k/\mu_k$, $k = 1, 2, \ldots, K$. Let $\rho = \rho_1 + \ldots + \rho_K$. We assume that $\rho < 1$ so that the queueing system is stable.

Exercise 4.9.1 Roughly speaking, $\lambda_k t$ is the number of type k customers arrived in $[0, t]$, which needs approximately $\lambda_k t/\mu_k$ units of time to be served, $k = 1, 2, \ldots, K$. Explain why condition $\rho < 1$ ensures system stability.

We first analyze the busy period of the queueing system. A busy period begins at the arrival epoch of a customer, regardless of its type, to an empty system, and ends when the system becomes empty again. The number of customers served in a busy period and the length of a busy period are investigated.

Let $\mathbf{n} = (n_1, n_2, \ldots, n_K)$, where $\{n_1, \ldots, n_K\}$ are nonnegative integers, and $\mathbf{z} = (z_1, z_2, \ldots, z_K)$, where $0 \leq z_1, \ldots, z_K \leq 1$. Let τ_B be the length of a busy period. For $k, k' = 1, \ldots, K$, define $G_{k,k'}(x, \mathbf{n})$ to be an $m_a \times m_a$ matrix with the (i, j)-th element being the conditional probability that

(i) The length of a busy period is less than or equal to x;
(ii) There are n_1 type 1, ..., and n_K type K customers served during the busy period;
(iii) The last arrival in the busy period is of type k'; and
(iv) $I_a(\tau_B+) = j$,

given that the first customer of the busy period is of type k and $I_a(0+) = j$. Denote by $G^*_{k,k'}(s, \mathbf{z})$ the joint transform of $G_{k,k'}(x, \mathbf{n})$, for $s \geq 0$, $0 \leq z_1, \ldots, z_K \leq 1$,

$$G^*_{k,k'}(s, \mathbf{z}) = \sum_{\mathbf{n} \geq 0} \mathbf{z}^{\mathbf{n}} \int_0^{\infty} e^{-sx} G_{k,k'}(\mathrm{d}x, \mathbf{n}), \tag{4.134}$$

where $\mathbf{z}^{\mathbf{n}} = z_1^{n_1} \cdots z_K^{n_K}$. Define $G^*(s, \mathbf{z})$ as a matrix with the (k, k')-th block $G^*_{k,k'}(s, \mathbf{z})$, $k, k' = 1, 2, \ldots, K$.

Theorem 4.9.1 (He (1996)) *For the MMAP[K]/G[K]/1 queue, the matrix $G^*(z, s)$ satisfies equation, for $s \geq 0, 0 \leq z_1, \ldots, z_K \leq 1$,*

$$G^*(s, \mathbf{z}) = \int_0^{\infty} \begin{pmatrix} z_1 F_1(\mathrm{d}x) I & & \\ & \ddots & \\ & & z_K F_K(\mathrm{d}x) I \end{pmatrix} \exp((-sI + T_0 + T_1 G^*(s, \mathbf{z}))x), \tag{4.135}$$

where

$$T_0 = \begin{pmatrix} D_0 & & \\ & \ddots & \\ & & D_0 \end{pmatrix}, \quad T_1 = \begin{pmatrix} D_1 & \cdots & D_K \\ \vdots & \vdots & \vdots \\ D_1 & \cdots & D_K \end{pmatrix}. \tag{4.136}$$

Exercise 4.9.2 If $K = 1$, show that Eq. (4.135) is reduced to Eq. (4.116).

Some implications and extensions of Theorem 4.9.1 are collected as follows.

1. Let $G = G^*(0, \mathbf{e}')$ and $G_{k,k'} = G^*_{k,k'}(0, \mathbf{e}')$, for $k, k' = 1, 2, \ldots, K$. Then the elements of $G_{k,k'}$ give the transition probabilities of the phase of the underlying Markov chain and the type of customers served, at the beginning and the end of the busy period. Consequently, the elements of G give the transition probabilities of the phase. It is clear that G is a stochastic matrix, given $\rho < 1$.
2. The elements of $-T_0^{-1}T_1G$ are the transition probabilities of the phase at the end of a busy period and the type of the last customer arrived (served) in the busy period. Let $\mathbf{u} = (\mathbf{u}(1), \ldots, \mathbf{u}(K))$ be the Perron-Frobenius eigenvalue of $-T_0^{-1}T_1G$. Then $\mathbf{u}$ is the joint distribution of the type of last customer served in a busy period and the phase of the arrival process at the end of a busy period. In addition, we have $\mathbf{x}(0) = c(\mathbf{u}(1) + \ldots + \mathbf{u}(K))$, where c is a positive constant and $\mathbf{x}(0)$ is the distribution of the phase of the arrival process at the end of an arbitrary busy period.

3. The mean numbers of different types of customers served in a busy period can be obtained (see He 1996).
4. Let $\Phi^*(x, s, z) = \exp((-sI + T_0 + T_1G^*(s, \mathbf{z}))x)$, for $x \geq 0$. Divide $\Phi^*(x, s, \mathbf{z})$ into $m_a \times m_a$ blocks. Then the (i, j)-th element of the block $\Phi_{k,k'}{}^*(x, s, \mathbf{z})$ is the joint transform of the numbers of different types of customers served in a busy period, and the length of a busy period, such that the busy period begins with serving a type k customer, and the service time is x, the phase of the underlying Markov chain is j right after the end of the busy period, given the phase at the beginning of the busy period is i.

Exercise 4.9.3 Define $G_k = G^*_{k,1}(0, \mathbf{e}') + \ldots + G^*_{k,K}(0, \mathbf{e}')$, $k = 1, 2, \ldots, K$. Explain G_k intuitively.

Exercise 4.9.4 Let $\Phi_k{}^*(x, s, \mathbf{z})$ be the sum of matrices $\{\Phi_{k,k'}{}^*(x, s, \mathbf{z}), k' = 1, 2, \ldots, K\}$. Explain $\Phi_k{}^*(x, s, \mathbf{z})$ intuitively. Let $\Phi_k(x) = \Phi^*_k(x, 0, \mathbf{e}')$, $k = 1, 2, \ldots, K$. Explain $\Phi_k(x)$ intuitively. Use the intuitive interpretation to argue that $\Phi_1(x) = \Phi_2(x) = \ldots = \Phi_K(x)$.

Next, we consider the waiting time distributions. Let V_t be the total amount of work in the system at time t, which is called the *virtual waiting time*. Define

$$\tau_B = \inf\{t:\ t \geq 0, V_t = 0\}, \tag{4.137}$$

which is the first passage time to the idle state after time 0. (Note: τ_B has been defined as the length of a busy period.) Let $\Phi(x)$ denote an $m_a \times m_a$ matrix whose (i, j)-th element is $P\{I_a(\tau_B) = j \mid I_a(0) = i, V_0 = x\}$, with boundary condition $\Phi(0+) = I$.

Exercise 4.9.5 Show that $\Phi(x) = \Phi_k(x)$, which is defined in Exercise 4.9.4.

Exercise 4.9.6 Similar to Lemma 4.8.1, show that $\Phi(x + y) = \Phi(x)\Phi(y)$, for $x, y > 0$.

Theorem 4.9.2 *The function* $\Phi(x)$ *is given by*

$$\Phi(x) = \exp(Qx);$$
$$Q = D_0 + \sum_{k=1}^{K} D_k \int_0^\infty \exp(Qx)\mathrm{d}F_k(x). \tag{4.138}$$

Proof. First note that, for *MMAP*[*K*], there can be only one arrival in a small interval, which leads to

$$\Phi(\delta x) = I + D_0\delta x + \sum_{k=1}^{K} D_k\delta x \int_0^\infty \Phi(x)\mathrm{d}F_k(x) + o(\delta x). \tag{4.139}$$

Replacing y in $\Phi(x + y) = \Phi(x)\Phi(y)$ with infinitesimally small $\delta x > 0$ leads to

$$\Phi(x+\delta x) = \Phi(x)\left(I + D_0\delta x + \sum_{k=1}^{K} D_k\delta x \int_0^{\infty} \Phi(x)\mathrm{d}F_k(x) + o(\delta x)\right). \quad (4.140)$$

Equation (4.140) leads to the following differential equation for $\Phi(x)$:

$$\frac{\mathrm{d}\Phi(x)}{\mathrm{d}x} = \Phi(x)\left(D_0 + \sum_{k=1}^{K} D_k \int_0^{\infty} \Phi(x)\mathrm{d}F_k(x)\right), \quad x \geq 0. \quad (4.141)$$

Denote by Q the part in the parentheses on the right hand side of Eq. (4.141), which is indeed a constant matrix. Then we obtain $\mathrm{d}\Phi(x)/\mathrm{d}x = \Phi(x)Q$. The solution in Eq. (4.138) is obtained directly from the differential equation with boundary condition $\Phi(0+) = I$. This completes the proof of Theorem 4.9.2.

Exercise 4.9.7 Show that the off-diagonal elements of Q are nonnegative.

Equation (4.138) indicates that matrix Q is an irreducible infinitesimal generator, which is for the underlying Markov chain after excising busy periods (see Takine 2001a). Let $\boldsymbol{\kappa} = (\kappa_1, \ldots, \kappa_{m_a})$ satisfying $\boldsymbol{\kappa}Q = 0$ and $\boldsymbol{\kappa}\mathbf{e} = 1$. Then $\boldsymbol{\kappa}$ is the stationary distribution of the phase during the idle periods after excising all busy periods, i.e., $\kappa_i = P\{I_a(t) = i \mid V_t = 0\}$, $i = 1, 2, \ldots, m_a$, at arbitrary time t.

Define $\mathbf{v}(x) = (\mathbf{v}_1(x), \ldots, \mathbf{v}_{m_a}(x))$, where $\mathbf{v}_i(x) = P\{I_a(t) = i,\ V_t \leq x\}$, $i = 1, 2, \ldots, K$, and $x \geq 0$, at arbitrary time t. Denote by $\mathbf{v}^*(s)$ the LST of $\mathbf{v}(x)$.

Exercise 4.9.8 By *Little's law* in queueing theory, we have $\mathbf{v}(0)\mathbf{e} = 1 - \rho$. Explain the result intuitively. Use the results to explain $\mathbf{v}(0) = (1 - \rho)\boldsymbol{\kappa}$ intuitively. (Note: The vector $\mathbf{v}(0)$ is $\mathbf{y}_0$ whose j-th element is the probability that the system is empty and the underlying Markov chain is in phase j, $j = 1, 2, \ldots, m_a$.)

Exercise 4.9.9 Show that $\mathbf{v}^*(0) = \mathbf{v}(\infty) = \boldsymbol{\theta}_a$.

Next, the classical Pollaczek-Khinchine formula is generalized to the *MMAP*[*K*]/*G*[*K*]/1 queue as follows.

Theorem 4.9.3 (He (1996) and Takine (2001a)) *The* LST $v^*(s)$ *is given by, for* $s > 0$,

$$\mathbf{v}^*(s)\left(sI + D_0 + \sum_{k=1}^{K} f_k^*(s)D_k\right) = (1-\rho)s\boldsymbol{\kappa}. \quad (4.142)$$

Based on Theorem 4.9.3, a number of results on the virtual waiting time and the actual waiting time can be obtained, which are collected in the following exercises. By taking derivatives of both sides of Eq. (4.142), we obtain

$$
\begin{aligned}
&\left.\frac{\mathrm{d}\mathbf{v}^*(s)}{\mathrm{d}s}\right|_{s=0} D + \boldsymbol{\theta}_{\mathrm{a}}\left(I - \sum_{k=1}^{K}\frac{D_k}{\mu_k}\right) = (1-\rho)\boldsymbol{\kappa};\\
&\left.\frac{\mathrm{d}^2\mathbf{v}^*(s)}{\mathrm{d}s^2}\right|_{s=0} D + 2\left.\frac{\mathrm{d}\mathbf{v}^*(s)}{\mathrm{d}s}\right|_{s=0}\left(I - \sum_{k=1}^{K}\frac{D_k}{\mu_k}\right) + \boldsymbol{\theta}_{\mathrm{a}}\left(\sum_{k=1}^{K}\frac{D_k}{\mu_k^{(2)}}\right) = 0.
\end{aligned}
\tag{4.143}
$$

where $\mu_k^{(2)}$ is the reciprocal of the second moments of the service time of type k customers.

Exercise 4.9.10 Find the mean virtual waiting time by the following steps.

1. Show that the matrix $D + \left(I - \sum_{k=1}^{K} D_k/\mu_k\right)\mathbf{e}\boldsymbol{\theta}_{\mathrm{a}}$ is invertible.
2. Use the equalities in Eq. (4.143) to show that

$$
\begin{aligned}
\left.\frac{\mathrm{d}\mathbf{v}^*(s)}{\mathrm{d}s}\right|_{s=0} = &\left((1-\rho)\boldsymbol{\kappa} - \boldsymbol{\theta}_{\mathrm{a}}\left(I - \sum_{k=1}^{K}\frac{D_k}{\mu_k}\right)\right)\left(D + \left(I - \sum_{k=1}^{K}\frac{D_k}{\mu_k}\right)\mathbf{e}\boldsymbol{\theta}_{\mathrm{a}}\right)^{-1}\\
&-\frac{\boldsymbol{\theta}_{\mathrm{a}}}{2(1-\rho)}\left(\sum_{k=1}^{K}\frac{D_k}{\mu_k^{(2)}}\right)\mathbf{e}\boldsymbol{\theta}_{\mathrm{a}}.
\end{aligned}
\tag{4.144}
$$

Let W_k be the (actual) waiting time of an arbitrary type k customer, and I_k be the phase of the underlying Markov chain right after the arrival of the type k customer, for $k = 1, 2, \ldots, K$. Denote by $\mathbf{w}_k(x)$ the joint distribution of (W_k, I_k) and by $\mathbf{w}_k^*(s)$ the LST of $\mathbf{w}_k(x)$.

Exercise 4.9.11 Use conditional probabilities to show, for $s \geq 0$,

$$
\mathbf{w}_k(x) = \frac{\mathbf{v}(x)D_k}{\lambda_k}, \quad \mathbf{w}_k^*(s) = \frac{\mathbf{v}^*(s)D_k}{\lambda_k}. \tag{4.145}
$$

Apparently, the moments of the actual waiting times can be obtained by Eq. (4.145) and Exercise 4.9.10.

Let S_k be the sojourn time of an arbitrary type k customer, and let $I_{\mathrm{a},k}$ be the phase of the underlying Markov chain, immediately after the arrival of a type k customer, for $k = 1, 2, \ldots, K$. Define $\mathbf{s}_k(x)$ as a $1\times m$ vector whose j-th element is $P\{S_k \leq x, I_{\mathrm{a},k} = j\}$. Define $\mathbf{s}^*_k(s)$ as the LST of $\mathbf{s}_k(x)$.

Exercise 4.9.12 Show, for $k = 1, 2, \ldots, K$,

$$
\begin{aligned}
\mathbf{s}_k(x) &= \frac{1}{\lambda_k}\int_0^x \mathbf{v}(x-y)D_k\mathrm{d}F_k(y);\\
\mathbf{s}_k^*(s) &= \frac{\mathbf{v}^*(s)D_k f_k^*(s)}{\lambda_k}.
\end{aligned}
\tag{4.146}
$$

Exercise 4.9.13 Consider the $M[2]/M[2]/1$ queue, where the arrival process consists of two independent Poisson processes and the service times are exponentially distributed. Find $\mathbf{v}^*(s)$, $\mathbf{w}_1^*(s)$, $\mathbf{w}_2^*(s)$, $\mathbf{s}_1^*(s)$, and $\mathbf{s}_2^*(s)$ explicitly.

Exercise 4.9.14 Consider an $MAP[2]/M[2]/1$ queue for which the two types of customers arrive cyclically:

$$D_0 = \begin{pmatrix} -2 & 0 \\ 0 & -10 \end{pmatrix},\ D_1 = \begin{pmatrix} 0 & 2 \\ 0 & 0 \end{pmatrix},\ D_2 = \begin{pmatrix} 0 & 0 \\ 10 & 0 \end{pmatrix}. \tag{4.147}$$

The service rates of the two types of customers are $\mu_1 = 0.5$ and $\mu_2 = 5$, respectively. Find the arrival rates λ_1 and λ_2. Find the mean (actual) waiting times of the two types of customers. Find the mean sojourn times of the two types of customers. Consider the $M[2]/M[2]/1$ queue defined in Exercise 4.9.13. Find the mean waiting times and mean sojourn times of the two types of customers using parameters $\{\lambda_1, \lambda_2, \mu_1, \mu_2\}$. Comment on the interaction between the two types of customers.

Finally, we look at the queue length distributions. Since all customers join a single queue, to know the service times of customers, we must keep track of the types of all customers in the queue. In Sect. 4.6, the K-ary tree is used for that purpose. If customers are served on an LCFS basis, the transitions of the Markov chain introduced in Sect. 4.6 possesses a tree structure. Then the Markov chain can be analyzed. Unfortunately, if the service discipline is FCFS, the Markov chain no longer has the tree structure. The approach developed in Sects. 3.6, 3.7, and 3.8 does not apply. A different approach, which is based on the relationship between the queue lengths and waiting times, is developed.

Let $L_k(t)$ be the number of type k customers in the system at an arbitrary time t, and let $L_{w,k}(t)$ be the number of type k customers waiting in the system at an arbitrary time t, for $k = 1, 2, \ldots, K$. Let $q_k(k')$ be the number of type k' customers in the system, and $I_{d,k}$ be the phase of the underlying Markov chain, immediately after the departure of a type k customer, for $k, k' = 1, 2, \ldots, K$.

In steady state, define the joint probability generating functions as follows: for $j = 1, 2, .., m_a$,

$$\begin{aligned} L_j^*(\mathbf{z}) &= E[z_1^{L_1(t)} \cdots z_K^{L_K(t)} 1_{\{I_a(t)=j\}}]; \\ L_{w,j}^*(\mathbf{z}) &= E[z_1^{L_{w,1}(t)} \cdots z_K^{L_{w,K}(t)} 1_{\{I_a(t)=j\}}]; \\ q_{k,j}^*(\mathbf{z}) &= E[z_1^{q_k(1)} \cdots z_K^{q_k(K)} 1_{\{I_{d,k}=j\}}]. \end{aligned} \tag{4.148}$$

Let $\mathbf{L}^*(\mathbf{z}) = (L_1^*(\mathbf{z}), L_2^*(\mathbf{z}), \cdots, L_K^*(\mathbf{z}))$, $\mathbf{L}_w^*(\mathbf{z}) = \left(L_{w,1}^*(\mathbf{z}), L_{w,2}^*(\mathbf{z}), \cdots, L_{w,K}^*(\mathbf{z})\right)$, and $\mathbf{q}_k^*(\mathbf{z}) = \left(q_{k,1}^*(\mathbf{z}), q_{k,2}^*(\mathbf{z}), \cdots, q_{k,K}^*(\mathbf{z})\right)$, for $k = 1, 2, \ldots, K$.

Theorem 4.9.4 (Takine (2001a)) *For the MMAP*[*K*]/*G*[*K*]/1 *queue, we have*

$$\mathbf{L}^*(\mathrm{z})\left(D_0 + \sum_{k=1}^{K} z_k D_k\right) = \sum_{k=1}^{K} \lambda_k (z_k - 1)\, \mathbf{q}_k^*(\mathbf{z}). \qquad (4.149)$$

Theorem 4.9.4 generalizes the relationship between the queue lengths at an arbitrary time and at departure epochs, given in Exercise 4.8.16.

Exercise 4.9.15 For the $M[2]/M[2]/1$ queue, simplify Eq. (4.149) and explain the resulting equations intuitively.

Next, the relationships between the waiting times, sojourn times, and queue lengths are established.

Theorem 4.9.5 (Takine (2001a))

$$\mathbf{L}_\mathrm{w}^*(\mathbf{z})\left(D_0 + \sum_{k=1}^{K} z_k D_k\right) = \sum_{k=1}^{K} \lambda_k (z_k - 1) \int_{0-}^{\infty} \mathbf{w}_k(\mathrm{d}x) \exp\left(\left(D_0 + \sum_{u=1}^{K} z_u D_u\right) x\right);$$

$$\mathbf{L}^*(\mathbf{z})\left(D_0 + \sum_{k=1}^{K} z_k D_k\right) = \sum_{k=1}^{K} \lambda_k (z_k - 1) \int_{0-}^{\infty} \mathbf{s}_k(\mathrm{d}x) \exp\left(\left(D_0 + \sum_{u=1}^{K} z_u D_u\right) x\right). \qquad (4.150)$$

Note that, in Eq. (4.150), $\int_{0-}^{\infty} \cdot \mathrm{d}x = \int_{0-}^{0+} \cdot \mathrm{d}x + \int_{0+}^{\infty} \cdot \mathrm{d}x$.

Exercise 4.9.16 For the $M[2]/M[2]/1$ queue, simplify Eq. (4.150).

Let ξ denote the maximum absolute value of the diagonal elements of matrix D_0. Define the following vectors, for $u = 0, 1, 2, \ldots$,

$$\mathbf{v}^{(u)}(\xi) = \int_0^{\infty} e^{-\xi x} \frac{(\xi x)^u}{u!} \mathrm{d}\mathbf{v}(x);$$

$$\left(I + \xi^{-1}\left(D_0 + \sum_{k=1}^{K} z_k D_k\right)\right)^u = \sum_{\mathbf{n}:\, \mathbf{n} \geq 0,\, \mathbf{ne} \leq u} z_1^{n_1} \cdots z_K^{n_K} \mathbf{F}_u(\mathbf{n}). \qquad (4.151)$$

Lemma 4.9.1 (Takine (2001a)) *The vectors* $\{\mathbf{F}_u(\mathbf{n}), \mathbf{n} \geq 0$ *and* $u = 1, 2, \ldots\}$ *can be calculated recursively as follows:*

$$\mathbf{F}_{u+1}(0) = (I + \xi^{-1} D_0)^{u+1};$$

$$\mathbf{F}_{u+1}(\mathbf{n}) = \mathbf{F}_u(\mathbf{n})(I + \xi^{-1} D_0) + \sum_{k=1:\, n_k \geq 1}^{K} \mathbf{F}_u(\mathbf{n} - \mathbf{e}_k)\xi^{-1} D_k, \qquad (4.152)$$

$$\text{for } \mathbf{ne} = 1, 2, \ldots, u;$$

$$\mathbf{F}_{u+1}((u+1)\mathbf{e}_k) = (\xi^{-1} D_k)^{u+1}, \quad \text{for } k = 1, 2, \ldots, K.$$

Lemma 4.9.2 (Takine (2001a)) *The vectors* $\{\mathbf{v}^{(u)}(\xi),\ u = 0, 1, 2, \ldots\}$ *are the stationary distribution of an M/G/1 type Markov chain whose transition probability matrix is given by*

$$\begin{pmatrix} B_0 + B_1 & B_2 & B_3 & B_4 & \cdots \\ & B_0 & B_1 & B_2 & B_3 & \cdots \\ & & B_0 & B_1 & B_2 & \cdots \\ & & & B_0 & B_1 & \cdots \\ & & & & \ddots & \ddots \end{pmatrix}, \tag{4.153}$$

where $B_0 = I + \xi^{-1}(D_0 + D^{(0)}(\xi))$, $B_n = \xi^{-1}D^{(n)}(\xi)$, $n = 1, 2, \ldots,$ *and*

$$\begin{aligned} D^{(n)}(\xi) &= \int_{0-}^{\infty} e^{-\xi x}\frac{(\xi x)^n}{n!}\mathrm{d}\left(\sum_{k=1}^{K} D_k F_k(x)\right); \\ D_k^{(n)}(\xi) &= \int_{0-}^{\infty} e^{-\xi x}\frac{(\xi x)^n}{n!}D_k \mathrm{d}F_k(x). \end{aligned} \tag{4.154}$$

Finally, the distributions of $(L_{w,1}(t), L_{w,2}(t), \ldots, L_{w,K}(t))$ and $(L_1(t), L_2(t), \ldots, L_K(t))$ are obtained as follows.

Theorem 4.9.6 (Takine (2001a)) *For the MMAP[K]/G[K]/1 queue, in steady state, we have*

$$\begin{aligned} \mathbf{L}_w(\mathbf{n}) &= \left(\sum_{k=1:\ n_k \geq 1}^{K} \mathbf{L}_w(\mathbf{n} - \mathbf{e}_k)D_k + \sum_{u=\mathbf{ne}}^{\infty} \mathbf{v}^{(u)}(\xi)D\mathbf{F}_u(\mathbf{n}) \right. \\ &\quad \left. - \sum_{k=1:\ n_k \geq 1}^{K} \sum_{u=\mathbf{ne}-1}^{\infty} \mathbf{v}^{(u)}(\xi)D_k\mathbf{F}_u(\mathbf{n} - \mathbf{e}_k) \right)(-D_0)^{-1}; \\ \mathbf{L}(\mathbf{n}) &= \left(\sum_{k=1:\ n_k \geq 1}^{K} \mathbf{L}(\mathbf{n} - \mathbf{e}_k)D_k + \sum_{u=\mathbf{ne}}^{\infty}\sum_{i=0}^{u} \mathbf{v}^{(u)}(\xi)D^{(u-i)}(\xi)\mathbf{F}_u(\mathbf{n}) \right. \\ &\quad \left. - \sum_{k=1:\ n_k \geq 1}^{K} \sum_{u=\mathbf{ne}-1}^{\infty}\sum_{i=0}^{u} \mathbf{v}^{(u)}(\xi)D_k^{(u-i)}(\xi)\mathbf{F}_u(\mathbf{n} - \mathbf{e}_k) \right)(-D_0)^{-1}. \end{aligned} \tag{4.155}$$

Exercise 4.9.17* For the $M[2]/M[2]/1$ queue, give all the necessary details for computing its queue length distributions.

Commentary In He (2001), the results on busy periods and waiting times are given to the batch arrival case, where customers of different types can arrive together. In Takine (2001a, b, c, d), the results on waiting times and queue lengths are given for the batch arrival case. In Takine (2001a), it has shown that Theorems 4.9.4 and 4.9.5 hold for queueing system with more general service disciplines.

The discovery of the algorithm for computing the joint distributions of the queue lengths of different types of customers (i.e., Theorems 4.9.4, 4.9.5, and 4.9.6) for the FCFS case is one important development in queueing theory.

Additional Exercises and Extensions

In queueing analysis, tail asymptotics analysis plays an important role. In Takine (2001a, 2004), a tail asymptotic analysis is done for the *MMAP*[*K*]/*G*[*K*]/1 queue. Distribution function $F(x)$ is called *heavy-tailed* if $1 - F(x) > 0$ for all $x > 0$, and, for all $y \geq 0$,

$$\lim_{x\to\infty} \frac{1-F(x+y)}{1-F(x)} = 1. \tag{4.156}$$

Denote by $F^{[2]}(x)$ the twofold convolution of $F(x)$ with itself, i.e., $F^{[2]}(x) = \int_0^x F(x-y)\mathrm{d}F(y)$. Heavy-tailed distribution function $F(x)$ is called *subexponential* if

$$\lim_{x\to\infty} \frac{1-F^{[2]}(x)}{1-F(x)} = 2. \tag{4.157}$$

Exercise 4.9.18* (Takine (2001a)) For the *MMAP*[*K*]/*G*[*K*]/1 queue, define $F_{\text{all}}(x) = \sum_{k=1}^{K} \lambda_k F_k(x)/\lambda$ and

$$F_{\text{e}}(x) = \frac{\lambda}{\rho}\int_0^x (1 - F_{\text{all}}(y))\mathrm{d}y, \tag{4.158}$$

which is the equilibrium distribution of $F_{\text{all}}(x)$. Assume that $F_{\text{e}}(x)$ is subexponential. Show that

$$\begin{aligned} \lim_{x\to\infty} \frac{\mathbf{e}' - \mathbf{v}(x)}{1-F_{\text{e}}(x)} &= \frac{\rho}{1-\rho}\boldsymbol{\theta}_{\text{a}}; \\ \lim_{x\to\infty} \frac{\mathbf{e}' - \mathbf{w}_k(x)}{1-F_{\text{e}}(x)} &= \left(\frac{\rho}{1-\rho}\right)\frac{\boldsymbol{\theta}_{\text{a}} D_k}{\lambda_k}. \end{aligned} \tag{4.159}$$

4.10 Additional Topics

Numerous queueing models have been analyzed using matrix-analytic methods. Although some of the queueing models are quite complicated, fairly explicit results have been obtained for various performance measures. In addition, efficient algorithms have been developed for numerical computation. We note that the literature on queueing models investigated by applying matrix-analytic methods is enormous, which is reflected by the articles published by M.F. Neuts from 1971 to 2006. Due to space limitation, only a small part of the papers written by Neuts and other researchers using matrix-analytic methods is referenced in this book. The following discussion can only highlight a small portion of the works appeared in the last forty years or so (from 1971 to 2012).

4.10.1 Generalization to *GI/PH/c* Queue

In theory, generalization from the *GI/PH/*1 queue to the *GI/PH/c* queue is straightforward. On the other hand, the state space of the corresponding Markov chain becomes complex, especially the boundary states. Thus, it is challenging to develop efficient algorithms for performance analysis. Related references are Asmussen (1992), Asmussen and O'Cinneide (1998) and Asmussen and Møller (2001). Other generalizations, such as the *SM/PH/*1, *BSM/PH/*1, and *MAP/SM/*1 queues, have been investigated by Ramaswami (1982), Sengupta (1990b) and Lucantoni and Neuts (1994).

Example 4.10.1 (The *MAP/PH/n* queue) Assume that customers arrive according to *MAP* (D_0, D_1) of order m_a. The system has n identical servers with *PH*-service time $(\boldsymbol{\beta}, S)$ of order m_s. Let $q(t)$ be the total number of customers in the system at time t, $I_a(t)$ the phase of the arrival process, and $I_j(t)$ the phase of the service process of server j (if server j is idle, then $I_j(t) = 0$), for $j = 1, 2, \ldots, n$. Then $\{(q(t), I_a(t), I_1(t), \ldots, I_n(t)), t \geq 0\}$ is a Markov chain. In fact, it is a QBD process, which can be analyzed using the matrix-analytic methods developed in Chap. 3.

There has been little progress in the study of *GI/G/n* queue. Given that fact, it is significant to see that the *MAP/PH/n* queue can be solved analytically and numerically. However, the size of the (matrix) transition blocks involved in the analysis can become very large. In fact, the size of the transition block for the QBD process $\{(q(t), I_a(t), I_1(t), \ldots, I_n(t)), t \geq 0\}$ is $m_a(m_s)^n$, if $q(t) \geq n$. Suppose $m_a = m_s = 2$, and $n = 10$. Then the size of the transition block is $2^{11} = 2048$, which implies that each transition block has more than four million elements. Computational efficiency is greatly compromised.

Some special methods have been used for space reduction in matrix-analytic methods. For Example 4.10.1, let $n_j(t)$ be the number of servers whose service phase is j at time t, for $j = 1, 2, \ldots, m_s$. Then $\{(q(t), I_a(t), n_1(t), \ldots, n_{m_s}(t)), t \geq 0\}$ is a QBD process. The size of the transition blocks is $m_a(m_s + n - 1)!/((n - 1)!m_s!)$. For the case with $m_a = m_s = 2$, and $n = 10$, the size is 110, which is significantly smaller than 2,048. This example demonstrates that, on the one hand, the effectiveness of matrix-analytic methods is affected by the "curse of dimensionality". On the other hand, special methods might be available to reduce the state space.

4.10.2 Transient Solutions to the *BMAP/G/1* Queue

Lucantoni et al. (1994) and Lucantoni (1998) study the queue length distribution at time t. The idea is to consider both busy periods and idle periods first. By applying the theory of the regenerative process, explicit results are obtained for the queue length distribution at any time.

4.10.3 Retrial Queues, Vacation Queues, Priority Queues, Discrete Time Queues, and Other Special Types of Queues

Matrix-analytic methods have been applied to study a number of different types of queueing models extensively. The main work on retrial queues can be found in two monographs: Falin and Templeton (1997) and Artalejo and Gomez-Corral (2008). Tian and Zhang (2006) give a comprehensive treatment of a number of vacation queues by using matrix-analytic methods. Alfa (2010) collects results for some discrete queues, including vacation queues, priority queues, and queues with multiclass of customers. Other works inlcude Alfa (1998), Diamond and Alfa (1999), Alfa et al. (2003), Artalejo (1999), Chakravarthy (2009), He et al. (2000), He (2003, 2004), Hsu and He (1994), Tian and Zhang (2003), van Houdt and Blondia (2002, 2004), Zhang and Tian (2001), Li and Zhao (2009), and Liu et al. (2011).

4.10.4 Queues with Multiple Types of Customers

Such queueing models can be categorized into two groups: queues without priority and queues with priority.

For queues without priority, He (1996, 2001) and van Houdt and Blondia (2002, 2004) study the *MMAP*[*K*]/*G*[*K*]/1/FCFS queue without priority. The focus is on the busy cycles, workload, and waiting times of individual types of customers. Takine (2001a, b, c, d, 2004) also investigate the *MMAP*[*K*]/*G*[*K*]/1/FCFS queue. A highlight of Takine (2001a, b, c, d, 2004) is about the queue length distributions of individual types of customers. The relationship between the queue lengths at departures, queue lengths at arrivals, queue length at an arbitrary time, and waiting time distributions are explored extensively. For *MMAP*[*K*]/*G*[*K*]/1/LCFS queues, Yeung and Sengupta (1994), He and Alfa (1998), and He (2000) look at the queue length distributions and found solutions of the matrix-geometric type.

For queues with priority, Takine and Hasegawa (1994) and Takine (1996, 1999, 2002) find the distributions of queue lengths and waiting times of individual types of customers. We would like to point out that simpler queues with priority have been investigated extensively (see Jaiswal (1968), Miller (1981), and references therein.)

4.10.5 Fluid Processes and Fluid Queues

An active research area in matrix-analytic methods is fluid processes and fluid queues. Some works in this area include Asmussen (1995a, b), Ahn and Ramaswami (2003), da Silva Soares and Latouche (2002, 2006, 2009), and Bean et al. (2005a, b). We refer to the conference proceedings of the *International Conferences on Matrix-analytic Methods in Stochastic Models* (1995, 1998, 2000, 2002, 2005, 2008, and 2011) for more details on fluid queues.

4.10.6 Queueing Networks

There has been no matrix-geometric solution found for general queueing networks. On the other hand, for some simple queueing networks (e.g., two nodes or three nodes), matrix-analytic results have been obtained for the tail distributions of queue lengths and waiting times. Some references on tail asymptotics of queueing networks are Foley and McDonald (2001), Takahashi et al. (2001), Kroese et al. (2004), Miyazawa and Zhao (2004), Tang and Zhao (2008), and Miyazawa (2009).

References

Ahn S, Ramaswami V (2003) Fluid flow models and queues – a connection by stochastic coupling. Stoch Models 19:325–348

Alfa AS (1998) Matrix geometric solution of discrete time *MAP/PH/*1 priority queue. Nav Res Logist 45:23–50

Alfa AS (2010) Queueing theory for telecommunications: discrete time modelling of a single node system. Springer, New York

Alfa AS, Liu B, He QM (2003) Discrete-time analysis of *MAP/PH/*1 multiclass general preemptive priority queue. Nav Res Logist 50:662–682

Artalejo JR (1999) Accessible bibliography on retrial queues. Math Comput Model 30:1–6

Artalejo JR, Gomez-Corral A (2008) Retrial queueing systems: a computational approach. Springer, Berlin

Asmussen S (1992) Phase-type representation in random walk and queueing problems. Ann Prob 20:772–789

Asmussen S (1995a) Stationary distributions for fluid flow models with or without Brownian noise. Stoch Models 11:21–49

Asmussen S (1995b) Stationary distributions via first passage times. In: Dshalalow J (ed) Advances in queueing: models, methods and problems. CRC Press, Boca Raton, pp 79–102

Asmussen S (2003) Applied probability and queues, 2nd edn. Springer, New York

Asmussen S, Møller JR (2001) Calculation of the steady state waiting time distribution in *GI/PH/*c and *MAP/PH/c* queues. Queueing Syst 37:9–29

Asmussen S, O'Cinneide C (1998) Representation for matrix-geometric and matrix-exponential steady-state distributions with applications to many-server queues. Stoch Models 14:369–387

Bean NG, O'Reilly MM, Taylor PG (2005a) Algorithms for return probabilities for stochastic fluid flows. Stoch Models 21:149–184

Bean NG, O'Reilly MM, Taylor PG (2005b) Hitting probabilities and hitting times for stochastic fluid flows. Stoch Processes Appl 115:1530–1556

Breuer L, Baum D (2005) An introduction to queueing theory and matrix-analytic methods. Springer, Dordrecht
Brill PH (2008) Level crossing methods in stochastic models. Springer, New York
Buzacott JA, Shanthikumar JG (1993) Stochastic models of manufacturing systems. Prentice-Hall, New York
C'inlar E (1969) Markov renewal theory. Adv Appl Prob 1:123–187
Chakravarthy S (2009) A disaster queue with Markovian arrivals and impatient customers. Appl Math Comput 214:48–59
Chaudhry ML, Templeton JGC (1983) A first course in bulk queues. Wiley, New York
Chen H, Yao DD (2001) Fundamentals of queueing networks: performance, asymptotics and optimization, vol 46, Applications of mathematics. Springer, New York
Cohen JW (1982) The single server queues. North-Holland, Amsterdam
da Silva Soares A, Latouche G (2002) Further results on the similarity between fluid queues and QBDs. In: Latouche G, Taylor P (eds) Matrix-analytic methods theory and applications – proceedings of the 4th international conference on matrix-analytic methods. World Scientific, New Jersey, pp 89–106
da Silva Soares A, Latouche G (2006) Matrix-analytic methods for fluid queues with finite buffers. Perform Eval 63:295–314
da Silva Soares A, Latouche G (2009) Fluid queues with level dependent evolution. Eur J Oper Res 196:1041–1048
Diamond JE, Alfa AS (1999) Matrix analytical methods for a multi-server retrial queue with buffer. TOP 7:249–266
Erlang AK (1917) Solution of some problems in the theory of probabilities of significance in automatic telephone exchange. Post Office Elec Eng J 10:189–197
Falin GI, Templeton JGC (1997) Retrial queues. Chapman & Hall, New York
Foley R, McDonald D (2001) Join the shortest queue: stability and exact asymptotics. Ann Appl Prob 11:569–607
Gross D, Harris CM (1985) Fundamentals of queueing theory, 2nd edn. Wiley, New York
Hassin R, Haviv M (1999) To queue or not to queue: equilibrium behavior in queueing systems. Kluwer Academic, Boston
He QM (1996) Queues with marked customers. Adv Appl Prob 28:567–587
He QM (2000) Quasi-birth-and-death Markov processes with a tree structure and the *MMAP*[*K*]/*PH*[*K*]/1 queue. Eur J Oper Res 120:641–656
He QM (2001) The versatility of *MMAP*[*K*] and the *MMAP*[*K*]/*G*[*K*]/1 queue. Queueing Syst 38:397–418
He QM (2003) Age process, total workload, sojourn times, and waiting times in a discrete time *SM*[*K*]/*PH*[*K*]/1/FCFS queue. Queueing Syst 49:363–403
He QM (2004) Workload process, waiting times, and sojourn times in a discrete time *MMAP*[*K*]/*SM*[*K*]/1/FCFS queue. Stoch Models 20:415–437
He QM, Alfa AS (1998) The *MMAP*[*K*]/*PH*[*K*]/1 queue with a last-come-first-served preemptive service discipline. Queueing Syst 28:269–291
He QM, Jewkes EM (1996) A level crossing analysis of the *MAP*/*G*/1 queue. In: Chakravarthy SR, Alfa AS (eds) Matrix-analytic methods in stochastic models. Marcel Dekker, New York, pp 107–116
He QM, Neuts MF (2002) Two *M*/*M*/1 queues with transfers of customers. Queueing Syst 42:377–400
He QM, Li H, Zhao Q (2000) Ergodicity of the *BMAP*/*PH*/*s*/*s* + *K* retrial queue with *PH*-retrial times. Queueing Syst 35:323–347
Hsu GH (1988) Stochastic service systems. Academic, Beijing
Hsu GH, He QM (1994) The matched queueing system *GI*∘*PH*/*PH*/1. Acta Math Appl Sin 17:34–47
Jaiswal NK (1968) Priority queues. Academic, New York

Kendall DG (1953) Stochastic processes occurring in the theory of queues and their analysis by the method of the imbedded Markov chain. Ann Math Stat 24:338–354

Kleinrock L (1975) Queueing systems, volume I: theory. Wiley Interscience, New York

Kroese DP, Scheinhardt WRW, Taylor PG (2004) Spectral properties of the tandem Jackson network, seen as a quasi-birth-and-death process. Ann Appl Prob 14:2057–2089

Li H, Zhao YQ (2009) Exact tail asymptotics in a priority queue – characterization of the preemptive model. Queueing Syst 63:355–381

Liu B, Wang X, Zhao YQ (2011) Tail asymptotics for *M*/*M*/*c* retrial queues with non-persistent customers. Oper Res Int J 12:173–188

Lucantoni DM (1991) New results on the single server queue with a batch Markovian arrival process. Stoch Models 7:1–46

Lucantoni DM (1998) Further transient analysis of the *BMAP*/*G*/1 queue. Stoch Models 14:461–478

Lucantoni DM, Neuts MF (1994) Some steady-state distributions for the *MAP*/*SM*/1 queue. Stoch Models 10:575–598

Lucantoni DM, Meier-Hellstern KS, Neuts MF (1990) A single server queue with server vacations and a class of non-renewal arrival processes. Adv Appl Prob 22:676–705

Lucantoni DM, Choudhury GL, Whitt W (1994) The transient *BMAP*/*G*/1 queue. Stoch Models 10:145–182

Miller DG (1981) Computation of steady-state probabilities for *M*/*M*/1 priority queues. Oper Res 29:945–958

Miyazawa M (2009) Tail decay rates in double QBD processes and related reflected random walks. Math Oper Res 34:547–575

Miyazawa M, Zhao YQ (2004) The stationary tail asymptotics in the *GI*/*G*/1 type queue with countably many background states. Adv Appl Prob 36:1231–1251

Neuts MF (1978) Markov chains with applications in queueing theory, which have a matrix-geometric invariant probability vector. Adv Appl Prob 10:185–212

Neuts MF (1981) Matrix-geometric solutions in stochastic models – an algorithmic approach. The Johns Hopkins University Press, Baltimore

Neuts MF (1986a) The caudal characteristic curve of queues. Adv Appl Prob 18:221–254

Neuts MF (1986b) Generalizations of the Pollaczek-Khinchin integral method in the theory of queues. Adv Appl Prob 18:952–990

Neuts MF (1989a) The fundamental period of the queue with Markov-modulated arrivals. In: Probability, statistics and mathematics: papers in honor of Samuel Karlin. Academic, Boston, pp 187–200

Neuts MF (1989b) Structured stochastic matrices of *M*/*G*/1 type and their applications. Marcel Dekker, New York

Ozawa T (2006) Sojourn time distributions in the queue defined by a general QBD process. Queueing Syst 53:203–211

Prabhu NU (1998) Stochastic storage processes: queues, insurance risk, dams, and data communication, 2nd edn. Springer, New York

Ramaswami V (1980) The *N*/*G*/1 queue and its detailed analysis. Adv Appl Prob 12:222–261

Ramaswami V (1982) The busy period of queues which have a matrix-geometric steady state probability vector. Opsearch 19:265–281

Ramaswami V, Taylor PG (1996) Some properties of the rate operators in level dependent quasi-birth-and-death processes with a countable number of phases. Stoch Models 12:143–164

Ross SM (2010) Introduction to probability models, 10th edn. Academic, Amsterdam/Boston

Sengupta B (1989) Markov processes whose steady state distribution is matrix-exponential with an application to the *GI*/*PH*/1 queue. Adv Appl Prob 21:159–180

Sengupta B (1990a) Phase-type representations for matrix-geometric solutions. Stoch Models 6:163–167

Sengupta B (1990b) The semi-Markovian queue: theory and applications. Stoch Models 6:383–413

Takahashi Y (1981) Asymptotic exponentiality of the tail of the waiting time distribution in a *PH*/*PH*/*c* queue. Adv Appl Prob 13:619–630

Takahashi Y, Fujimoto K, Makimoto N (2001) Geometric decay of the steady-state probabilities in a quasi-birth-and-death process with a countable number of phases. Stoch Models 17:1–24

Takine T (1996) A nonpreemptive priority *MAP*/*G*/1 queue with two classes of customers. J Oper Res Soc Jpn 39:266–290

Takine T (1999) The nonpreemptive priority *MAP*/*G*/1 queue. Oper Res 47:917–927

Takine T (2001a) A recent progress in algorithmic analysis of FIFO queues with Markovian arrival streams. J Korean Math Soc 38:807–842

Takine T (2001b) Subexponential asymptotics of the waiting time distribution in a single-server queue with multiple Markovian arrival streams. Stoch Models 17:429–448

Takine T (2001c) Distributional form of Little's law for FIFO queues with multiple Markovian arrival streams and its application to queues with vacations. Queueing Syst 37:31–63

Takine T (2001d) Queue length distribution in a FIFO single-sever queue with multiple arrival streams having different service time distributions. Queueing Syst 39:349–375

Takine T (2002) Matrix product-form solution for an LCFS-PR single-server queue with multiple arrival streams governed by a Markov chain. Queueing Syst 42:131–151

Takine T (2004) Geometric and subexponential asymptotics of Markov chains of *M*/*G*/1 type. Math Oper Res 29:624–648

Takine T, Hasegawa T (1994) The workload in a *MAP*/*G*/1 queue with state-dependent services: its applications to a queue with preemptive resume priority. Stoch Models 10:183–204

Tang J, Zhao YQ (2008) Stationary tail asymptotics of a tandem queue with feedback. Ann Oper Res 160:173–189

Tian NS, Zhang ZG (2003) Stationary distributions of *GI*/*M*/*c* queue with *PH* type vacations. Queueing Syst 44:183–202

Tian NS, Zhang ZG (2006) Vacation queueing models: theory and applications. Springer, New York

van Houdt B, Blondia C (2002) The delay distribution of a type *k* customer in a FCFS *MMAP*[*K*]/*PH*[*K*]/1 queue. J Appl Prob 39:213–222

van Houdt B, Blondia C (2004) The waiting time distribution of a type *k* customer in a discrete time *MMAP*[*K*]/*PH*[*K*]/*c* ($c = 1, 2$) queue using QBDs. Stoch Models 20:55–69

Whitt W (2002) Stochastic-process limits. Springer, New York

Yeung RW, Sengupta B (1994) Matrix product-form solutions for Markov chains with a tree structure. Adv Appl Prob 26:965–987

Zhang ZG, Tian NS (2001) Discrete time *Geo*/*G*/1 queue with multiple adaptive vacations. Queueing Syst 38:419–429

Zhao YQ, Grassmann WK (1990) The shortest queue model with jockeying. Nav Res Logist 37:773–787

Zhao YQ, Grassmann WK (1995) Queueing analysis of a jockeying model. Oper Res 43:520–529

Chapter 5
Applications in Inventory and Supply Chain Management

Abstract This chapter studies a number of inventory and supply chain models. Using matrix-analytic methods, algorithms are developed for computing performance measures such as the expected total cost per unit time. For two inventory models, algorithms are developed for computing the optimal policy.

Inventory can be found everywhere and is a "loyal" companion of many human activities/events. Books in a bookstore, food in a refrigerator, goods in a supermarket, cars to be sold, and spare parts to be used, are all inventory of some kind. Inventory takes up space and ties up with cash/resource, which might be scarce or can be used somewhere else. Consequently, inventory management becomes an issue of interest. We refer readers to Hadley and Whitin (1963), Bartmann and Beckmann (1992), Axsäter (2000), Buzacott and Shanthikumar (1993), Zipkin (2000), and Porteus (2002) for basic reading on inventory management.

Inventory models usually consist of a demand process, goods in a warehouse, and a replenishment process of ordered goods. Inventory management is about the control of inventory in order to satisfy customer demands in an efficient and/or economic manner. For that purpose, two fundamental issues to be addressed are (i) when to order and (ii) how much to order. Usually, an inventory control policy answers the two questions. Thus, in inventory management, finding the optimal policy is the most important issue.

To get familiar with the structure of inventory models, the terminologies in inventory management, and inventory policies, we take a brief look at two simple but important inventory models.

Exercise 5.1 The economic order quantity (EOQ) model is the simplest inventory model: (1) *demand rate d* is constant; (2) an *order-up-to policy* is applied: any order, if it is placed, is to bring the inventory to fixed level q; and (3) orders are fulfilled after constant *lead time L*. Figure 5.1 shows the change of inventory level in an EOQ model.

DOI 10.1007/978-1-4614-7330-5_5,

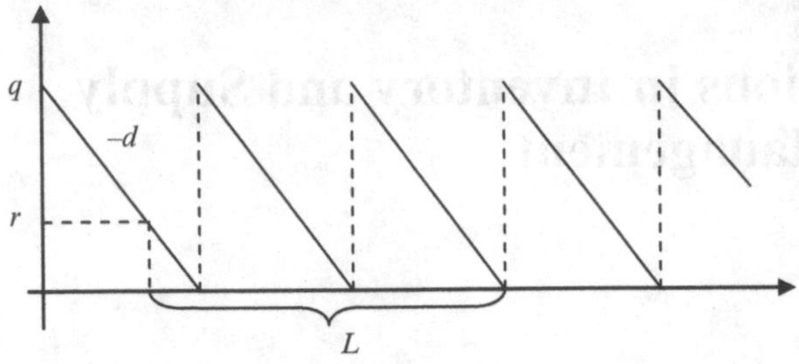

Fig. 5.1 Inventory level in an EOQ model

1. Determine the *reorder point* so that the inventory level is zero right before orders arrive. Note that a reorder point is the threshold such that an order is placed whenever the inventory level drops to or below it.
2. For an EOQ model with *ordering cost* K per order, and *holding cost* C_h per unit of goods per unit time, show that the average total cost per unit time is $Kd/q + C_h q/2$. Show that the optimal order size (or the optimal *order-up-to level*) that minimizes the *average total cost per unit time* is given by $q^* = EOQ = \sqrt{2Kd/C_h}$.

There is no randomness in the EOQ model, but the EOQ model is useful for introducing basic concepts. Next, we consider the Newsvendor model.

Exercise 5.2 (The newsvendor model) The newsvendor model is the simplest stochastic inventory model. The newsvendor orders goods to satisfy customer demands. The model is defined as follows: (1) the probability distribution of demand is known; (2) place at most one order (or one decision to make) per period; (3) lead time is zero; and (4) *leftover* has no value and cannot be used in the next period; and (5) *shortage* incurs a penalty cost.

1. Explain intuitively why it is not a good idea to order too few or too many products.
2. Assume that *overage cost* is c_o (i.e., a unit of leftover costs c_o) and *underage cost* is c_u (i.e., a unit of lost demand costs c_u). Let d be the demand, a continuous random variable, and x be the order size in a period. Show that the total cost incurred in the period is $c_o\max\{0, x–d\} + c_u\max\{0, d–x\}$.
3. For given *demand distribution* $F(x)$, show that the optimal order size x^* that minimizes the expected total cost per period satisfies $F(x^*) = c_u/(c_u + c_o)$. (Note: To be consistent with the convention in inventory theory, in this chapter, we use "the expected value" for the mathematical expectation of random variables.)

In this chapter, we present a few examples to demonstrate the use of matrix-analytic methods in analyzing inventory and supply chain models. In Sects. 5.1 and 5.2, we introduce Markovian models for performance analysis of a few classical inventory models. In Sects. 5.3 and 5.4, we develop methods for computing the optimal policy for two inventory and supply chain models. We refer to Axsäter

(2000), Zipkin (2000), and Porteus (2002) for basic concepts on inventory control and supply chain management: *inventory level, inventory position, inventory on hand, lead time, reorder point, order-up-to level, base stock policy, (s, S) policy, (r, q) policy, holding cost, ordering cost, penalty cost, expected cost per unit time, expected cost per product*, etc.

5.1 An Inventory System with an (*s*, *S*) Policy

In this section, we consider a stochastic inventory system with a simple structure. The inventory system consists of a warehouse with a single type of products. Demands arrive to the warehouse in batches. If there is enough inventory in the warehouse to meet a demand, then the demand is satisfied immediately. Otherwise, the demand is backlogged. Next, we define the inventory model explicitly.

1. Demands arrive according to a *batch Markovian arrival process* (*BMAP*) with matrix representation (D_n, $n = 0, 1, 2, \ldots$), where $\{D_0, D_1, D_2, \ldots\}$ are $m_a \times m_a$ matrices with nonnegative elements, except for the diagonal elements of D_0, which are negative numbers, and m_a is a positive integer. Let $D = D_0 + D_1 + D_2 + \ldots$. Then D is the infinitesimal generator of the (continuous time) underlying Markov chain for the demand arrival process. We assume that D is irreducible. Denote by $I_a(t)$ the phase of the underlying Markov chain at time t. Denote by $\boldsymbol{\theta}_a$ the stationary distribution of D, i.e., $\boldsymbol{\theta}_a D = 0$ and $\boldsymbol{\theta}_a \mathbf{e} = 1$. The (average) demand rate is given by $\lambda = \boldsymbol{\theta}_a(D_1 + 2D_2 + \ldots + nD_n + \ldots)\mathbf{e}$.
2. We assume that the lead time for product replenishment is zero. Thus, any order is immediately fulfilled. Then there is no *outstanding order* in the system.
   To specify the inventory management policy, we need to define inventory level and inventory position. *Inventory level* is inventory on-hand less the amount of backordered demands. If there is no backorder, inventory level equals inventory on-hand; otherwise, inventory level is negative and its absolute value equals the amount of backordered demands. *Inventory position* is the sum of all outstanding orders and the inventory level. For many inventory systems, replenishment policy is based on inventory position, rather than inventory level. Since the lead time is zero, there is no outstanding order and the inventory level equals the inventory position for the model of interest.
3. Inventory in the warehouse is reviewed continuously. Replenishment decision is made based on inventory position. The warehouse adopts an (s, S) policy for its inventory management (Note: Usually, the policy is called an (r, S) policy for inventory systems with a periodic review scheme.) That is, whenever the inventory position reaches s or drops below s, an order is placed to bring up the inventory position to S. Note that s can be any integer and S has to be a nonnegative integer. The constant s is called the *reorder point* and S is called the *order-up-to level*.

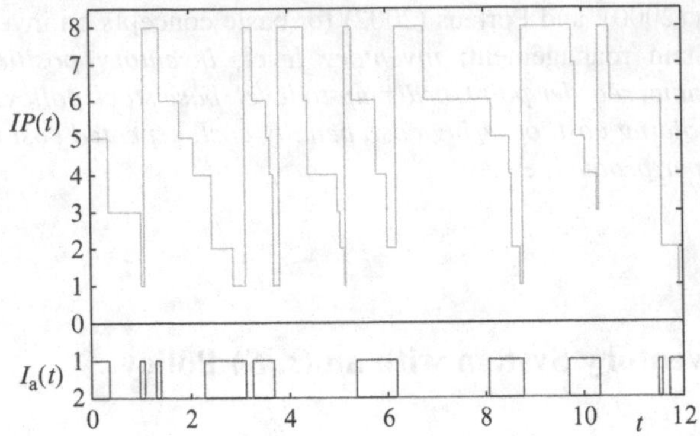

Fig. 5.2 A sample path of the (*s*, *S*) system

We are interested in the stationary distribution of the inventory position. Since lead times are zero, the inventory position and inventory level are equal. We focus on the inventory position, instead of the inventory level because the former is the typical variable used in inventory policy, and it is convenient for generalization to models with nonzero lead times.

Let $IP(t)$ be the inventory position at time t. Notice that the inventory position, in general, has nothing to do with when a backorder is filled or an outstanding order arrives. The inventory position only depends on the arrival times of demands and the placement time and quantity of orders. Since the (s, S) policy has been specified, for our model, $IP(t)$ is always between $s + 1$ and S and, depends on when the next demand arrives. Since $I_a(t)$ gives us enough information on when the next demand arrives probabilistically, it is easy to see that $\{(IP(t), I_a(t)), t \geq 0\}$ is a continuous time Markov chain with state space $\{s + 1, s + 2, \ldots, S\} \times \{1, 2, \ldots, m_a\}$. Sample paths of both $IP(t)$ and $I_a(t)$ are plotted in Fig. 5.2. The infinitesimal generator of the process $\{(IP(t), I_a(t)), t \geq 0\}$ is given by

$$Q_{\mathrm{IP}} = \begin{matrix} s+1 \\ s+2 \\ \vdots \\ \vdots \\ S \end{matrix} \begin{pmatrix} D_0 & & & & \hat{D}_1 \\ D_1 & D_0 & & & \hat{D}_2 \\ \vdots & \ddots & \ddots & & \vdots \\ D_{S-s-2} & \cdots & D_1 & D_0 & \hat{D}_{S-s-1} \\ D_{S-s-1} & \cdots & \cdots & D_1 & D_0 + \hat{D}_{S-s} \end{pmatrix}_{((S-s)m_a)\times((S-s)m_a)}, \tag{5.1}$$

where $\hat{D}_n = \sum_{k=n}^{\infty} D_k$, for $n = 0, 1, 2, \ldots$. The infinitesimal generator in Eq. (5.1) can be interpreted as follows: If a transition is not accompanied by the arrival of a demand (corresponding to D_0), the level variable does not change. If a transition is accompanied by the arrival of a demand and the inventory position drops to s or below s, then the level variable becomes S since an order is placed and fulfilled immediately. If a transition is accompanied by the arrival of a demand and the inventory position is still above s after the demand is satisfied, then the level variable decreases by the size of the batch.

Denote by $\boldsymbol{\pi}_{\mathrm{IP}} = (\boldsymbol{\pi}_{s+1}, \boldsymbol{\pi}_{s+2}, \ldots, \boldsymbol{\pi}_S)$, where $\boldsymbol{\pi}_n = (\pi_{n,1}, \ldots, \pi_{n,m_a})$, for $n = s + 1, \ldots, S$, the stationary distribution of the continuous time Markov chain $\{(IP(t), I_a(t)), t \geq 0\}$. Then $\boldsymbol{\pi}_{\mathrm{IP}}$ satisfies $\boldsymbol{\pi}_{\mathrm{IP}} Q_{\mathrm{IP}} = 0$ and $\boldsymbol{\pi}_{\mathrm{IP}}\mathbf{e} = 1$. It is easy to obtain $\boldsymbol{\pi}_{S-1} = \boldsymbol{\pi}_S(-D_1 D_0^{-1}) \equiv \boldsymbol{\pi}_S R(S-1)$. In a similar way, we can obtain $\boldsymbol{\pi}_n \equiv \boldsymbol{\pi}_S R(n)$ for some nonnegative matrix $R(n)$, for $n = s + 1, \ldots, S - 1$.

Theorem 5.1.1 *Assume that the continuous time Markov chain $\{(IP(t), I_a(t)), t \geq 0\}$ is irreducible. Then the stationary distribution of the inventory position is given by $\boldsymbol{\pi}_n = \boldsymbol{\pi}_S R(n)$, for $n = s + 1, \ldots, S - 1$, where*

$$R(n) = -D_{S-n} D_0^{-1} + \sum_{k=n+1}^{S-1} R(k)(-D_{k-n} D_0^{-1}), \quad n = s+1, \ldots, S-1;$$

$$\boldsymbol{\pi}_S \left(D_0 + \hat{D}_{S-s} + \sum_{k=s+1}^{S-1} R(k) \hat{D}_{k-s} \right) = 0; \tag{5.2}$$

$$\boldsymbol{\pi}_S \left(I + \sum_{k=s+1}^{S-1} R(k) \right) \mathbf{e} = 1.$$

Exercise 5.1.1 Show that $\boldsymbol{\pi}_S + \boldsymbol{\pi}_{S-1} + \ldots + \boldsymbol{\pi}_{s+1} = \boldsymbol{\theta}_a$. In addition, explain the equality intuitively. (Note: This result can be used to check the accuracy of computation).

Usually, there are three types of costs under consideration: (i) ordering cost; (ii) holding cost; and (iii) penalty cost. Assume that ordering cost K is incurred each time an order is placed; holding cost C_h is incurred per unit inventory per unit time; and penalty cost C_p is incurred per unit backlog per unit time.

Theorem 5.1.2 *Assume that the lead times are zero. In steady state, the expected cost per unit time is given by*

$$K\left(\sum_{n=s+1}^{S} \boldsymbol{\pi}_n \hat{D}_{n-s} \mathbf{e} \right) + C_h \sum_{n=s+1}^{S} \max\{0, n\} \boldsymbol{\pi}_n \mathbf{e} + C_p \sum_{n=s+1}^{S} \max\{0, -n\} \boldsymbol{\pi}_n \mathbf{e}. \tag{5.3}$$

Proof. By definitions, $\boldsymbol{\pi}_n \hat{D}_{n-s}\mathbf{e}$ is the expected number of orders per unit time, if the inventory level is n. Thus, the first item in Eq. (5.3) is the expected ordering cost per unit time. Since lead times are zero, the inventory level and the inventory position are equal. It is easy to see that $\boldsymbol{\pi}_n\mathbf{e}$ is the probability that the inventory level is n. If the inventory position is n, then $\max\{0, n\}$ is the inventory on hand, $C_h \max\{0, n\} \boldsymbol{\pi}_n \mathbf{e}$ is the holding cost per unit time, $\max\{0, -n\}$ is the number of backlogged units, and $C_p \max\{0, -n\} \boldsymbol{\pi}_n \mathbf{e}$ is the penalty cost per unit time. Consequently, Eq. (5.3) gives the expected total cost per unit time. This completes the proof of Theorem 5.1.2.

Exercise 5.1.2 Outline an algorithm for computing the stationary distribution of the inventory position. Find the stationary distribution for a system with following parameters: $s = -5$, $S = 2$, and

$$D_0 = \begin{pmatrix} -3 & 1 \\ 0 & -5 \end{pmatrix}, \quad D_1 = \begin{pmatrix} 1 & 0 \\ 0 & 2 \end{pmatrix}, \quad D_2 = \begin{pmatrix} 1 & 0 \\ 1 & 1 \end{pmatrix}, \quad D_3 = \begin{pmatrix} 0 & 0 \\ 1 & 0 \end{pmatrix}. \tag{5.4}$$

Exercise 5.1.3 (A special case) Assume that the batch size of demands is always one, i.e., $D_n = 0$, for $n = 2, 3, \ldots$.

(1) Show that the stationary distribution of the inventory position is given explicitly as $\boldsymbol{\pi}_n = \boldsymbol{\theta}_a/(S-s)$, $n = s+1, \ldots, S$. Then the limiting distribution of $\{IP(t), t \geq 0\}$ is the discrete uniform distribution on $\{s+1, s+2, \ldots, S\}$. The expected total costs per unit time is given by

$$\frac{\lambda K}{S-s} + \frac{1}{S-s} \sum_{n=s+1}^{S} \left(C_{\mathrm{h}} \max\{0, n\} + C_{\mathrm{p}} \max\{0, -n\} \right). \tag{5.5}$$

Explain Eq. (5.5) intuitively.

(2) If $C_{\mathrm{h}} = 0$, find the (s, S) policy that minimizes Eq. (5.5). If $C_{\mathrm{p}} = 0$, find the (s, S) policy that minimizes Eq. (5.5).

Exercise 5.1.3 shows that for a model with demands arriving one at a time, explicit results can be obtained and the optimal (s, S) policy can be found by using Eq. (5.5). On the other hand, if the demands arrive in batches, the solution is more complex.

Commentary An important part of the literature on inventory systems focuses on finding optimal replenishment policy that minimizes costs or optimizes a certain performance measure. It has been shown that the optimal policy for many inventory systems is of (s, S) type (e.g., Scarf (1960), Iglehart (1963), Zheng (1991)). Variants of the (s, S) policy, such as the (r, Q) policy, have been studied extensively.

Additional Exercises and Extensions

We consider two generalizations where the lead times are not zero. If lead times are not zero, the issue of *order crossovers* may occur. That is, an order placed early may arrive later than an order placed after it. Order crossovers make it difficult to analyze the inventory system. In our generalizations, we assume that the demands arrive one at a time. Then the sizes of outstanding orders are the same. For such a case, order crossovers do not affect the analysis.

Exercise 5.1.4 (Generalization 1: exponential lead time) For the (s, S) model, we further assume that (i) the demand batch size is always one, i.e., $MAP\ (D_0, D_1)$; and (ii) the production times of ordered products have independent exponential distributions with parameter μ. Note that the production of ordered products is independent of each other (e.g., if there are infinitely many production machines). Let $B(t)$ be the number of outstanding products at time t. For this case, the inventory level and inventory position can be different.

(1) Explain intuitively why the process $\{(B(t), IP(t), I_a(t)), t \geq 0\}$ is a continuous time Markov chain.

(2) Show that the infinitesimal generator of $\{(B(t), IP(t), I_a(t)), t \geq 0\}$ is given by

$$\begin{array}{c} \\ \\ \end{array}\begin{matrix} 0 & 1 & \cdots & S-s-1 & S-s & S-s+1 & \cdots \end{matrix}$$

$$\begin{pmatrix} Q_{IP,0} & 0 & \cdots & 0 & Q_{IP,1} & & \\ \mu I & Q_{IP,0}-\mu I & 0 & \cdots & 0 & Q_{IP,1} & \\ & 2\mu I & Q_{IP,0}-2\mu I & 0 & \cdots & 0 & Q_{IP,1} \\ & & \ddots & \ddots & \ddots & \ddots & \ddots \\ & & & \ddots & \ddots & \ddots & \ddots \end{pmatrix}, \tag{5.6}$$

where

$$Q_{IP,0} = \begin{pmatrix} D_0 & & & & \\ D_1 & D_0 & & & \\ & \ddots & \ddots & & \\ & & D_1 & D_0 & \\ & & & D_1 & D_0 \end{pmatrix}, \quad Q_{IP,1} = \begin{pmatrix} 0 & & & & D_1 \\ & 0 & & & \\ & & \ddots & & \\ & & & 0 & \\ & & & & 0 \end{pmatrix}, \tag{5.7}$$

(3) Is the Markov chain $\{(B(t), IP(t), I_a(t)), t\geq 0\}$ ergodic? Why?

(4) Denote by $\boldsymbol{\pi}_B = (\boldsymbol{\pi}_{B,0}, \boldsymbol{\pi}_{B,1}, \ldots, \boldsymbol{\pi}_{B,n}, \ldots)$ the stationary distribution of $\{(B(t), IP(t), I_a(t)), t \geq 0\}$, where $(IP(t), I_a(t))$ is defined as the phase variable. Prove $\boldsymbol{\pi}_{B,0} + \boldsymbol{\pi}_{B,1} + \ldots + \boldsymbol{\pi}_{B,n} + \ldots = \boldsymbol{\pi}_{IP}$ (obtained in Theorem 5.1.1). Explain the result intuitively. (Note: This result can be useful for checking the accuracy of computation if one uses an approximation method to find the stationary distribution.)

In general, there is no explicit solution to the stationary distribution of the Markov chain given in Eq. (5.6). Since transitions of the Markov chain are level dependent, it is not easy to develop a simple algorithm for computing the stationary distribution. Approximation methods must be utilized for this case.

Exercise 5.1.5 Explain intuitively why sometimes a replenishment policy is better based on inventory position than on inventory level.

Exercise 5.1.6 (Generalization 2) For the (s, S) model, we further assume that (i) the demand batch size is always one, i.e., MAP (D_0, D_1); and (ii) ordered products are produced one at a time and the production time has an exponential distribution with parameter μ. Let $B(t)$ be the number of outstanding products at time t.

(1) Explain intuitively why the process $\{(B(t), IP(t), I_a(t)), t \geq 0\}$ is a continuous time Markov chain.

(2) Show that the infinitesimal generator of $\{(B(t), IP(t), I_a(t)), t \geq 0\}$ is given by

$$\begin{pmatrix} Q_{IP,0} & 0 & \cdots & 0 & Q_{IP,1} & & & \\ \mu I & Q_{IP,0}-\mu I & 0 & \cdots & 0 & Q_{IP,1} & & \\ & \mu I & Q_{IP,0}-\mu I & 0 & \cdots & 0 & Q_{IP,1} & \\ & & \ddots & \ddots & \ddots & \ddots & \ddots & \ddots \\ & & & \ddots & \ddots & \ddots & \ddots & \ddots \end{pmatrix}, \quad (5.8)$$

where $Q_{IP,0}$ and $Q_{IP,1}$ are given in Eq. (5.7).

(3) Show that the Markov chain $\{(B(t), IP(t), I_a(t)), t \geq 0\}$ is ergodic if Q_{IP} is irreducible and $\lambda < \mu$. (Hint: Use Neuts condition for *M*/*G*/1 type Markov chains.)

(4) Denote by $\boldsymbol{\pi}_B = (\boldsymbol{\pi}_{B,0}, \boldsymbol{\pi}_{B,1}, \ldots, \boldsymbol{\pi}_{B,n}, \ldots)$ the stationary distribution of $\{(B(t), IP(t), I_a(t)), t \geq 0\}$, where $(IP(t), I_a(t))$ is defined as the phase variable. Prove $\boldsymbol{\pi}_{B,0} + \boldsymbol{\pi}_{B,1} + \ldots + \boldsymbol{\pi}_{B,n} + \ldots = \boldsymbol{\pi}_{IP}$. Explain the result intuitively.

(5) Outline an algorithm for computing the stationary distribution of $\{(B(t), IP(t), I_a(t)), t \geq 0\}$. (Hint: You can treat the process as an *M*/*G*/1 type Markov chain or reblock it into a QBD process.)

(6) Find the stationary distribution for a system with the following parameters: $s = 1$, $S = 6$, $\mu = 5$, and

$$D_0 = \begin{pmatrix} -3 & 1 \\ 0 & -5 \end{pmatrix}, \quad D_1 = \begin{pmatrix} 2 & 0 \\ 3 & 2 \end{pmatrix}. \quad (5.9)$$

(7) Assume that the production times have common *PH*-distribution $(\boldsymbol{\alpha}, T)$. Redo problems (1) to (6).

(8) Assume that there are in total N identical production facilities. The production times are exponentially distributed with parameter μ. Redo problems (1) to (6).

Exercise 5.1.7 Write a simulation program to visualize the inventory position process and the phase process for the system given in Exercise 5.1.2 for the (s, S) system with $s = 1$ and $S = 8$.

5.2 A Multi-Item Assemble-to-Order System

We consider an assemble-to-order system in which two types of items are stored to satisfy customer demands that consist of one or more items. The items are supplied by production facilities dedicated to each type. Specifically, the assemble-to-order system is defined as follows.

1. There are two different items {1, 2} stored in the system to meet customer demands.
2. Assembly times are negligible (as compared to the production times).

3. There are three types of customer demands: $\{1\}$, $\{2\}$, and $\{1, 2\}$. The three types of demands arrive according to three independent Poisson processes with parameters λ_1, λ_2, and λ_{12}, respectively. (Note: According to the superposition and decomposition of Poisson processes, the demand process can be defined as a Marked Poisson process with total arrival rate $\lambda = \lambda_1+\lambda_2+\lambda_{12}$ and marking probabilities λ_1/λ, λ_2/λ, and λ_{12}/λ.)
4. Demands are filled on a first-come-first-served (FCFS) basis. If a demand arrives and there are enough items in stock, the demand is filled immediately; otherwise, the demand is queued in some backlog queues. If item 1 (and/or 2) is out of stock, the demand is queued in queue 1 (and/or 2). The demand will be filled once all items become available. Each backlog queue has a limited capacity: b_1 and b_2, which are positive integers. Thus, if a demand finds that an item is out of stock and its queue is longer than its capacity, the entire demand is lost (lost sales case). This assumption is called *total order service* (TOS).
5. Items are replenished independently according to two *base-stock* policies: s_1 and s_2. That is, whenever a unit of item 1 (or 2) is shipped out, an order for a unit of item 1 (or 2) is placed. We assume that the inventory levels of the two items are s_1 and s_2, respectively, at time 0.
6. Replenishment orders for item 1 (or 2) are sent to single machine production facility 1 (or 2), where they are processed on a FCFS basis. The processing times are independent exponential random variables with parameters μ_1 and μ_2, respectively. Based on the above assumptions, each facility can be considered as an $M/M/1$ queue with a finite buffer size, $s_1 + b_1$ or $s_2 + b_2$. The arrival rates to the two $M/M/1$ queues are $\lambda_1+\lambda_{12}$ and $\lambda_2+\lambda_{12}$, respectively. We remark that the two queues are not independent because of the demands of the type $\{1, 2\}$.

To represent the inventory system, we define two random variables for outstanding orders. We first find the joint distribution of the two random variables. Then we establish relationships between outstanding orders and other system random variables such as inventory position and the amount of backorders. Performance measures shall then be obtained. Define

$IO_1(t)$: the number of outstanding units of item 1, i.e., the number of units of item 1 either waiting to be produced or in production.

$IO_2(t)$: the number of outstanding units of item 2.

Consider the process $\{(IO_1(t), IO_2(t)), t \geq 0\}$. Since the demand processes are Poisson processes and the production times are exponential, it is easy to see that $\{(IO_1(t), IO_2(t)), t \geq 0\}$ is a continuous time Markov chain. Consider $IO_1(t)$ as a level variable and $IO_2(t)$ as a phase variable. For any state of $(IO_1(t), IO_2(t))$, there can be arrivals of type $\{1\}$, $\{2\}$, and $\{1, 2\}$, and production of items 1 and 2. The arrival of a type $\{1\}$ (or $\{2\}$) demand increases $IO_1(t)$ (or $IO_2(t)$) by one. The arrival of a type $\{1, 2\}$ demand increases both $IO_1(t)$ and $IO_2(t)$ by one. A production completion of a type 1 (or 2) item decreases $IO_1(t)$ (or $IO_2(t)$) by one. Then the state space of $\{(IO_1(t), IO_2(t)), t \geq 0\}$ can be organized as $\{0, 1, 2, \ldots, s_1 + b_1\} \times \{0, 1, 2, \ldots, s_2 + b_2\}$ and its infinitesimal generator can be written as

$$Q = \begin{array}{c} 0 \\ 1 \\ \vdots \\ s_1+b_1-1 \\ s_1+b_1 \end{array} \begin{pmatrix} A_1 & A_0 & & & \\ \mu_1 I & A_1-\mu_1 I & A_0 & & \\ & \ddots & \ddots & \ddots & \\ & & \mu_1 I & A_1-\mu_1 I & A_0 \\ & & & \mu_1 I & A_2-\mu_1 I \end{pmatrix}, \tag{5.10}$$

where

$$A_1 = \begin{pmatrix} -\lambda & \lambda_2 & & & \\ \mu_2 & -\lambda-\mu_2 & \lambda_2 & & \\ & \ddots & \ddots & \ddots & \\ & & \mu_2 & -\lambda-\mu_2 & \lambda_2 \\ & & & \mu_2 & -\lambda_1-\mu_2 \end{pmatrix}_{(s_2+b_2+1)\times(s_2+b_2+1)}, \tag{5.11}$$

$$A_0 = \begin{pmatrix} \lambda_1 & \lambda_{12} & & \\ & \lambda_1 & \lambda_{12} & \\ & & \ddots & \ddots \\ & & & \lambda_1 & \lambda_{12} \\ & & & & \lambda_1 \end{pmatrix}, \quad A_2 = \begin{pmatrix} -\lambda_2 & \lambda_2 & & & \\ \mu_2 & -\lambda_2-\mu_2 & \lambda_2 & & \\ & \ddots & \ddots & \ddots & \\ & & \mu_2 & -\lambda_2-\mu_2 & \lambda_2 \\ & & & \mu_2 & -\mu_2 \end{pmatrix}.$$

Note that $\lambda = \lambda_1+\lambda_2+\lambda_{12}$. Also note that in matrix A_2, there is no arrival of type $\{1\}$ or type $\{1, 2\}$ demand since queue 1 is full in these states. In A_1 and A_2, there is no type $\{2\}$ nor type $\{1, 2\}$ demand associated with the last state since queue 2 is full at these states.

Denote by $\boldsymbol{\pi} = (\boldsymbol{\pi}_0, \boldsymbol{\pi}_1, \ldots, \boldsymbol{\pi}_{s_1+b_1})$, where $\boldsymbol{\pi}_n = (\pi_{n,0}, \ldots, \pi_{n,s_2+b_2})$, for $n = 0, 1, \ldots, s_1 + b_1$, the stationary distribution of the Markov chain $\{(IO_1(t), IO_2(t)), t\geq 0\}$. Then $\boldsymbol{\pi}$ satisfies $\boldsymbol{\pi}Q = 0$ and $\boldsymbol{\pi}\mathbf{e} = 1$. Solving the linear system, we obtain the following matrix-geometric solution.

Theorem 5.2.1 *Assume that the continuous time Markov chain $\{(IO_1(t), IO_2(t)), t \geq 0\}$ is irreducible. Then the joint stationary distribution of the queues is given by $\boldsymbol{\pi}_n = \boldsymbol{\pi}_0 R^{(1)} \ldots R^{(n)}$, for $n = 1, 2, \ldots, s_1 + b_1$, where*

$$\begin{aligned} R^{(s_1+b_1)} &= -A_0(A_2 - \mu_1 I)^{-1}; \\ R^{(n)} &= -A_0(A_1 - \mu_1 I + \mu_1 R^{(n+1)})^{-1}, \quad n = 1, \ldots, s_1 + b_1 - 1; \\ \boldsymbol{\pi}_0 &\left(A_1 + \mu_1 R^{(1)}\right) = 0; \\ \boldsymbol{\pi}_0 &\left(I + \sum_{k=1}^{s_1+b_1} \left(\prod_{j=1}^{k} R^{(j)}\right)\right)\mathbf{e} = 1. \end{aligned} \tag{5.12}$$

Exercise 5.2.1 Find the stationary distributions of the number of outstanding units for the following system: $\lambda_1 = 1$, $\lambda_2 = 2$, $\lambda_{12} = 1.5$, $\mu_1 = 2$, $\mu_2 = 4$, $b_1 = 5$, $b_2 = 5$, $s_1 = 2$, and $s_2 = 7$.

Next, performance measures can be obtained. We list the results as follows.

1. The stationary distribution of $\{IO_1(t), t \geq 0\}$ is given by $\{\boldsymbol{\pi}_n \mathbf{e}, n = 0, 1, \ldots, s_1 + b_1\}$;
2. The stationary distribution of $\{IO_2(t), t \geq 0\}$ is given by $\sum_{n=0}^{s_1+b_1} \boldsymbol{\pi}_n$.
3. The distribution of inventory on hand can be found using the relationship $I_i = \max\{0, s_i - IO_i\}, i = 1, 2$.
4. The distribution of back orders can be found using the relationship $B_i = \max\{0, IO_i - s_i\}, i = 1, 2$.
5. The *fill rate* of type {1} demands can be found by $F^{\{1\}} \equiv P\{I_1 > 0\} = P\{IO_1 < s_1\}$; the fill rate of type {2} demands can be found by $F^{\{2\}} \equiv P\{I_2 > 0\} = P\{IO_2 < s_2\}$; and the fill rate of type {1, 2} demands can be found by $F^{\{12\}} \equiv P\{I_1 > 0, I_2 > 0\} = P\{IO_1 < s_1, IO_2 < s_2\}$.
6. The *service level* of type {1} orders is $SL^{\{1\}} \equiv P\{B_1 < b_1\} = P\{IO_1 < s_1 + b_1\}$; the service level of type {2} demands is $SL^{\{2\}} \equiv P\{B_2 < b_2\} = P\{IO_2 < s_2 + b_2\}$; and the service level of type {1, 2} demands is $SL^{\{12\}} \equiv P\{B_1 < b_1, B_2 < b_1\} = P\{IO_1 < s_1 + b_1, IO_2 < s_2 + b_2\}$.
7. The fill rate of item 1 (which includes item 1 from type {1} demands and from type {1, 2} demands) is given by

$$F_1 \equiv \frac{\lambda_1}{\lambda_1 + \lambda_{12}} P\{IO_1 < s_1\} + \frac{\lambda_{12}}{\lambda_1 + \lambda_{12}} P\{IO_1 < s_1,\ IO_2 < s_2 + b_2\}; \quad (5.13)$$

   The fill rate of item 2 (which includes item 2 from type {2} demands and from type {1, 2} demands) can be obtained similarly.
8. The service level of item 1 is given by

$$SL_1 \equiv \frac{\lambda_1}{\lambda_1 + \lambda_{12}} SL^{\{1\}} + \frac{\lambda_{12}}{\lambda_1 + \lambda_{12}} SL^{\{12\}}; \quad (5.14)$$

   The service level of item 2 can be obtained similarly.
9. The *system fill rate* is given by $F \equiv \frac{\lambda_1}{\lambda} F^{\{1\}} + \frac{\lambda_2}{\lambda} F^{\{2\}} + \frac{\lambda_{12}}{\lambda} F^{\{12\}}$.
10. The *system service level* is given by $SL \equiv \frac{\lambda_1}{\lambda} SL^{\{1\}} + \frac{\lambda_2}{\lambda} SL^{\{2\}} + \frac{\lambda_{12}}{\lambda} SL^{\{12\}}$.

Exercise 5.2.2 (Exercise 5.2.1 continued) For the model in Exercise 5.2.1, compute all the above performance measures.

Exercise 5.2.3 Similar to Eqs. (5.13) and (5.14), find expressions for fill rate of item 2 (i.e., F_2) and the service level of item 2 (i.e., SL_2).

Commentary The model considered in this section is a special case of the model investigated in Song et al. (1999). We refer to Song et al. (1999) for the model with any number of items and any number of product configurations.

Additional Exercises and Extensions

Exercise 5.2.4* The waiting time of an order is the total time for the entire demand to be satisfied. The waiting time of an item is the time for the item to be filled. Find the distributions of the waiting times of demands and items.

Exercise 5.2.5* Carry out an analysis on the inventory model with Markovian demand process (D_0, D_1, D_2, D_{12}).

5.3 A Make-to-Order Inventory-Production System

The model analyzed in this section consists of both inventory control and production management. The inventory control policies considered are more general than the (s, S) policy described in Sect. 5.1 and the based stock policy used in Sect. 5.2. The analysis consists of three parts: (i) model definition and notation; (ii) performance analysis; and (iii) the optimal policy. A highlight of this section is an algorithm for computing the optimal policy, which is given in Sect. 5.3.3.

5.3.1 *Model Definition and Replenishment Policy*

The inventory-production system of interest consists of a workshop and a warehouse (Fig. 5.3). The workshop produces products to satisfy customer demands. Raw materials used in production are stored in the warehouse. The warehouse places orders to an outside supplier for raw materials. More specifically, the model is defined as follows.

The workshop Customer demands arrive one at a time to the workshop according to a Poisson process with parameter λ. All demands are processed in the workshop by a single machine. Production (or processing) times of products have a common exponential distribution with parameter μ. When the machine is ready to produce a product, a call for a unit of raw materials is sent to the warehouse. A unit of raw materials is immediately sent to the workshop and production on that unit begins. The transportation time between the warehouse and the workshop is assumed to be negligible. We further assume that the shortage of raw materials is not allowed.

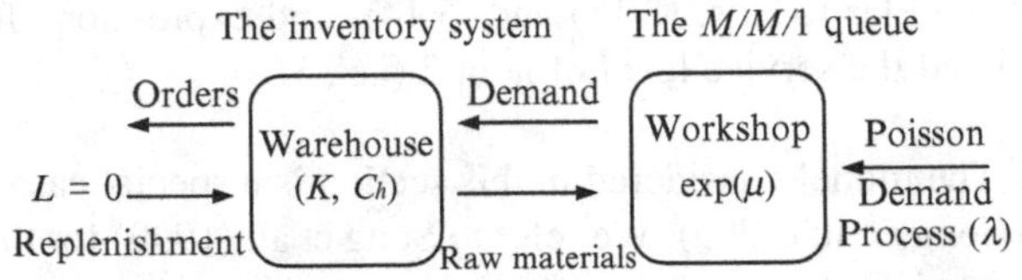

Fig. 5.3 The inventory-production system

Exercise 5.3.1 Explain why the workshop can be modeled as an $M/M/1$ queue.

The warehouse Raw materials in the warehouse are replenished from a supplier according to a continuous review replenishment policy. Lead times of raw materials are assumed to be zero so that production in the workshop can occur whenever there are demands. The replenishment policy for raw materials is such that it does not allow raw material shortages.

Let $q(t)$ be the number of waiting customers (i.e., the unfilled demands) in the workshop. By Exercise 5.3.1, $q(t)$ is the queue length of an $M/M/1$ queue. Let $I(t)$ be the total number of units of raw materials in the inventory-production system at time t, i.e., the number of units of raw materials in the warehouse plus the one unit of raw materials in the workshop, if the workshop is working. Then the status of the inventory-production system can be represented by $(q(t), I(t))$ at time t. (We note that $I(t)$ is used for the phase of the underlying Markov chain in other sections and chapters).

Replenishment policy Raw material replenishment in the warehouse is controlled according to a replenishment policy defined on system status $(q(t), I(t))$. Since lead times for raw material replenishment are zero, there is no need to place an order if $I(t)$ is positive. Thus, orders for raw materials may be placed only if $I(t) = 0$. Consequently, a replenishment policy can be expressed as a function of $q(t)$ only and can be represented as vector $\boldsymbol{\xi} = (\xi(0), \xi(1), \xi(2), \ldots)$, where $\xi(n)$ is the order size if *the inventory level is zero* and the number of waiting customers is n. At time t, if $I(t) = 0$ and $\xi(q(t)) > 0$, an order of size $\xi(q(t))$ is issued and filled; otherwise, no action takes place. If an order of the size $\xi(q(t))$ is filled at time t, the inventory level becomes $\xi(q(t))$, i.e., $I(t+) = \xi(q(t))$. If $q(t) = 0$, there is no need to order raw materials. Thus, we assume $\xi(0) = 0$ for all feasible policies. The set of feasible replenishment policies under consideration in this model is then defined as

$$\Pi = \{\boldsymbol{\xi} : \xi(0) = 0, \xi(q) \geq 1, \quad q = 1, 2, \ldots\}. \tag{5.15}$$

Inventory Costs The fixed ordering cost associated with each order from the warehouse to the supplier, regardless of the order size, is K. The holding cost is C_h per unit of raw materials held in the system per unit time. We note that, since shortage of raw materials in the workshop is not allowed, no penalty cost is incurred.

Objective The objective is to find policy $\boldsymbol{\xi}^*$ in $\boldsymbol{\Pi}$, called the optimal policy, such that the long-run average (i.e., expected) total inventory costs per product is minimized.

Example 5.3.1 Consider an inventory-production system with system parameters $\lambda = 0.4, \mu = 1, K = 60$, and $C_h = 1.5$. Using the methods developed in Sects. 5.3.2 and 5.3.3, the optimal replenishment policy $\boldsymbol{\xi}^*$ can be obtained and is presented in Table 5.1

Table 5.1 The optimal replenishment policy of Example 5.3.1

q	0	1	2	3	4	5	6	7	8	9	10	(≥)11
$\xi^*(q)$	0	5	6	7	8	9	9	10	10	8	8	9

If the queue is very long, the demands for raw materials from the workshop to the warehouse are similar to that of a Poisson process with parameter μ. To reduce the total cost, the order size $\xi(q)$ should be EOQ(μ) if q is sufficiently large. This observation is confirmed by Example 5.3.1 (also see Exercise 5.3.2) and it leads to the following characterization of the optimal policy $\boldsymbol{\xi}^*$.

Theorem 5.3.1 (He et al. (2002a)) *The optimal policy* $\boldsymbol{\xi}^*$ *is in the following subset of feasible policies*

$$\mathbf{\Pi}_{\text{EOQ}} = \left\{ \boldsymbol{\xi} : \ \boldsymbol{\xi} \in \Pi, \ \xi(q) \le 2\sqrt{\frac{K\mu}{C_{\text{h}}}} + 2, \ \text{for } q \ge 0; \ \exists\, q_{\xi} \ge 1, \ \xi(q) = EOQ(\mu), \ \text{for } q \ge q_{\xi} \right\}, \tag{5.16}$$

where

$$EOQ(\mu) = \arg \min_{S=1,2,\ldots} \left\{ \frac{K}{S} + \frac{(S+1)C_{\text{h}}}{2\mu} \right\}. \tag{5.17}$$

Exercise 5.3.2 Assume that $q(t)$ is always positive, which can be true if $\lambda \ge \mu$.

(i) Show that (a) the demand process from the workshop to the warehouse is a Poisson process with parameter μ; and (b) the inventory process in the warehouse is the same as that of the model introduced in Sect. 5.1.
(ii) Argue that the optimal replenishment policy is independent of the queue length and should be an order-up-to policy, i.e., an (s, S) policy with $s = 0$. (Hint: Lead times are zero and shortage is not allowed.)
(iii) Under the order-up-to policy $(0, S)$, show that the process $\{I(t), t \ge 0\}$ is a continuous time Markov chain with infinitesimal generator

$$Q = \begin{matrix} 1 \\ \vdots \\ S-1 \\ S \end{matrix} \begin{pmatrix} -\mu & \mu & & \\ & \ddots & \ddots & \\ & & -\mu & \mu \\ \mu & & & -\mu \end{pmatrix}. \tag{5.18}$$

(iv) Under the order-up-to policy $(0, S)$, show that the expected total cost per replenishment cycle (i.e., the period between two consecutive replenishments) is $K + S(S + 1)C_{\text{h}}/(2\mu)$, the expected length of a cycle is S/μ, and the number of demands satisfied in a cycle is S. Explain the difference between this formulation and that of Theorem 5.1.2.
(v) Show that the optimal (s, S) policy, which minimizes the expected total cost per unit time, is $s = 0$ and $S = EOQ(\mu)$, where $EOQ(\mu)$ is given by Eq. (5.17).

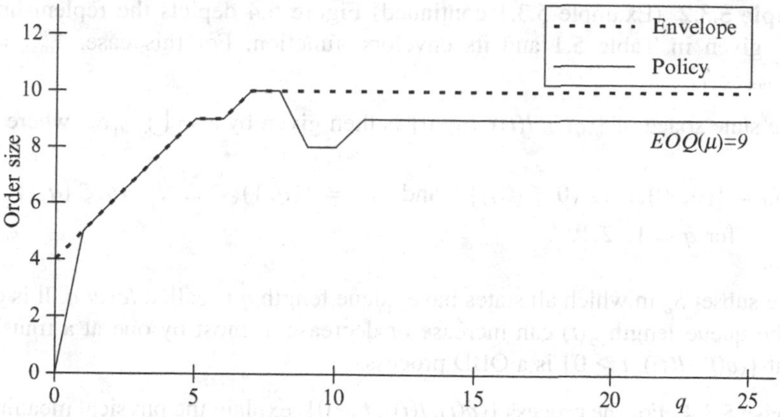

Fig. 5.4 A replenishment policy and its envelope function

If $\lambda < \mu$, the inventory-production system is not equivalent to a simple inventory system. In Sect. 5.3.2, for any policy in $\mathbf{\Pi}_{\text{EOQ}}$, two methods are developed for computing performance measures, especially the expected total costs per product. In Sect. 5.3.3, an algorithm is developed for finding the optimal policy ξ^*.

5.3.2 Performance Analysis

Consider an inventory-production system with replenishment policy $\xi \in \mathbf{\Pi}_{\text{EOQ}}$. It is easy to see that $\{q(t), t \geq 0\}$ is the birth-and-death process defined in Exercise 3.1.8, and $\{(q(t), I(t)), t \geq 0\}$ is a QBD process. To construct the state space and the infinitesimal generator of the Markov chain $\{(q(t), I(t)), t \geq 0\}$, function ξ^e, called the *envelope function* of ξ, is introduced first. Let

$$\xi_{\max} = \max_{\{q \geq 0\}}\{\xi(q)\} \quad \text{and} \quad q_{\max} = \min_{\{q \geq 0\}}\{q : \ \xi(q) = \xi_{\max}\}. \tag{5.19}$$

Note that $q_\xi \geq q_{\max}$. The envelope function ξ^e is defined recursively as

$$\xi^e(q) = \begin{cases} \xi_{\max}, & q \geq q_{\max}; \\ \max\{\xi(0), \xi(1), \cdots, \xi(q), \xi^e(q+1) - 1\}, & 0 \leq q < q_{\max}. \end{cases} \tag{5.20}$$

The function ξ may fluctuate (before settling down at EOQ(μ)). The function ξ^e is a nondecreasing function (see Fig. 5.4).

Exercise 5.3.3 Show that, if $f(q)$ is a nondecreasing function and $f(q) \geq \xi(q)$, for $q \geq 0$, then $f(q) \geq \xi^e(q)$, for $q \geq 0$. That is, ξ^e is the minimum nondecreasing function that is no less than ξ elementwise.

Example 5.3.2 (Example 5.3.1 continued) Figure 5.4 depicts the replenishment policy given in Table 5.1 and its envelope function. For this case, $\xi_{max} = 10$ and $q_{max} = 7$.

The state space of $\{(q(t), I(t)), t \geq 0\}$ is then given by $S = \bigcup_{q=0}^{\infty} S_q$, where

$$S_0 = \{(0,\ 0), \ldots,\ (0, \xi^e(0))\} \quad \text{and} \quad S_q = \{(q, 1), \quad \ldots, \quad (q, \xi^e(q))\},$$
$$\text{for } q = 1,\ 2, \ldots.$$

The subset S_q in which all states have queue length q is called *level* q. It is clear that the queue length $q(t)$ can increase or decrease at most by one at a transition so that $\{(q(t), I(t)), t \geq 0\}$ is a QBD process.

Exercise 5.3.4 For the process $\{(q(t), I(t)), t \geq 0\}$, explain the physical meaning of the following transitions: (1) a transition from state (q, i) to $(q + 1, i)$; (2) a transition from (q, i) to $(q - 1, i - 1)$; and (3) a transition from $(q, 1)$ to $(q - 1, \xi(q - 1))$.

Exercise 5.3.5 Show that $\{(q(t), I(t)), t{\geq}0\}$ is a level independent QBD process and is irreducible for any feasible policy $\boldsymbol{\xi}$ in $\boldsymbol{\Pi}_{\mathrm{EOQ}}$.

The infinitesimal generator of $\{(q(t), I(t)), t \geq 0\}$ is given by

$$Q = \begin{pmatrix} A_{0,0} & A_{0,1} & & & & & \\ A_{1,0} & A_{1,1} & A_{1,2} & & & & \\ & \ddots & \ddots & \ddots & & & \\ & & A_{q_\xi,q_\xi-1} & A_{q_\xi,q_\xi} & A_{q_\xi,q_\xi+1} & & \\ & & & A_2 & A_1 & A_0 & \\ & & & & A_2 & A_1 & A_0 \\ & & & & & \ddots & \ddots & \ddots \end{pmatrix}. \tag{5.21}$$

It is easy to find

$$A_0 = \lambda I,\ A_1 = -(\lambda + \mu)I, \text{ and } A_2 = \begin{matrix} 1 \\ 2 \\ \vdots \\ \xi_{max} \end{matrix} \begin{pmatrix} 0 & \cdots & \mu & \cdots & 0 \\ \mu & \ddots & & & \\ & \ddots & \ddots & & \\ & & \ddots & \ddots & \\ & & & \mu & 0 \end{pmatrix}, \tag{5.22}$$

where A_0, A_1, and A_2 are $\xi_{max} \times \xi_{max}$ matrices. The elements in the first row of A_2 are zero, except for the (1, $EOQ(\mu)$)-th element, which is μ. The level zero S_0 has $\xi^e(0) + 1$ phases. For all policies $\boldsymbol{\xi} \in \boldsymbol{\Pi}_{\mathrm{EOQ}}$, we actually have $\xi^e(0) + 1 = \xi^e(1)$. Then we have $A_{0,0} = -\lambda I$, $A_{1,0} = \mu I_{\xi^e(1)\times(\xi^e(0)+1)}$, $A_{0,1}$ is a $(\xi^e(0) + 1)\times\xi^e(1)$ matrix with all elements being zero except for the $(1, \xi(1))$-th element, which is λ, and for the $(i, i-1)$-st element, which is λ, for $i = 2, \ldots, \xi^e(0) + 1$.

Exercise 5.3.6 Explain A_0, A_1, and A_2 intuitively. Find $A_{q,q}$, $A_{q,q+1}$, and $A_{q+1,q}$ explicitly. What are the sizes of these matrices?

Let $\boldsymbol{\pi} = (\boldsymbol{\pi}_0, \boldsymbol{\pi}_1, \ldots)$ be the stationary distribution of $\{(q(t), I(t)), t \geq 0\}$, which satisfies $\boldsymbol{\pi}Q = 0$ and $\boldsymbol{\pi}\mathbf{e} = 1$. Next, we use a series of exercises to lay out a procedure to analyze the inventory-production system.

Exercise 5.3.7 Show that the inventory-production system can reach steady state if and only if $\lambda<\mu$ (i.e., the Markov chain $\{(q(t), I(t)), t \geq 0\}$ is ergodic if and only if $\lambda<\mu$). Explain the system stability condition intuitively. (Note: Although the Markov chain is irreducible, $A = A_0 + A_1 + A_2$ can be reducible.)

Exercise 5.3.8 Let $\rho = \lambda/\mu$. We assume $\rho < 1$. Prove $\boldsymbol{\pi}_0\mathbf{e} = 1-\rho$. Explain the result intuitively.

Exercise 5.3.9 Consider feasible replenishment policy $\boldsymbol{\xi} = (l, l, \ldots)$, where l is a positive integer. This replenishment policy is special since the order size is always l when the inventory level is zero (independent of the queue length at the order epochs). (Note: The policy $\boldsymbol{\xi}$ is not in $\boldsymbol{\Pi}$.)

(1) Show that the stationary distribution is given by $\boldsymbol{\pi}_q = (1-\rho)\rho^q(1/l, \ldots, 1/l)$, $q \geq 0$, if $\rho = \lambda/\mu < 1$.
(2) Show that the stationary distribution of the inventory level is given as $\sum_{q=0}^{\infty} \boldsymbol{\pi}_q = (1, 1, \cdots, 1)/l.$. That is, the inventory level $I(t)$ is uniformly distributed over $\{1, 2, \ldots, l\}$.
(3) Let $g(l)$ be the expected total cost per product with respect to order size l. Prove $g(l) = K/l + (l+1)C_h/(2\lambda)$. Find the optimal order size l^* that minimizes the expected total cost per product.
(4) Find the expected total cost per unit time.

For policy $\boldsymbol{\xi}$ in $\boldsymbol{\Pi}$, in general, no explicit result can be obtained. The system can be analyzed using the matrix-geometric solution of the QBD process. We call it an R-based analysis. Since the R-based analysis is routine and standard (see Chaps. 3 and 4, and Sect. 5.1), we omitted the details.

Exercise 5.3.10 (An R-Based Analysis) Assume that $\boldsymbol{\xi}$ is in $\boldsymbol{\Pi}$ and $\lambda<\mu$. For the QBD process $\{(q(t), I(t)), t \geq 0\}$ defined by Eq. (5.21), find its matrix-geometric solution and use it to find performance measures.

1. Find the equation for the rate matrix R. Find the equations for rate matrices $\{R_q, q = 0, 1, \ldots, q_\xi\}$ for the boundary levels.
2. Find the equations for $\{\boldsymbol{\pi}_q, q = 0, 1, \ldots, q_\xi\}$. Find $\{\boldsymbol{\pi}_q, q = q_\xi+1, \ldots\}$.
3. Find the stationary distribution of the inventory level process $\{I(t), t \geq 0\}$.
4. Explain why the expected total cost per unit time is the product of λ and the expected total cost per product.

5. Let $\boldsymbol{\alpha}$ be the stationary distribution of $\{I(t), t \geq 0\}$. Express $\boldsymbol{\alpha}$ in terms of $\{\boldsymbol{\pi}_q, q = 0, 1, 2, \ldots\}$. Show that the long-run average total costs per product can be given by

$$\frac{K}{\lambda}((\alpha_1 - \pi_{1,1} - \pi_{0,1})\mu + \pi_{0,0}\lambda) + \frac{C_{\text{h}}}{\lambda}\sum_{i=1}^{\xi_{\max}} i\alpha_i. \tag{5.23}$$

6. Use the algorithm you developed to find the expected total cost per product for Example 5.3.1. (Answer: $g(\boldsymbol{\xi}) = 20.7961$)

Next, we conduct a cost analysis using the matrix G, which is called a *G-based analysis*. We present details on this approach since it is related to the *policy iteration algorithm* for the optimal replenishment policy to be introduced in Sect. 5.3.3.

Denote by $\tau(q, i)$ the sojourn time in (q, i). Let $C(q, i)$ be the total cost incurred during the stay in (q, i), including any possible ordering cost at the end of the stay.

Exercise 5.3.11 Let u and v be random variables with exponential distributions with parameters λ and μ, respectively. For $q, i \geq 1$, show

$$\begin{aligned}
\tau(q,i) &= \min\{u, v\};\\
E[\exp(-s\tau(q,i))] &= \frac{s}{s+\lambda+\mu};\\
C(q,i) &= \begin{cases} iC_{\text{h}}\tau(q,i), & \text{if } i > 1,\ q \geq 1;\\ C_{\text{h}}\tau(q,1), & \text{if } i = 1,\ u,\ q \geq 2;\\ C_{\text{h}}\tau(q,1) + K, & \text{if } i = 1,\ u \geq v,\ q \geq 2. \end{cases}
\end{aligned} \tag{5.24}$$

In addition, we have $\tau(0, i) = u$, for $i \geq 1$, $C(1, 1) = C_{\text{h}}\tau(1, 1)$, $C(0, 0) = K$, $C(0, i) = iC_{\text{h}}\tau(0, i)$, for $i \geq 1$.

Exercise 5.3.12 Conditioning on the state of the Markov chain, the Laplace Stieltjes Transform (LST) of the cost function $C(q, i)$ is given as, for $q > q_\xi$,

$$\begin{aligned}
&E[\exp(-sC(q,i)) :\ q(\tau(q,i)+) = q+1,\ I(\tau(q,i)+) = j\ |q(0) = q, I(0) = i]\\
&= \frac{(-A_1)_{i,i}}{(iC_{\text{h}}s - (A_1)_{i,i})}\frac{(A_0)_{i,j}}{(-A_1)_{i,i}} = \frac{(A_0)_{i,j}}{iC_{\text{h}}s - (A_1)_{i,i}},
\end{aligned} \tag{5.25}$$

and

$$\begin{aligned}
&E[\exp(-sC(q,i)) :\ q(\tau(q,i)+) = q-1,\ I(\tau(q,i)+) = j\ |q(0) = q, I(0) = i]\\
&= \begin{cases} \frac{(-A_1)_{i,i}}{(iC_{\text{h}}s-(A_1)_{i,i})}\frac{(A_2)_{i,j}}{(-A_1)_{i,i}} = \frac{(A_2)_{i,j}}{iC_{\text{h}}s-(A_1)_{i,i}}, & i \geq 2;\\ e^{-sK}\frac{(-A_1)_{1,1}}{(C_{\text{h}}s-(A_1)_{1,1})}\frac{(A_2)_{1,j}}{(-A_1)_{1,1}} = e^{-sK}\frac{(A_2)_{1,j}}{C_{\text{h}}s-(A_1)_{1,1}}, & i = 1. \end{cases}
\end{aligned} \tag{5.26}$$

Exercises 5.3.11 and 5.3.12 find the cost incurred during the stay of the Markov chain $\{(q(t), I(t)), t \geq 0\}$ in a state. Next, we find the cost incurred in a fundamental period (see Chap. 3). Define the following function associated with the costs incurred in a fundamental period, for $(q-1, j) \in S_{q-1}$ and $(q, i) \in S_q$,

$\Phi_{q,(i,j)}(x)$ = The probability that the total cost incurred in a fundamental period is less than x, and the Markov chain reaches level $q-1$ for the first time in state $(q-1, j)$, given that it starts in state (q, i).

The LST of this function is defined as

$$\Phi^*_{q,(i,j)}(s) = \int_0^\infty e^{-sx} \mathrm{d}\Phi_{q,(i,j)}(x), \quad s > 0. \tag{5.27}$$

Let $\Phi^*_q(s)$ be a matrix with elements $\Phi^*_{q,(i,j)}(s)$. For $q > q_\xi$, $\Phi^*_q(s)$ is the same and is denoted as $\Phi^*(s)$. We also define $\Phi^*(2, s) = (\Phi^*_{i,j}(2, s))$ as the LST of the total cost incurred during the first passage time from level $q+1$ to level $q-1$ (i.e., the fundamental periods from $q+1$ to q, and from q to $q-1$).

Combining Exercise 5.3.12 and the definition in Eq. (5.27), for $1 \leq i, j \leq \xi_{\max}$, we obtain, for $q > q_\xi$,

$$\Phi^*_{i,j}(s) = \frac{(-A_1)_{i,i}}{(iC_h s - (A_1)_{i,i})} \times \left[\frac{\exp(-sKI_{\{i=1\}})(A_2)_{i,j}}{(-A_1)_{i,i}} + \sum_{k=1}^{\xi_{\max}} \frac{(A_0)_{i,k}}{(-A_1)_{i,i}} \Phi^*_{k,j}(2, s) \right]. \tag{5.28}$$

Recall that $I_{\{.\}}$ is the indicator function. Note that the LST of the sum of two independent random variables is the product of the LSTs of the two random variables. In matrix form, after some algebra, Eq. (5.28) leads to

$$\begin{aligned}(sC_h \mathrm{diag}(1, \cdots, \xi_{\max}) - A_1)\Phi^*(s) &= \mathrm{diag}(e^{-sK}, 1, \cdots, 1)A_2 \\ &\quad + A_0\Phi^*(2, s),\end{aligned} \tag{5.29}$$

where diag(.) is a diagonal matrix. Note that, in the left-hand side of Eq. (5.29), A_1 should be diag(A_1). Since A_1 is a diagonal matrix (see Eq. (5.22)), we have $A_1 = \mathrm{diag}(A_1)$. For convenience, we use A_1 in Eq. (5.29). In fact, in this section, we use $A_{q,q}$ for diag($A_{q,q}$) whenever diag($A_{q,q}$) should be used since all matrices $\{A_{q,q}, q = 0, 1, \ldots, q_\xi\}$ are diagonal matrices.

Exercise 5.3.13 For $s > 0$, show $\Phi^*(2, s) = (\Phi^*(s))^2$. Explain the relationship intuitively.

The LST of the total cost incurred in a fundamental period is obtained as follows.

Theorem 5.3.2 *For $s > 0$ and $q > q_\xi$, $\Phi^*(s)$ satisfies*

$$(sC_{\text{h}}\text{diag}(1,\cdots,\xi^{\max}) - A_1)\Phi^*(s) = \text{diag}(e^{-sK},1,\cdots,1)A_2 + A_0(\Phi^*(s))^2. \tag{5.30}$$

For $2 \le q \le q_\xi$,

$$\Phi_q^*(s) = \left(sC_{\text{h}}\text{diag}(1,\cdots,\xi^{\text{e}}(q)) - A_{q,q} - A_{q,q+1}\Phi_{q+1}^*(s)\right)^{-1}\text{diag}(e^{-sK},1,\cdots,1)A_{q,q-1}. \tag{5.31}$$

For $q = 1$,

$$\Phi_1^*(s) = \left(sC_{\text{h}}\text{diag}(1,\cdots,\xi^{\text{e}}(1)) - A_{1,1} - A_{1,2}\Phi_2^*(s)\right)^{-1}A_{1,0}. \tag{5.32}$$

For $q = 0$, $\Phi_0^*(s)$ is defined as the LST of the total cost incurred in a cycle (i.e., the time period begins with zero inventory and ends when the inventory becomes zero again). Then

$$\Phi_0^*(s) = \left(sC_{\text{h}}\text{diag}(0,1,\cdots,\xi^{\text{e}}(0)) - A_{0,0}\right)^{-1}\text{diag}(e^{-sK},1,\cdots,1)A_{0,1}\Phi_1^*(s). \tag{5.33}$$

Exercise 5.3.14 Explain the equations for $\Phi_q^*(s)$ intuitively (without any calculations).

With the LSTs of the total costs incurred during fundamental periods, some performance measures such as the distributions and moments of the costs can be determined. We focus on the expected total cost incurred in a fundamental period. As a result, the expected total cost incurred in a busy period and expected total cost per product/unit time can be obtained. Again, the analysis starts with level $q > q_\xi$.

Let $G_q = \Phi_q^*(0)$, for $q \ge 0$. For $q > q_\xi$, denote by $G = \Phi^*(0)$. Then matrix G is the minimal nonnegative solution to the equation

$$A_2 + A_1G + A_0G^2 = 0. \tag{5.34}$$

Other matrices can be obtained: $G_{q_\xi+1} = G$,

$$\begin{aligned} G_q &= -\left(A_{q,q} + A_{q,q+1}G_{q+1}\right)^{-1}A_{q,q-1}, \quad 1 \le q \le q_\xi; \\ G_0 &= (-A_{0,0}^{-1}A_{0,1})G_1. \end{aligned} \tag{5.35}$$

Exercise 5.3.15 If $\rho < 1$, show that the matrix G_q has unit row sums, and the matrix $A_{q,q} + A_{q,q+1}G_{q+1}$ is invertible, for $q \ge 1$. Show that G_0 is a stochastic matrix.

Define, for $q = 0, 1, 2, \ldots$,

$$\Phi_q^{(1)} = -\left.\frac{\mathrm{d}\Phi_q^*(s)}{\mathrm{d}s}\right|_{s=0} \quad \text{and } \mathbf{u}_{c,q} = \Phi_q^{(1)}\mathbf{e}. \tag{5.36}$$

The element $(\mathbf{u}_{c,q})_i$ is the expected total cost incurred during a fundamental period started in state (q, i). If $q > q_\xi$, $\Phi_q^{(1)}$ is denoted as $\Phi^{(1)}$, and $\mathbf{u}_{c,q}$ is denoted as $\mathbf{u}_c$. The next theorem summarizes formulas for computing the expected costs.

Theorem 5.3.3 *Assume $\rho < 1$. We have*

$$\begin{aligned}
\Phi^{(1)} &= \Phi_{\infty,0}^{(1)} + (-A_1^{-1}A_0)(G\Phi^{(1)} + \Phi^{(1)}G);\\
\Phi_q^{(1)} &= \Phi_{q,0}^{(1)} + (-A_{q,q}^{-1}A_{q,q+1})(G_{q+1}\Phi_q^{(1)} + \Phi_{q+1}^{(1)}G_q), \quad 1 \le q \le q_\xi;\\
\Phi_0^{(1)} &= \Phi_{0,0}^{(1)} + (-A_{0,0}^{-1}A_{0,1})\Phi_1^{(1)},
\end{aligned} \tag{5.37}$$

where

$$\begin{aligned}
\Phi_{\infty,0}^{(1)} &= -A_1^{-1}(C_h\mathrm{diag}(1,2,\cdots,\xi_{\max})G + \mathrm{diag}(K,0,\cdots,0)A_2);\\
\Phi_{q,0}^{(1)} &= -A_{q,q}^{-1}\big(C_h\mathrm{diag}(1,2,\cdots,\xi^e(q))G_q + \mathrm{diag}(K,0,\cdots,0)A_{q,q-1}\big), \; 2 \le q \le q_\xi;\\
\Phi_{1,0}^{(1)} &= -A_{1,1}^{-1}C_h\mathrm{diag}(1,2,\cdots,\xi^e(1))G_1;\\
\Phi_{0,0}^{(1)} &= -A_{0,0}^{-1}\big(C_h\mathrm{diag}(0,1,\cdots,\xi^e(0))G_0 + \mathrm{diag}(K,0,\cdots,0)A_{0,1}\big).
\end{aligned} \tag{5.38}$$

Proof. The results are obtained from Theorem 5.3.2 by routine calculations. This completes the proof of Theorem 5.3.3.

Exercise 5.3.16 Use the direct-sum of matrices (see Sect. 1.4) and Eq. (5.37) to find $\Phi^{(1)}$ explicitly. Develop a method for computing $\Phi_q^{(1)}$, for $0 \le q \le q_\xi$.

Exercise 5.3.17 Use Theorem 5.3.3 to show that

$$\begin{aligned}
\mathbf{u}_c &= \Phi^{(1)}\mathbf{e} = (I - (-A_1^{-1}A_0)(I+G))^{-1}\Phi_{\infty,0}^{(1)}\mathbf{e};\\
\mathbf{u}_{c,q} &= \Phi_q^{(1)}\mathbf{e} = \left(I - (-A_{q,q}^{-1}A_{q,q+1})G_{q+1}\right)^{-1}\left(\Phi_{q,0}^{(1)}\mathbf{e} + (-A_{q,q}^{-1}A_{q,q+1})\mathbf{u}_{c,q+1}\right), \; 1 \le q \le q_\xi;\\
\mathbf{u}_{c,0} &= \Phi_0^{(1)}\mathbf{e} = \Phi_{0,0}^{(1)}\mathbf{e} + (-A_{0,0}^{-1}A_{0,1})\mathbf{u}_{c,1}.
\end{aligned} \tag{5.39}$$

To find the expected total cost per product, consider the expected total cost incurred in a busy cycle. The idea is to consider the embedded Markov chain at the ending points of busy periods. By definition, it is known that this embedded Markov chain has transition matrix G_0. It is easy to see that G_0 is an irreducible stochastic matrix when the Markov chain $\{(q(t), I(t)), t \geq 0\}$ is ergodic. Let $\boldsymbol{\beta}_0$ be the left invariant vector of G_0, i.e., $\boldsymbol{\beta}_0 G_0 = \boldsymbol{\beta}_0$, $\boldsymbol{\beta}_0 \geq 0$, and $\boldsymbol{\beta}_0 \mathbf{e} = 1$. Conditioning on the initial state of a busy cycle, the expected total cost incurred in a busy cycle can be obtained as

$$\boldsymbol{\beta}_0 \mathbf{u}_{c,0} = \sum_{i=1}^{\xi^e(0)+1} (\boldsymbol{\beta}_0)_i (\mathbf{u}_{c,0})_i. \tag{5.40}$$

Since the total number of products produced in a busy cycle is the total number of customers served in a busy cycle for an $M/M/1$ queue, the expected number of products produced in a busy cycle is $\mu/(\mu-\lambda)$ (see Cohen (1982)). Then the expected total cost per product is given by, when the policy $\boldsymbol{\xi}$ is applied,

$$g(\boldsymbol{\xi}) = \boldsymbol{\beta}_0 \mathbf{u}_{c,0}(\mu - \lambda)/\mu. \tag{5.41}$$

In summary, an algorithm for computing performance measures related to busy cycles can be developed as follows.

Algorithm 5.3.1: A *G*-Based Approach

Step 0 Input parameters: λ, μ, K, C_h, $\boldsymbol{\xi}$, and q_{ξ}.
Step 1 Construct the envelope function $\boldsymbol{\xi}^e$ by using Eqs. (5.19) and (5.20), and construct transition blocks in Q.
Step 2 Compute the matrix G (Eq. (5.34)).
Step 3 Compute the following groups of matrices and vectors:
(i) $\{G_q, q = 0, 1, \ldots, q_{\xi}\}$ using Eq. (5.35);
(ii) $\Phi^{(1)}_{\infty,0}$ and $\{\Phi^{(1)}_{q,0}, q = 0, 1, \ldots, q_{\xi}\}$ using Eq. (5.38);
(iii) $\Phi^{(1)}$ and $\{\Phi^{(1)}_q, q = 0, 1, \ldots, q_{\xi}\}$ using Eq. (5.37);
(iv) $\mathbf{u}_c$ and $\{\mathbf{u}_{c,q}, q = 0, 1, \ldots, q_{\xi}\}$ using Eq. (5.39);
(v) Vector $\boldsymbol{\beta}_0$.

Step 4 Compute the performance measures (e.g., $g(\boldsymbol{\xi})$ in Eq. (5.41)).

Exercise 5.3.18 (Example 5.3.1 continued) For the inventory-production system defined in Example 5.3.1, use the *G*-based algorithm to calculate the expected total cost per product. (Answer: $g(\boldsymbol{\xi}) = 20.7961$)

Exercise 5.3.19 Assume that $\rho < 1$. Choose $\boldsymbol{\xi} = (0, 1, 1, \ldots)$, $K = 0$, and $C_h = 1$. Show that (i) $\mathbf{u}_{c,1}$ is the expected length of a busy period of the $M/M/1$ queue, which is $1/(\mu - \lambda)$; and (ii) $\boldsymbol{\beta}_0 \mathbf{u}_{c,0}$ is the expected length of a busy cycle of the $M/M/1$ queue, which is $1/\lambda + 1/(\mu - \lambda)$.

5.3.3 *Optimal Replenishment Policy: An MDP Approach*

We use a Markov decision process (MDP) approach to develop an algorithm for computing the optimal policy in $\mathbf{\Pi}$. We refer to Puterman (1994) and Sennott (1999) for the theory on MDPs. By Theorem 5.3.1, it is sufficient to find the optimal policy in $\mathbf{\Pi}_{\text{EOQ}}$. We further restrict the set of policies to, for $k \geq 0$,

$$\mathbf{\Pi}_{\text{EOQ}}[k] = \{\boldsymbol{\xi}: \ \boldsymbol{\xi} \in \mathbf{\Pi}, \quad \xi(q) \equiv EOQ(\mu), \quad \text{for} \quad q = k, k+1, \ldots\}. \quad (5.42)$$

For any $\boldsymbol{\xi} \in \mathbf{\Pi}_{\text{EOQ}}[k]$, we can choose q_ξ to be k. Assume that $\boldsymbol{\xi} \in \mathbf{\Pi}_{\text{EOQ}}[k]$ is applied in inventory control. For $0 \leq q, i \leq n$, and $n \geq 1$, define

$V^\xi(q, i, n) =$ The expected total cost to produce n products, given that there are q demands and i units of raw materials in the system initially.

The long-run average total costs per product is defined as

$$g(\boldsymbol{\xi}) \equiv \lim_{n\to\infty} \frac{V^\xi(q,i,n)}{n}. \quad (5.43)$$

The limit in Eq. (5.43) exists and is independent of the initial state if $\rho = \lambda/\mu < 1$. Define the *relative cost function*

$$h^\xi(q,i,n) = V^\xi(q,i,n) - V^\xi(1,1,n), \quad q \geq 0, \quad i \geq 0, \quad n \geq 1. \quad (5.44)$$

By Theorem 3.1 in Tijms (1986), the limit of $\{h^\xi(q, i, n), n = 1, 2, \ldots\}$ exists and is finite for each state (q, i). Based on the theory of MDPs, we formulate equations for the limits of $\{h^\xi(q, i, n), n = 1, 2, \ldots\}$ without rigorous mathematical proofs. Denote by, for $q \geq 0$ and $i \geq 0$,

$$h^\xi(q,i) = \lim_{\{n\to\infty\}} h^\xi(q,i,n) = \lim_{\{n\to\infty\}} \left(V^\xi(q,i,n) - V^\xi(1,\ 1,n)\right). \quad (5.45)$$

Then, for $q \geq 0$, $i \geq 0$, $n \geq 0$,

$$V^\xi(q,i,n) = ng(\boldsymbol{\xi}) + h^\xi(q,i) + \varepsilon(q,i,n), \quad (5.46)$$

where $\varepsilon(q, i, n) \to 0$ as $n \to \infty$. By conditioning on the events during the production time of a product, it can be proved that $\{g(\boldsymbol{\xi}), h^\xi(q, i), q, i \geq 1\}$ satisfy

$$
\begin{aligned}
& g(\boldsymbol{\xi}) + h^{\xi}(1,1) = \tfrac{C_h}{\mu} + K + \omega h^{\xi}(1,\xi(1)) + \sum_{j=1}^{\infty} \omega(1-\omega)^j h^{\xi}(j,\xi(j));\\
i \geq 2, \quad & g(\boldsymbol{\xi}) + h^{\xi}(1,i) = \frac{iC_h}{\mu} + \omega\left(\frac{(i\text{-}1)C_h}{\lambda} + h^{\xi}(1,i-1)\right)\\
& \qquad + \sum_{j=1}^{\infty} \omega(1-\omega)^j h^{\xi}(j,i-1);\\
q \geq 2, \quad & g(\boldsymbol{\xi}) + h^{\xi}(q,1) = \tfrac{C_h}{\mu} + \sum_{j=0}^{\infty} \omega(1-\omega)^j \left(K + h^{\xi}(q-1+j,\xi(q-1+j))\right);\\
i,q \geq 2, \quad & g(\boldsymbol{\xi}) + h^{\xi}(q,i) = \tfrac{iC_h}{\mu} + \sum_{j=0}^{\infty} \omega(1-\omega)^j h^{\xi}(q-1+j,i-1),
\end{aligned}
\tag{5.47}
$$

where $\omega = \mu/(\lambda+\mu)$.

Theoretically, when $\lambda < \mu$, an algorithm can be developed for computing the expected total cost per product $g(\boldsymbol{\xi})$ using Eq. (5.47) and the value iteration method (see Puterman (1994)). Furthermore, an algorithm for computing the optimal replenishment policy can be developed by using Eq. (5.47) and the policy iteration method (also see Puterman (1994)). A difficulty that has much to do with the infinite state space of $\{(q(t), I(t)), t \geq 0\}$ arises. There is no direct way to evaluate the summations with infinite items in Eq. (5.47). Fortunately, since the tail order size of the optimal policy is fixed and known, the problem can be transformed into a finite semi-Markov decision process with a finite state space using the G-based analysis developed in Sect. 5.3.2. Consequently, an algorithm for computing the optimal replenishment policy in a finite number of iterations can be developed.

Next, we establish a relationship between the functions $\{g(\boldsymbol{\xi}), h^{\xi}(q, i), q, i = 1, 2, \ldots\}$ and the matrices $\{G, G_q, \mathbf{u}_c, \mathbf{u}_{c,q}, q = 1, 2, \ldots, k\}$ defined in Sect. 5.3.2 for $\boldsymbol{\xi}$ in $\boldsymbol{\Pi}_{\text{EOQ}}[k]$. The relationship between the two sets of measures is interesting since it brings Markov decision processes and matrix-analytic methods together. For $q = 1, 2, \ldots$, denote by

$$
\mathbf{h}^{\xi}(q) = \left(h^{\xi}(q,\ 1), \ldots,\ h^{\xi}(q,\ \xi^{e}(q))\right)', \tag{5.48}
$$

a column vector of order $\xi^{e}(q)$.

Theorem 5.3.4 *Assume that $\rho < 1$. Then the inventory-production system with replenishment policy $\boldsymbol{\xi}$ in $\boldsymbol{\Pi}_{\text{EOQ}}[k]$ can reach its steady state, and we have*

$$
\begin{aligned}
g(\boldsymbol{\xi})\mathbf{e}\frac{\mu}{\mu-\lambda} + \mathbf{h}^{\xi}(q+1) &= \mathbf{u}_c + \mathbf{G}\mathbf{h}^{\xi}(q), \quad q = k+1,\ k+2,\ \ldots;\\
g(\boldsymbol{\xi})\mathbf{e}\frac{\mu}{\mu-\lambda} + \mathbf{h}^{\pi}(q) &= \mathbf{u}_{c,q} + G_q\mathbf{h}^{\xi}(q-1), \quad q = 2, \ldots, k;\\
g(\boldsymbol{\xi})\mathbf{e}\frac{\mu}{\mu-\lambda} + \mathbf{h}^{\xi}(1) &= \mathbf{u}_{c,1} + G_1\left(\boldsymbol{\Phi}_{0,0}^{(1)}\mathbf{e} + (-A_{0,0}^{-1}A_{0,1})\mathbf{h}^{\xi}(1)\right).
\end{aligned}
\tag{5.49}
$$

Proof. We consider the numbers of products produced in fundamental periods of the Markov chain $\{(q(t), I(t)), t \geq 0\}$. Similar to the function $G_{q,(i,j)}$ defined in Sect. 5.3.2, define

$G_{i,j}(q, n)$: the probability that, when n products are completed during the first passage time, the Markov chain transits from level q to level q–1 into state $(q-1, j)$, given that the process started in state (q, i).

Define $G(q, n)$ as an $\xi^{e}(q)\times\xi^{e}(q-1)$ matrix with elements $G_{i,j}(q, n)$. Clearly, $\sum_{i=1}^{n} G(q,i) \leq G_q$ and $\lim_{n\to\infty}\sum_{i=1}^{n} G(q,i) = G_q$.

Similar to the cost function $\Phi_{q,(i,j)}(x)$ defined in Sect. 5.3.2, define

$\Phi_{i,j}^{(1)}(q,n)$: the expected total cost incurred before or when n products are completed during the first passage time from level q to level $q-1$ into state $(q-1, j)$, given that the process started in state (q, i).

Note that $\Phi_{i,j}^{(1)}(q,n)$ only includes the total cost of producing n products, not all products produced during the first passage time. Define $\Phi^{(1)}(q, n)$ as an $\xi^{e}(q)\times\xi^{e}(q-1)$ matrix with elements $\Phi_{i,j}^{(1)}(q,n)$. Clearly, $\Phi^{(1)}(q,n) \leq \Phi_q^{(1)}$, and $\{\Phi^{(1)}(q,n), n = 1,2,\ldots\}$ is a nondecreasing sequence and $\lim_{\{n\to\infty\}}\Phi^{(1)}(q,n) = \Phi_q^{(1)}$. Consequently, $\lim_{\{n\to\infty\}}\Phi^{(1)}(q, n)\mathbf{e} = \mathbf{u}_{c,q}$, for $q = 1, 2, \ldots$.

Consider the expected total cost vector $\mathbf{V}^{\xi}(q, n)$. It is clear that $\mathbf{V}^{\xi}(q, n)$ can be decomposed into two parts: the total cost incurred during the first passage from level q to level $q-1$ and the total cost incurred thereafter. Conditioning on the initial inventory level and the queue length at the first transition from level q to level $q-1$, it has, for $q \geq 0$,

$$V^{\xi}(q,i,n) = \sum_{j=1}^{\xi^{e}(q-1)} \Phi_{i,j}(q,n) + \sum_{j=1}^{\xi^{e}(q-1)} \sum_{m=0}^{n-1} G_{i,j}(q,m)V^{\xi}(q-1,j,n-m). \quad (5.50)$$

Note that, if $m \geq n$, production completes before the first passage ends. Writing the above equation into a matrix form, yields

$$\mathbf{V}^{\xi}(q,n) = \Phi(q,n)\mathbf{e} + \sum_{m=0}^{n-1} G(q,m)\mathbf{V}^{\xi}(q-1,n-m). \quad (5.51)$$

Combining Eqs. (5.47) and (5.51), yields, for $q \geq 0$,

$$\begin{aligned} ng(\boldsymbol{\xi})\mathbf{e} + \mathbf{h}^{\xi}(q) &= \Phi(q,n)\mathbf{e} + \sum_{m=0}^{n-1} G(q,m)\left((n-m)g(\boldsymbol{\xi})\mathbf{e} + \mathbf{h}^{\xi}(q-1)\right) + o(1) \\ &= \Phi(q,n)\mathbf{e} + g(\boldsymbol{\xi})n\sum_{m=0}^{n-1} G(q,m)\mathbf{e} - g(\boldsymbol{\xi})\sum_{m=0}^{n-1} mG(q,m)\mathbf{e} \quad (5.52) \\ &\quad + \left(\sum_{m=0}^{n-1} G(q,m)\right)\mathbf{h}^{\xi}(q-1) + o(1). \end{aligned}$$

Thus, to prove the first two equalities in Eq. (5.49), we need to find $\sum_{m=0}^{\infty} mG(q,m)\mathbf{e}$, which is the expected number of demands satisfied in a fundamental period. Since the workshop is an $M/M/1$ queue, it is clear that the elements in $\sum_{m=0}^{\infty} mG(q,m)\mathbf{e}$ are the expected numbers of customers served in a busy period in an $M/M/1$ queue. By Cohen (1982), we have $\sum_{m=0}^{\infty} mG(q,m)\mathbf{e} = \mathbf{e}\mu/(\mu-\lambda)$.

We also need to prove

$$\lim_{n\to\infty} n\left(\mathbf{e} - \sum_{m=0}^{n-1} G(q,m)\mathbf{e}\right) = 0. \tag{5.53}$$

Note that $G_q = \sum_{m=0}^{n-1} G(q,m) + \sum_{m=n}^{\infty} G(q,m)$. Since $G_q\mathbf{e} = \mathbf{e}$, we have

$$0 \le n\left(\mathbf{e} - \sum_{m=0}^{n-1} G(q,m)\mathbf{e}\right) = \sum_{m=n}^{\infty} nG(q,m)\mathbf{e} \le \sum_{m=n}^{\infty} mG(q,m)\mathbf{e} \xrightarrow{n\to\infty} 0. \tag{5.54}$$

To prove the third equality in Eq. (5.49), the level 0 is considered. Note that

$$\mathbf{V}^{\xi}(0,n) = \mathbf{\Phi}_{0,0}^{(1)}\mathbf{e} + (-A_{0,0}^{-1}A_{0,1})\mathbf{V}^{\xi}(1,n). \tag{5.55}$$

Since $(-A_{0,0}^{-1}A_{0,1})\mathbf{e} = \mathbf{e}$, Eq. (5.55) leads to

$$\mathbf{h}^{\xi}(0) = \mathbf{\Phi}_{0,0}^{(1)}\mathbf{e} + (-A_{0,0}^{-1}A_{0,1})\mathbf{h}^{\xi}(1), \tag{5.56}$$

which proves the third equality in Eq. (5.49). This completes the proof of Theorem 5.3.4.

Using Theorem 5.3.4, we calculate $\{\mathbf{h}^{\xi}(q),\ q = 1, 2, \ldots\}$ from $\{g(\boldsymbol{\xi}),\ G,\ G_q,\ \mathbf{u}_c,\ \mathbf{u}_{c,q},\ q = 0, 1, \ldots, k\}$ for any replenishment policy $\boldsymbol{\xi}$ in $\mathbf{\Pi}_{\text{EOQ}}[k]$. To use the recursive formulas in Eq. (5.49), we first find $\mathbf{h}^{\xi}(1)$. Let

$$\mathbf{c} = \left(I - G_1(-A_{0,0}^{-1}A_{0,1}) + \mathbf{e}\boldsymbol{\theta}\right)^{-1}\left(\mathbf{u}_{c,1} + G_1\mathbf{\Phi}_{0,0}^{(1)}\mathbf{e} - g(\boldsymbol{\xi})\frac{\mu}{\mu-\lambda}\mathbf{e}\right), \tag{5.57}$$

where $\boldsymbol{\theta}$ is the left invariant vector of the matrix $G_1(-A_{0,0})^{-1}A_{0,1}$.

Exercise 5.3.20 Show that $G_1(-A_{0,0})^{-1}A_{0,1}$ is an irreducible stochastic matrix. Show that $I - G_1(-A_{0,0})^{-1}A_{0,1} + \mathbf{e}\boldsymbol{\theta}$ is invertible. Show that $\boldsymbol{\theta} = \boldsymbol{\beta}_0(-A_{0,0})^{-1}A_{0,1}$, where $\boldsymbol{\beta}_0$ is defined in Sect. 5.3.2.

By the last equality in Eq. (5.49) (see Theorem 5.3.4), it can be proved that

$$\mathbf{h}^{\xi}(1) = \mathbf{c} + \left(\boldsymbol{\theta}\mathbf{h}^{\xi}(1)\right)\mathbf{e}. \tag{5.58}$$

Thus, it is possible to determine $\mathbf{h}^{\xi}(1)$ first, which is equivalent to determining the product $\mathbf{h}^{\xi}(1)\mathbf{e}$. This can be done by using the original definition of $\{\mathbf{h}^{\xi}(q), q = 1, 2, \ldots, k\}$. Since the state $(q, i) = (1, 1)$ was chosen as the base state to determine the relative cost functions, then $h^{\xi}(1, 1) = 0$ (see Eq. (5.44)). The product $\mathbf{h}^{\xi}(1)\mathbf{e}$ is determined by setting $h^{\xi}(1, 1) = 0$ in Eq. (5.58), which yields

$$\boldsymbol{\theta}\mathbf{h}^{\xi}(1) = -(\mathbf{c})_1. \tag{5.59}$$

Then $\mathbf{h}^{\xi}(1)$ can be obtained by Eq. (5.58).

Now, we are ready to state an algorithm for computing the optimal policy in $\boldsymbol{\Pi}_{\mathrm{EOQ}}(k)$, which minimizes the expected total cost per product,

$$\boldsymbol{\xi}^{*(k)} = \arg \min_{\{\boldsymbol{\xi}\in\boldsymbol{\Pi}_{\mathrm{EOQ}}[k]\}} \{g(\boldsymbol{\xi})\}. \tag{5.60}$$

Algorithm 5.3.2: A *G*-Based MDP Algorithm

1. Initialize the policy iteration process by choosing $\boldsymbol{\xi}$ as $\xi(0) = 0$,

$$\begin{aligned} \xi(q) &= 2\sqrt{\frac{K\mu}{C_h}} + 2, \quad \text{for } q = 1, 2, \ldots, k; \\ \xi(q) &= EOQ(\mu), \qquad \text{for } q = k+1, k+2, \ldots. \end{aligned} \tag{5.61}$$

2. For policy $\boldsymbol{\xi}$, calculate $\{g(\boldsymbol{\xi}), \boldsymbol{\Phi}_1^{(1)}, \boldsymbol{\Phi}_{0,0}^{(1)}, \mathbf{u}_c, \mathbf{u}_{c,q}, G_q, q = 1, \ldots, k\}$ by using Algorithm 5.3.1 presented in Sect. 5.3.2.
3. Compute $\mathbf{c}$ using Eq. (5.57); $\boldsymbol{\theta}\mathbf{h}^{\xi}(1)$ using Eq. (5.59); and $\mathbf{h}^{\xi}(1)$ using Eq. (5.58).
4. Compute $\{\mathbf{h}^{\xi}(q), q = 1, 2, \ldots, k\}$ using Eq. (5.49).
5. Determine new policy $\boldsymbol{\xi}'$ using, $\xi'(0) = 0$,

$$\begin{aligned} \xi'(q) &= \arg \min_{1\le i\le \xi^{e}(q)} \{h^{\xi}(q, i)\}, \quad 1 \le q \le k; \\ \xi'(q) &= EOQ(\mu), \qquad \text{for } q > k. \end{aligned} \tag{5.62}$$

6. If $\boldsymbol{\xi} = \boldsymbol{\xi}'$, stop; otherwise, set $\boldsymbol{\xi} = \boldsymbol{\xi}'$, and repeat steps 2–5.

Using Algorithm 5.3.2, the optimal policy $\boldsymbol{\xi}^{*(k)}$ can be found for $k = 1, 2, \ldots$. Since $\boldsymbol{\Pi}_{\mathrm{EOQ}}(k) \subset \boldsymbol{\Pi}_{\mathrm{EOQ}}(k + 1)$, if the optimal policy in $\boldsymbol{\Pi}_{\mathrm{EOQ}}$ (or equivalently $\boldsymbol{\Pi}$) exists, then it is $\boldsymbol{\xi}^{*(k)}$ if k is sufficiently large. Thus, we can use the following procedure to find the optimal policy in $\boldsymbol{\Pi}_{\mathrm{EOQ}}$. We give k a moderate value first and calculate the optimal policy $\boldsymbol{\xi}^{*(k)}$ by Algorithm 5.3.2. Then we set $k =: 2k$ and apply Algorithm 5.3.2 to find the optimal policy $\boldsymbol{\xi}^{*(2k)}$. If $\boldsymbol{\xi}^{*(k)} = \boldsymbol{\xi}^{*(2k)}$, we may consider to stop the iteration process; otherwise, we find $\boldsymbol{\xi}^{*(3k)}, \boldsymbol{\xi}^{*(4k)}, \ldots$, until $\boldsymbol{\xi}^{*(jk)} = \boldsymbol{\xi}^{*((j+1)k)}$. Then $\boldsymbol{\xi}^{*(jk)}$ can be considered to be the optimal policy in $\boldsymbol{\Pi}_{\mathrm{EOQ}}$. Note that it is not guaranteed that $\boldsymbol{\xi}^{*(jk)}$ is the optimal policy. On the other hand, $\boldsymbol{\xi}^{*(jk)}$ must be optimal if j is sufficiently large.

Table 5.2 The optimal replenishment policies for Example 5.3.3

$k \backslash q$	0	1	2	3	4	5	6	7	($\geq$)8	$g(\xi^{(k)})$
1	0	10	10	10	10	10	10	10	10	19.1429
2	0	8	10	10	10	10	10	10	10	18.8787
4	0	8	9	9	10	10	10	10	10	18.8440
6	0	8	9	9	10	10	10	10	10	18.8440
8	0	8	9	9	10	10	10	11	10	18.8438

Example 5.3.3 Consider an inventory-production system with $\lambda = 0.7$, $\mu = 10$, $C_h = 1.5$, and $K = 40$.

Results for Example 5.3.3 are presented in Table 5.2.

Commentary Models with exponential lead times are considered in He (1996) and He et al. (2002b). The inventory-production system can be analyzed by the R-based analysis or the G-based analysis. The details are more tedious, though. On the other hand, it is much more complicated to develop an algorithm for computing the optimal policy. It is clear from the proof of Theorem 5.3.4 that the relationship between $\{g(\boldsymbol{\xi}), \mathbf{h}^{\pi}(q), q = 1, 2, \ldots\}$ and $\{G, G_q, \mathbf{u}_c, \mathbf{u}_{c,q}, q = 1, 2, \ldots, k\}$ may hold for more general stochastic models where the cost structure is defined appropriately and the model can be represented by a QBD Markov process (with level dependent transitions).

Additional Exercises and Extensions

Exercise 5.3.21 For Example 5.3.3, find the optimal policy in $\boldsymbol{\Pi}$ for $\lambda = 0.1, 0.2, 0.4, 0.618$, and 0.9.

Exercise 5.3.22 Write a simulation program to compute the expected total cost per product for the inventory-production system. Run the simulation program for Example 5.3.3 and Exercise 5.3.21 and compare the results. In addition, use the R-based algorithm developed in Exercise 5.3.10 to verify the results in Example 5.3.3 and Exercise 5.3.21.

Exercise 5.3.23* Analyze the inventory-production system with Markovian demand process (D_0, D_1) and phase-type service time $(\boldsymbol{\alpha}, T)$.

Exercise 5.3.24* Analyze the inventory-production system with a compound Poisson demand process.

5.4 An Inventory System with a Markovian Demand Process

We consider a discrete time inventory system in this section. Demands arrive according to a discrete time Markovian arrival process. Demands are satisfied from on-hand inventory with complete backlogging (that is, all demands are eventually satisfied). Inventory is replenished from an outside supplier. The replenishment lead time is constant, i.e., an order placed in period t arrives in period $t + L$, where L is a nonnegative integer.

Two types of costs are considered. At the end of each period, the holding costs are assessed based on the on-hand inventory at a constant rate of C_h per unit, and the penalty costs are assessed based on the backordered demands at a constant rate of C_p per unit.

In the system, the phase of the underlying Markov chain and the state of inventory (inventory position) are observed at the beginning of a period. Then a replenishment order is placed, if it is necessary. Finally, a shipment is received (if any) and demands arrive. At the end of a period, holding and backordering costs are assessed.

The objective is to find an inventory control policy that minimizes the long-run average costs. The rest of this section focuses on developing an algorithmic method to compute the optimal policy. The idea is to find a lower bound on the long-run average costs, which turns out to be the minimum long-run average costs.

Assume that $(D_0, D_1, D_2, \ldots)$ is a matrix representation of the discrete time Markovian arrival process, where $\{D_0, D_1, D_2, \ldots\}$ are nonnegative matrices of size m_a. Let $D = D_0 + D_1 + D_2 + \ldots$. Then D is a stochastic matrix, which is the transition probability matrix of the underlying Markov chain of the Markovian arrival process. We assume that D is irreducible. Denote by $\boldsymbol{\theta}_a$ the stationary distribution of D, i.e., $\boldsymbol{\theta}_a = \boldsymbol{\theta}_a D$ and $\boldsymbol{\theta}_a \mathbf{e} = 1$. The conditional probability generating function of the demand per period is defined as $D^*(z) = D_0 + zD_1 + z^2 D_2 + \ldots$, for $0 \leq z \leq 1$. Define $d_{n,(i,j)}(k)$ as the probability that the total demands in k consecutive periods is n, and the phase of the underlying Markov chain is j at the end of period k, given that the underlying Markov chain is in phase i at the beginning of period 1. Let $D_n(k) = (d_{n,(i,j)}(k))$, for $n = 0, 1, 2, \ldots$. It is easy to see that $D_n(1) = D_n$, for $n = 0, 1, 2, \ldots$.

We remark that, since arrivals of demand occur during a period, it is necessary to distinguish the phases at the beginning and at the end of a period, as they can be different. For discrete time stochastic models, one has to pay more attention to the sequence of event occurrence. For the model of interest, the phase of the underlying Markov chain is the same at the end of one period and at the beginning of the next period.

Exercise 5.4.1 Show that the conditional probability generating function of the total demands in $L + 1$ consecutive periods is $D^*(L + 1, z) = (D^*(z))^{L+1}$.

Based on Exercise 5.4.1, the conditional distributions of the total demands in $L + 1$ consecutive periods can be computed recursively as follows.

Exercise 5.4.2 Show that, for $n = 0, 1, 2, \ldots$,

$$D_n(L+1) = \sum_{k=0}^{n} D_k D_{n-k}(L). \tag{5.63}$$

Define $f_{i,L+1}(n)$ as the probability that the total demand in $L + 1$ periods is n, given that the underlying Markov chain is in phase i at the beginning of period 1, $i = 1, 2, \ldots, m_a$. Let $\mathbf{f}_{L+1}(n)$ be a column vector with elements $\{f_{i,L+1}(n), i = 1, 2, \ldots, m_a\}$. Then $\mathbf{f}_{L+1}(n) = D_n(L + 1)\mathbf{e}$, for $n = 0, 1, 2, \ldots$.

Suppose that the inventory position in period t is y (i.e., inventory level plus all outstanding orders) and the total demands in the periods $t, t + 1, \ldots$, and $t + L$ is x. Then all the outstanding orders included in y must have been realized in period $t + L$, and any new orders are not yet realized in period $t + L$. Thus, the inventory level in period $t + L$ is $y - x$. The on-hand inventory is $\max\{0, y - x\}$ and the backlogs is $\max\{0, x - y\}$. Then the total inventory cost in period $t + L$ can be calculated as $C_h\max\{0, y - x\} + C_p\max\{0, x - y\}$. A common practice in inventory management is to charge the cost determined by $y - x$ to period t, instead of period $t + L$. Let

$C^1(i, y)$ = the expected total costs charged to period t, given that the underlying Markov chain is in phase i and the inventory position is y at the beginning of period t.

It can be shown that the shift of costs in time does not affect the long-run average costs. Without rigorous mathematical proofs, one can use simulation (see Exercise 5.4.12) and performance analysis (see Exercises 5.4.13 and 5.4.14) to verify the conclusion numerically.

Exercise 5.4.3 Show that, for $i = 1, 2, \ldots, m_a$, and $y = 0, 1, 2, \ldots$

$$C^1(i,y) = \sum_{x=0}^{\infty} f_{i,L+1}(x)\left(C_h \max\{0,\ y-x\} + C_p \max\{0,\ x-y\}\right). \tag{5.64}$$

In matrix form, Eq. (5.64) can be rewritten as

$$\begin{pmatrix} C^1(1,y) \\ \vdots \\ C^1(m_a,y) \end{pmatrix} = \sum_{x=0}^{\infty} \mathbf{f}_{L+1}(x)\left(C_h \max\{0,\ y-x\} + C_p \max\{0,\ x-y\}\right). \tag{5.65}$$

The set of functions $\{C^1(i, .), i = 1, 2, \ldots, m_a\}$ determines the expected costs charged to each period, given the phase of the underlying Markov chain and the inventory position in that period. That is, if the system is in state (i, y) at the beginning of a period, then the cost incurred in that period is given by $C^1(i, y)$. We say $\{C^1(i, .), i = 1, 2, \ldots, m_a\}$ defining a cost-accounting scheme, denoted by $\mathcal{G}^1$.

Let $I(t)$ be the phase of the underlying Markov chain at the beginning of period t; and $IP(t)$ be the inventory position at the beginning of period t. Then the long-run average costs per period for any feasible policy is given by, under the cost-accounting scheme $\mathcal{G}^1$,

$$C\{C^1(i,\cdot),\ i = 1, \ldots,\ m_a\} = \lim_{N\to\infty} \frac{\sum_{t=1}^{N} C^1(I(t),\ IP(t))}{N}, \tag{5.66}$$

if the limit exists.

Our goal is to find a policy that minimizes the long-run average costs. An algorithm is to be developed. To make some technical preparation for that purpose, first, we find the expected total costs while the underlying Markov chain stays in a subset of phases. Without loss of generality, we assume that the subset is $\{1, 2, \ldots, m_1\}$ with $m_1 < m_a$.

Exercise 5.4.4 Suppose that the cost charged per period, if the underlying Markov chain is in phase i, is $C(i)$. Let $r(i)$ be the expected total costs incurred before the underlying Markov chain leaves the set $\{1, 2, \ldots, m_1\}$, given that the Markov chain is in phase i in period 1. Show that $\{r(i), i = 1, 2, \ldots, m_1\}$ satisfy the following equation:

$$r(i) = C(i) + \sum_{j=1}^{m_1} \left(\sum_{n=0}^{\infty} d_{n,(i,j)} \right) r(j), \quad i = 1, 2, \ldots, m_1. \tag{5.67}$$

Let $D^{[m_1]}$ be the north-western $m_1 \times m_1$ part of D. Show that

$$\begin{pmatrix} r(1) \\ \vdots \\ r(m_1) \end{pmatrix} = \left(I - D^{[m_1]} \right)^{-1} \begin{pmatrix} C(1) \\ \vdots \\ C(m_1) \end{pmatrix}. \tag{5.68}$$

In Eq. (5.68), the cost charged per period is independent of the inventory position. If the cost charged per period does depend on the inventory position, then we have the following result.

Exercise 5.4.5 Suppose that an order is placed to bring the inventory position to zero, whenever the inventory position is negative. Suppose that the cost charged per period, if the underlying Markov chain is in phase i and the inventory position is y, is $C(i, y)$. Let $R(i, y)$ be the total costs incurred before the underlying Markov chain leaves the set $\{1, 2, \ldots, m_1\}$, given that the Markov chain is in phase i and the inventory position is y in period 1. Let $\mathbf{R}(y) = (R(1, y), \ldots, R(m_1, y))'$. Show that $\{\mathbf{R}(y), y = 0, 1, 2, \ldots\}$ satisfy the following equations:

$$\mathbf{R}(0) = \left(I - D^{[m_1]} \right)^{-1} \begin{pmatrix} C(1,0) \\ \vdots \\ C(m_1,0) \end{pmatrix};$$

$$\mathbf{R}(y) = \left(I - D_0^{[m_1]} \right)^{-1} \left(\begin{pmatrix} C(1,y) \\ \vdots \\ C(m_1,y) \end{pmatrix} + \sum_{n=1}^{y-1} D_n^{[m_1]} \mathbf{R}(y-n) + \left(\sum_{n=y}^{\infty} D_n^{[m_1]} \right) \mathbf{R}(0) \right), \quad y \geq 1, \tag{5.69}$$

where $D_n^{[m_1]}$ is the north-western $m_1 \times m_1$ part of D_n, for $n = 0, 1, 2, \ldots$.

Now, we begin the development of an algorithm for computing the optimal policy. Since there is no ordering cost, the optimal policy is an order-up-to policy for which the order-up-to level depends on the phase of the underlying Markov chain. Suppose that the optimal policy is $\{s^*(1^*), s^*(2^*), \ldots, s^*(m_a^*)\}$, where $\{1^*, 2^*, \ldots, m_a^*\}$ is a permutation of $\{1, 2, \ldots, m_a\}$, $s^*(i^*)$ is the order-up-to level if the underlying Markov chain is in phase i^*, and the order-up-to levels are arranged as $s^*(1^*) \le s^*(2^*) \le \ldots \le s^*(m_a^*)$. In any period, if the underlying Markov chain is in phase i^* and the inventory position is less than $s^*(i^*)$, then an order is placed to bring the inventory position to $s^*(i^*)$. If the inventory position is greater than or equal to $s^*(i^*)$, do nothing.

Roughly speaking, the steps to find the optimal order-up-to levels are:

(i) To find $s^*(1^*)$ by using $\{C^1(i, .), i = 1, 2, \ldots, m_a\}$;
(ii) To construct cost functions $\{C^k(i, .), i = 1, 2, \ldots, m_a\}$ from $\{C^{k-1}(i, .), i = 1, 2, \ldots, m_a\}$; and
(iii) To find $s^*(k^*)$ by using $\{C^k(i, .), i = 1, 2, \ldots, m_a\}$

Repeat steps (ii) and (iii) for $k = 2, 3, \ldots, m_a$ to complete the iteration process. To clearly present the iteration steps, we first introduce some notation.

Define $U^1 = \phi$ (an empty set) and $V^1 = \{1, 2, \ldots, m_a\}$. One period cost functions $\{C^1(i, y), i = 1, 2, \ldots, m_a\}$ have been defined and can be computed using Eq. (5.64).

Exercise 5.4.6 Show that the functions $\{C^1(i, y), i = 1, 2, \ldots, m_a\}$ are convex in y. Show that $C^1(i, y) \to +\infty$, if $y \to +\infty$, for $i = 1, 2, \ldots, m_a$.

By Exercise 5.4.6, function $C^1(i, y)$ is minimized at a finite point $y = s^1(i)$, for $i = 1, 2, \ldots, m_a$. Then $s^1(i)$ can be considered as the optimal order-up-to level for phase i, if all other phases are censored. However, $\{s^1(i), i = 1, 2, \ldots, m_a\}$ may not be the optimal policy. Define $1^* = \arg\min_{i \in V^1}\{s^1(i)\}$ and $s^*(1^*) = s^1(1^*)$. Then we consider $s^*(1^*)$ the optimal order-up-to level for phase 1^*. We define a new cost function for phase 1^* as follows:

$$\underline{C}^1(1^*, y) = \begin{cases} C^1(1^*, s^*(1^*)), & \text{if } y \le s^*(1^*); \\ C^1(1^*, y), & \text{if } y > s^*(1^*). \end{cases} \tag{5.70}$$

The relationship between functions $C^1(1^*, .)$ and $\underline{C}^1(1^*, .)$ is shown in Fig. 5.5.

Next, we want to find phase 2^* and $s^*(2^*)$. Set $U^2 = \{1^*\} = \{1^*\} \cup U^1$ and $V^2 = \{1, 2, \ldots, m_a\} \backslash \{1^*\} = V^1 \backslash \{1^*\}$. Note that we have $U^1 \cup V^1 = U^2 \cup V^2 = \{1, 2, \ldots, m_a\}$ and $U^1 \cap V^1 = U^2 \cap V^2 = \phi$. Then we use cost functions $\{C^1(i, .), i \in V^2, \underline{C}^1(1^*, .)\}$ to define new cost-accounting scheme $\mathcal{G}^1$. Under $\mathcal{G}^1$, the functions used in Eq. (5.66) are replaced with $\{C^1(i, .), i \in V^2, \underline{C}^1(1^*, .)\}$. Also under $\mathcal{G}^1$, if the phase is 1^* and the inventory position is smaller than $s^*(1^*)$, then an order is placed to bring inventory position to $s^*(1^*)$. Since $C^1(1^*, y) \ge \underline{C}^1(1^*, y)$ for all y, we must have

$$C\{C^1(i, .),\ i = 1,\ 2, \ldots,\ m\} \ge C\{C^1(i, .),\ i \ne 1^*,\ \underline{C}^1(1^*, .)\}. \tag{5.71}$$

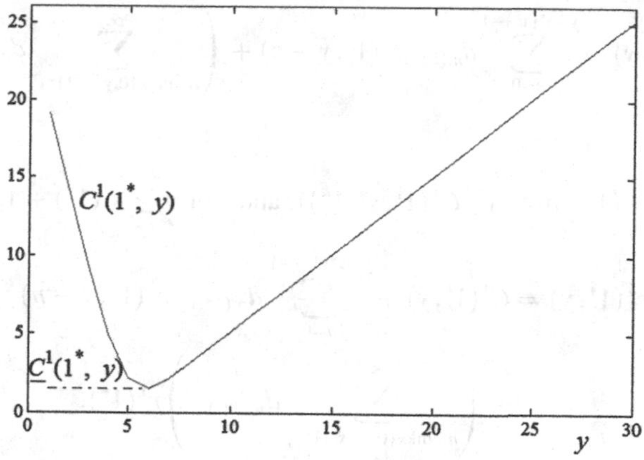

Fig. 5.5 Functions $C^1(1^*, .)$ and $\underline{C}^1(1^*, .)$ with $s^*(1^*) = 6$

We need to update the cost functions. Set $C^2(1^*, y) \equiv 0$, which means that the cost to be charged is zero, if the underlying Markov chain is in phase 1^*. In order to keep the total cost unchanged, we spread the cost $C^1(1^*, y)$ of phase 1^* to other phases as follows.

At the beginning of a period, the underlying Markov chain is either in a phase in U^2 or in a phase in V^2. Define a type $\mathcal{U}$ interval as an interval with consecutive periods in which the Markov chain is always in U^2. A type $\mathcal{V}$ interval is defined similarly for V^2. Then the underlying Markov chain can be seen alternating between the two types of intervals.

We allocate the cost incurred in a type $\mathcal{U}$ interval under $\underline{\mathcal{g}}^1$ to the last phase in the last $\mathcal{V}$ interval. That is, the total costs incurred while the underlying Markov chain is in U^2 (i.e. phase 1^*) is reallocated to phases in V^2 by considering the first passage from one phase to phases in V^2. For $i \in V^2$, define $C^2(i, y)$ as the expected total cost incurred during the first passage time from i back to V^2, given that the underlying Markov chain is in phase i and the inventory position is y initially, and the cost-accounting scheme is $\underline{\mathcal{g}}^1$.

Example 5.4.1 Consider a case with $m_a = 4$. Suppose $V^2 = \{1, 2, 4\}$ and $U^2 = \{3\}$. A sample path of the underlying Markov chain $\{I(t), t = 1, 2, 3, \ldots\}$ is $\{2, 1, 1, 3, 3, 4, 3, 2, 2, 1, 1, 3, 2, 4, 1, \ldots\}$. We can divide the entire interval into intervals each beginning with a phase in V^2: $\{2\}$, $\{1\}$, $\{1, 3, 3\}$, $\{4, 3\}$, $\{2\}$, $\{2\}$, $\{1\}$, $\{1, 3\}$, $\{2\}$, $\{4\}$, and $\{1\}$. Then the cost incurred in each interval is calculated. The costs incurred in periods 4 and 5 are added to that of period 3, whose phase is 1; the cost incurred in period 7 is added to period 6, whose phase is 4; and the cost incurred in period 12 is added to period 11, whose phase is 1.

Use results in Exercises 5.4.4 and 5.4.5 to prove the following results.

Exercise 5.4.7 Under $\underline{\mathcal{g}}^1$, show, for $i \in V^2$,

$$C^2(i,y) = C^1(i,y) + \sum_{n=0}^{y-s^*(1^*)-1} d_{n,(i,1^*)} r^1(1^*, y-n) + \left(\sum_{n=\max\{0,y-s^*(1^*)\}}^{\infty} d_{n,(i,1^*)} \right) r^1(1^*), \tag{5.72}$$

where, $r^1(1^*) = \left(1 - d_{1^*,1^*}\right)^{-1} C^1(1^*, s^*(1^*))$, and, for $y \geq s^*(1^*) + 1$,

$$r^1(1^*,y) = C^1(1^*,y) + \sum_{n=0}^{y-s^*(1^*)-1} d_{n,(1^*,1^*)} r^1(1^*, y-n) + \left(\sum_{n=\max\{0,y-s^*(1^*)\}}^{\infty} d_{n,(1^*,1^*)} \right) r^1(1^*). \tag{5.73}$$

From Eq. (5.72), it is clear that $C^2(i, y)$, for $i \in V^2$, is the cost incurred in the current period, which is $C^1(i, y)$, plus all the cost incurred in the subsequent consecutive periods for which the phase of the Markov chain is in U^2.

Now, we move to the second iteration to determine 2^* and $s^*(2^*)$ by utilizing the cost functions $\{C^2(i, .), i = 1, 2, \ldots, m_a\}$. Note that $C^2(i, .) \equiv 0$, for $i \in U^2$.

Exercise 5.4.8 Explain intuitively why $C\{C^1(i, .), i \in V^2, \underline{C}^1(1^*, .)\} = C\{C^2(i, .), i = 1, 2, \ldots, m_a\}$. Show that functions $\{C^2(i, y), i = 1, 2, \ldots, m_a\}$ are convex in y. Show that $C^2(i, y) \to +\infty$, if $y \to +\infty$, for $i \in V^2$.

We use functions $\{C^2(i, .), i = 1, 2, \ldots, m_a\}$ to define a cost-accounting scheme, denoted by $\mathcal{G}^2$. By Exercise 5.4.8, $\mathcal{G}^2$ and $\underline{\mathcal{G}}^1$ have the same long-run average costs. Also by Exercise 5.4.8, function $C^2(i, y)$ is minimized at a finite point $y = s^2(i)$, for $i \in V^2$. Then $s^2(i)$ can be considered as the optimal order-up-to level for phase i, if all other phases are censored. Define $2^* = \arg\min_{i \in V^2}\{s^2(i)\}$ and $s^*(2^*) = s^2(2^*)$. Then $s^*(2^*)$ is considered to be the optimal order-up-to level for phase 2^*.

Exercise 5.4.9 Show that $s^*(1^*) \leq s^*(2^*)$.

Next, we assume that the k-th iteration has completed and $\{U^k, V^k, C^k(i, .), i = 1, 2, \ldots, m_a\}$ are found. We must have $U^k \cup V^k = \{1, 2, \ldots, m_a\}$ and $U^k \cap V^k = \phi$. The functions $C^k(i, .) \equiv 0$, if $i \in U^k$; and $C^k(i, y)$ is convex in y, if $i \in V^k$. Suppose that $C^k(i, y)$ is minimized at $s^k(i)$, for $i \in V^k$. Define $k^* = \arg\min_{i \in V^k}\{s^k(i)\}$ and $s^*(k^*) = s^k(k^*)$. Then $s^*(k^*)$ is considered to be the optimal order-up-to level if the underlying Markov chain is in phase k^*. Define

$$\underline{C}^k(k^*,y) = \begin{cases} C^k(k^*, s^*(k^*)), & \text{if } y \leq s^*(k^*); \\ C^k(k^*,y), & \text{if } y > s^*(k^*). \end{cases} \tag{5.74}$$

We use cost functions $\{C^k(i, .), i = 1, 2, \ldots, m_a\}$ to define cost-accounting scheme $\mathcal{G}^k$ and $\{C^k(i, .), i \neq k^*, \underline{C}^k(k^*, .)\}$ to define cost-accounting scheme $\underline{\mathcal{G}}^k$. Since $C^k(k^*, y) \geq \underline{C}^k(k^*, y)$ for all y, we must have

$$C\{C^k(i,.),\ i=1,\ 2,\ldots,\ m_a\} \geq C\{C^k(i,.),\ i\neq k^*,\ \underline{C}^k(k^*,.)\}. \tag{5.75}$$

Set $U^{k+1} = \{k^*\}\cup U^k$ and $V^{k+1} = V^k\backslash\{k^*\}$. Again, we have $U^{k+1}\cup V^{k+1} = \{1, 2, \ldots, m_a\}$ and $U^{k+1}\cap V^{k+1} = \phi$. We set $C^{k+1}(i, .) \equiv 0$, for $i \in U^{k+1}$. In order to keep the total cost unchanged, we spread the cost $C^k(k^*, y)$ to the phases in V^{k+1}. That is, the total costs incurred while the underlying Markov chain is in U^{k+1} is reallocated to phases in V^{k+1}. For $i \in V^{k+1}$, define $C^{k+1}(i, y)$ as the expected total cost incurred during the first passage time back to V^{k+1}, given that the underlying Markov chain is in phase i and the inventory position is y initially. Intuitively, for $i \in V^{k+1}$, $C^{k+1}(i, y)$ includes $C^k(i, y)$ and subsequent costs incurred in phases in U^{k+1}, before the underlying Markov chain returns to V^{k+1}.

Proposition 5.4.1 *Under $\{C^k(i,.), i\neq k^*, \underline{C}^k(k^*,.)\}$, we have, for $i \in V^{k+1}$,*

$$\begin{aligned} C^{k+1}(i,y) = C^k(i,y) &+ \sum_{n=0}^{y-s^*(k^*)-1} \sum_{j\in U^{k+1}} (D_n)_{i,j} \left(\mathbf{R}^{(k+1)}(y-n)\right)_j \\ &+ \sum_{j\in U^{k+1}} \left(\sum_{n=\max\{0,y-s^*(k^*)\}}^{\infty} D_n \right)_{i,j} \left(\mathbf{R}^{(k+1)}\right)_j, \end{aligned} \tag{5.76}$$

or, in matrix form,

$$\begin{aligned} \left(C^{k+1}(i,y)\right)_{i\in V^{k+1}} = \left(C^k(i,y)\right)_{i\in V^{k+1}} &+ \sum_{n=0}^{y-s^*(k^*)-1} D_n^{[V^{k+1},\, U^{k+1}]}\mathbf{R}^{(k+1)}(y-n) \\ &+ \left(\sum_{n=\max\{0,y-s^*(k^*)\}}^{\infty} D_n^{[V^{k+1},\, U^{k+1}]} \right)\mathbf{R}^{(k+1)} \end{aligned} \tag{5.77}$$

where

$$\mathbf{R}^{(k+1)} = \left(I - D^{[U^{k+1},U^{k+1}]}\right)^{-1} \begin{pmatrix} 0 \\ \vdots \\ 0 \\ C^k(k^*, s^*(k^*)) \end{pmatrix}, \tag{5.78}$$

and, for $y \geq s^*(k^*)+1$,

$$\begin{aligned} \mathbf{R}^{(k+1)}(y) = &\left(I - D_0^{[U^{k+1},U^{k+1}]}\right)^{-1} \left(\begin{pmatrix} 0 \\ \vdots \\ 0 \\ C^k(k^*, y) \end{pmatrix} + \sum_{n=1}^{y-s^*(k^*)-1} D_n^{[U^{k+1},U^{k+1}]}\mathbf{R}^{(k+1)}(y-n) \right) \\ &+ \left(I - D_0^{[U^{k+1},U^{k+1}]}\right)^{-1} \left(\sum_{n=\max\{1,y-s^*(k^*)\}}^{\infty} D_n^{[U^{k+1},U^{k+1}]} \right)\mathbf{R}^{(k+1)}, \end{aligned} \tag{5.79}$$

where $D_n^{[V^{k+1},U^{k+1}]}$ *is a submatrix of* D_n *corresponding to the rows indexed by* V^{k+1} *and the columns indexed by* U^{k+1}*, and* $D_n^{[U^{k+1},U^{k+1}]}$ *and* $D^{[U^{k+1},U^{k+1}]}$ *are submatrices of* D_n *and* D *corresponding to the rows and columns indexed by* U^{k+1}*, respectively. The cost functions* $\{C^{k+1}(i,y),\ i=1,2,\ldots,m_a\}$ *are convex in* y*, and are minimized at finite points.*

Proof. Since an order is placed whenever the inventory position drops to or below $s^*(k^*)$, the corresponding cost is then determined by $\mathbf{R}^{(k+1)}$. The results are obtained by applying results in Exercises 5.4.4 and 5.4.5. This completes the proof of Proposition 5.4.1.

The cost-accounting scheme associated with $\{C^{k+1}(i, y),\ i = 1, 2, \ldots, m_a\}$ is denoted as $\mathcal{G}^{k+1}$. We have $C\{C^k(i, .), i \in V^2, \underline{C}^k(k^*, .)\} = C\{C^{k+1}(i, .), i = 1, 2, \ldots, m_a\}$. Repeat the above procedure m_a times to obtain $\{s^*(1^*), s^*(2^*), \ldots, s^*(m_a{}^*)\}$, and $C^{m_a}(m_a^*, y)$, which is minimized at $s^*(m_a{}^*)$.

Lemma 5.4.1 (Chen and Song (2001)) *The functions* $\{C^{k+1}(i,y),\ i=1,2,\ldots,m_a\}$ *are convex. The base-stock levels are arranged as* $s^*(1^*) \le s^*(2^*) \le \ldots \le s^*(m_a^*)$*. In addition, we have*

$$C\{C^1(i,.),\ i=1,\ 2,\ldots,\ m_a\} \ge \cdots \ge C\{C^{m_a}(i,.),\ i=1,\ 2,\ldots,\ m_a\}. \quad (5.80)$$

Proof. See Chen and Song (2001) for a complete proof. By the construction of the functions $\{C^{k+1}(i, y), i \in V^{k+1}\}$, it is clear that they are decreasing for $y \le s^*(k^*)$. The functions $\{C^{k+1}(i, y),\ i = 1, 2, \ldots, m_a\}$ are convex. Thus, we must have $s^*(k^*) \le s^*((k+1)^*)$. This completes the proof of Lemma 5.4.1.

Now, we are ready to state the main result of this section.

Theorem 5.4.1 (Chen and Song (2001)) *The state-dependent base-stock policy* $\{s^*(1^*), s^*(2^*), \ldots, s^*(m_a^*)\}$ *is optimal. The corresponding minimum expected total cost per unit time is given by* $(\boldsymbol{\theta}_a)_{m_a^*} C^{m_a}(m_a^*, s^*(m_a^*))$.

Proof. In Chen and Song (2001), the theorem is proved for the case with

$$D_n = \mathrm{diag}(f_1(n), \cdots, f_{m_a}(n))D, \text{ for } n = 0, 1, 2, \ldots, \quad (5.81)$$

where $f_i(.)$ is a probability distribution on $\{0, 1, 2, \ldots\}$, for $i = 1, 2, \ldots, m_a$. The basic idea of the proof is to show that $(\boldsymbol{\theta}_a)_{m_a^*} C^{m_a}(m_a^*, s^*(m_a^*))$ is a lower bound of the expected total cost per unit time associated with any feasible policy, and $(\boldsymbol{\theta}_a)_{m_a^*} C^{m_a}(m_a^*, s^*(m_a^*))$ is attained at the policy $\{s^*(1^*), s^*(2^*), \ldots, s^*(m_a{}^*)\}$. Thus, the policy $\{s^*(1^*), s^*(2^*), \ldots, s^*(m_a{}^*)\}$ is optimal. The same proof is valid for the case with a *BMAP* demand process. Details are omitted. This completes the proof of Theorem 5.4.1.

We note that $C^{m_a}(m_a^*, s^*(m_a^*))$ can be considered as the expected total costs incurred during the first passage from phase m^*_a to m^*_a, if the cost-accounting scheme is $\underline{\mathcal{G}}^k$ with $k = m_a^*$. Recall that $1/(\boldsymbol{\theta}_a)_{m_a^*}$ can be interpreted as the expectation

Table 5.3 The optimal policies for Example 5.4.2

L	$(1^*, s^*(1^*))$	$(2^*, s^*(2^*))$	$(3^*, s^*(3^*))$	ETC
1	(3, 1)	(2, 3)	(1, 4)	1.8221
3	(3, 2)	(2, 6)	(1, 7)	4.0933
5	(3, 4)	(2, 9)	(1, 10)	5.9252
10	(3, 10)	(2, 15)	(1, 16)	9.9842

ETC expected total cost per unit time

of the first passage time from phase m^*_{a} to m^*_{a}. Then $(\boldsymbol{\theta}_{\text{a}})_{m^*_{\text{a}}} C^{m_{\text{a}}}(m^*_{\text{a}}, s^*(m^*_{\text{a}}))$ is the expected cost per unit time.

Now, we are ready to state an algorithm for computing the optimal policy $\{s^*(1^*), s^*(2^*), \ldots, s^*(m_{\text{a}}^*)\}$.

Algorithm 5.4.1 Find the optimal phase-dependent base-stock levels. System parameters are L, C_{h}, C_{p}, m_{a}, and $\{D_0, D_1, \ldots\}$.

1. Compute the conditional distributions of the demands per period $\{\mathbf{f}_1(n), n = 0, 1, 2, \ldots\}$, the conditional distributions of the lead time demands $\{D_n(L + 1), n = 0, 1, 2, \ldots\}$ using Eq. (5.63), and $\{\mathbf{f}_{L+1}(n), n = 0, 1, 2, \ldots\}$,
2. Compute cost functions $\{C^1(i, .), i = 1, 2, \ldots, m_{\text{a}}\}$ by Eq. (5.65). Set $k = 1$, $U^1 = \phi$, and $V^1 = \{1, 2, \ldots, m_{\text{a}}\}$.
3. Use $\{C^k(i, .), i \in V^k\}$ to find $\{s^k(i), i \in V^k\}$. Find $k^* = \arg\min_{i \in V^k}\{s^k(i)\}$ and $s^*(k^*) = s^k(k^*)$. If $k = m_{\text{a}}$, stop; otherwise, go to step 4.
4. Update $U^{k+1} = \{k^*\} \cup U^k$ and $V^{k+1} = V^k \backslash \{k^*\}$. Compute $\mathbf{R}^{(k+1)}$ by Eq. (5.78); Compute $\mathbf{R}^{(k+1)}(y)$ recursively using Eq. (5.79) for $y = s^*(k^*) + 1, s^*(k^*) + 2, \ldots$; and functions $\{C^{k+1}(i, .), i \in V^{k+1}\}$ by Eq. (5.77). (Note: By definition, we set $C^{k+1}(i, .) \equiv 0$, for $i \in U^{k+1}$.) Set $k =: k + 1$, and go to step 3.

Comments on implementation: All computations in Algorithm 5.4.1 can be carried out by matrix operations. For convenience, we permute rows and columns in matrices $\{D_0, D_1, \ldots\}$ so that $U^{k+1} = \{1, 2, \ldots, k\}$ throughout the iteration process. More specifically, if k^* is identified among phases $\{k, k + 1, \ldots, m_{\text{a}}\}$, switch the k-th row (column) and the k^*-th row (column) in matrices $\{D_0, D_1, \ldots\}$. In this way, the formulas developed in Eqs. (5.68) and (5.69) can be used directly. In order to keep track of the original indexes of phases, a vector can be introduced to track the permutations.

Example 5.4.2 We consider a system with $C_{\text{h}} = 1$, $C_{\text{p}} = 5$, $m_{\text{a}} = 3$, and

$$D_0 = \begin{pmatrix} 0.1 & 0.0 & 0.1 \\ 0.0 & 0.1 & 0.1 \\ 0.0 & 0.1 & 0.8 \end{pmatrix}, \quad D_1 = \begin{pmatrix} 0.0 & 0.0 & 0.0 \\ 0.3 & 0.1 & 0.0 \\ 0.0 & 0.0 & 0.1 \end{pmatrix},$$
$$D_2 = \begin{pmatrix} 0.7 & 0.1 & 0.0 \\ 0.2 & 0.1 & 0.1 \\ 0.0 & 0.0 & 0.0 \end{pmatrix}. \tag{5.82}$$

For different lead time L, the optimal policies are given in Table 5.3.

Commentary The material in this section comes mainly from Chen and Song (2001). We consider a slightly more general model for which the distribution of demands per period is modulated by both the initial and ending phases of the underlying Markov chain. In Chen and Song (2001), the results are generalized to multi-echelon models. The model considered in this section has no ordering cost. If ordering costs are involved, intuitively, the optimal policy should be a phase-dependent (s, S) one. In Song (1991), Song and Zipkin (1993), Beyer and Sethi (1997), and Sethi and Cheng (1997), inventory models with ordering costs are considered. The optimality of the phase-dependent (s, S) policy is confirmed under fairly general conditions.

Additional Exercises and Extensions

Exercise 5.4.10 If $C_h = 0$, what does the optimal policy look like? If $C_p = 0$, what does the optimal policy look like?

Exercise 5.4.11 Implement Algorithm 5.4.1. Verify the results in Example 5.4.2.

Exercise 5.4.12 Develop the following two simulation programs and run them for Example 5.4.2. Compare the results. (Hint: In the simulation program, pay special attention to the sequence of events in a period.)

1. Write a simulation program based on the inventory position to calculate the long-run average costs per unit time by using Eqs. (5.64) and (5.66).
2. Write a simulation program based on the inventory level to calculate the long-run average costs per unit time by using the following equation. Let $IL(t)$ be the inventory level at the end of period t. Given policy $(s_1, \ldots, s_{m_a})$ the long-run average costs per unit time is defined as

$$C\{s_1, \cdots, s_{m_a}\} = \lim_{N\to\infty} \frac{\sum_{t=1}^{N} (C_h \max\{0, IL(t)\} + C_p \max\{0, -IL(t)\})}{N}. \tag{5.83}$$

Exercise 5.4.13 Consider a system with state dependent base-stock policy $\{s(1), s(2), \ldots, s(m_a)\}$ satisfying $s(1) \le s(2) \le \ldots \le s(m_a)$. Consider the process $\{(I(t), IP(t)), t = 0, 1, 2, \ldots\}$.

(1) Show that $\{(I(t), IP(t)), t = 0, 1, 2, \ldots\}$ is a discrete time Markov chain.
(2) Show that the state space of $\{(I(t), IP(t)), t = 0, 1, 2, \ldots\}$ is $\{(i, j): i = 1, 2, \ldots, m, j = s(j), s(j + 1), \ldots, s(m_a)\}$.
(3) Construct the transition probability matrix P for $\{(I(t), IP(t)), t = 0, 1, 2, \ldots\}$.
(4) Find the limiting probabilities for P, i.e., find $\boldsymbol{\pi} = (\boldsymbol{\pi}_1, \boldsymbol{\pi}_2, \ldots, \boldsymbol{\pi}_{m_a}), \boldsymbol{\pi}_j = \left(\pi_{j,s(j)}, \pi_{j,s(j)+1}, \cdot s, \pi_{j,s(m_a)}\right)$, satisfying $\boldsymbol{\pi} = \boldsymbol{\pi}P$ and $\boldsymbol{\pi}\mathbf{e} = 1$.
(5) Show that the long-run average costs per unit time can be obtained as

$$C\{(s(1), s(2), \ldots, s(m_a))\} = \sum_{i=1}^{m_a} \sum_{y=s(i)}^{s(m_a)} \pi_{i,y} \left(\sum_{n=0}^{\infty} f_{L+1,i}(n) \left(C_h \max\{0, y-n\} + C_p \max\{0, n-y\} \right) \right). \tag{5.84}$$

Exercise 5.4.14 (Exercise 5.4.13 continued) Consider a system with state dependent base-stock policy $\{s(1), s(2), \ldots, s(m_a)\}$ satisfying $s(1) \leq s(2) \leq \ldots \leq s(m_a)$. Show that the process $\{(IL(t), I(t), IP(t)), t = 0, 1, 2, \ldots\}$ is a discrete time Markov chain. Use this process to develop a method for computing the long-run average costs per unit time based on the inventory level.

Exercise 5.4.15* Consider systems with special demand processes: (i) geometric demand; (ii) compound geometric demand; and (iii) *DMAP* $\{D_0, D_1\}$. Try to simplify Algorithm 5.4.1 for the special cases.

5.5 Additional Topics

Applications of matrix-analytic methods are versatile. We briefly discuss a few key application areas and provide references for further reading.

5.5.1 Inventory Control and Supply Chain Management

In general, the use of matrix-analytic methods in the study of inventory and supply chain models is limited. An early work of this kind is Ramaswami (1981), where an algorithm is developed for performance analysis of an (*s*, *S*) model. Similar analyses of inventory models can be found in, for example, Chakravarthy and Alfa (1996), He (1996), He and Jewkes (2000), He et al. (2002a, b), Lian and Liu (2001, 2005), Chakravarthy and Daniel (2004), Isotupa (2006), Boute et al. (2007a, b), and Zhao and Lian (2011).

5.5.2 Design of Telecommunications Systems

Matrix-analytic methods are natural tools for the analysis and design of telecommunications systems. Traditionally, queueing theory and queueing models are used in the analysis and design of telecommunications systems. Thus, matrix-analytic methods have been used in the study of, for example, ATM and multiplexing of packets (e.g., Chandramouli et al. (1989), Neuts and Li (1996), Ramaswami and Wang (1996), Ost (2001)). Recently, matrix-analytic methods are used in the investigation of wireless systems, call centers, etc. (e.g., Alfa and Liu (2004), van Houdt and Blondia (2004, 2005), Le et al. (2006)).

5.5.3 Risk/Insurance Analysis

The use of matrix-analytic methods in risk and insurance analysis dates back to Asmussen and Rolski (1992), Asmussen (1995a, b), Asmussen and Bladt (1996). As mentioned in Chaps. 1 and 2, phase-type distributions and Markovian arrival processes are suitable for modeling the claim amounts and claim arrival processes (e.g., Asmussen (2000), Drekic et al. (2004), Drekic and Wilmot (2005)). Recent research work demonstrates that the algorithmic approach of matrix-analytic methods is also effective in analyzing risk/insurance models (e.g., Alfa and Drekic (2007), Asmussen et al. (2002), Avram and Usabel (2004), Badescu et al. (2007a, b), Stanford et al. (2011), Ren et al. (2009), etc.) This is a fertile area for further exploration.

5.5.4 Reliability Models

Matrix-analytic methods are suitable for analyzing reliability models for two reasons: (i) phase-type distributions can be used to approximate life times of units in a system; and (ii) the set of phase-type distributions is closed under a number of operations on random variables (see Sect. 1.3). Thus, the time to failure and reliability of many reliability systems can be found explicitly. The following papers demonstrate the usefulness of matrix-analytic methods in the study of reliability systems: Assaf and Levikson (1982), He et al. (1990), Chakravarthy et al. (2001), and Montoro-Cazorla and Pérez-Ocón (2006), and Ruiz-Castro et al. (2008).

References

Alfa AS, Drekic S (2007) Algorithmic analysis of the Sparre Andersen model in discrete time. ASTIN Bull 37:293–317

Alfa AS, Liu B (2004) Performance analysis of a mobile communication network: the tandem case. Comput Commun 27:208–221

Asmussen S (1995a) Stationary distributions for fluid flow models with or without Brownian noise. Stoch Models 11:21–49

Asmussen S (1995b) Stationary distributions via first passage times. In: Dshalalow J (ed) Advances in queueing: models, methods & problems. CRC Press, Boca Raton, pp 79–102

Asmussen S (2000) Ruin probabilities. World Scientific, Hong Kong

Asmussen S, Bladt M (1996) Phase-type distributions and risk processes with state-dependent premiums. Scand Actuarial J 1996:19–36

Asmussen S, Rolski T (1992) Computational methods in risk theory: a matrix algorithmic approach. Insur Math Econ 10:259–274

Asmussen S, Avram F, Usabel M (2002) Erlangian approximations for finite-horizon ruin probabilities. ASTIN Bull 32:267–281

Assaf D, Levikson B (1982) Closure of phase type distributions under operations arising in reliability theory. Ann Prob 10:265–269

Avram F, Usabel M (2004) Ruin probabilities and deficit for the renewal risk model with phase-type interarrival times. ASTIN Bull 34:315–332

Axsäter S (2000) Inventory control. Kluwer, London

Badescu AL, Drekic S, Landriault D (2007a) Analysis of a threshold dividend strategy for a *MAP* risk model. Scand Actuarial J 2007:227–247

Badescu AL, Drekic S, Landriault D (2007b) On the analysis of a multi-threshold Markovian risk model. Scand Actuarial J 2007:248–260

Bartmann D, Beckmann MJ (1992) Inventory control: models and methods. Springer-Verlag, New York

Beyer D, Sethi S (1997) Average cost optimality in inventory delays with Markovian demands. J Optimiz Theory Appl 92:497–526

Boute RN, Disney SM, Lambrecht MR, van Houdt B (2007a) An integrated production and inventory model to dampen upstream demand variability in the supply chain. Eur J Oper Res 178:121–142

Boute RN, Lambrecht MR, van Houdt B (2007b) Performance evaluation of a production/inventory system with periodic review and endogenous lead times. Nav Res Log 54:462–473

Buzacott JA, Shanthikumar JG (1993) Stochastic models of manufacturing systems. Prentice Hall, New York

Chakravarthy SR, Alfa AS (1996) Matrix-analytic methods in stochastic models. Marcel Dekker, New York

Chakravarthy SR, Daniel JK (2004) A Markovian inventory system with random shelf time and back orders. Comput Ind Eng 47:315–337

Chakravarthy SR, Krishnamoorthy A, Ushakumari PV (2001) A k-out-of-n reliability system with an unreliable server and phase type repairs and services: the (N, T) policy. J Appl Math Stoch Anal 14:361–380

Chandramouli Y, Neuts MF, Ramaswami V (1989) A queueing model for meteor burst packet communication systems. IEEE Trans Commun 31:1024–1030

Chen F, Song JS (2001) Optimal policies for multi-echelon inventory problems with Markov modulated demand. Oper Res 49:226–234

Cohen JW (1982) The single server queue, North-Holland series in applied mathematics and mechanics. North-Holland, Amsterdam

Drekic S, Willmot GE (2005) On the moments of the time of ruin with applications to phase-type claims. North Am Actuarial J 9:17–30

Drekic S, Dickson DCM, Stanford DA, Willmot GE (2004) On the distribution of the deficit at ruin when claims are phase-type. Scand Actuarial J 2004:105–120

Hadley G, Whitin T (1963) Analysis of inventory systems. Prentice-Hall, Englewood Cliffs

He QM (1996) The value of information used in inventory replenishment. Ph.D. thesis, University of Waterloo, Waterloo

He QM, Jewkes EM (2000) Performance measures of a make-to-order inventory-production system. IIE Trans 32:409–419

He QM, Wang YD, Zhang DJ (1990) On the first failure time for a two-unit system in two environments. Microelectron Reliab 30:1095–1097

He QM, Jewkes EM, Buzacott J (1998) An efficient algorithm for computing the optimal replenishment policy for an inventory-production system. In: Advances in matrix-analytic methods for stochastic models, (proceedings of the second international conference on matrix-analytic methods, Winnipeg, 1998), Notable Publications, pp 381–402

He QM, Jewkes EM, Buzacott J (2002a) Optimal and near-optimal inventory control policies for a make-to-order inventory-production system. Eur J Oper Res 141:113–132

He QM, Jewkes EM, Buzacott J (2002b) The value of information used in inventory control of a make-to-order inventory-production system. IIE Trans 34:999–1013
Iglehart D (1963) Optimality of (s, S) policies in the infinite horizon dynamic inventory problem. Manag Sci 9:259–267
Isotupa KPS (2006) An (s, Q) Markovian inventory system with lost sales and two demand classes. Math Comput Model 43:687–694
Le BL, Hossain E, Alfa AS (2006) Delay statistics and throughput performance for multi-rate wireless networks under multiuser diversity. IEEE Trans Wireless 5:3234–3243
Lian Z, Liu L (2001) Continuous review perishable inventory systems: models and heuristics. IIE Trans 33:809–822
Lian Z, Liu L (2005) A discrete-time model for common lifetime inventory systems. Math Oper Res 30:718–732
Montoro-Cazorla D, Pérez-Ocón R (2006) Reliability of a system under two types of failures using a Markovian arrival process. Oper Res Lett 34:525–530
Neuts MF (1981) Matrix-geometric solutions in stochastic models: an algorithmic approach. The Johns Hopkins University Press, Baltimore
Neuts MF (1989) Structured stochastic matrices of $M/G/1$ type and their applications. Marcel Dekker, New York
Neuts MF, Li JM (1996) Conditional overflow probability and profile curve for ATM congestion detection. Proc IEEE 3:970–977
Ost A (2001) Performance of communication systems: a model-based approach with matrix-geometric methods. Springer, New York
Porteus E (2002) Foundations of stochastic inventory theory. Stanford University Press, Stanford
Puterman M (1994) Markov Decision Processes: Discrete Stochastic Dynamic Programming. Wiley, New York
Ramaswami V (1981) Algorithms for a continuous-review (s, S) inventory system. J Appl Prob 18:461–472
Ramaswami V, Wang JL (1996) A discrete time queueing analysis of ATM systems with heterogeneous traffic sources. GLOBECOM '96 1:623–628
Ren J, Breuer L, Stanford D, Yu K (2009) Perturbed risk processes analyzed as fluid flows. Stoch Models 25:522–544
Ruiz-Castro JE, Pérez-Ocón R, Fernández-Villodre G (2008) Modelling a reliability system governed by discrete phase-type distributions. Reliab Eng Syst Saf 93:1650–1657
Scarf H (1960) The optimality of (s, S) policies in dynamic inventory problems. In: Arrow K, Karlin L, Suppes P (eds) Mathematical methods in the social sciences. Stanford University Press, Stanford
Sennott LI (1999) Stochastic dynamic programming and the control of queueing systems. Wiley, New York
Sethi SP, Cheng F (1997) Optimality of (s, S) policies in inventory models with Markovian demand. Oper Res 45:931–939
Song J (1991) Inventory management in a fluctuating environment. Ph.D. Dissertation. Graduate School of Business, Columbia University, New York
Song JS, Zipkin P (1993) Inventory control in a fluctuating demand environment. Oper Res 41:351–370
Song JS, Xu S, Liu B (1999) Order-fulfillment performance measures in an assemble-to-order system with stochastic leadtimes. Oper Res 47:131–149
Stanford DA, Yu K, Ren J (2011) Erlangian approximation to finite time ruin probabilities in perturbed risk models. Scand Actuarial J 2011:38–58
Tijms HC (1986) Stochastic modeling and analysis: a computational approach. Wiley, New York
van Houdt B, Blondia C (2004) Robustness of Q-ary collision resolution algorithms in random access systems. Perform Eval 57:357–377

van Houdt B, Blondia C (2005) Throughput of Q-ary splitting algorithms for contention resolution in communication networks. Commun Inform Syst 4:135–164

Zhao N, Lian Z (2011) A queueing-inventory system with two classes of customers. Int J Prod Econ 129:225–231

Zheng YS (1991) A simple proof for optimality of (s, S) policies in infinite-horizon inventory systems. J Appl Prob 28:802–810

Zipkin PH (2000) Foundations of inventory management. McGraw Hill, Boston

Printed in the United States
By Bookmasters

Printed in the United States
By Bookmasters